Microwave Solutions for Ceramic Engineers

Microwave Solutions for Ceramic Engineers

Editors
David E. Clark
Diane C. Folz
Carlos E. Folgar
Morsi M. Mahmoud

Published by

The American Ceramic Society

735 Ceramic Place, Suite 100
Westerville, Ohio 43081
www.ceramics.org

For information on ordering titles published by The American Ceramic Society, or to request a publications catalog, please call 614-794-5890, or visit www.ceramics.org

ISBN 1-57898-224-9

Contents

Preface

The papers presented in this book, for the most part, have been published previously by The American Ceramic Society (ACerS), either in proceedings from microwave symposia and congresses or in the American Ceramic Society Bulletin. Those not previously published by ACerS are noted and permission was received from the respective publishers to reproduce the work in this book. It is important to note that the papers published by ACerS from the microwave symposia and congresses have been through the a review process. All have been reviewed by at least three recognized experts in microwave technologies. Papers that "round out" the topics, such as Hutcheon's paper on dielectric property measurements, Gerling's paper on microwave safety and Cresko's paper on dielectric heating, were solicited specifically for this publication.

The papers selected for this book are representative of a larger collection of papers on microwave processing of ceramics that address a wide range of issues, such as fundamentals, equipment, economics, modeling and applications. They present the materials in a manner that will help to introduce technical managers, engineers and scientists to the potential for applying microwave technologies to their specific manufacturing processes.

The topics have been broken up in such a manner as to build upon the previous section, from the "The Basics" that provides an overview of fundamental knowledge regarding microwave processing to "Profiting from Microwaves" that provides steps to commercialization in this area as well as case studies.

The appendices provide information about the editors as well as the authors of the papers included in this volume. Also included is a summary list of all the papers published in the six microwave proceedings published by ACerS. We hope this summary will lead you to further useful information in your specific area of interest. In addition to the topics, titles and authors, this list gives information about the organizations and countries currently conducting microwave technology research and development. Finally, data from this summary has been organized into several figures to provide an overview of microwave processing work presented at ACerS meetings and the first three World Congresses.

While the papers in this text were selected because of their applicability to ceramics processing and because of their quality, there is one paper that is worthy of special recognition. Willard Sutton authored perhaps the seminal paper in microwave processing of ceramics. This paper will give you a sound understanding of the reasons many of us currently work in this field became interested in conducting microwave research. Sutton presents experimental data as well as a fundamental introduction to the practical ways in which microwave energy interacts with materials. He also provides an excellent reference section that will lead you to the early work conducted in this field of study.

David E. Clark
Diane C. Folz
Carlos E. Folgar
Morsi M. Mahmoud

What is Microwave Processing?

D.E. Clark and D.C. Folz

Introduction

The purpose of this book is to present a basic introduction to microwave processing, with a particular emphasis on ceramic materials. The reader also is encouraged to refer to *Microwave Processing of Materials,*[1] published by the National Academy Press in 1994, which provides an excellent overview of the subject and includes materials other than ceramics. Some of the material in this chapter has been adapted from Clark and Sutton[2] which contains an extensive set of references up to 1996.

Applications of microwave energy can be divided into several categories: (1) communications and information transfer; (2) processing/manufacturing; (3) diagnostics/analyses; (4) medical treatment; and (5) weapons. By far, the largest uses fall into the first two categories. In fact, most of us are familiar with microwaves primarily through the use of our home microwave ovens and cell phones. Many of the microwave applications involve the use of specialized ceramics for their successful operation. However, the primary focus of this book is on how to use microwave energy to process and manufacture ceramics. The principles and equations that govern the interaction of electromagnetic energy with materials (which includes microwave energy) have been known for over 50 years. In the early 1950s, these were used to develop equipment and methods for processing food at the industrial level as well as within our homes. These same principles and methods now are being used to guide the development of microwave processing of ceramics in the laboratory and, to a more limited extent, in ceramics plants.

Microwave energy can revolutionize the way we, as ceramic engineers and production managers, approach product manufacturing. While microwave energy alone will not be the answer to every technical barrier in materials processing, it can give us an alternative to the high energy consumption resistance heating techniques that are commonplace in industry. This book will provide you with a more complete understanding of the significant advantages microwave processing can offer and of some of the barriers that have prevented more rapid, wide-spread industrial use of this technology in the United States.

The question always arises as to the types of materials that are suitable for microwave processing. As shown in the bar charts and tables that follow, there has been significant interest in microwave processing of ceramics since the mid-1980s. These figures present data based on the numbers and distribution of papers in the six proceedings published by the American Ceramic Society for three international microwave symposia and the first three World Congresses focusing on microwave and radio frequency processing and applications.[3,4,5,6,7,8] As clearly indicated in Figure 1, out of 383 total papers published, nearly half are focused on ceramics and glasses. The use of microwaves for sintering accounts for a significant fraction of the research (Figure 2). This point is interesting in light of the fact that most microwave ovens are designed to operate at low temperatures.

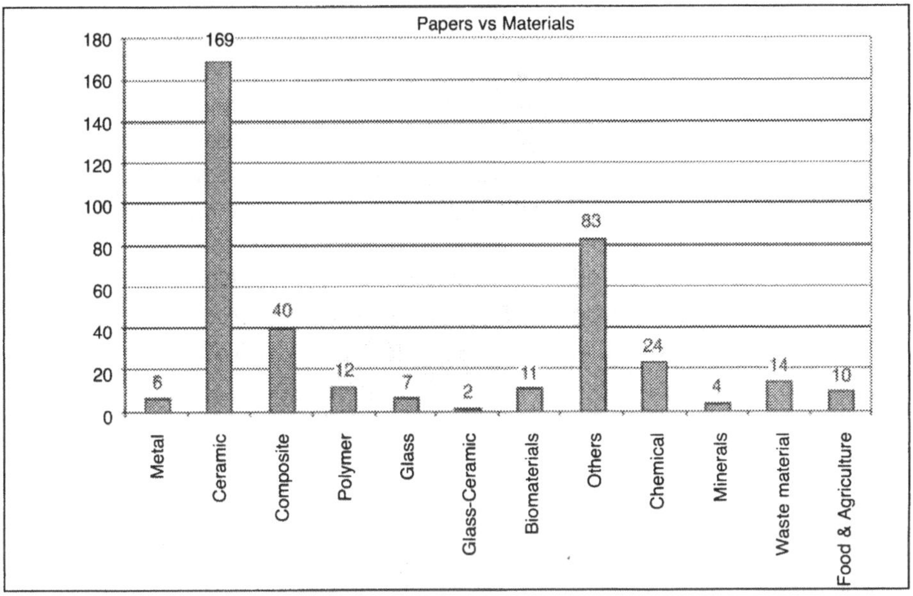

Figure 1. Distribution of materials topical papers published by the American Ceramic Society in the microwave symposia and conference proceedings from 1991 to 2003.

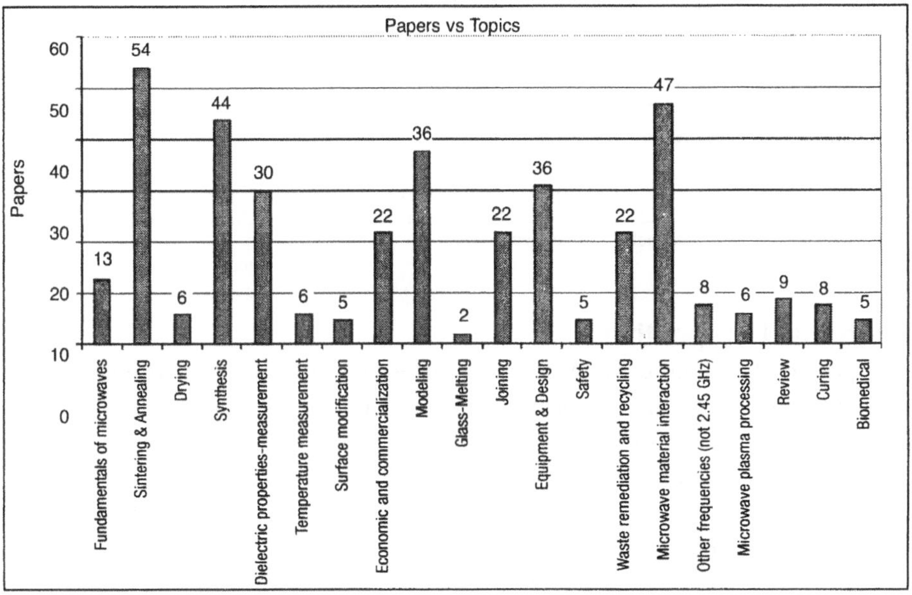

Figure 2. Distribution of papers with respect to research topics published by the American Ceramic Society in the microwave symposia and conference proceedings from 1991 to 2003.

The information presented in Table 1 illustrates the distribution of interest in microwave processing of ceramics throughout industry, academia and government labs as presented in these six proceedings. Much of the work in microwave processing is being conducted through collaboration between two or more of these research groups. It is not uncommon to see interagency and interdisciplinary research teams working together to attempt to work through some of the practical and theoretical issues associated with implementing microwave processing.

TABLE 1. THE DISTRIBUTION OF CERAMICS (DOES NOT INCLUDE GLASS OR GLASS-CERAMICS) PAPERS ACCORDING TO RESEARCHER INSTITUTION.

Topics	Ceramics papers							
	G	A	I	IA	GA	GI	GIA	Total
Fundamentals of microwaves		5	1					6
Sintering & annealing	7	29	4	4	1	2		47
Drying				1				1
Synthesis	1	14		4				19
Dielectric properties-measurement	2	8	3	3	1			17
Temperature measurement	1	3						4
Surface modification		3				1		4
Economic and commercialization			3					3
Modeling		15			2			17
Joining	1	7		2	2	1	2	15
Equipment & design	2	4		1		1		8
Microwave material interaction	1	12		6				19
Other frequencies (not 2.45GHz)	1	1		1	1			4
Microwave plasma processing		2		1				3
Review articles		2						2
Total	16	105	11	23	7	5	2	169

G=government laboratory; I=industry; A=academia

Historical Perspective

The possibility of processing ceramic materials with microwave energy was known by the 1950s and had been investigated by Tinga *et al,*[9] Levinson,[10] and Bennett *et al*[11] on a limited basis by the 1960s. In 1975, while investigating microwave drying of high alumina castables, Sutton[12] observed that, in addition to removing water, microwaves heated the ceramic. Using an industrial microwave oven lined with non-microwave-absorbing refractories, temperatures in excess of 1400°C were attained. This study demonstrated that microwave energy could be used for processing full-scale ceramic products (low- and high-temperature). It also showed microwave energy to be faster and more cost effective, as well as capable of producing products with equal or superior performance when compared with products resulting from conventional gas heating. It was fortuitous that the particular ceramic used by Sutton was a good microwave absorber at 2.45 GHz (the most commonly used microwave frequency), as many ceramic materials are transparent to microwave energy at this frequency. On the other hand, it was unfortunate that, comparatively, the low cost of fossil fuels at the time limited implementation of microwave energy for industrial applications.

By the 1980s, quite a number of researchers in North America were investigating microwave processing of ceramics, most likely due to three factors: 1) many laboratories were equipped with inexpensive home-model microwave ovens that could be modified for high-temperature research, 2) researchers at national and foreign laboratories were reporting successes and unusual effects with microwave processing, and 3) an international symposium on microwave processing was held in 1988. As a consequence of factor (1), most of the studies have been and still are performed at 2.45 GHz. This fact has resulted in some creative engineering in developing microwave-conventional hybrid heating techniques and microwave-absorbing additions to compositions in order to overcome the issue of low microwave absorption by many ceramics at room temperature. Regarding factor (2), scientists in the Ukraine and Russia used high frequency (37 and 84 GHz) microwaves to achieve very rapid heating and processing for applications, such as plastic coating of oil piping; glazing ceramics; soldering and brazing; treatment of polymers, composites and semiconductors; and, synthesis of materials. Researchers at Los Alamos and Oak Ridge National Labs used 2.45 GHz and higher frequencies (28 and 60 GHz) to prepare ceramic-glass seals; sinter Al_2O_3, ZrO_2, B_4C and TiB_2; heat treat Si_3N_4; fabricate composites; and, form ceramic joints. The laboratories also developed unique microwave susceptors and casketing techniques. Scientifically, one of the most important discoveries was reported by Janney *et al*[13] at Oak Ridge National Laboratory. They demonstrated that alumina (Al_2O_3) could be sintered at much lower temperatures with microwave energy than with electric resistance heating. This phenomenon is referred to as a "microwave effect" and has been reported by numerous investigators. Enhancements in the rates of activated processes involving material transport (e.g., sintering, ion exchange and chemical reactions) are considered to be microwave effects because a reduction in the activation energy appears to be required, and investigators have been unable to provide a scientific basis for this behavior. In addition to enhanced rates, reaction pathways and reaction products found in microwave processes, but not in conventional processing, also are considered as microwave effects.

As microwave processing becomes more widely used, it also will be necessary to understand the differences in heating schedules, reaction pathways and phase transformations that occur in a microwave processing field. It is anticipated that new diagnostic/analytical tools, such as microwave thermogravimetric analyzers (MTGAs) and microwave differential scanning calorimeters (MDSCs) will need to be developed to provide this understanding. One such tool, scanning dielectric analysis (SDA), is described by Hutcheon *et al* in this book.

Since 1988, the Materials Research Society (MRS) and The American Ceramic Society (ACerS) have held seven major symposia in the United States on microwave processing. The Microwave Working Group (MWG) (www.microwave-rf.org) was formed in 1993 to foster microwave research and manufacturing in the United States. In addition to being active in the MRS and ACerS symposia, the MWG has organized four World Congresses on microwave processing.

Other professional societies worldwide (the International Microwave Power Institute, the American Chemical Society, the Institute of Electromagnetic Wave Application – Japan, the Association of Microwave Power in Europe for Research and Education) regularly sponsor microwave processing conferences, short courses and workshops. These symposia have brought together internationally recognized experts in materials processing, microwave technology and equipment designers, and provided an excellent forum for exchanging information, sparking fresh ideas and stimulating new interest. As a result of the increased interest among scientists and engineers, many advances are being made in dielectric property measurements, modeling and processing. A better understanding of microwave-material interactions and the economics of microwave processing is evolving that will provide the basis for future industrial applications. In parallel, new equipment with variable frequencies, high-temperature capabilities, and better temperature monitoring and control systems are being designed and manufactured.

The Basics

Microwaves (frequencies of 0.3 GHz to 300 GHz and wavelengths of 1 m to 1 mm) lie between radio wave frequencies (RF) and infrared (IR) frequencies in the electromagnetic (EM) spectrum (Figure 3). Microwaves can be reflected, transmitted and/or absorbed. The absorbed microwave energy is converted into heat within the material, resulting in an increase in temperature. Gases, liquids and solids can interact with microwaves and be heated. Under certain conditions, gases can be excited by microwaves to form plasmas that also can be useful for processing.

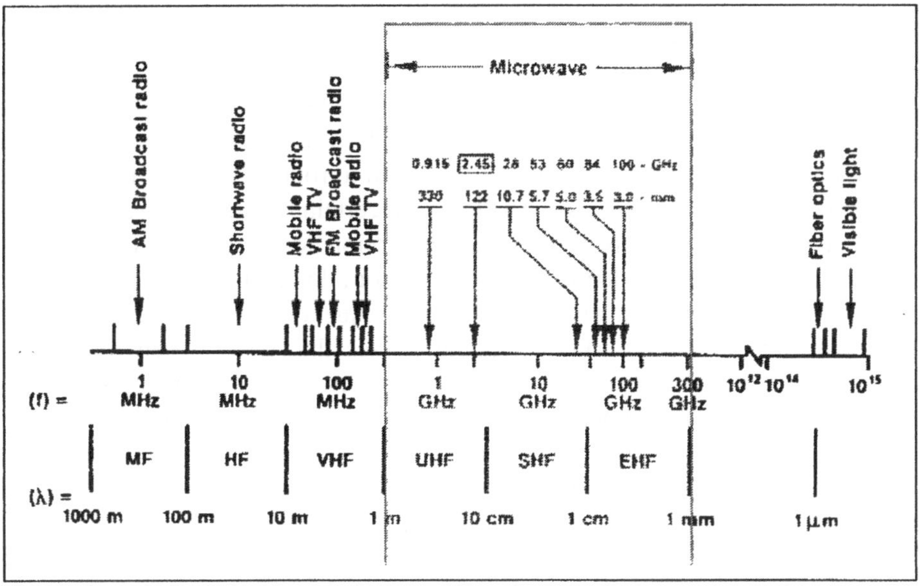

Figure 3. The electromagnetic (EM) spectrum and the microwave range. The most commonly used frequency is 2.45 GHz with a wavelength of 12.2 cm.

Nearly all microwave ovens operate at 2.45 GHz due to the fact that they were designed to process foods and the water in the foods is a good absorber at this frequency. Unfortunately, at room temperature, many ceramics do not absorb 2.45 GHz microwave energy appreciably. Their absorption may be increased by adding absorbing constituents (e.g., silicon carbide, carbon, organic binders), altering their microstructures and defect structures, changing their form (e.g., bulk vs. powder) or changing the frequency of the incident radiation. The latter often is not feasible due to the relative unavailability of equipment. Fortunately,

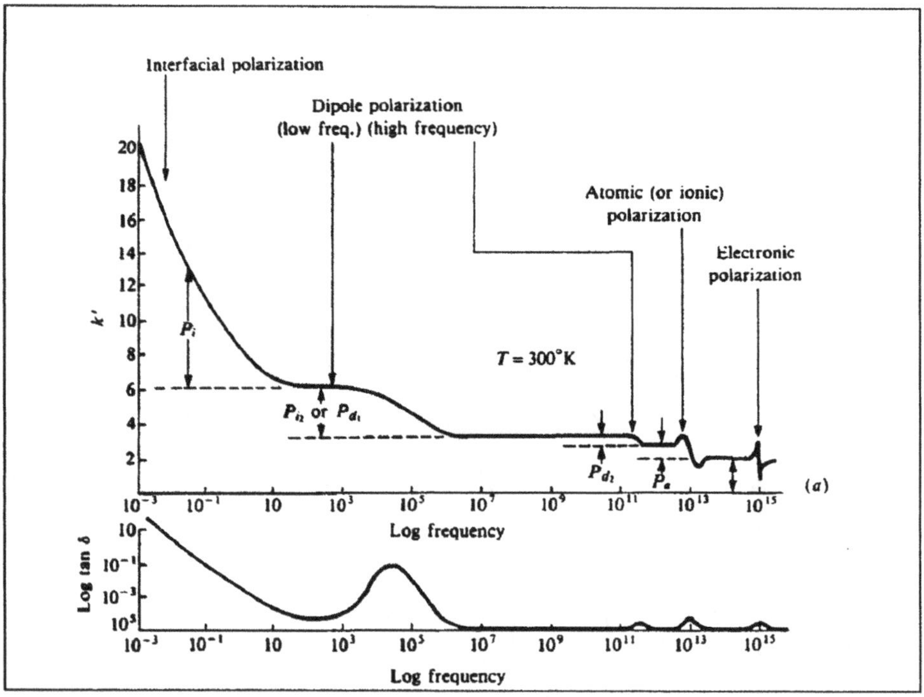

Figure 5. Frequency dependence of the polarization mechanisms in dielectrics: (a) contribution to the charging constant (representative values of ε'); (b) contribution to the loss angle (representative of ε").[15]

In addition to dipole rotation, other widely recognized polarization processes include interfacial, ionic and electronic polarization. These processes are discussed in detail by Hench and West[15] for typical ceramics. Because several processes can contribute to the losses, and it is not always easy to differentiate experimentally between the loss mechanisms, losses typically are reported as effective losses, ε''_{eff}. To adequately characterize the response of ceramics to the electromagnetic spectrum, measurements should be made over a broad frequency range. Frequency ranges, where each of these can contribute to losses, vary considerable over the EM spectrum, depending on the ceramic (Figure 5). It is because of the frequency dependence of absorption that a single frequency, such as 2.45 GHz, cannot be used easily to heat all ceramics at room temperature. To efficiently heat a material that does not have an absorption mechanism close to 2.45 GHz requires changing either the frequency of the radiation or the composition of the material, or the use of hybrid heating. Likewise, ceramics can be designed to be microwave safe, such as with microwave cookware. These materials have insufficient absorption to be self-heating at the operating frequency of the microwave oven.

Whether or not a ceramic can be heated by microwave energy depends on how much and how fast that energy can be absorbed. The rate of energy absorption is expressed in terms of power per unit volume as

$$P_a\ (\mathrm{W/m^3}) = \omega\varepsilon_0\, \varepsilon''_{eff}\, E^2_{rms} + \omega\, \mu_0\, \mu''_{eff}\, H^2_{rms} \qquad (\mathrm{watts/m^3}) \qquad (1)$$

where the terms on the right side of the equation refer to electric and magnetic losses, respectively. Generally, the magnetic losses are negligible, and only the electric losses contribute to the absorbed power. Here

ω = angular frequency = $2\pi f$ (f = operating frequency in hertz);

ε_o = permittivity of free space = 8.85×10^{-12} F/m;

ε''_{eff} = effective relative dielectric loss (dissipation) factor (unitless); and,

E_{rms} = root mean square of the internal electric field (V/m).

$$\varepsilon''_{eff} = \varepsilon''_c + \varepsilon''_s + \varepsilon''_d + \varepsilon''_i + \varepsilon''_e \qquad (2)$$

where, ε''_c is the loss due to dc conductivity and is negligible at microwave frequencies for most ceramics, as shown in Figure 4.[14] The last four terms represent dielectric losses due to polarization mechanisms, where ε''_s = space charge or interfacial; ε''_d = dipolar; ε''_i = ionic, and ε''_e = electronic. For ceramic materials in the microwave frequency range, ε''_s and ε''_d are the most important. Since these terms are dependent on both frequency and temperature, they must be measured in order to know how well the ceramic will heat at a specified frequency (i.e., 2.45 GHz and room temperature). Sometimes, ε''_{eff} is replaced by $\varepsilon' \tan\delta$, where δ is the loss angle, $\tan\delta$ is the dissipation factor and ε' is the dielectric constant. Values of dielectric properties are provided in Table 2 for a few ceramics. A rule of thumb is that ceramics with loss factors between the limits $10^{-2} < \varepsilon''_{eff} < 5$ are good candidates for microwave heating. Ceramics with $\varepsilon''_{eff} < 10^{-2}$ would be difficult to heat, while those with $\varepsilon''_{eff} > 5$ would experience most of the heating in the surface and not the bulk.[16]

The rate of temperature rise ($\Delta T/\Delta t$) in a ceramic when microwave energy is absorbed is given by[16]

$$\frac{\Delta T}{\Delta t} = \frac{P_a}{\rho(C_p)} \qquad (^\circ\mathrm{C/sec}) \qquad (3)$$

where, ρ = density of the ceramic (kg/m^3), Cp = specific heat (kJ/(kg ·°C)).

It can be seen from equations (1) and (3) that, in addition to the dielectric properties, the internal electric field also must be known in order to predict the power absorbed and heating efficiency of the ceramic. It is not trivial to determine the internal electric field, since it can vary from point to point within the ceramic. The internal field is less than the external field and is controlled by the dielectric constant, ε', as shown by,[2]

$$E_{\text{int}} = \left[1 - \frac{N(\varepsilon' - 1)}{1 + N(\varepsilon' - 1)}\right] E_{ext} \quad \text{(V/m)} \tag{4}$$

where N is a depolarization factor that depends on shape or geometry. For an alumina sample sandwiched between parallel electrodes, N=1, ε'=9 and $E_{\text{int}} = E_{\text{ext}}/9$. Thus high values of ε' can reduce the internal field and rates of absorption and heating. The internal electric field for a given point within the material can be estimated by measuring $\Delta T/\Delta t$ and knowing C_{p} and $\varepsilon''_{\text{eff}}$ by[16]

$$E_{\text{int}}\,(\text{rms}) = \left[\frac{\rho C_{\text{p}} \dfrac{\Delta T}{\Delta t}}{\omega \varepsilon''_{\text{eff}}}\right]^{1/2} \quad \text{(V/m)} \tag{5}$$

From this, the E_{ext} (rms) can be estimated using equation (4). It should be pointed out that the external electric field is not uniform in most multimode microwave ovens, hence neither is the internal electric field. This is why many home microwave ovens require turntables or mode stirrers. The field uniformity can be improved by increasing the size of the oven cavity. The cavity should be much larger than the wavelength of the microwaves.

At the macroscopic level, the dielectric properties control the microwave processibility of a wide range of non-magnetic materials through the quantitative relationships of absorbed power per unit volume (P_{a}) and depth of penetration (D_{p}, defined as the distance from the surface into the material at which P_{a} drops to e^{-1} (~37%) of the surface value)[16]

$$D_p = \frac{\lambda_{\text{o}}}{2\pi(2\varepsilon')^{1/2}} \left[\left(1 + \left(\frac{\varepsilon''_{\text{eff}}}{\varepsilon'}\right)^2\right)^{1/2} - 1\right]^{-1/2} \quad \text{(cm)} \tag{6}$$

where, λ_{o} = free space wavelength of the microwave radiation (λ_{o}= 12.2cm for 2.45 GHz). The penetration depths given in Table 2 were calculated using equation (6).

TABLE 2. DIELECTRIC PROPERTIES, PENETRATION DEPTHS AND CRITICAL TEMPERATURES (T_c) FOR SOME TYPICAL CERAMICS AND OTHER COMMONLY USED MATERIALS. THE DIELECTRIC PROPERTIES CAN BE VERY SENSITIVE TO THE PURITY AND PHASE OF THE MATERIAL.THEREFORE, THE DATA SHOWN SHOULD BE USED ONLY AS A GUIDE.

CERAMIC MATERIALS	Frequency (GHz)	Temp. (°C)	ε'	ε''	tan δ	Depth of Penetration, D_p (cm)	Critical Temp. T_c
Alumina[1]	2.45	+25	8.9	0.009	0.00010	1.2	800°C (3.89-3.61(GHz))[5]
Silicon carbide[2]	2.45	+200	105	110	1.048	0.28	
Mullite[3]	0.001	+25	5.8	0.012	0.002	3674.4	
Fused Quartz[1]	2.45	+25	3.8	0.0001	0.00003	1.9	
Soda Lime Glass[4]	2.45	+20	6.0	1.20	0.20	1.5	
Pyrex [1]	2.45	+25	4.0	0.005	0.0013	1.8	
Boron Nitride[5]	8.52	+25	4.37	0.01311	0.00300	0.5	750-800°C (5.17-4.96 (GHz))[5]
Aluminum Nitride[7]	2.45	+100	9	0.008	0.000889	1.18	
Silicon Nitride[5]	8.52	+25	5.54	0.01994	0.00360	0.44	
	2.45	+25	0.68	0.015	0.022059	8.13	
Water (Distilled)[4]	2.45	+25	78.0	12.00	0.1565	0.39	
Lithium Disilicate glass[7]	2.45	+25	7.5	0.17152	0.023	1.30	
Lithium Disilicate glass-ceramics[7]	2.45	+25	5	0.02038	0.004	1.63	
Teflon [1]	2.45	+25	2.1	0.0003	0.00014	2.73	
Natural Rubber[4]	2.45	+25	2.2	0.01	0.005	2.65	
Polyethylene[4]	2.45	+20	2.3	0.001	0.0004	2.57	
Beef (cooked)[4]	2.45	+30	30.5	9.6	0.31	0.62	
Frozen Beef [5]	2.45	-20	4.4	0.53	0.12	1.74	
Potato (Mashed)[4]	2.45	+30	72.5	24.0	0.33	0.40	

1: Data provided by J. Gerling, Gerling Applied Engineering, Inc., Modesto, CA. 2: K. Lesier, "Microwave Behavior of Silicon Carbide/High-Alumina Cement Composites," Doctoral dissertation, University of Florida (2001). 3: www.accuratus.com 4: F. R. Schiffmann, "Principles of Industrial Microwave and RF Heating," *Microwave: Theory and Application in Materials Processing IV,* Ceramic Transactions, American Ceramic Society, **80** (1997). 5: *Metaxas, Industrial Microwave Heating,* Peter Peregrinus Ltd., London, United Kingdom (1988). 6: Committee on Microwave Processing of Materials: "An Emerging Industrial Technology," Commission on Engineering and Technical Systems, National Research Council, *Microwave Processing of Materials,* National Materials Advisory Board, National Academy Press, Washington, D.C. (1994). 7: Dielectric data provided by Dr. R. Hutcheon, Microwave Properties North, Canada.

The depth of penetration has important consequences when processes are scaled up to manufacture components with large cross-sections or multiple samples where stacking is required. As an example, the depths of penetration for many ceramics in the microwave range usually are much greater (many cm) than their cross-sections. Thus multiple samples stacked inside a microwave furnace will all "see" the microwaves and uniform heating can occur. On the other hand, most of the radiant heat in conventional furnaces is absorbed in the first few microns of the outer most samples, producing shielding effects for samples that are not directly exposed to the radiation. Therefore, long times are required to bring all the samples to the same temperature and each sample experiences a different thermal history. This phenomenon increases production times and can reduce product consistency.

In some materials, the magnetic dipoles may be able to couple with the magnetic component of the electromagnetic field and provide an additional heating mechanism. Similar to the dielectric properties, the magnetic permeability, μ', and loss, μ'', must be considered. There have been relatively few investigations on microwave processing of magnetic materials and this area appears sufficiently promising so as to warrant further investigation.

Historically, there has been a serious lack of data on dielectric properties of most materials as a function of temperature over the microwave range. The data that are beginning to emerge indicate that there is a complex interdependence of these properties on temperature and frequency. The microwave processing community has begun to address these complexities through improved dielectric measurement techniques and processing equipment design. Several groups have established laboratories for measuring the dielectric properties over broad ranges of microwave frequency and temperatures.

Many ceramics exhibit an abrupt increase in ε''_{eff} with increasing temperature, as shown in Figure 6. Since P_a is directly proportional to ε''_{eff}, there is a corresponding increase in the power absorbed and the heating rate of the ceramic as the temperature is increased. The reason for this abrupt change in ε''_{eff} most likely is due to easier dipole rotation that allows the peak shown in Figure 4 to move closer to the microwave frequency being used. The temperature where the abrupt change in ε''_{eff} occurs is referred to as the critical temperature, T_c. Below T_c, an external energy source is required to heat the ceramic. Above T_c, the sample becomes self-heating in the microwave field. Very few T_c values have been reported for ceramics thus far. Collection of more T_c values for these materials would be very useful. There is no equation relating T_c to fundamental materials values, thus the critical temperature values must be measured. For all calculations and measurements, it is important to identify specifically the materials characteristics (i.e., composition, phase, form) and assumed or measured process parameters (i.e., temperature, pressure, atmosphere). For example, 99% alumina differs from 96% alumina and α-SiC differs from β-SiC.

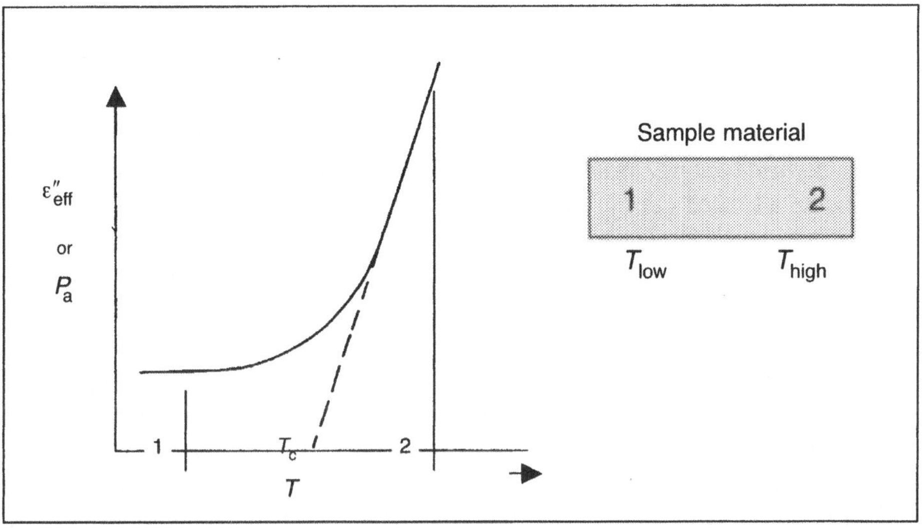

Figure 6. The effect of temperature on ε''_{eff} and P_a. Points 1 and 2 in the sample material represent positions in or on the sample with different electric field strengths. The electric field is greater at point 2.

A potential consequence of the rapid increase in ε'' with temperature ($d\varepsilon''/dT$) is that, unless the sample is heated uniformly by an external source, localized hot spots as indicated by point 2 in Figure 6, can develop. The development of hot spots usually is referred to as "thermal runaway" (unstable accelerated heating). This terminology is unfortunate since it suggests a process that cannot be controlled. In fact, it is this very property (large $d\varepsilon''/dT$) that permits the microwave heating of many materials that are poor absorbers at ambient temperatures, as long as the entire sample is heated uniformly by an external source and the internal field is fairly constant.

Some excellent quantitative models of thermal runaway have been developed in the last few years. The models incorporate thermal conductivity, dielectric properties and their nonlinear dependence on temperature. The effects of the electric field intensity, sample size and insulation were evaluated thoroughly by Johnson using various purities of Al_2O_3 as a model system. These investigators delineated regions of stable and unstable heating and demonstrated that insulation normally used to contain the heat inside the cavity exacerbates the problem of thermal runaway. However, the data also suggested that thermal runaway could be minimized by hybrid heating (heating the article with radiant heat at the same time that microwave energy is deposited into the sample) and with proper control of the electric field. This is consistent with the experimental results of many investigations using hybrid heating and controlling the electric field with various techniques.

Many different aspects of the microwave process have been modeled: drying and heating rates, temperature profiles, electric field distributions, power absorption and hybrid heating. Most research has been focused on modeling the electric field inside a loaded cavity, and then coupling these data with equation (1) and various heat transfer equations. Huang *et al* used a variable mesh Finite-Difference-Time-Domain (FDTD) analysis of electromagnetic power deposition in combination with a 3-D heat transfer program to model sintering in a multimode cavity. This model accounts for external stimuli, such as SiC heating rods (hybrid heating) and insulation effects. When the dielectric and thermophysical properties are known as a function of temperature, the model can accurately predict the heating rate of the sample, as shown in Figure 7. Modeling can greatly help to understand the quantitative effects of microwave-material interactions and will play an increasingly important role in the automation and control of processing parameters.

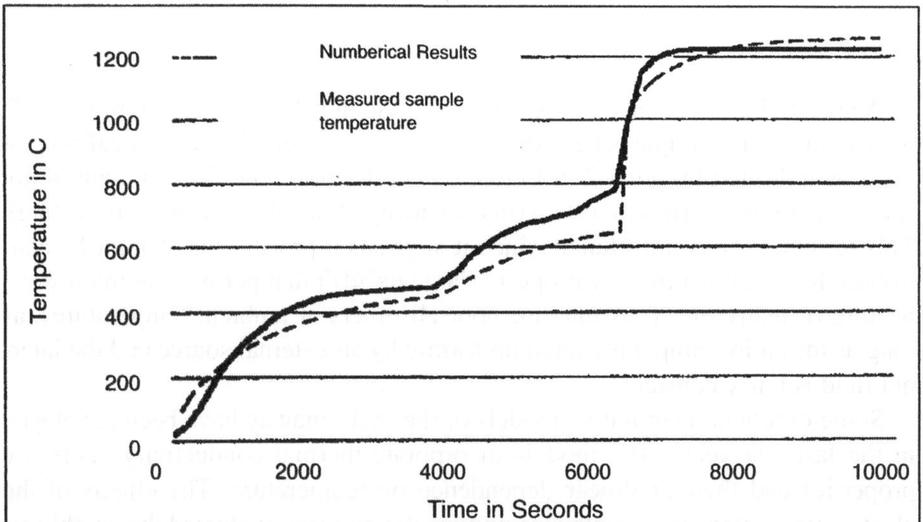

Figure 7. Comparison between experimental and numerical results of a zirconia sample's temperature while heating in a multimode microwave cavity.[26]

Equipment

A microwave system typically consists of a generator to produce the microwaves, a waveguide to transport the microwaves and an applicator (usually a cavity) to manipulate microwaves for a specific purpose and a control system (tuning, temperature, power, etc.). Until recently, only fixed frequency single-mode or multimode systems (home microwave ovens) were readily available. Single-mode systems have had limited (but successful) applications in industry because of a limited processing volume over which the electric field is useful, but have been particularly effective in plasma processing, joining and fiber curing. In a multimode system, the fixed frequency microwaves yield resonant modes over a narrow frequency band around the operating frequency. The modes result in regions of high and low electric fields (i.e., non-uniformity) within the cavity. In general, the uniformity of the field increases as the cavity size increase, but the uniformity also is dependent on the overall cavity dimensions.

Magnetrons, klystrons, gyrotrons and traveling wave tubes (TWTs) are used to generate microwaves. Each has its advantages. For example, klystrons offer precise control in amplitude, frequency and phase. Gyrotrons offer the possibility of providing much higher power output (megawatts) and beam focusing. The TWTs can provide variable and controlled frequencies of microwave energy. Magnetrons are by far the most widely used microwave source for home microwave ovens and industrial microwave systems, due to their availability and low cost. Solid state devices also are available for generating microwaves, but typically have been limited in power (few tens of watts).

The total cost of microwave processing equipment usually is greater than conventional equipment designed to perform the same tasks. However, it is not easy to do a one-to-one comparison between conventional and microwave processing equipment because, in some cases, a hybrid system may be best. Certainly, as microwave energy becomes more widely used for manufacturing ceramics, the costs will come down. Shown in Figure 8 is a microwave furnace designed by Cober Electronics, Inc. for manufacturing ceramics. In addition to the substantial amount of microwave power, the equipment includes a hot air mode to provide high efficiency heating in conjunction with the microwave power. The control systems are highly automated and precise with minimum operating labor. Equipment of this type costs approximately $500,000 installed. Panasonic manufactures a microwave kiln that will achieve 1250°C and may be used for both bisque and final firing of traditional ceramic ware. The system sells for about $15,000 in Japan (Figure 9). Additional systems cost and efficiency information is provided in Table 3. While the costs have changed significantly since Sheppard's table was produced, the relative data remains the same.

TABLE 3. CAPITAL AND OPERATING COSTS OF MICROWAVE EQUIPMENT. ADAPTED FROM SHEPPARD, 1988.

Component	Typical Costs
Complete System	$1,000 – 5,000/kW
Generator	< 50% system cost
Applicator	> 50% system cost
Power Transmission	$1,000 – 3,000; < 5% system cost
Instrumentation	$1,000 – 3,000; < 5% system cost
External Materials Handling	$1,000 – 3,000; <5% systems cost
Installation, Start-up	5 - 15% system cost
Magnetron Replacement	$0.01- 0.12/kWh
Electric Energy	$0.05 – 0.12/kWh
Plug-to-Product Efficiency	50 – 65%
Routine Maintenance	5 – 10%

Figure 8. Hybrid microwave oven manufactured by Cober Electronics, Inc. for drying technical ceramics (0.915GHz; 200kW).

Figure 9. Microwave kiln manufactured by Panasonic for firing ceramics (2.45 GHz, 6kW).

Certain features of microwave equipment have most likely hindered more widespread use of microwave energy for industrial processes;(i.e, limited and fixed frequencies, low temperature capabilities, field non-uniformity and complexity, and inability to monitor or control the internal temperatures). Although microwave energy covers a broad frequency range, in the United States, the Federal Communications Commission (FCC) allocates frequency bands centered at 2.45 GHz and 0.915 GHz for industrial, scientific and medical (ISM) use. These two frequencies provide a good compromise between penetration depth, absorption and equipment costs for food processing. Water and other polar molecules couple well with these frequencies. Unfortunately for the rest of the materials processing community, they are not always the optimum frequencies for heating polymers, ceramics, composites and other materials. Most of the equipment is designed for low- temperature (few hundred °C) cooking and drying. Consequently, the equipment is not designed or properly insulated for the high temperatures (>1000°C) required to process most ceramics and composites. Field uniformity also has been a problem, especially for the lower frequency microwaves (2.45 and 0.915 GHz), where the wavelengths are several inches. As discussed previously, the power absorbed by the material is strongly dependent on the internal electric field. In turn, the internal field is controlled by the field inside the cavity, which can vary widely in most cavities. Mode stirrers and moveable platforms provide some improvement in the field uniformity as "seen" by the sample, through time averaging.

Potential users and equipment suppliers have, in many cases, avoided implementing microwave technologies dues to (1) the complexity of the equipment, (2) the need to perform expensive feasibility studies, and (3) the requirement for custom designs. In the last few years, microwave equipment and component companies have begun to more fully address these obstacles. Many equipment companies are alleviating some of the complexities and research costs by allocating facilities and technical assistance for customers to use for feasibility studies and prototype stage evaluations. If prototyping proves successful, the equipment manufacturers are prepared to construct, install and service the equipment, and to provide technical assistance during product manufacture. As an example, Lambda Technologies, Inc. is commercializing variable frequency microwave (VFM) equipment under exclusive license from Oak Ridge National Laboratory for materials processing. This new generation of microwave oven is computer controlled and takes advantage of broadband microwave sources (TWTs) to tune the incident frequency to the materials to be processed. Furthermore, the VFM system can sweep over several frequencies to achieve uniform distribution of the EM field. These applicators are available now for research and industrial applications.

Many researchers use home microwave ovens that have been modified to suit their needs. For example, when sintering ceramics, the oven must be well-insulated to protect the interior from high temperatures. In most cases, granules or rods of highly absorbing materials are attached to or embedded in the insulation to create susceptor caskets. The ceramics to be heated are put into these caskets and then placed inside the microwave oven. The microwaves heat the susceptors which, in turn, heat the ceramics. Using this hybrid system, the ceramics are heated by both the radiant energy from the susceptor and the microwave energy from the oven. In order to control and monitor the temperature, shielded thermocouples as well as radiation pyrometers are used. These devices provide feedback information to the control panel that controls the power to the magnetron (Figure 10). Most home microwave ovens provide "power level" control through a variable duty cycle. For example, a power level setting of 50 for a 1000 watt oven might result in the magnetron being on "full power" for 50% of the specified "cooking" time. The magnetron is cycled on/off using a specified time base (i.e., 15 seconds on/15 seconds off) to essentially achieve a 500 watt power level. In more expensive industrial microwave ovens, the power level can be controlled without pulsing. You can build as complex or as simple a system as you like. However, the real key to success is to understand the basics with respect to microwave heating.

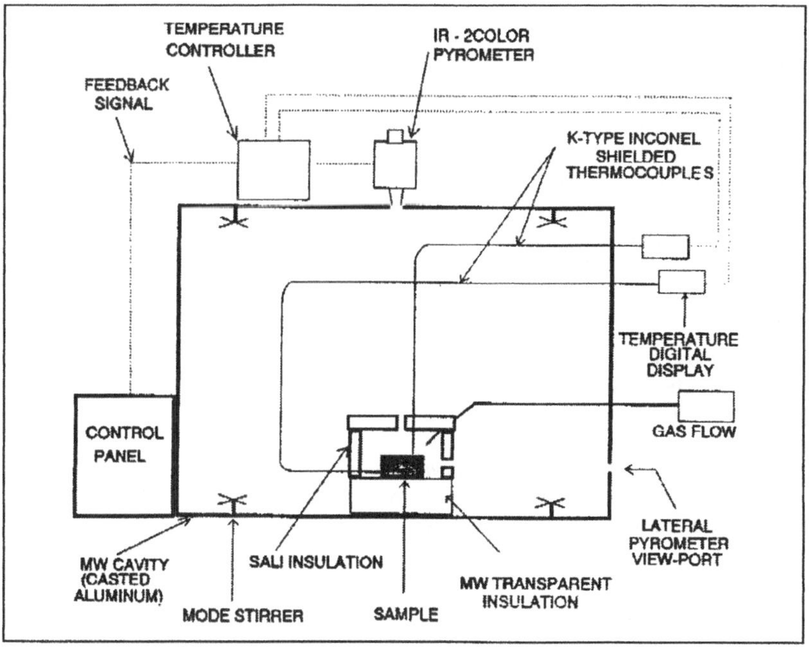

Figure 10. Schematic of a research system using a multimode microwave oven.[17] (The SALI is transparent to microwave, but is a thermal insulator. To make the SALI into a susceptor casket, it is lined with a microwave absorber, such as SiC.)

Why Use Microwave Energy?

Based on research performed by many engineers and scientists, a list of the potential benefits as well as challenges and needs associated with microwave processing has been compiled (Table 4).

TABLE 4. BENEFITS, CHALLENGES AND NEEDS OF MICROWAVE PROCESSING.[18]

Benefits
• Cost savings (time and energy, reduced floor space)
• Rapid heating of thermal insulators (most ceramics and polymers)
• Precise and controlled heating (instantaneous on/off heating)
• Selective heating
• Volumetric and uniform heating (due to deep energy penetration)
• Short processing times
• Improved quality and properties
• Synthesis of new materials
• Processing not possible with conventional means
• Reduction of hazardous emissions
• Increased product yields
• Environmentally friendly (clean and quiet)
• Self-limiting heating in some materials
• Power supply can be remote
• Clean power and process conditions
Challenges
• Heating low-loss poorly absorbing materials
• Controlling accelerated heating (thermal runaway)
• Exploiting inverted temperature profiles
• Eliminating arcing and controlling plasmas
• Efficient transfer of microwave energy to workpiece
• Compatibility of the microwave process with the rest of the process line
• Reluctance to abandon proven technologies
• Timing
• Economics
Needs
• Availability of affordable equipment and supporting technologies
• Kiln furniture, thermal insulation, and other processing support hardware
• Development of compositions and processes tailored specifically for microwave processing
• Better fundamental understanding and modeling of microwave-material interactions
• Better process controls, electronic tuning and automation (smart processing)
• Better communication among equipment manufacturers, technology developers, researchers and commercial users
• More emphasis on microwave processing of magnetic materials

Cost Savings

A Canadian study has shown there can be considerable energy savings when ceramics are manufactured with microwaves. Table 5 compares conventional drying and firing with microwave drying and firing for a wide range of ceramic products. It can be seen that there is a factor of about two in energy savings for microwave drying and about a factor of ten in energy savings for microwave sintering. Similar savings have been reported by E.A. Technologies for microwave hybrid heating, as discussed in more detail in Wroe's paper in this book.

TABLE 5. COMPARISON OF ENERGY SAVINGS FOR CONVENTIONAL AND MICROWAVE DRYING AND FIRING OF CERAMICS.[19]

	Energy savings (× 10^6 kW • h/yr)				
	Conventional Drying	Microwave Drying	Conventional Firing	Microwave Firing	Total Energy Savings
Brick and Tile	56.10	28.05	198.90	19.90	207.06
Electrical Porcelain	3.52	1.76	12.48	1.25	12.99
Glazes	16.63	8.30	58.97	5.89	61.37
Pottery	1.96	0.98	6.94	0.69	7.23
Refractories	10.87	5.40	38.53	3.85	40.08
Sanitary Ware	25.04	12.52	88.76	8.88	92.40
Advanced Ceramics	1.30	0.65	4.60	0.46	4.79
Total (× 10^6 kW · h/yr)	115.42	57.66	409.18	40.92	425.92
(× 10^6 PJ/yr)	0.42	0.21	1.47	0.15	1.50

*Source: Ontario Ministry of Energy, Canada.

A few examples of other benefits in ceramic processing are discussed below. Many other examples can be found throughout this book.

Selective and Self-Limiting Heating

As discussed previously, many factors can influence the rate at which ceramics heat in the microwave field (i.e., dielectric loss, sample volume). Leiser[20] observed that other factors, such as particle size and particle size distribution, sample geometry, and material phase(s) present can influence the heating rate significantly. Leiser loaded an alumina cement (microwave transparent) matrix with varying amounts of silicon carbide (absorber) to determine the effects of the range of composite composition on heating rates. The cement alone could not be heated much above room temperature. In contrast, pure SiC could be heated rapidly and to very high temperatures. This research showed (1) that the ratio of microwave-absorbing material to microwave-transparent materials was significant when evaluating heating rate and maximum temperature, and (2) that self-limiting heating also can be achieved by adjusting this ratio (Figure 11).

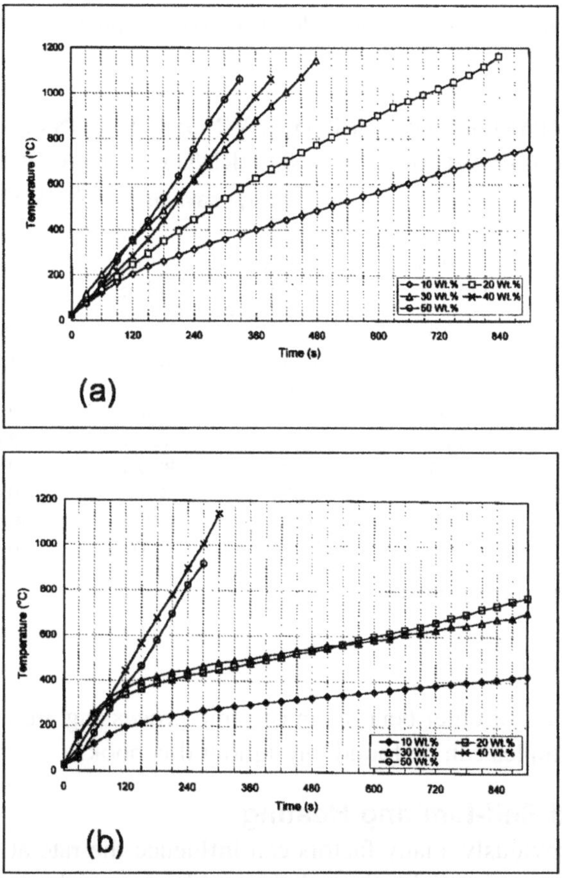

Figure 11. Comparison of heating rates of susceptors as a function of particle size and weight percent of silicon carbide: (a) 1000μm SiC; (b) 85μm SiC.

Volumetric and Uniform Heating

One of the most important characeristics associated with the use of microwave hybrid heating is the potential to achieve uniform heating throughout the cross-section of a material. In the late 1980s, Deã *et al* showed that microwave hybrid heating could result in samples with no significant density gradient throughout the cross-section and that this phenomenon was enhanced with increased sample size (Figure 12).[21] In addition, he noted that the density was higher for the microwave hybrid heated samples as compared to those processed with conventional fast firing techniques.

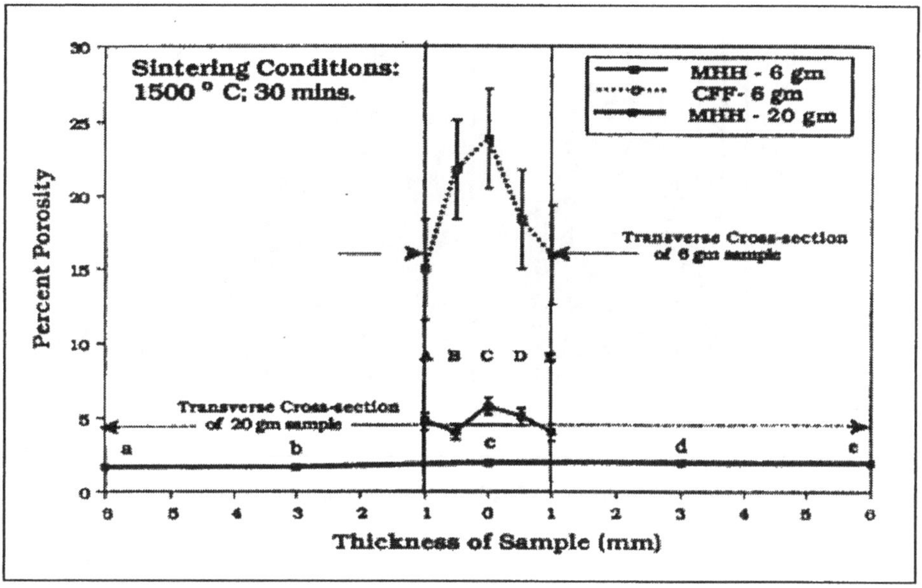

Figure 12. Percent porosity versus sample thickness for alumina processed using microwave hybrid heating and conventional fast firing methods.

Rapid Heating and Short Processing Times

As shown in Figure 13, microwave processing of materials offers the potential for reducing production times for ceramic materials. The material for the study presented was yttrium-barium-copper oxide superconductor and the sintering time was reduced from over 900 minutes to under 200 minutes. Also note that the sintering temperature to achieve 80% of theoretical density was reduced by over 100°C. This work, performed by Ahmad *et al*, supported the data presented in Table 5, showing the potential for significant cost savings in processing of ceramic materials.[22]

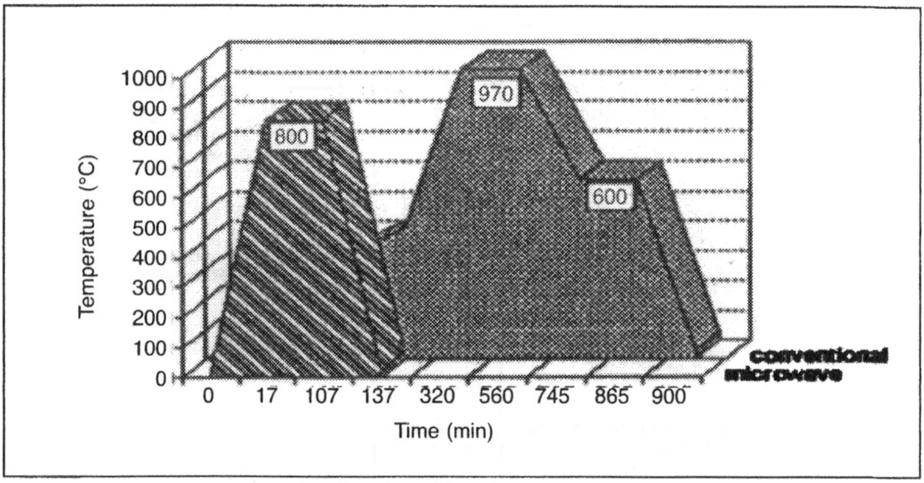

Figure 13. Typical sintering schedule for $YBa_2Cu_3O_{7-x}$ superconductors to achieve 80% theoretical density.

Waste Treatment and Environmentally Friendly Technologies

In addition to offering the potential for advantages in ceramics fabrication, microwave processing has been shown to reduce significantly the emissions generated during processing operations. In a study by Earl *et al* presented in this book, it was shown that microwave drying of tile resulted in an average reduction in flourine emissions of approximately 33%. The possibility of significant emissions reduction warrants further research.

In the 1990s, a large-scale research project conducted by the University of Florida and Westinghouse Savannah River Company showed that microwave energy could be used as a method for remediating waste materials (hazardous and nonhazardous) and recycling precious metals. Schulz *et al* designed and built a tandem microwave system that accomplished high-temperature incineration of composite feed stock (electronic circuit boards), recycling of the valuable metal components and treatment of hazardous emissions resulting from the incineration process. A schematic illustrating the basic design of the system is provided in Figure 14.

While the value-added process of recovering recyclables during the volume reduction experiments was a significant achievement, perhaps the most valuable results were obtained in the emissions treatment portion of the work. As can be seen in Table 6, hazardous emissions from the polymer incineration process were reduced by orders of magnitude and, in some cases, were no longer detected using gas chromatography mass spectroscopy (GCMS). This work indicates the value of microwave processing as a "back end" technology as well as a "front end" production method.

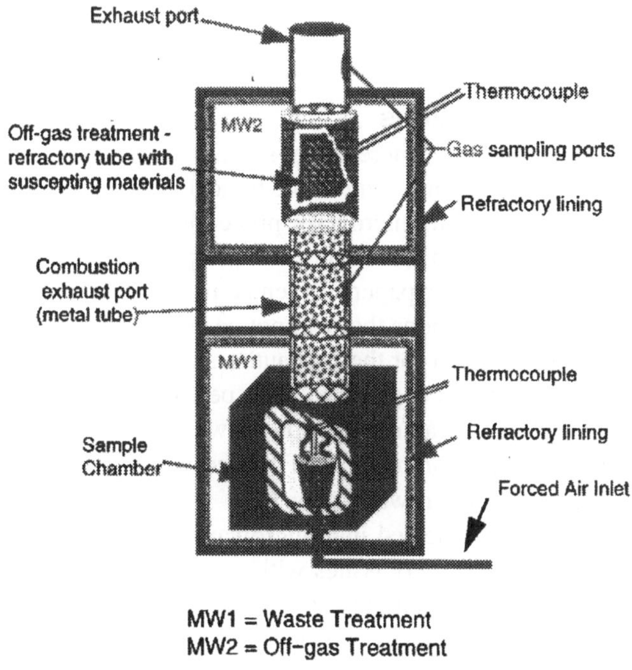

Figure 14. Tandem microwave waste treatment and recycling system designed and patented by the University of Florida and Westinghouse Savannah River Company.[23,24]

TABLE 6. SUMMARY OF GC MASS SEPCTROSCOPY RESULTS OF EMISSIONS RESULTING FROM COMBUSION OF PRINTED CIRCUIT BOARDS (A = EMISSION LEVEL BEFORE MICROWAVE TREATMENT; B = EMISSION LEVEL AFTER MICROWAVE TREATMENT)[25]

Compound	SR-4 Emissions (ppb)		SR-5 Emissions (ppb)		SR-6 Emissions (ppb)		SR-7 Emissions (ppb)	
	A	B	A	B	A	B	A	B
Benzene	115.3	5.2	119.6	8.1	176.6	14.7	165.4	13.5
Toluene	67.5	6.1	78.7	6.9	159.1	18.1	115.7	5.9
Ethylbenzene	13.9	nd*	26.7	nd	142.9	5.0	91.8	nd
Styrene	165.2	2.9	167.7	2.6	472.3	27.2	482.9	6.5
Napthalene	75.1	1.3	35.2	1.3	6.8	3.4	47.6	2.4
m/p Xylene	27.5	nd	23.8	nd	53.3	1.6	60.0	nd
1,3,5 Trimethylbenzene	15.6	1.6	18.4	nd	12.8	2.4	46.2	1.7
1.2.4 Trimethylbenzene	nd	nd	nd	nd	15.1	nd	6.1	1.8

*nd = not detected <1ppb

Frequently Asked Questions

As with any new technology, there always seem to be more questions than answers. The list below contains many of the typical questions asked regularly of those of us actively involved in microwave processing research and/or manufacturing. Some of the answers to these questions are provided in this Introduction, while others are answered in the papers presented in this collection. You will most certainly have additional questions that will not be answered in this book. For this reason, the contact information for the first authors is provided in the appendices.

1. Can I use a home model microwave to perform research?
2. Can I get high temperatures in a low-power microwave oven?
3. What are the major components of a microwave oven?
4. How efficient is a microwave oven?
5. Can all materials be heated in a microwave oven?
6. What are the major safety issues with using microwave ovens?
7. How are microwaves generated?
8. What are the differences in microwave and radio frequency heating?
9. Why do some materials absorb and others reflect microwaves?
10. What is the difference between microwaves used in communications and in processing?

11. How do you measure temperature in the microwave oven?
12. How do you measure electric field in a microwave oven?
13. What is the difference between single and multimode microwave cavities?
14. What is a susceptor?
15. What is the microwave effect?
16. What are magnetrons, klystrons and gyrotrons?
17. Can microwaves be used for processes involving ceramics, other than drying?
18. What is the largest use of microwave energy for processing?
19. Can microwave ovens be scaled up to manufacture ceramics?
20. Is there an economic advantage in using microwave energy to manufacture ceramics?
21. Are the properties of microwave and conventionally processed ceramics the same?
22. What kinds of ceramics can be heated with microwave energy?
23. What is the cost of microwave equipment capable of manufacturing ceramics on a large scale?
24. Why are the dielectric properties of ceramics important when microwaves are being considered for heating them?
25. What is the operating frequency of a home model microwave oven and what are some of the other frequencies that have been used to process ceramics?
26. What is the significance of an Industrial, Scientific and Medical (ISM) band?
27. Why do materials typically heat from the inside out in a microwave oven?
28. What is thermal runaway?
29. What are modes?
30. Why are some ceramics "microwave safe" and others not?
31. Where does all that energy go in an empty microwave oven when it is turned on?

Recommendations

What needs to be done in order to generate more widespread use of microwave energy in the ceramics industry? The National Materials Advisory Board convened a committee in 1992 that heard presentations from a number of experts in the field. The purpose of this committee was to recommend to the Department of Defense and the National Aeronautics and Space Administration future directions for research and development in microwave processing. While their recommendations do not specifically address the question posed above, they do provide some specific direction for researchers and manufacturers to move this technology forward. Following is a synopsis of the committee's recommendations as they pertain to ceramics.[2]

1. Define the conditions under which microwaves provide uniform, stable heating for ceramics. Hybrid heating schemes should be considered.
2. Establish multidisciplinary teams to properly develop microwave processes and procedures.
3. Provide training and education in the fundamentals to ensure a knowl edgeable and sustainable workforce for the engineering technology.
4. Compile existing ceramic property information in dielectric, magnetic and thermal properties in the range useful for processing ceramics.
5. Make better use of numerical modeling to design equipment and processes.
6. Develop practical methods for monitoring and controlling microwave processing.
7. Determine the validity of the "microwave effect."

The papers in this book are meant to help you to gain the insight that will lead to successful evaluation of microwave energy as a solution to your processing needs. We especially emphasize that you should read carefully John Gerling's paper, "Equipment Safety for Microwave and Radio Frequency Processing of Ceramics." This overview of safety issues will help prepare you to work with your microwave system and to separate truth and fiction with respect to microwave oven safety.

If you prefer, there is a great deal of expertise in research on microwave processing of a wide variety of materials that can be tapped into at academic and government research laboratories. Many times it is better to work with an experienced microwave research team than to "re-invent the wheel" in your own facility. However, if you want to start the process from scratch, the information provided in this book should be sufficient to get you started.

REFERENCES

1 Materials Research Advisory Board, Microwave Processing of Materials, National Research Council, Publication NMAB-473, National Academy Press (1994).

2 Clark, D.E. and Sutton, W.H., "Microwave Processing of Materials," *Annu. Rev. Mater. Sci.*, **26**:299-331 (1996).

3 Ceramic Transactions - Microwaves: Theory and Application in Materials Processing, D.E. Clark, F.D. Gac and W.H. Sutton, eds., American Ceramic Society, Inc., Vol. 21, Westerville, OH (1991).

4 Ceramic Transactions - Microwaves: Theory and Application in Materials Processing II, D.E. Clark, W.R. Tinga and J.R. Laia, Jr. eds., American Ceramic Society, Inc., Vol. 36, Westerville, OH, (1993).

5 Ceramic Transactions - Microwaves: Theory and Application in Materials Processing III, D.E. Clark, D.C. Folz and R. Silberglitt, eds., American Ceramic Society, Inc., Vol. 59, Westerville, OH, (1995).

6 Ceramic Transactions - Microwaves: Theory and Application in Materials Processing IV, Proceedings from the First World Congress on Microwave and RF Processing, D.E. Clark, W. Sutton and D.L. Lewis, eds., American Ceramic Society, Inc., Vol. 80, Westerville, OH (1997).

7 Ceramic Transactions – Microwaves: Theory and Application in Materials Processing V, Proceedings from the Second World Congress on Microwave and RF Processing, D.E. Clark, J.G.P. Binner and D.A. Lewis, eds., American Ceramic Society, Inc., Vol. 111, Westerville, OH (2001).

8 Microwave and Radio Frequency Applications, Proceedings from the Third World Congress on Microwave and RF Processing, D.C. Folz, J.H. Booske, D.E. Clark and J.F. Gerling, eds., American Ceramic Society, Westerville, OH (2003).

9 Tinga, W.R. and Voss, W.A.G., Microwave Power Engineering. 1968. New York: Academic

10 Levinson, M.L., 1971. U.S. Patent No. 3585258.

11 Bennett, C.E.G., McKinnon, N.A. and Williams, L.S., 1968. Nature 217:1287-88.

12 Sutton, W.H., 1988. Microwave Processing of Materials, 124. Reno: Mater. Res. Soc. 399 pp.

13 Janney, M.A. and Kimrey, H. 1988. Ceramic Powder Science II, Ceram. Trans. Westerville, OH: Am. Ceram. Soc. 1:919-24.

14 Metaxas, A.C. 1993. See ref. 16.

15 Hench, L.L. and West, J.K. 1990. Principles of Electronic Ceramics. New York: Wiley & Sons. 546 pp.

16 Metaxas, A.C. and Meredith, R.J., Industrial Microwave Heating, Peter Peregrinus Ltd., London, United Kingdom (1988).

17 Fathi, Z., Surface Modification of Sodium Aluminosilicate Glasses Using Microwave Energy. Doctoral dissertation, University of Florida (1994).

18 Clark, D.E. and Sutton, W.H., "Microwave Processing of Materials," *Annu. Rev. Mater. Sci.*, **26**:299-331 (1996).

19 Sheppard, L.M., *Ceramic Bulletin,* 67 [10] 1988. 1656-61.

20 Leiser, K.S., Microwave Behavior of Silicon Carbide/High Alumina Cement Composites. Doctoral dissertation, University of Florida (2001).

21 Deã, A. Ultra-Rapid Sintering of Alumina with Microwave Energy at 2.45 GHz. Masters thesis. University of Florida (1990).

22 Ahmad, I., Chandler, G.T. and Clark, D.E., Microwave Processing of Materials Proceedings, W. Sutton, I. Chabinsky and M. Brooks, eds., Vol. 124, Materials Research Society, Pittsburgh, 239-246 (1988).

23 Wicks, G.G., Schulz, R.L. and Clark, D.E., Tandem Microwave Waste Remediation and Decontamination System, U.S. Patent # 5,968,400, issued 10/19/99.

24 Wicks, G.G., Schulz, R.L. and Clark, D.E., Microwave Waste Treatment and Decontamination System, U.S. Patent # 6,262,405, issued 6/17/2001.

25 Clark, D.E., Folz, D.C., Schulz, R.L. Boonyapiwat, A., DiFiore, R.R., Darby, G. and Leiser, K., (1997) See Ref. 6.

26 Huang, Z., Iskander, M.F., Tucker, J., Kimrey, H.D., 1994, Microwave Processing of Materials IV, Mat. Res. Soc., 347: 331-45.

Basics

Microwave Processing of Ceramic Materials

Willard H. Sutton

Although still in an early stage of development, the use of microwave energy to process a wide variety of ceramic materials offers many new and exciting opportunities.[1-9] There are several reasons for the growing interest in microwave processing over conventional processing methods, including the potential for (1) significant reductions in manufacturing costs due to energy savings and shorter processing times, (2) improved product uniformity and yields, (3) improved or unique microstructures and properties, and (4) synthesis of new materials.

Microwave heating is fundamentally different from conventional processes. In the microwave process, heat is generated internally within the material instead of originating from external heating sources. As a result of this internal and volumetric heating, the thermal gradients and the flow of heat in microwave-processed materials are the reverse of those in materials processed by conventional heating, as indicated in Figure 1. Consequently, microwave processing makes it possible to heat both small and large shapes very rapidly and uniformly; to efficiently remove volatile constituents (binders, moisture, etc.) from thick sections; and to reduce thermal stresses that cause cracking during processing. Since the absorption of microwave energy varies with the composition and structure of different phases, selective heating is also possible and presents new opportunities. These and other characteristics and advantages of microwave processing are summarized in Table 1 and will be discussed further in this paper.

In the past, factors that have hindered the acceptance of microwave processing included unfavorable socioeconomic conditions; a natural reluctance to accept an entirely new, unproven process; a lack of experience and lack of understanding of the efficient use of microwaves; and the unavailability of positive results from successful research and development investigations because of proprietary restrictions. In addition, fundamental data are lacking on dielectric properties at elevated temperatures and on the heating response of ceramic and ceramic-composite materials at various microwave frequencies. Such data are necessary for developing optimized processing parameters and microwave equipment. In the past, many microwave processors used in research and development programs

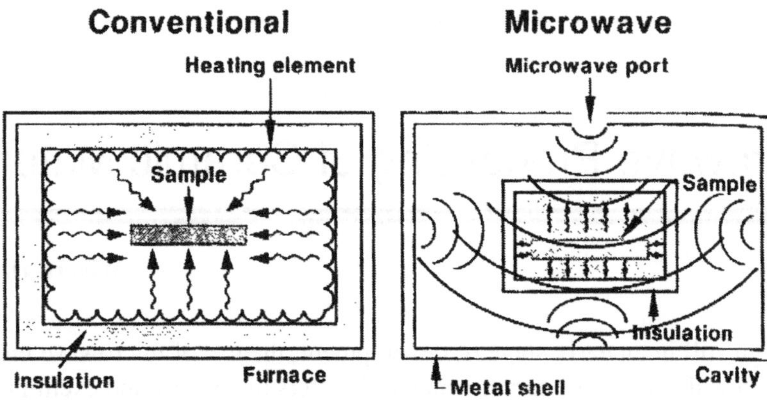

Figure 1. Heating patterns in conventional and microwave furnaces.

were not designed for optimum or effective use and did not take advantage of the latest developments in microwave and instrument technology. However, rapid progress is now being made in many of these areas.

Microwave–Material Interactions

Microwaves are electromagnetic waves that have a frequency range of 0.3 to 300 GHz and corresponding wavelengths ranging from 1 m to 1 mm. In contrast with visible waves, except for lasers, microwaves are coherent and polarized. Microwaves also obey the laws of optics and can be transmitted, absorbed, or reflected, depending on the material type, as shown in Figure 2. Metals are opaque to microwaves and thus are good reflectors. However, in this paper, the materials of interest include a broad range of ceramic dielectric (electrically insulating) materials (e.g., Al_2O_3, MgO, SiO_2, most glasses), many of which are transparent to microwaves at ambient temperature.

When heated to above a critical temperature (T_{crit}), these materials begin to absorb and couple more efficiently with microwave radiation. Other ceramic materials (e.g., CO_2O_3, MnO_2, NiO, and CuO) absorb microwaves efficiently at room temperature. The absorption of microwaves by transparent ceramics can be greatly enhanced by adding conductive or magnetic phases in the form of fibers, particles, or other additives, as shown in Figure 2. In these materials, the conductive or magnetic phases will absorb microwave energy far more rapidly than the matrix and, hence, will be heated selectively and rapidly.

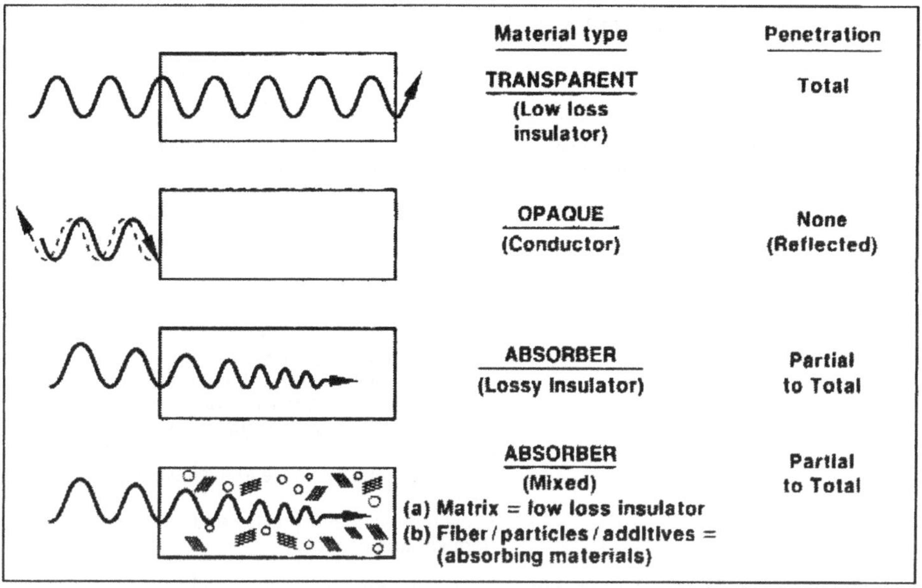

Figure 2. Interaction of microwaves with materials.

The degree of interaction (or absorption) of microwaves by a dielectric material is related to the material's complex permittivity, ε^* (F/m), which is composed of a real part (ε', dielectric con stant) and an imaginary part (ε'', dielectric loss factor) by

$$\varepsilon^* = \varepsilon' - j\varepsilon'' = \varepsilon_0(\varepsilon'_r - j\varepsilon''_{eff}) \tag{1}$$

where $j = (-1)^{1/2}$, ε_0 is the permittivityof free space ($\varepsilon_0 = 8.86 \times 10^{-12}$ F/m), ε' is the relative dielectric constant, and ε''_{eff} is the effective relative dielectric loss factor.

When microwaves penetrate and propagate through a dielectric material, the internal electric fields generated within the affected volume induce translational motions of free or bound charges (e.g., electrons or ions) and rotate charge complexes such as dipoles. The resistance of these induced motions due to inertial, elastic, and frictional forces, which are frequently dependent, causes losses and attenuates the electric field. As a consequence of these losses, volumetric heating results. For convenience, the loss mechanisms are all combined together in one loss parameter, ε''_{eff}. However, the loss tangent (tan δ) is commonly used to describe these losses as

$$\tan\delta = \frac{\varepsilon''_{eff}}{\varepsilon'_r} = \frac{\sigma}{2\pi f \varepsilon_0 \varepsilon'_r} \tag{2}$$

TABLE 1. CHARACTERISTICS AND ADVANTAGES FOR MICROWAVEPROCESSING CERAMICS

Characteristic	Advantages over conventional heating processes
Direct coupling (absorbing) of microwave creates volumetric (bulk) heating	• Potential to heat large sections uniformly • Reversed thermal gradients: surface cooler than interior • Process materials at lower surface temperatures • Rapid removal of water, binders, and gases without rupture or cracking • Internal stresses reduction by lower thermal gradients • Heat in clean (pure) environment; air, controlled atmospheres, vacuum, or pressure • Control partial pressure of reactive gases for selective oxidation/reduction of certain phases • Improvement of product quality, uniformity, and yields • Instantaneous response to microwave power changes • Low thermal mass; precise and automated temperature control
Dielectric losses (and heating) accelerate very rapidly with increasing temperatures above Tcrit	• Ability to heat "transparent" materials above Tcrit • Very rapid processing (2 to 50 times faster than conventional) • Densify materials rapidly with minimum grain growth (accelerated sintering) • Reduce process costs (time, energy, and labor) • Ability to heat ceramics well above 2000°C (in air, vacuum, or controlled atmospheres)
Microwaves are polarized and coherent; location of maximum electric and magnetic fields can be controlled	• Capability of high~energy concentration in short times and in selected regions • Frequency and power level optimization for given material, size, and shape • Potential for process automation, flexibility, efficiency, and energy savings • Precise heating of selected regions, ie., brazing or sealing of joints, fiber drawing, and plasma generation • Acceleration of sintering and diffusion due to high electrical fields; thus densification at lower temperatures
Differential microwave coupling of phases, additives, and constituents leads to selective heating	• Synthesis of new materials and microstructures • Heating of selected zones (brazing and sealing) • Enhanced coupling of microwave transparent materials • Use of fugitive coupling materials for preheating of otherwise transparent materials • Use of microwave-coupling materials as shapes or containers to heat the more transparent materials • Superior control over state of oxidation through selective heating of phases and control over oxygen partial pressure

where σ is the total effective conductivity (S/m) caused by conduction and displacement currents and f is the frequency (GHz).

The power absorbed per unit volume P (W/m^3), provides the following basis for heating:

$$P=\sigma|E|^2=2\pi f\varepsilon_0\varepsilon_r' \tan\delta |E|^2 \quad (3)$$

where E (V/m) is the magnitude of the internal field. Equation (3) shows that the power absorbed varies linearly with the frequency, the relative dielectric constant, and tan δ, and varies with the square of the electric field. It is assumed that P is uniform throughout the volume and that thermal equilibrium has been achieved. This is rarely the case. In addition, f, ε_r', tan δ, and E are all interdependent. Furthermore E is dependent on the size, geometry, and location of the material within a microwave cavity and on the design and volume of the cavity. However Eq. (3) provides a useful approximation of P and describes the basic relationships between the four variables.

Since the electric fields are attenuated as the microwaves penetrate and propagate through an absorbing material, a useful parameter describing this is D, which is the penetration depth[8] at which the incident power is reduced by one half.

$$D=\frac{3\lambda_0}{8.686\pi \tan\delta(\varepsilon_r'/\varepsilon_0)^{1/2}} \quad (4)$$

where λ_0 is the incident or free-space wavelength. For greater wavelengths (and lower frequencies), the penetration depth is also greater, as shown in Figure 3 for water. The 0.915 and 2.450 GHz frequencies shown in Figure 3 are the most commonly used of the four allocated by the Federal Communications Commission for industrial, scientific, and medical (ISM) applications.[10] Although low frequencies result in greater penetration depths, the heating does not necessarily increase, since the internal field E could be low, depending on the properties of the material.[10]

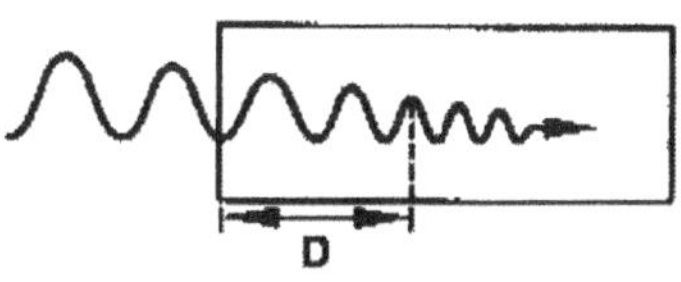

D_{H_2O} (cm)	f (GHz)	λ_0 (cm)
7.6 (3")	0.915	32.8 (12.8")
1.3 (0.5")	2.450	12.3 (4.8")

$$D=\frac{3\lambda_0}{(8.686\pi)(\tan\delta)\sqrt{\epsilon_r'/\epsilon_0}}$$

D = penetration depth at 1/2 power

λ_0 = free space wavelength

Tan δ = dielectric loss tangent

ϵ_r'/ϵ_0 = dielectric constant

Figure 3. Absorption of microwaves to half-power depth, *D*.

The relative dielectric constant (ε_r') and the loss tangent (tan δ) are the two most widely used and measured parameters that describe the behavior of a dielectric under the influence of a microwave field. They both affect the power absorbed (Eq. (3)) and the half-power depth (Eq. (4)), and thus, they influence the volumetric heating behavior of a given material. The value of ε_r' is a measure of the polarizability of a material in an electric field, whereas the value of tan δ is a measure of the loss (or absorption) of the microwave energy within the material.

During heating ε_r' and tan δ change with temperature, and a knowledge of these changes is important for process control. Figure 4 shows that ε_r' of fused SiO_2, Si_3N_4, hot-pressed BN, and a glass-ceramic increase slowly from room temperature to 1400°C and that ε_r' of Al_2O_3, increases at a greater rate. Ho[12] has shown that the increase in ε_r' with temperature is due to an increase in the polarizability caused by volumetric expansion. This could explain the greater increase of ε_r' for Al_2O_3 over that for the other materials shown in Fig. 4, since Al_2O_3 has a significantly greater coefficient of thermal expansion. In addition, the composition and the density have a major influence on the value of ε_r' which is evident from the curves in Fig. 4.

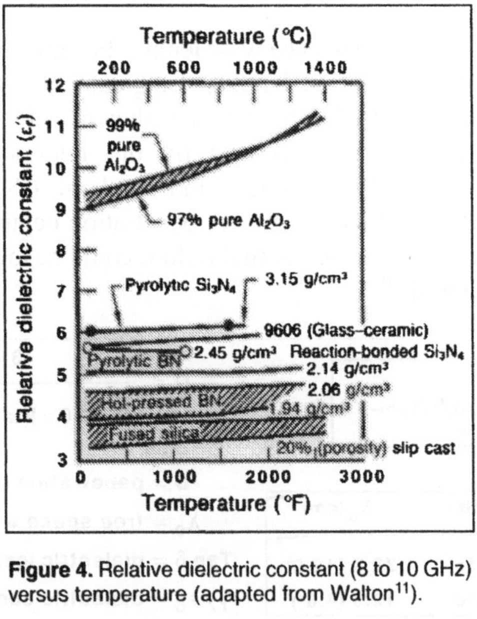

Figure 4. Relative dielectric constant (8 to 10 GHz) versus temperature (adapted from Walton[11]).

In contrast to ε_r', tan δ is far more affected by temperature, as shown in Figure 5. In general, tan δ initially rises slowly with increasing temperature, until some critical point (T_{crit}) is reached, beyond which tan δ rises rapidly. Compositional additives and impurities usually account for the rapid rise in tan δ. Ho[12] reports that the rapid increase in tan δ in polycrystalline ceramics is associated with the softening of intergranular, amorphous phases, which causes an increase in the local conductivity σ (in Eq. (2)). This could explain, for example, why tan δ of the 99% pure Al_2O_3 sample was less affected by temperature than that for the less pure (97% pure Al_2O_3) material.

Thermal Runaway

At temperatures above T_{crit} tan δ rises very rapidly, which causes a condition of thermal runaway in a microwave-heated material. As tan δ begins to increase rapidly, the material begins to absorb microwave energy more efficiently, which also raises the temperature. This causes tan δ to rise even faster. The net result is an exponential increase (runaway) in temperature. The rate of temperature rise and the triggering temperature (T_{crit}) vary widely for different materials.

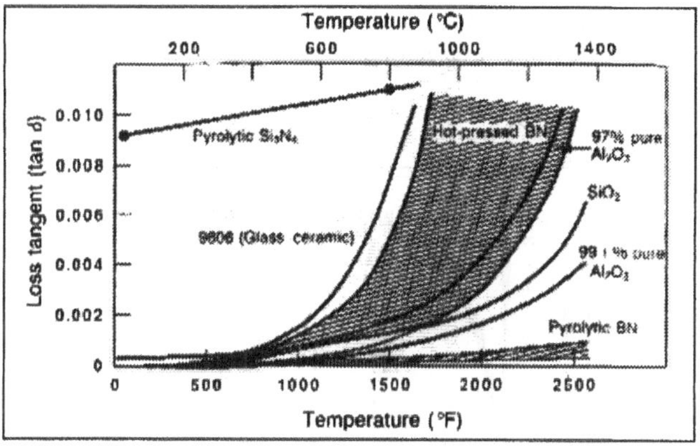

Figure 5. Loss tangent (8 to 10 GHz) versus temperature (adapted from Walton[11]).

The thermal runaway condition is indicated in Figure 6 for two different materials. In material B, microwaves are absorbed efficiently at room temperature, and thermal runaway occurs as soon as the sample is irradiated by microwaves at a given power level. In material A, which is less absorptive (low tan δ), thermal runaway does not occur at the same power level used for material B. However, when the power is increased sufficiently, the initial rate of heating is increased to the point where thermal runaway can be triggered. Figure 7 shows that Fe_3O_4 is similar initially to material B in Figure 6 and that thermal runaway was initiated at all the indicated input power levels.

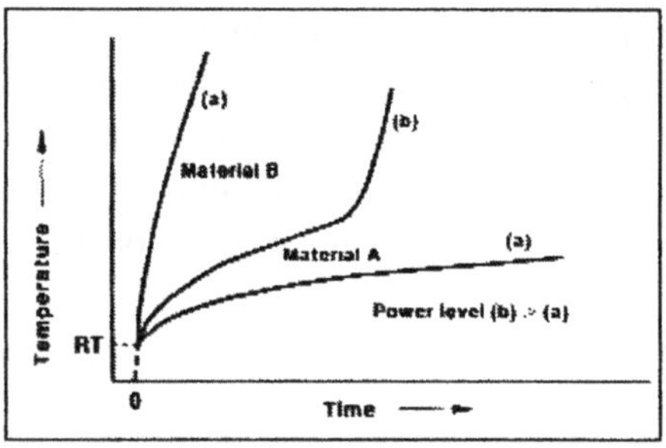

Figure 6. Response of two different materials to microwave heating.

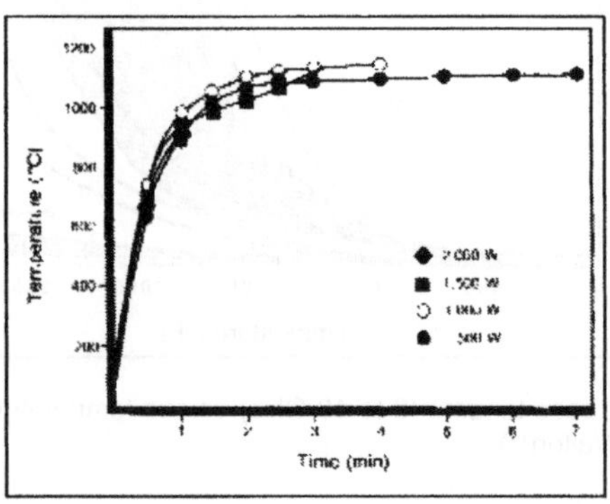

Figure 7. Microwave (2.45 GHz) heating of Fe_3O_4 (McGill et al.[13]).

Since Fe_3O_4 is ferromagnetic, a high value of tan δ_μ (magnetic equivalent of tan δ) is probably responsible for triggering thermal runaway at room temperature. Above 588°C (the Curie point), the rate of heating begins to decrease, and further heating to 1100°C is due to electrical losses only. Since Fe_3O_4 couples (heats) readily with microwaves at room temperature, it can be added to a microwave transparent material, such as Al_2O_3, to increase the overall rate of heating at the lower temperatures.[14] Figure 8 shows the effect of 10 and 50 wt% of Fe_3O_4 additions to Al_2O_3 on the heating curves of 300-g powder mixtures. The Fe_3O_4, addition is selectively heated in the microwave field. The Al_2O_3 behaves like material A in Figure 6, and the mixture of 50 wt% Al_2O_3 and 50 wt% Fe_3O_4, behaves like material B.

Figure 9 shows that Cr_2O_3 behaves like material A in Figure 6. At low-power inputs, the buildup of heat in the material requires more time to trigger thermal runaway. Because the initial temperature rise is slower at 500-W input power than at 1000 W, a significant fraction of the heat can be dissipated from the material to the surrounding environment, thereby increasing the time for the triggering temperature to be reached.

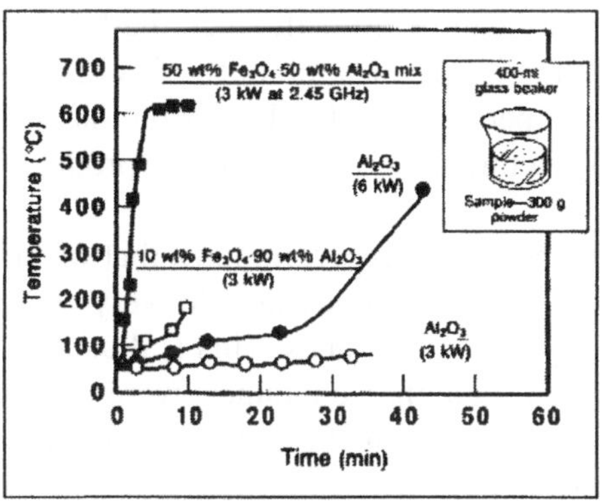

Figure 8. Effect of Fe_3O_4 additions on microwave heating of alumina powders (Sutton[14]).

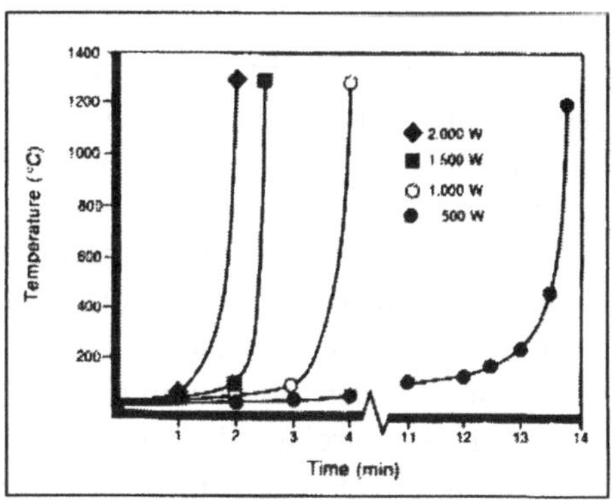

Figure 9. Microwave (2.45 GH_z) heating of Cr_2O_3 (McGill et al.[13]).

Thermal runaway is an important aspect of microwave heating. It can cause undesirable hot spots within a material; it can also be used to heat materials at rapid rates. There are several means to control or prevent thermal runaway, such as by controlling or pulsing the microwave powder level (where changes in the power cause an instantaneous response in the material) or by designing the microwave system and applicator to deposit the microwave energy within the material or product in a prescribed manner.

Applicator Design

The applicator is a device that provides a means for applying the microwave energy from the generator to the workpiece (material to be heated) The most common type of applicator is the multimode cavity, of which thc home oven is an example. The cavity is typically a closed metal box, such as the one shown in Figure 1; is simple mechanically; and can accept a wide range of loads. Its chief drawback is that the nonuniform electric field distribution can cause inhomogeneous heating and hot spots.

When microwaves are introduced into a multimode cavity, they are reflected off of the cavity walls to form a complex pattern of standing waves, which produce regions of high and low field intensity. This means that the volume and placement of the workpiece within the cavity can be critical, since the heating efficiency and uniformity will vary, depending on the location of the field gradients. Using mode stirrers to continuously perturb the field distribution and moving the workpiece through the fields via a turntable or other device produces a more uniform heating pattern.

The filling factor, which is the ratio of the workpiece volume to that of the cavity, also affects the heating efficiency of the cavity. For example, very small workpieces tend to be heated less effectively than larger ones. This is illustrated in Figure 10, where aluminum oxide powders (in a glass container) were placed in the center of a relatively large (1 m^3) microwave oven.[14] Figure 10 shows that the heating effectiveness increases significantly for a given power level as the quantity of Al_2O_3 was increased from 50 to 500 g.

One approach for achieving a uniform electric-field distribution is to design a cavity such that at least one of the major dimensions is 100 times greater than the wavelength of the microwave incident radiation. This results in an untuned cavity in which there is a continuous overlapping of the individual mode resonances of the microwaves. By changing the frequency from the standard 2.45 to 28 GHz, Swain[3] and Kimrey and Janney[15] were able to increase the ratio of the microwave wavelength to the size of a 0.56-m^3 cavity from 1:3 to 1:100. The result was an extremely uniform distribution of the electromagnetic fields throughout the cavity, where the total variation in the field strength was less than ±4%.

For precise and predictable control over the fields within the workpiece, the single mode applicator is used. Such applicators maximize the electrical field strength at the workpiece location so that rapid and uniform heating can be achieved. The design of the applicator can also be changed to provide localized heating in specific regions of the workpiece, which is desired in the case of joining materials together, drawing glass fibers, or creating plasmas to rapidly sinter various ceramic materials. Figure 11 shows a single-mode applicator used by Kemer and Johnson[16] to generate a plasma that induced rapid sintering in Al_2O_3 and in other materials.

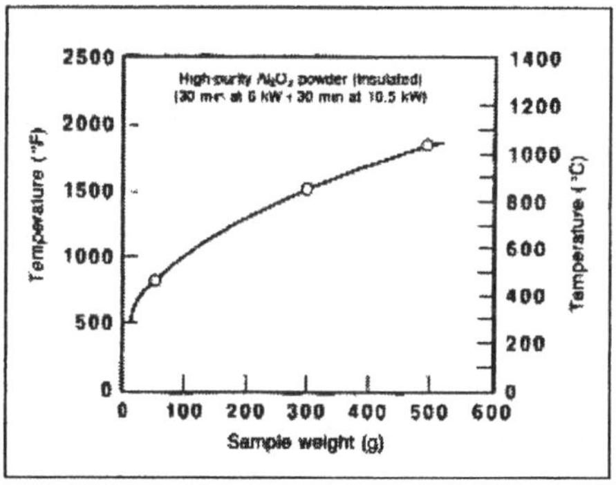

Figure 10. Effect of mass on heating effectiveness of alumina powder (Sutton[14]).

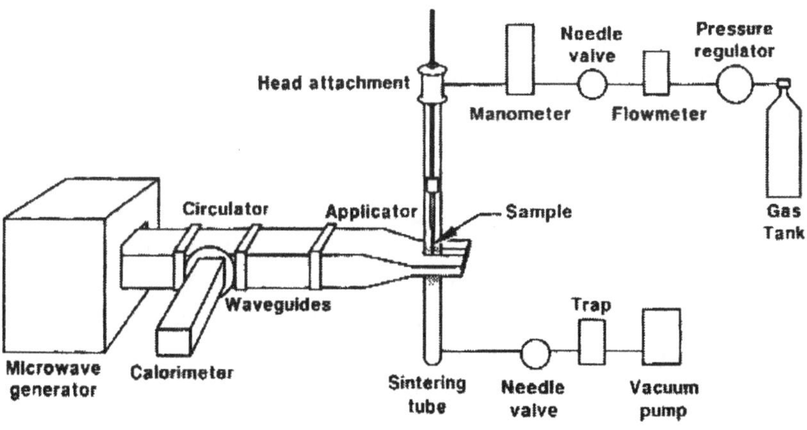

Figure 11. Microwave system used for plasma-induced sintering (Kemer and Johnson[16]).

The design of applicators[17,18] for specific uses is a broad subject and is beyond the scope of this paper. However, it is an important aspect of microwave processing. Some examples of the multimode and single-mode applicators will be given in the next section.

Applications of Microwave Processing

Most of the early work on the processing of ceramic materials dates back to the early 1960s and has led to some commercial applications in the foundry and investment casting industry. Work on high-temperature applications, such as melting and sintering, began in the late 1960s and early 1970s.[19-22] Much of the theory and some data on microwave-material interactions, which provided a valuable background for microwave heating, had been reported earlier by von Hippel in 1954.[23] Although considerable interest was directed toward the heating of ceramics (drying, calcining, sintering, etc.) through the 1970s, several other applications, such as accelerated slip casting and acid digestion of ceramics and minerals, were developed. Up to the present time, many types of potential applications have been investigated, and these are categorized in Figure 12, which was adapted from an original chart by Roy *et al.*[24] Each of these categories will be discussed with some examples, which are listed in Table 2.

Process Control

Since many ceramic materials are transparent to microwaves at ambient temperature, microwaves can be used to detect moisture, defects, pores, etc., that have different absorption characteristics. Thus, microwaves can be used for both process control and nondestructive evaluation (NDE). Campbell and Shivers[25] have discussed the use of microwaves for ceramic process control, citing several examples of moisture measurements, contact-free thickness gaging, flaw detection, and vibration measurements. Barr[26] investigated the use of microwaves to detect flaws in dense Si_3N_4 and indicated that there were many opportunities for this type of application.

Plasma Processing

Figure 12 indicates that microwave-generated plasmas can be used in different ways: (1) in microwave-plasma chemical vapor deposition (MPCVD) and (2) in microwave-plasma-induced sintering (MPIS). The MPCVD process assists in decomposing gaseous species to produce highly excited electronic states of the elements that recombine to form metastable elements. Diamond and diamondlike films have been produced using this process.[27,28]

The MPIS method generates plasmas that provide the intense heat to very rapidly sinter ceramic materials. This process will be discussed in a later section.

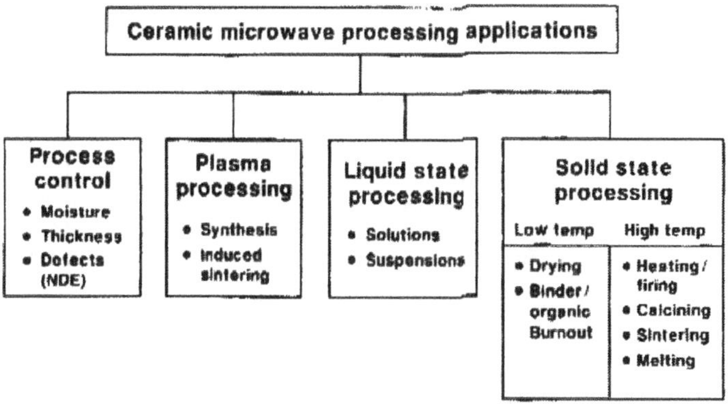

Figure 12. Application of microwave energy to process ceramics (adapted from Roy et al.[24]).

Liquid-State Processing

The application of microwaves to liquids in the form of solutions or suspensions provides a new avenue to process, synthesize, or analyze materials.

Solutions. The U.S. Bureau of Mines has developed a means to greatly reduce the time for the wet-chemical analysis of minerals, ceramics, alloys, etc., by greatly increasing their dissolution rates in acids.[29,30] Microwaves generate the heat in which the temperature and pressure are both increased in a sealed vessel so that the digestion times can be reduced from several hours to a few minutes for many materials. Specially designed, microwave-transparent Teflon® reaction vessels were designed for this process and can be used to temperatures up to 180° to 250°C and pressures as high as 8.2 MPa.[31] Another advantage of using sealed vessels is that volatile elements can be quantitatively retained.[32]

The vitrification of high-level radioactive wastes by microwave energy has been demonstrated on a pilot scale. The evaporation of solutions, the calcination of uranyl salts and radioactive wastes with glass-forming additives, and the final vitrification were all accomplished using microwaves.[33]

Another demonstrated advantage of microwave processing was the direct heating of a product without contamination from heating elements or other devices. Hall Chemical, for example, has replaced tank immersion heaters which corrode and contaminate tank contents, with wraparound microwave heaters.[34]

Suspensions. Microwaves have been used successfully on a pilot scale in casting, setting, and drying slip-cast whiteware and refractory bodies. An early patent in 1973[35] showed that microwaves could greatly speed up the process of slip casting sanitary ware by heating the slip to 65°C and then hardening the ware in the molds after casting.

TABLE 2. EXAMPLES OF MICROWAVE-HEATED CERAMICS

Process	Material†	Sample size, quantity, or weight
Combined		
Dry, calcine, BB,* sinter	PZT ceramics	10-mm-dia. pellets
	PLZT ceramics	10-mm-dia. pellets
Dry, heat	Mullite, SiO_2, Al_2O_3, gels	2-26 g
Calcine, anneal	$YBa_2Cu_3O_{7-x}$	10-mm rod
BB, sinter	Ba-ferrite	12.7-mm disks
BB, sinter	Al_2O_3 spark plug insulator	186 insulators
Dry, fire	Al_2O_3 castable shapes	1-m (68-kg) long
Clinker	Portland Cement	10-mm cylinders
Sinter	Al_2O_3	Pellet
	99.995% pure Al_2O_3	10-mm pellet
	99% pure Al_2O_3	Pellet
	Al_2O_3 +0.1MgO	100-g pellet
	Al_2O_3 +2SiC (powder)	Pellet
	Al_2O_3 +10SiC (platelets)	Pellet
	Al_2O_3 +10SiC (whiskers)	Pellet
	Al_2O_3 +10SiC (whiskers)	10-mm pellet
	(Al_2O_3+0.2MgO)+30TiC	4-mm rod
	99% pure Y_2O_3	Pellet
	Y_2O_3 +2ZrO_2	Pellet
	Y_2O_3 +10Zr-oxynitrate	Pellet
	ZrO_2+4CaO	Pellet
	ZrO_2+3Y_2O_3	50-g disk
	ZrO_2+8Y_2O_3	Pellet
	ZrO_2 (PSZ)	Pellet
	ZrO_2+10Zr-oxynitrate	Pellet
	ZrO_2+10SiC (whiskers)	Pellet
	ZrO_2+10Si_3N_4 (whiskers)	Pellet
	$SrTiO_3$	Pellet
	B_4C (pure)	10-mm cylinder
	SiC+0.6B+4C	4.5-mm rod
	Glass+SiC-fiber composite	76x76x3 mm
Plasma-induced sintering	Al_2O_3+0.25MgO	4.8-mm rod
	Al_2O_3+0.25MgO	4.8-mm rod
	SiC+0.6C+2B_4C	Rod
	Si_3N_4+5MgO	Rod
Melt	UO_2	103 g
	U_3O_8	55 g
Join	Al_2O_3+sealing glass	50x50x0.5 mm plates
	99% pure Al_2O_3 + flux	3-mm rod: butt join
	92% or 96% pure Al_2O_3, no flux	3-mm rod: butt join
	Mullite, no flux	9.5-mm rod: butt join
	Si_3N_4+flux	3-mm rod: butt join
Draw fibers	SiO_2 (pure)	Rod
	(ultrapure)	Tube/rod

*BB=Binder burnout. †All proportions are in wt%, except where purity percentages are indicated.

‡Single-mode applicator (otherwise, multimode). §Heating time from room temperature to temperature indicated.

Microwave frequency (GHz)	Atmosphere	Temperature (°C)	Hold time (s)	Theoretical density (%)	Reference
2.45	Air	1310	300	Dense	Harrison[44]
2.45	Air	1200	300	Dense	Harrison[44]
2.45	Air	1500+			Komarneni[48]
2.45	Air/O_2	1800	750/500		Ahmad[47]
2.45	Air	1205	2700		Krage[63]
2.45	Air	1600	1740	99.9	Schubring[21]
2.45	Air	1200	2.9x104[§]		Sutton[14]
2.45[±]	Air	>1400	420[§]		Quéméneur[49]
2.45	Air	1500		95	Varadan[51]
60		1700	370[§]	95.7	Meek[53]
2.45	Air			91.7	Meek[52]
28	Vacuum	1200	3600	98.2	Janney[54]
2.45	Air			96.9	Meek[52]
2.45	Air	1600		94	Katz[56]
2.45	Air	1600		72	Katz[56]
60		1450	320[§]	77	Meek[53]
2.45	1.1 MPa	1850		95	Tian[57]
2.45	Air			57.4	Meek[52]
2.45	Air			93.3	Meek[52]
2.45	Air		1800[§]	99.6	Meek[52]
2.45	Air			67.5	Meek[52]
28	Vacuum	1150	3600	99+	Kimrey[59]
2.45	Air			97.6	Meek[52]
2.45	Air			98	Wilson[60]
2.45	Air			99.6	Meek[52]
2.45	Air		>1700		Blake[61]
2.45	Air		>1700		Blake[61]
2.45	Air	1300	430[§]	98	Varadan[51]
2.45	Air	2000	450	95	Katz[65]
2.45	0.8 MPa N_2	2200			Tian[57]
2.45	Air	1200	120[§]		Sutton[58]
2.45[‡]	4 kPa N_2	1600	<120	99.0	Kemer[16]
2.45[‡]	4 kPa N_2	1660	600	99.9	Kemer[16]
2.45[‡]	N_2				Hsu[66]
2.45[‡]	N_2				Hsu[66]
2.45	Air	>2600	900[§]	Melted	Haas[67]
2.45	Air		1200[§]	Melted	Haas[67]
2.45	Air	1400	1200		Meek[69]
2.45[‡]	Air	1800	180		Fukushima[72]
2.45[‡]	Air	1750	180		Fukushima[72]
2.45[‡]	Air	1300	<600		Palaith[71]
2.45[‡]	N_2	1720	180		Fukushima[72]
2.45[‡]	Air	>2000			Berteaud[22]
2.45[‡]	Air	>2000			Hässler[73]

Chabinsky and Eves[8] worked with Tobin (Tobin process[36]) to develop a total production scheme (from casting to firing) for using microwaves in several of the processing steps for producing slip-cast sanitary ware. They investigated the use of microwaves to assist in (1) removing water after the slip had been poured into molds, (2) hardening the cast green ware after the excess slip had been drained, (3) drying the molds, (4) heating after glazing, and (5) firing the glazed ware.

Experiments on production castings of the first three process steps demonstrated the following: (1) microwaves accelerated the movement of water from the slip through and to the exterior surfaces of the molds where evaporation took place; (2) the drained molds retained an excellent surface and could be reused immediately without the need for additional drying; and (3) the development of green strength in the cast ware and the drying process were accelerated. The time required for the total casting cycle was reduced from 3 or 4 h to 0.5 h. Microwave processing improved product quality and yields, reduced the number of required molds by two-thirds, doubled the throughput for the same number of workers, and greatly reduced the floor-space requirements.

Oda and Balbaa[37] reported similar benefits from using microwaves in slip-casting clay bodies, in which the casting process time was reduced from 1.5 h to 20 min, and the drying of castings from 24 h to 8 min.

Pike and Binner[38] described another advantage for using microwaves during the slip casting process, in which the normal solids content (60% to 70%) in the slip could be reduced significantly and thus could improve the microstructural uniformity of the cast parts.

Solid-State Processing

The use of microwaves for processing solid ceramics can be divided into low- (<500°C) and high- (>500°C) temperature applications. Several commercial applications involve the use of microwaves at low temperatures. However, applications for high temperatures are primarily in the research stage, although some commercial processes have been investigated on a pilot scale. Melting was included in this section, since most of the examples only involve partial melting of the solid material (e.g., joining and fiber drawing). Table 2 lists several examples of various types of solid-state processing cited in the literature.

Low-Temperature Processing

Drying. Water strongly absorbs 0.915 and 2.450 GHz microwaves and, thus, is very efficiently heated at these frequencies. However, for water contents above 5% to 10% in solids, conventional heating processes are more economical since microwave energy is more expensive. Only about 50% of the electrical input energy in microwave generators is converted to microwave energy. However, at low water contents (<5%), the removal of water becomes very inefficient for conventional processes, and microwave drying becomes economical.[17] A hybrid system, in which both conventional and microwave heating are used, is often the most efficient. This system is often the most rapid drying process[17,34] for materials that have high water contents (>10%) and that must be thoroughly dried.

Microwave drying is particularly effective in removing low water contents from thick sections, where the volumetric heating drives the water away from the interior, hotter sections. One of the first commercial applications of microwave heating was the curing and drying of foundry cores,[39,40] where the time was reduced to one-half to one-tenth of that required by conventional heating. Another foundry application for microwaves was the drying of ceramic wash coatings applied to molds for use in the lost-foam casting process.[6]

The microwave drying of plaster molds used in casting large aluminum patterns has had a dramatic impact on improving the quality of the casting,[41,42] since the surface of the microwave-dried molds are free from cracks and other flaws that are often found in conventionally dried molds. Microwaves have also been used to dry space-shuttle tiles[9] and large fibrous silica insulating blocks.[43]

Microwaves have been used to precisely control the moisture content in clays prior to shipment and to dry pigments and other powders. The microwave-processed powders do not tend to form clusters that require additional processing.[8]

Organic Material Burnout. The binders normally used in forming ceramic shapes can be removed by microwave heating, either by direct coupling (if they are lossy materials) or by the heat generated in the ceramic part. For example, binders have been burned out of spark plug insulators[21] and ferroelectric pellets of polycrystalline lead zirconate titanate (PZT) and lead-based lanthanum-doped zirconate titanate (PLZT)[44] during the early stages of the microwave-firing process (Table 2).

The use of microwave energy to remove waxes from ceramic shell molds in the investment-casting industry has been a commercial success[18,45] and has demonstrated many advantages over conventional dewaxing, including recovering wax for reuse as pattern wax; using only one-tenth the energy of the autoclave process; burning very little of the wax that is essentially smoke free and minimizes polychlorobiphenyl (PCB) contamination; reducing labor and storage space; and reducing significantly the breakage of shells. This process is an excellent example of selective heating, since the shell and not the wax is heated first. In the conventional process, both are heated to a higher temperature, and the greater expansion of the wax often causes the shell mold to crack.

High-Temperature Processing

Several processing steps, such as calcining, firing, sintering, and melting with microwave energy, have been investigated. Examples of each of these steps will be presented to illustrate the progress being made and to indicate the versatility of the microwave process. Table 2 gives examples of each of the processing steps that have been investigated either singly or in combination.

Combined Process Steps. Harrison *et al.*[44] were the first to demonstrate the feasibility of using microwave energy for all of the thermal treatment steps (drying, calcining, binder burnoff, and sintering) required for preparing PZT and PLZT ferroelectric materials. They further showed that the microwave process was 6 to 12 times faster and required only 5% of the energy used by the conventional sintering process. In addition, the use of microwave processing made it possible to tailor the microstructure to obtain unique properties in PZT.

Schurbring[21] investigated the microwave sintering (including binder burnout) of Al_2O_3 spark plug insulators. He showed that sintering temperatures of 1600°C could be achieved in a multimode cavity, and by controlling the power input, thermal runaway was prevented. As many as 186 insulators were heated to 1600°C in 2.6 h, and the quality was comparable with the gas-fired product. The microwave energy required per insulator for the complete firing cycle was about 50% of the energy required in the standard manufacturing process. The use of a microwave technique was suggested for on-line thermal analysis of parts during the heating process to assist in quality control and for automatic control of the firing profile [21]

Sutton and Johnson[14,46] used microwave heating to dry and fire production shapes of high Al_2O_3 castables weighing as much as 68 kg and having lengths of up to 1 m. A two-stage firing schedule was developed (Fig. 13) in which the free water (remaining after the casting process) was removed in the first stage. In the second stage, the power was increased until the peak temperature was reached. Ceramic-fiber insulating blankets transparent to microwaves were wrapped around the shape before the second stage was initiated to minimize radiation losses. Figure 14 indicates that trade-offs can be made in firing time versus microwave power. For example, a castable shape heated at a power concentration of 550 W/kg (250 W/lb) can be fired 3.6 times faster than one heated at 220 W/kg (100 W/lb). By optimizing firing time versus microwave power input, the microwave-heating process was 50% faster and 20% to 30% lower in energy costs than the gas-fired production process.

Ahmad *et al.*[47] used microwaves to calcine, sinter, and anneal $YBa_2Ca_3O_{7-x}$-superconducting pellets. The microwave-processed pellets exhibited a more refined microstructure, lower porosity, improved oxygen content, and higher

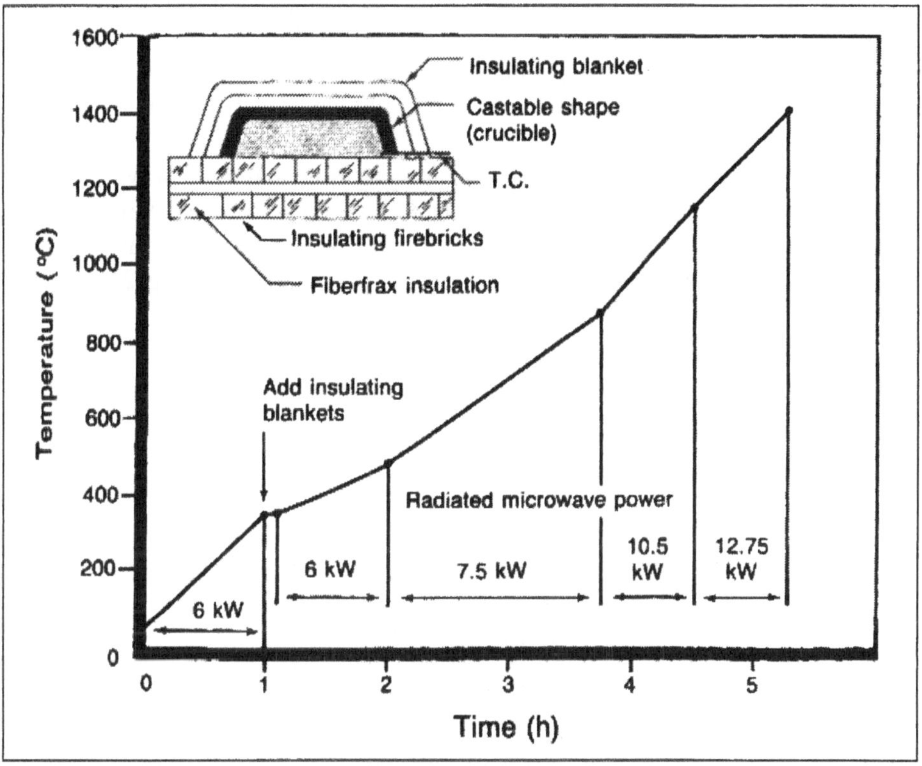

Figure 13. Microwave firing schedule for alumina castable shape (Sutton[14]).

superconductivity transition temperature (93 K versus 90 K) over conventionally processed samples.

Microwaves have also been used to dry, to heat, and, in some cases, to melt gels.[24] Komarneni *et al.*[48] prepared crystalline mullite powders by microwave-heating single-phase gels. Gels heated for 5 min after thermal runaway showed euhedral grains of approximately 100 to 200 nm. Microwave heating appears to be a feasible way to produce fine mullite powders that are usually difficult to make by normal grinding or milling procedures.

Clinkering. Quéméneur *et al.*[49] studied the clinkering of portland cement using a grooved, resonant single-mode applicator. They demonstrated for the first time that microwaves could produce a clinker with characteristics similar to the industrial product.

Sintering. Table 2 shows some of the materials investigated in microwave-sintering experiments. Tinga[50] and Varadan *et al.*[51] have recently reviewed the fundamentals of the microwave-sintering process. Tinga described some simplified models that could give the process designer insight into the response of materials to microwaves during sintering. Varadan discussed how a rigorous application of the multiple scattering theory could lead to a better understanding of the microwave-sintering process.

The influence of composition, specimen preparation, and microwave-processing conditions (microwave frequency, applicator type, power level, vacuum versus pressure, etc.) on the sintering behavior of Al_2O_3 compositions has been studied by several investigators[51-57] (Table 2). Meek *et al.*[52] using 2.450-GHz

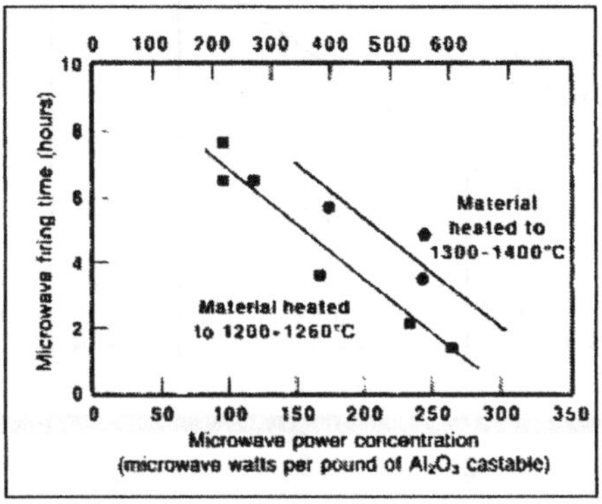

Figure 14. Firing time versus microwave power for alumina castable shapes (Sutton[14]).

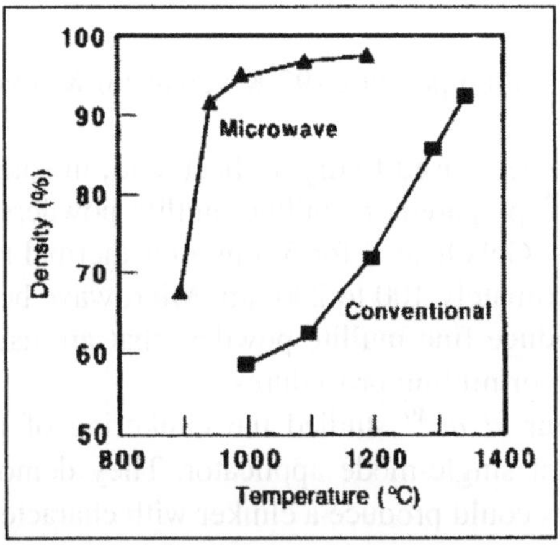

Figure 15. Density versus temperature of microwave (28 GHz) and conventionally sintered Al_2O_3 (Janney and Kimrey[55]).

radiation heated powdered compacts of 99.99% pure Al_2O_3, to a final sintered theoretical density of 91.7% and an average pore size of 0.3 μm. Additions of 2% of TaC, SiC, or $MoSi_3$ increased the final density to about 97% and resulted in an average pore size of 1.2 to 1.8 μm. Meek *et al.*[53] observed that high-purity, submicrometer powder compacts of pure Al_2O_3 could be densified from an initial value of 50% to a final value of 95.7% by heating from room temperature to 1700°C using 60-GHz radiation. They noted that conventional sintering of high-purity submicrometer Al_2O_3 powder compacts (50% initial density) required 20 h at 1600°C to achieve the same final density.

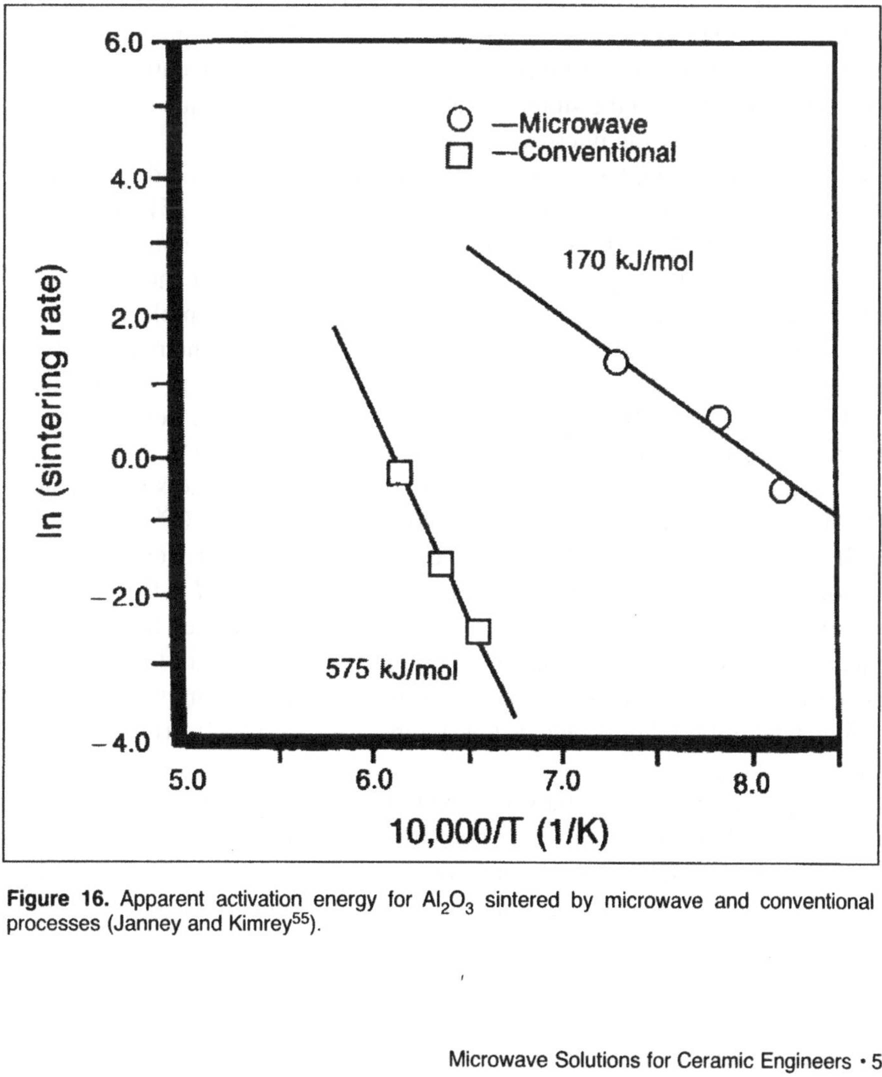

Figure 16. Apparent activation energy for Al_2O_3 sintered by microwave and conventional processes (Janney and Kimrey[55]).

Janney and Kimrey[54,55] sintered compacts of Al_2O_3 + 0.1 wt% MgO under vacuum using 28-GHz radiation. By processing at 28 GHz instead of at 2.45 GHz, they suggested that about 25 times more power can be dissipated in high-purity Al_2O_3 (according to Eq. (3)), because of the higher frequency and the greater value of tan δ at 28 GHz. They also observed that microwave heating densifies Al_2O_3 far more rapidly than by conventional heating, as shown in Fig. 15, in which the material, the initial density (55%), the heating rate (50°C/min), and the pressure (102 to 103 Pa) were the same for both processes. After 1-h exposure at 950°C, sintering had begun in the microwave-heated sample, whereas an additional 200°C at 1 h was required to achieve the same density in the conventionally heated material. After 1 h at 1200°C, the microwave-heated sample was 98.2% dense compared with 71.4% for the conventionally sintered Al_2O_3. Janney and Kimrey[55] calculated the apparent activation energy for sintering Al_2O_3 by the two processes. Figure 16 shows that the apparent activation energy for microwave sintering is about one-third that of conventional sintering. The large differences in the sintering rate were thought to be due to higher diffusion rates induced by the microwave field.

Below the triggering temperature for thermal runaway, Al_2O_3 does not absorb microwaves effectively because of the low value of tan δ (Fig. 5). However, the heating rate of Al_2O_3 can be greatly increased by adding a conductive phase, such as SiC particles or fibers. The final sintered density of these mixtures is influenced by the shape and size of the added phases. For example, a 50% dense Al_2O_3 compact containing 10 vol% SiC whiskers could be sintered to only 77% theoretical density after being microwave heated from room temperature to 1450°C in 320 s in a 60 GHz-microwave cavity.[53] No further densification was observed when the sintering temperature was raised to 1700°C. Similar results were obtained when compacts of Al_2O_3 + 10 vol% SiC whiskers were heated in a 2.450-GHz field to 1600°C, where the final density was 72%.[56]

The sintering of Al_2O_3 –30 wt% TiC composites under nitrogen gas pressures of about 1 MPa in a microwave field was studied by Tian *et al.*[57] Composite rods were rapidly and uniformly heated to 1850°C and were sintered to 95% of theoretical density without cracking. These composites are difficult to densify using conventional sintering methods, which require high temperatures (1950°C), high heating rates (400°C/min), and a low temperature (1100°C) prior to the high-heating-rate step.

Many glass-ceramic materials also exhibit low loss tangent values below 600° to 700°C (see Fig. 5) and, thus, are poor microwave absorbers below these temperatures. By adding continuous fibers of silicon carbide to a lithium aluminosilicate (LAS) glass ceramic, the low-temperature microwave absorption

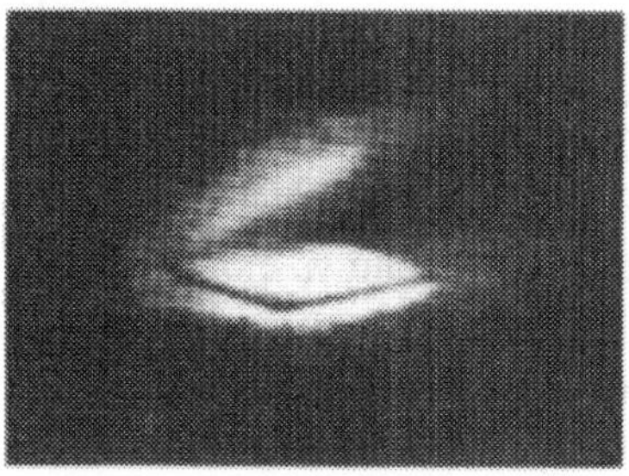

Fig. 17. Nicalon (SIC) fiber-glass composite specimen microwave heated to 1200°C (upper layer of insulation raised, specimen size: 76 by 76 by 3 mm[58]).

was increased dramatically A composite of LAS containing SiC fibers was heated to 1200°C in about 120 s in a 2.45-GHz multimode cavity.[58] Figure 17 shows this composite at about 1200°C just after the upper layer of a noncoupling ceramic fiber insulation was raised to expose the composite.

Many other oxide and composite oxide systems, as shown in Table 2, have been sintered by using microwaves.[51,52,59-61] Varadan *et al.*[51] rapidly sintered $SrTiO_3$ to a density of 98% by microwave heating to 1300°C at 3°C/s. Conventional heating (0.08°C/s) resulted in a sintered density of 80%. They also found that the amount of initial porosity in the $SrTiO_3$ compacts influenced both the time to reach maximum temperature and the maximum rate of temperature rise (thermal runaway). They noted that the initial porosity in compacts of Al_2O_3 also had a similar effect. Meek[62] suggested that a cold-pressed green ceramic body will begin to heat uniformly if the initial porosity is distributed uniformly throughout, since it appears that the electric fields and power density arc greatly intensified in thc porous regions between the grains. When dense regions begin to form, the microwave field will begin to decouple from these regions and then to couple more strongly to the regions of low density and low ε_r'. The net effect is a uniformly densified composite.

Magnetic oxides such as ferrites have been rapidly sintered in microwave fields, and the physical and magnetic properties were comparable with those of conventionally sintered ferrites.[63,64] Krage[63] indicated that in addition to dielectric losses, magnetic losses were also operative at temperatures below the Curie temperature.

The covalently bonded carbides and nitrides are not effectively densified by conventional pressureless sintering, without the addition of sintering aids. However, Katz *et al.*[65] have shown that B_4C can be sintered without the benefit of sintering aids to a theoretical density of 95% in less than 12 min by microwave heating to 2000°C. To achieve a comparable density using conventional sintering processes, additions of sintering aids (C, TiB_2, or AlF_3) and temperatures in excess of 2100°C were required.

Microwave-Plasma-Induced Sintering (MPIS). As was discussed earlier, high-temperature plasmas generated by microwave excitation of gases can cause rapid heating and sintering of many ceramic materials. Kemer and Johnson[16] were able to heat thin rods of porous Al_2O_3, containing 0.25 wt% MgO, at a rate of about 100°C/s in a nitrogen plasma. Less than 2 min were required to increase the density of the starting rods from 50% to 99% at about 1600°C. Densities of 99.9% could be achieved in less than 10 min at 1660°C. A grain size of about 2.5 μm was achieved in a 99.5% dense sample, which is typically 1 order or magnitude smaller than that found in conventionally sintered samples of comparable density. Hsu and Johnson[66] found that other materials, such as SiC (+0.6 wt% C + 2 wt% B_4C) or Si_3N_4 (+5 wt% MgO), would not sinter in a plasma. They concluded that the partial pressure of oxygen was too high in the plasma and that a reducing plasma would be beneficial to sintering.

Microwave Melting. Many examples of microwave-melted ceramics are cited in the literature.[22,24,67-73] Some examples are presented in Table 2 and will be discussed. Haas[67] observed that pure gels of UO_2 and U_3O_8, would continue to heat after all moisture had been removed. Pellets of UO_2 and U_3O_8 could be heated to their melting points using 2.45 GHz. However, no significant heating was observed with dried samples of UO_3, which indicates that reduced states of oxidation below U^{6+} were necessary.

Roy *et al.*[24] found that silica gels could be melted with microwave energy. Komarneni and Roy[68] noted that zeolites absorbed microwaves quite efficiently, which was believed due to a loss mechanism caused by ionic “rattling” in a rigid molecular cage. Both rapid heating and melting were observed in Na, K, and Cs containing zeolites.

Joining. Meek and Blake[69] bonded two plates of Al_2O_3 together by microwave heating a mixture of glass-sealing powders applied as a thin layer between the plates. The powders contained a microwave-coupling material[70] so that they could be selectively heated and rapidly fused. The microwave process proved to be 8 times faster than conventional sealing processes and used about 5% of the energy.

High-quality joints have been produced in alumina, mullite, and silicon nitride by microwave heating the ends of two rods butted together.[71,72] A specially designed single-mode cavity was used to maximize the electric field and maximize heating in the area to be joined. Palaith *et al.*[71] butt-joined two mullite rods together without any sintering aids (fluxes) at 1300°C in about 10 min. A novel feature of the process was the use of acoustic waves to continuously monitor the reflections from the butt interface. When these reflections disappeared, the joining was completed. Microscopic examination of the joined zone revealed a lower porosity and greater hardness, but no evidence of the original interface was found. The strength and toughness of the joint was superior to that of the original rods. The joining of

Si_3N_4 rods required an Ar atmosphere.

Fukushima *et al.*[72] also used a single-mode rectangular cavity to self-join rods of Al_2O_3 and Si_3N_4. Alumina rods of 92% and 96% purity could be directly bonded without sintering aids or fluxes. Maximum joint strengths, which were equal to the strength of the original material, were achieved at temperatures of 1750°C and above for 3 min. The bending strength of joined 92% alumina rods was 420 MPa, which was retained at temperatures of up to 800°C. The joining of high-purity Al_2O_3 or Si_3N_4 rods required a thin layer of a lower-purity sheet placed in the joint prior to microwave heating. The silicon nitride specimens required a nitrogen atmosphere for strong joints to occur.

Fiber Drawing. Hassler and Johansen[73] used microwave heating in a specially designed single-mode applicator to draw ultrapure quartz optical fibers at temperatures above 2000°C. The process involved preheating hollow tube quartz preforms to 1000°C using an oxygen-hydrogen torch to increase the microwave absorption ($> T_{crit}$). This was done so that microwave energy could then be used to collapse thc tubes and to initiate the fiber-drawing process. The microwave process reduced the amount of material blown away by the torching during tube collapse from 10% to 15% to 1% and required about 5% of the energy cost. Optical fiber quality was comparable with the standard process.

From Research and Development to Production

The microwave processing of ceramics is a relatively new development and is fundamentally different from the traditional heating processes used in the ceramic industry. The chief difference is that microwave energy is deposited instantaneously and volumetrically within the bulk of the ceramic material, which is typically a (dielectric) insulator.

The reasons for the growing interest in the use of microwave energy include rapid processing and reduced production costs and energy; improved product quality and uniformity; development of novel microstructures and properties; and synthesis of new materials. These and other characteristics of the microwave process were summarized in Table I.

Many examples were cited in the text and in Table 2, which indicate that considerable progress has been made in processing a wide variety of ceramic materials. The results also show the versatility and adaptability of the microwave process to applications involving liquid, gaseous (plasma), and solid states.

Since microwave processing is new to most of the ceramics industry, considerable time may be required to develop and implement it as a new production process. In a new application, the use of microwaves may be difficult to justify on energy costs alone,[74] since microwave energy is more expensive than fossil fuel energy. However, other benefits from the microwave process, such as much faster throughput of materials, improved quality and yields, less floor space, lower inventory, and new or unique properties, may show that the total economics of the new process is not only viable but is highly desirable.

Unfortunately, many of the benefits are not realized at the initiation of an economic analysis and do not become evident until after the laboratory tests have been completed. Thus, careful planning, execution, and extrapolation of economic analyses and laboratory results to a production scale are crucial to accepting and implementing a production process.[18,34] In some cases, major savings in processing time, energy, and costs can be achieved by hybridizing the process (i.e., by using both conventional and microwave processing,[18,34] or possibly by using microwaves for more than one process step[8,14,44]).

Thus far, commercial application for microwave processing of ceramics have been in the low-temperature (<500°C) region, such as the curing, drying, and dewaxing of ceramic molds and shells used in the metal-casting industries. However, recent investigations on the use of microwaves during the slip casting of production shapes have shown that significant cost savings were possible in terms of energy, time, labor, floor space, mold inventory, etc., but also that mold surface quality and product uniformity and yields were significantly improved over the conventional process.

The industrial applications for high-temperature processing are yet to be developed, although some pilot-scale and experimental studies are showing encouraging results. For example, several studies have shown the advantages of microwave heating over conventional heating processes, including accelerated sintering with finer grain sizes; lower densification temperatures; densification of materials such as B_4C without the use of sintering aids: the development of high-quality, high-strength joints with fluxes, and in some cases direct self-joining without fluxes; and the synthesis of new materials (diamond, powders from gels, and superconducting and ferroelectric materials).

In recent years, much progress has been made in microwave-processing technology. Many problems still need to be solved, ranging from basic scientific studies on microwave-material interactions and loss mechanisms to engineering application programs involving both process and equipment design. There is a critical need for a broad data base on dielectric properties to high temperatures at various frequencies for a wide range of ceramic materials. Such data are needed to develop process models that will predict the internal fields and the heating patterns and rates, so that optimum processing parameters can be developed to meet material and product requirements.

Because of the interdisciplinary nature (e.g., chemistry, physics, electrical and mechanical engineering, process engineering, materials science) of microwave processing, there is increasing need for scientists and engineers trained in these fields for future research and development and manufacturing programs. There is a need also for greater sophistication in processing equipment and applicators designed to meet specific product requirements and to provide high-power densities to achieve maximum use of energy, heating uniformity, and more efficient heating of low loss (low tan δ) materials. This includes the need to further develop sensors for continuously monitoring and controlling the process parameters to provide feedback for in-process control.

During the past 10 years, there has been a significant increase in the number of research and development programs concerned with the microwave processing of ceramics. In the next 10 to 20 years, major advances will develop in the science and technology of microwave processing and in the growth of production microwave processes in the ceramic industry.

ACKNOWLEDGMENTS

The author is grateful to the Materials Research Society for permission to use Figures 7 to 10, 13 and 14 from Ref. 1.

REFERENCES

1. W.H. Sutton, M.H. Brooks, and I.J. Chabinsky, eds., Microwave Processing of Materials, Symposium Procecdings, Vol. 124. Materials Research Society, Pittsburgh, PA, 1988.
2. L. Sheppard, "Manufacturing Ceramics with Microwaves: The Potential for Economic Production," *Am. Ceram. Soc. Bull.*, **67** [10] 1156-61 (1988).
3. B. Swain, "Microwave Sintering of Ceramics," *Met. Prog.*, **134** [3] 76-82 (1988).
4. R.G. Ford "Microwave Processing Symposium Report," *Materials in Processing Report,* **3** [4] 1-4 (1988).
5. R.G. Ford, "Microwave Processing: A New Materials Processing Technology," *Materials in Processing Report,* **1** [11] 1-2 (1987).
6. J. Zygmont, "Industry Warms Up to Microwave Ovens," *High Technol.*, [Mar.] 62-63 (1987).
7. A.J. Klein, "Processing with Microwaves," *Met. Prog.* **1** [4] 36-39 (1985).
8. I.J. Chabinsky and E.E. Eves III, ?The Application of Microwave Energy in Drying, Calcining. and Firing of Ceramics," *InterCeram,* **6** [Dec.] 30-35 (1986).
9. G. Fisher ?Ceramic Producers Explore Microwave Processing," *Ceram. Ind.,* **121**[1] 40-42 (1983).
10. J.M. Osepchuk, "A History of Microwave Heating Applications," *IEEE Trans., Microwave Theory Tech.,* **MIT-32** [9] 1200-24 (1984).

11. J.D. Walton, Jr., "Inorganic Randomes"; pp. 229-344 in Randome Engineering Handbook. Edited by J.D. Walton, Jr. Marcel Dekker, Inc., New York, 1970.

12. W.W. Ho, "High-Temperature Dielectric Properties of Polycrystalline Ceramics"; pp. 137-48 in Microwave Processing of Materials, Vol. 124. Edited by W.H. Sutton, M.H. Brooks, and I.J. Chabinsky. Materials Research Society, Pittsburgh, PA, 1988

13. S.L. McGill, J. Walkiewicz, and G.A. Smyres, "The Effect of Power Level on the Microwave Heating of Selected Chemical Minerals"; pp. 247-52 in Microwave Processing of Materials, Vol. 124. Edited by W.H. Sutton, M.H. Brooks, and I.J. Chabinsky. Materials Research Society, Pittsburgh, PA, 1988.

14. W.H. Sutton, "Microwave Firing of High Alumina Ceramics"; pp. 287-95 in Microwave Processing of Materials, Vol, 124. Edited by W.H. Sutton, M.H. Brooks, and I.J. Chabinsky. Materials Research Society, Pittsburgh, PA, 1988.

15. H.D. Kimrey and M.A. Janney, "Design Principles for High Frequency Microwave Cavities"; pp. 367-72 in Microwave Processing of Materials, Vol. 124. Edited by W.H. Sutton, M.H. Brooks, and I.J. Chabinsky. Materials Research Society, Pittsburgh, PA, 1988.

16. E.L. Kemer and D.L. Johnson, "Microwave Plasma Sintering of Alumina," *Am. Ceram. Soc. Bull.,* **64** [8] 1132-36 (1985).

17. R.B. Smith, ed., Industrial Applications of Microwave Energy. Transactions of International Power Institute, **2,** 112 pp. (1974).

18. A.C. Metaxas and R.J. Meredith, Industrial Microwave Heating. Peter Peregrinus Ltd., London, 357 pp., 1983.

19. W.R. Tinga; Thesis. Department of Electrical Engineering, University of Edmonton, Edmonton, Alberta, Canada, 1969.

20. W.R. Tinga and W.A.G. Voss, Microwave Power Engineering. Academic Press, New York, 1968.

21. N.W Schubring, "Microwave Sintering of Alumina Spark Plug Insulators," Tech. Rept. GMR-4471 General Motors Research Laboratory, 1983, pp. 1-27.

22. A.J. Berteaud and J.C. Badot, "High Temperature Microwave Heating in Refractory Materials," *J. Microwave Power,* **11** [4] 315-20 (1976).

23. A. von Hippel ed., Dielectric Materials and Applications. Technoiogy Press of MIT and Wiley, New, York, 1954, 422 pp.

24. R. Roy, S. Komarneni, and L.J. Yang. "Controlled Microwave Heating and Melting of Gels," *J. Am. Ceram. Soc.,* **68** [7] 392-95 (1985).

25. W.B. Campbell and J.V. Shivers, "Microwaves for Ceramic Process Control," *Am. Ceram. Soc. Bull.,* **52** [3] 260-62 (1973).

26. A. J. Barr, "Microwave Techniques for Nondestructive Evaluation of Ceramics-Phase II," Tech. Rept. AMMRC TR79-9, Contract No. DAAG46-76C-0048, Stanford Research Institute, Menlo Park, CA, Feb., 1979, pp. 1-90.

27. M. Kamo, Y Sato, S. Matsumoto, and N. Setaka "Diamond Synthcsis From Gas Phase in Microwave Plasma," *J. Cryst. Growth,* **62**, 642-44 (1983).

28. B.E. Williams, J.T. Glass, T Horiuchi, and K. Kobashi, "Electron Microscopy of Diamond Films Grown by MPCVD," Diamond and Diamond-like Materials Synthesis, Extended

Abstract EA-15, Spring Meeting of the Materials Research Society, Reno, NV, April 5-6 1988. Materials Research Society, Pittsburgh, PA 1988.

29. U.S. Bureau of Mines, "A Microwave System for the Acid Dissolution of Metal and Mineral Samples," *Technol. News,* No. **185** (Oct., 1983).

30. S.A. Matthes R.F. Farrell, and A.J. Mackie, "A Microwave System for the Acid Dissolution of Metal and Mineral Samples," *Tech. Prog. Rept.,* TPR No. 120, U.S. Bureau of Mines, Albany, OR, 1983.

31. H.M. Kingston and L.B. Jassie, "Fundamental Relationships in Acid Decomposition of Samples for Elemental Analysis Using Microwave Energy"; pp. 121-28 in Microwave Processing of Materials, Vol, 124. Edited by W.H. Sutton, M.H. Brooks, and I.J. Chabinsky. Materials Research Society, Pittsburgh, PA, 1988.

32. K.Y Patterson, C. Veillon, and H.M. Kingston, "Microwave Digestion of Biological Samples for Selenium Analysis of Electrothermal Atomic Absorption Spectroscopy"; published in Introduction to Microwave Sample Preparation: Theory and Practice. Edited by H.M. Kingston and L.B. Jassie. American Chemical Society Reference Book Series, Washington, DC.

33. R. Gaylor and W.H. Hartwick, "The Vitrification of High Level Radioactive Waste Using Microwaves," Colloquium on Techniques and Applications of Microwave Power, IEE Digest No. 1979/65 (Nov, 1969).

34. F.J. Smith, "Microwave Processing Is Increasing, But It Needs Special Knowledge," *Res. Dev.,* **30** [1] 54-58 (1988).

35. M.H. Guerga and B.L.D. Hailler, "Apparatus for Casting of Ceramics," U.S. Pat. No. 3 732 048, May 8, 1973.

36. L.W. Tobin Jr., "Ceramic Material Processing," U.S. Pat. No. 4 292 262, Sept. 18, 1981.

37. S.J. Oda and I.S. Balbaa, "Microwave Processing at Ontario Hydro Research Division"; pp. 303-309 in Microwave Processing of Materials, Vol. 124. Edited by W.H. Sutton, M.H. Brooks, and I.J. Chabinsky. Materials Research Society, Pittsburgh, PA, 1988

38. S. Pike and J.G.R Binner, "Microwave Slip Casting"; presented at the 89th Annual Meeting of the American Ceramic Society, Pittsburgh, PA, April 29, 1987 (Basic Science Division, Paper No. 232-B-87).

39. R.E. Schroeder and W.S. Hackett, "Microwave Energy in the Foundry," *Br. Foundryman,* [Aug.] 293-98 (1971).

40. J.M. Valentine, "Role of Microwaves in the Coreroom," *Foundry*, [Oct] 37-39 (1973).

41. J.M. Valentine, "Microwave Heating for Molds and Cores," *Industrial Models and Patterns,* [July-Aug.] (1973).

42. J.M. Valentine "Production of Plaster Molds by Microwave Treatment," U.S. Pat. No. 4 043 380, Aug. 23, 1977.

43. R.E. Stengel, "Microwaves Speed Drying of Ceramic Blocks," *Process Design News,* [Sept.] 24 (1974).

44. W.B. Harrison, M.R.B. Hanson, and B.G. Koepke, "Microwave Processing and Sintering of PZT and PLZT Ceramics"; pp. 279-86 in Microwave Processing of Materials, Vol. 124. Edited by W,H. Sutton, M.H. Brooks, and I.J. Chabinsky. Materials Research Society, Pittsburgh, PA, 1988.

45. Anonymous, "Microwave Energy Beams Benefits to Investment Casters," *Precision Met.,* **34** [5] 48-50 (1976).

46. W.H. Sutton and W.E. Johnson, "Method for Microwave Heating," U.S. Pat. No. 4 140 887, Feb. 20 1979,

47. I. Ahmad, G.T. Chandler, and D.E. Clark, "Processing of Superconducting Ceramics Using Microwave Energy", pp. 239-46 in Mierowave Processing of Materials, Vol. 124. Edited by W.H. Sutton, M.H Brooks and I.J. Chabinsky. Materials Research Society, Pittsburgh, PA, 1988.

48. S. Komarneni, E. Breval, and R. Roy, "Microwave Preparation of Mullite Powders"; pp. 235-38 in Microwave Processing of Materials, Vol. 124. Edited by W.H. Sutton, M.H. Brooks, and I.J. Chabinsky. Materials Research Society, Pittsburgh, PA, 1988.

49. L Quémenéur, J. Choisnet, B. Raveau J.M. Thiebaut, and G. Roussy, "Microwave Clinkermg wtth a Grooved Resonant Applicator, " *J. Am. Ceram. Soc.,* **66** [12] 855-59 (1983).

50. W.R. Tinga, "Fundamentals of Microwave-Material Interactions and Sintering"; pp. 33-43 in Microwave Processing of Materials, Vol. 124. Edited by W.H. Sutton, M.H. Brooks, and I.J. Chabtosky Materials Research Society. Pittsburgh, PA, 1988.

51. V.K. Varadan, Y. Ma, A. Lakhtakia, and V.V. Varadan, "Microwave Sintering of Ceramics"; pp. 45-57 in Microwave Processing of Materials, Vol. 124. Edited by W.H. Sutton, M.H. Brooks, and I.J. Chabinsky. Materials Research Society, Ptttsburgh, PA, 1988.

52. T.T. Meek, C.E. Holcomb, and N. Dykes, "Microwave Sintering of Some Oxide Materials Using Sintering Aids, " *J. Mater Sci. Lett.,* **6,** 1060-62 (1987).

53. T.T. Meek R.D. Blake, and J.J. Petrovic, "Microwave Sintering of Al_2O_3 and Al_2O_3-SiC-Whisker Composites," *Ceram. Eng. Sci. Proc.,* **8** [7-8] 861-71 (1987).

54. M.A. Janney and H.D. Kimrey, "Microwave Sintering of Alumina at 28 GHz"; pp. 919-24 in Ceramic Transactions, Ceramic Powder Science. II, B. Edited by G.L. Messing, E.R. Fuller, Jr., and H. Hausner American Ceramic Society, Westerville, OH, 1988.

55. M.A. Janney and H.D. Kimrey, "Microstructure Evolution in Microwave Sintered Alumina"; published in Advances in Sintering, American Ceramic Society, Westerville, OH.

56. J.D. Katz. R.D. Blake, and J.J. Petrovic, "Microwave Sintering of Alumina-Silicon Carbide Composites at 2.45 and 60 GHz," *Ceram. Eng. Sci. Proc.,* **9,** 725-34 (1988).

57. Y.L. Tian, M.E. Brodwin, and D.L. Johnson, "Microwave Sintering of Ceramics Under High Gas Pressure" pp. 213-18 in Microwave Processmg of Materials Vol. 124. Edited by W.H. Sutton, M.H. Brooks, and I.J. Chabinsky. Materials Research Society, Pittsburgh, PA, 1988.

58. W.H. Sutton; unpublished data.

59. H.D. Kimrey; private communication.

60. J. Wilson and S. Kunz, "Microwave Sintering of Partially Stabilized Zirconia"; presented at the 89th Annual Meeting of the American Ceramic Society, Pittsburgh, PA, April 30. 1987 (Engineering Ceramics Division Paper No. 172-C-87).

61. R.D. Blake and T.T. Meek, "Microwave Processed Composites," *J. Marer Sci. Lett.,* **5,** 1097-98 (1986).

62. T.T. Meek, "Proposed Model for the Sintering of a Dielectric in a Microwave Field," *J. Mater Sci. Lett.,* **6,** 638-40 (1987).

63.M.K. Krage, "Microwave Sintering of Ferrites," *Am. Ceram. Soc. Bull.,* **60** [11] 1232-34 (1981).

64.F. Okada, S. Tashiro, and M. Suzuku, "Microwave Sintering of Ferrites"; pp. 201-205 in Advances in Ceramics, Vol. 15, International Conference on Ferrites, Part I. Edited by F.Y. Wang. American Ceramic Society, Westerville, OH, 1986.

65.J.D. Katz, R.D. Blake, J.J. Petrovic, and H. Scheinberg, "Microwave Sintering of Boron Carbide", pp 219-26 in Microwave Processing of Materials, Vol. 124. Edited by W.H. Sutton, M.H. Brooks, and I.J. Chabinsky. Materials Research Society, Pittsburgh, PA 1988.

66.M. Hsu and D.L. Johnson, "Microwave-Induced Plasma Sintering of Ceramic Materials"; presented at the 88th Annual Meeting of the American Ceramic Society, Chicago, IL. April 29, 1986 (Basic Science Division Paper No. 112-B-86).

67.P.A. Haas, "Heating of Uranium Oxides in a Microwave Oven," *Am. Ceram. Soc. Bull.,* **58** [9] 873 (1979).

68.S. Komarneni and R. Roy, "Anomalous Microwave Melting of Zeolites," *Mater Lett.,* **4** [2] 107-110 (1986).

69.T.T. Meek and R.D. Blake, "Ceramic-Ceramic Seals by Microwave Heating," *J. Mater Sci. Lett.,* **5**, 270-74 (1986).

70.R.D. Blake and T.T. Meek, "Method for Producing Ceramic-Glass-Ceramic Seals by Microwave Heating," U.S. Pat. No. 4 606 748, Aug. 19, 1986.

71.D. Palaith, R. Silberglitt, C.C.M. Wu, R. Kleiner, and E.L. Libelo, "Microwave Joining of Ceramics", pp. 255-66 in Microwave Processing of Materials, Vol. 124. Edited by W.H. Sutton, M.H. Brooks. and I.J. Chabinsky. Materials Research Society, Pittsburgh, PA, 1988.

72.H. Fukushima, T Yamanka, and M. Matsui "Microwave Heating of Ceramics and Its Application to Joining", pp. 267-72 in Microwave Processing of Materials, Vol. 124. Edited by W.H. Sutton, M.H. Brooks, and I.J. Chabinsky. Materials Research Society, Pittsburgh, PA, 1988.

73.Y. Hässler and L. Johansen, "Microwave Heating of Fused Quartz to High Temperatures in the Fabrication Process of Optical Fibers"; pp. 273-78 in Microwave Processing of Materials, Vol. 124. Edited by W.H. Sutton, M.H. Brooks, and I.J Chabinsky Materials Research Society. Ptttsburgh, PA, 1988.

74.S. Das and T.R. Curlee, "Microwave Sintering of Ceramics: Can We Save Energy?," *Am. Ceram. Soc. Bull.,* **66** [7] 1093-97 (1987).

Principles of Industrial Microwave and RF Heating

Robert F. Schiffmann

Abstract

Microwave and RF heating are used throughout industry and provide products or components that are used daily. Despite a half century of successful commercial application, the principles of utilizing these electromagnetic energy forms are not commonly understood. This paper presents a largely non-mathematical introduction to the topic.

Introduction

It was over 50 years ago that Percy Spencer of Raytheon discovered the principles that led to the development of the microwave oven. Yet, the application of this unique method of heating materials has been slow to be adopted by industry. There are, however, approximately 1000 industrial installations operating in the food, pharmaceutical, automotive, rubber and other processing industries, responsible for the production of many tons of products daily.

Of even greater industrial application is RF heating which has been in use since the 1930's. Many thousands of installations are on stream in nearly every industry. It is so pervasive that it has become a "workhorse" providing unique heating and production benefits.

Background

The term "dielectric heating" can be applied logically to all electromagnetic frequencies up to and including at least the infrared spectrum. However, it is generally accepted that dielectric heating is done at frequencies between 1 and 100 MHz, whereas microwave heating occurs between 300 MHz and 300 GHz.

TABLE 1. INDUSTRIAL SCIENTIFIC AND MEDICAL (ISM) FREQUENCY BANDS USED FOR INDUSTRIAL HEATING.

Frequency (MHz)	Frequency tolerance	Wavelength (meters)	Area permitted
RF Heating			
13.56	0.05%	22.1	Worldwide
27.12	0.6%	13.6	"
40.68	0.05%	7.5	"
MicrowaveHeating			
896	10 MHz	0.32	Great Britain
915	13 MHz	0.33	N & S America
2,450	50 MHz	0.122	Worldwide

Table 1 [1] lists the ISM (industrial, scientific and medical) bands established by international agreement which are used for industrial heating applications. While a great deal of industrial RF heating is done at frequencies outside of these bands this will have to change in Europe due to E.C. regulations. The basic principles of RF and microwave heating are the same, but, the methods of generation, equipment and some applications are different as will be described below.

Fundamentals of Microwave and Dielectric Heating

Electromagnetic Waves

The environment is filled with electromagnetic waves at all times. Light, X-irradiation, TV, AM and FM radio waves, ultraviolet, infrared, and microwaves are some of the common manifestations of these waves. The relationship of these waves is found in the electromagnetic spectrum (see Figure 1).

All electromagnetic waves are characterized by their wavelength and frequency, and an illustration of a plane monochromatic electromagnetic wave is seen in Figure 2.

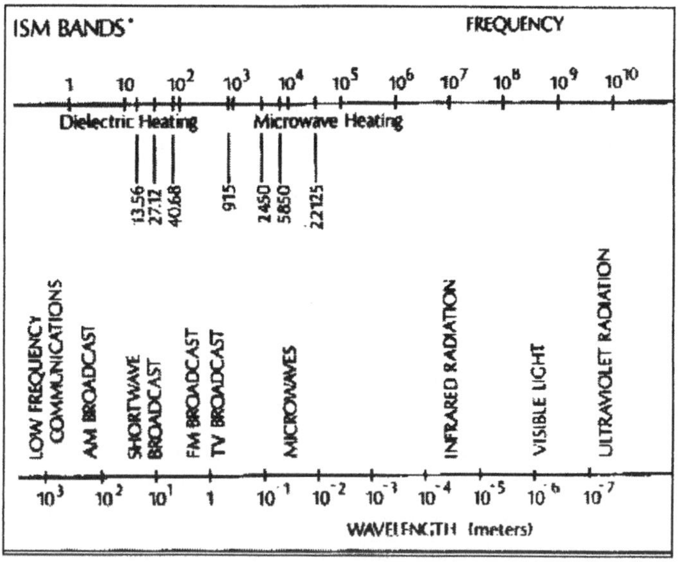

Figure 1. The electromagnetic spectrum [2].

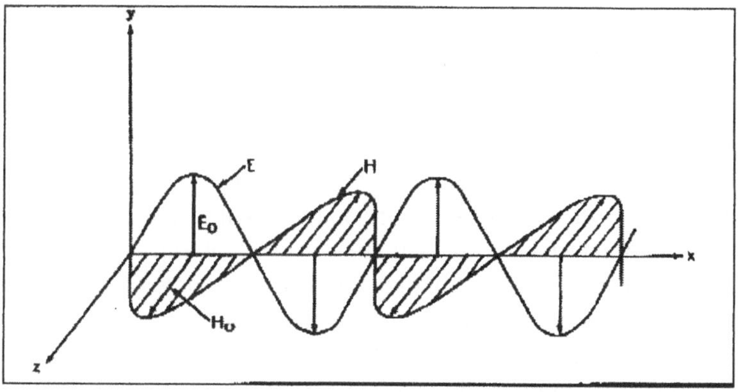

Figure 2. A plane electromagnetic wave. E and H represent the electrical and magnetic components of the wave; E_o and H_o are their respective amplitudes [2].

An electromagnetic wave is composed of an electric component E and a magnetic component H. Note that E and H are perpendicular to each other and are both perpendicular to the direction of travel. The wave is traveling in the X direction at the velocity of light in air or vacuum but slows as it passes through another medium, and this causes the wavelength to decrease, as indicated in Equation (1).

$$\lambda_m = \lambda_o / \sqrt{\varepsilon'} \tag{1}$$

where,
λ_o = free space wavelength
λ_m = wavelength in the material
$\sqrt{\varepsilon'}$ = relative dielectric constant of the material

An electromagnetic wave is an energy wave that changes its energy content and amplitude as it travels through a medium. For example, if we trace the E component we see that at some point it is zero; then it builds up to a maximum value, decays to zero, and again builds up to a maximum value with the opposite polarity before again decaying to zero. The same thing happens to the H component. The amplitude of the wave at any point along the X axis represents the electrical (E) or magnetic (H) field strength, which are measured as volts or amperes per unit distance, respectively. It is this periodic flip-flopping of the wave's polarity and its decay through zero that causes the stress upon ions, atoms, and molecules, which is converted to heat, and the greater the field strength, the greater will be the effect.

The Heating Mechanism

A crucial fact to keep in mind at all times is that microwave and RF are not forms of heat but rather forms of energy that are manifested as heat through their interaction with materials. It is as if they cause materials to heat themselves. There are many mechanisms for dielectric energy conversion but two are of major interest in industrial heating of most materials.

Ionic Conduction. Since ions are charged units they are accelerated by electric fields. In a solution of salt in water, for example, there are sodium, chloride, hydronium, and hydroxyl ions, all of which will be caused to move in the direction opposite to their own polarity by the electric field. In so doing they collide with unionized water molecules, giving up kinetic energy and causing them to accelerate and collide with other water molecules in billiard ball fashion, and when the polarity changes the ions accelerate in the opposite direction. Since this occurs many millions of times per second, large numbers of collisions and transfers of energy occur. Therefore, there is a two-step energy conversion: electric field energy is converted to induced ordered kinetic energy, which in turn is converted to disordered kinetic energy, at which point it may be regarded as heat. This type of heating is not dependent to any great extent upon either temperature or frequency. The power developed per unit volume (Pv) through ionic conduction is shown in Equation (2).

$$Pv = E^2qnu \tag{2}$$

where,
q = amount of electrical charge on each of the ions
n = ion density, the number of ions per unit volume
u = lepermittedvel of mobility of the ions

Dipolar Rotation. Many molecules, such as water, are dipolar in nature; that is, they possess an asymmetric charge center. Water is typical of such a molecule. Other molecules may become "induced dipoles" because of the stresses caused by the electric field. Dipoles are influenced by the rapidly changing polarity of the electric field. Although they are normally randomly oriented, the electric field attempts to pull them into alignment. However, as the field decays to zero (relaxes), the dipoles return to their random orientation only to be pulled toward alignment again as the electric field builds up to its opposite polarity. This buildup and decay of the field, occurring at a frequency of many millions of times per second, causes the dipoles similarly to align and relax millions of times per second. This causes an energy conversion from electrical field energy to stored potential energy in the material and then to stored random kinetic or thermal energy in the material. This temperature-dependent, molecular size-dependent time for buildup and decay defines a frequency known as the relaxation frequency. For small molecules, such as water and monomers, the relaxation frequency is already higher than the microwave frequency and rises further as the temperature increases, causing a slowing of energy conversion. On the other hand, large molecules, such as polymers, have a relaxation frequency that is much lower than the microwave frequency but that increases and approaches it as the temperature rises, resulting in better energy conversion into heat. This may lead to runaway heating in materials that at room temperature are nearly transparent to the microwave field. This must be superimposed upon the fact that such liquids as water and monomers are better absorbers of microwave energy than polymers. Since in drying or curing applications it is the liquids and monomers that require heating, not the existing polymeric substrate, it is possible to execute the process well, often at a lower temperature. In fact, it is even possible to dry materials such as foods and medicinals at cold or subfreezing temperatures. The power absorbed per unit volume by dipolar rotation (Equation 3) is:

$$Pv = k\,E^2\,f\,\varepsilon'\tan\delta\ ;\quad \text{or}\quad Pv = k\,E^2\,f\,\varepsilon'' \qquad (3)$$

where,
k = constant dependent upon the units of measurement used
E = electric field strength, in volts per unit distance
f = frequency, cycles per second (Hertz)
ε' = relative dielectric constant or permittivity, unitless
$\tan\delta$ = loss tangent or dissipation factor, unitless
ε'' = dielectric loss factor, unitless

The relative dielectric constant expresses the degree to which an applied electric field may be supported by a material while the dissipation factor indicates how much of it will be converted to heat. The loss factor combines these effects into a single term which expresses the degree to which a material is influenced by an applied electromagnetic field and heated. In a sense, E and f are functions of the equipment, whereas ε', ε'' and $\tan\delta$ are factors related to the material being heated. Another important point is that as frequency f is reduced it is necessary to increase the electric field strength E in order to maintain a particular level of absorbed energy. Since dielectric heating frequencies are much lower than microwave frequencies, this requires that the field strengths be much higher for comparable power absorption in a dielectric system. This may lead to voltage breakdown of air (arcing) or in the process material. The sparking threshold for air is about 30,000 V/cm (75,000 V/inch).

Interaction of Electromagnetic Fields with Materials

Materials and the way they interact with electromagnetic fields may be divided into four categories:

1. *Conductors*. Materials with free electrons, such as metals, are materials that reflect electromagnetic waves just as light is reflected by a mirror. These materials are used to contain and direct electromagnetic waves in the form of electrodes, applicators and waveguides.
2. *Insulators*. Electrically non-conductive materials, such as glass, ceramics, and air, are insulators which reflect and absorb electromagnetic waves to a negligible extent and primarily transmit them (that is, they are transparent

to the waves). They are therefore useful to support or contain materials to be heated by the electromagnetic field and may take the form of conveyor belts, support trays, dishes or others. These materials may also be considered "non-lossy dielectrics". Note that polymers and some inorganics fit this category but under certain conditions may become lossy dielectrics (see below). Heating a polymer past it's heat distortion temperature is an example. Some materials also may heat poorly at one frequency but heat well at another. Also, use of specially designed applicators may cause materials to heat well whereas they are virtually transparent in home microwave ovens.

3. *Dielectrics*. These are materials whose properties range from conductors to insulators. There is within this broad class of materials a group referred to as "lossy dielectrics," and it is this group that absorbs electromagnetic energy and converts it to heat. Examples of lossy dielectrics are water, oils, wood, food materials containing moisture, and the like.
4. *Magnetic compounds*. These are materials, such as ferrites, that interact with the magnetic component of the electromagnetic wave and as such will heat. They are often used as shielding or choking devices that prevent leakage of electromagnetic energy. They may also be used for heating in special devices.

The various dielectric properties of materials, and consequently their heating, are affected by moisture content, temperature, frequency and other factors. The amount of free moisture in a substance greatly affects its dielectric properties and heating rate, since water has a high dielectric loss factor, approximately 12 at room temperature; that of base materials is of the order of less than 1 (usually 0.01 to 0.001 at 2.45 GHz). Thus, with a larger percentage of water the loss factor generally increases, usually proportionally. It should be emphasized that very complex phenomena occur when different dielectrics are mixed. However, a few rules of thumb may be applied [2]. (See end of text for examples.)

1. The higher the moisture content, nearly always the higher is the loss factor.
2. Dielectric loss usually increases with increasing moisture content but levels off at values in the range of 20-30% and may decrease at still higher moisture.
3. The dielectric loss of a mixture usually lies between that of its components.

Various materials, including alcohols and some organic solvents, also exhibit dielectric properties that make them suitable for heating with microwave and

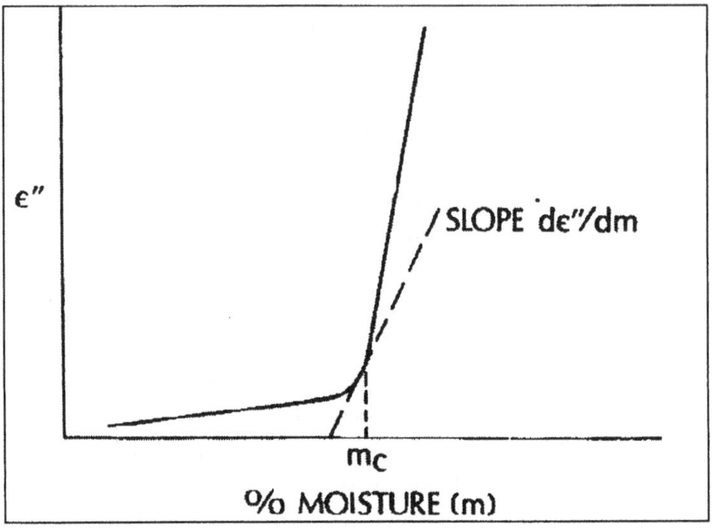

Figure 3. Variation of ε″ with moisture content [2,3].

dielectric energy and, so, behave similarly to water. Since drying is concerned with removal of water or a solvent, it is interesting to note that as these liquids are removed the dielectric loss decreases and, hence, the material heats less well. In many cases this leads to self-limitation of the heating as the material becomes relatively transparent at low moisture content. This has great value in obtaining moisture leveling, especially in sheet materials, in which the electromagnetic energy is likely to preferentially dry the wetter areas. Figure 3 shows a general graph of the variation in loss factor with moisture content [3]. Water exists in materials in different states, for example, bound or free, and these states may be ascribed to different regions on the graph, as indicated by the change of slope ($d\varepsilon''/dm$). Thus at low moisture contents, below the critical moisture content, we are dealing primarily with bound water; above it we encounter primarily free water.

Penetration Depth

The penetration depth is of utmost importance and is dependent upon various properties of a material. Since electromagnetic heating is, in effect, bulk heating, it is important that the energy penetrate as deeply as possible. If it does not, then the heating is limited to the surface. Those parameters affecting the depth of the field into the material are the wavelength, the dielectric constant, and the loss factor, as shown in Equation (4):

TABLE 2. PENETRATION DEPTH IN DOUGLAS FIR AT VARIOUS FREQUENCIES [2].

Frequency (MHz)	13.56	27.12	40.0	915	2,450
Penetration Depth (meters)	8.8	4.4	3.0	0.13	0.05

$$Dp = \lambda_0 \sqrt{\sqrt{\varepsilon'}} / 2 \pi \varepsilon'' \tag{4}$$

where D = penetration depth at which the available power in the material has dropped to about 37% (1/e) of its value at the surface; λ_0 = the free space wavelength; and π is pi (3.1416). (Note that this is a simplified version of the general equation and is useful in most cases.)

From this equation it is obvious that materials with high dielectric loss factors will have smaller depths of penetration than those with lower values. It is also apparent that the depth of penetration is greatly affected by the wavelength (and hence the frequency) of the applied field. This is illustrated in Table 2, where penetration depth into Douglas fir is shown for various wavelengths [2]. This demonstrates the very clear superiority of RF heating of very large materials with substantial dielectric properties.

A peculiarity of this type of heating is the unusual temperature gradients that may be generated. This is due to a number of factors. First, unless an auxiliary heat form is applied, the air in the system remains close to ambient temperature. Hence, the surface will be cooler than a zone somewhat below the surface. This is especially true in a drying system in which evaporative cooling of the surface will occur.

Another circumstance concerns the depth of penetration as it relates to the size of the piece being heated. If the piece is several times larger than the depth of penetration, then the temperature gradient will resemble conventional gradients, with a cooler interior and a warmer exterior. However if the piece is small in comparison with the penetration depth, for example, only one or two times greater, there may be a focusing of the electromagnetic field in the center of the piece. In this case, the center may be the hottest place, and in fact, the center may burn while the surface remains cool.

Process Advantages of Microwave and RF Systems

Heating and drying with microwave and dielectric energy is distinctly different from conventional means. Whereas conventional methods depend upon the slow march of heat from the surface of the material to the interior as determined by a differential in temperature from a hot outside to a cool inside, heating with dielectric and microwave energy is, in effect, bulk heating in which the electromagnetic field interacts with the material as a whole. The heating occurs nearly instantaneously and can be very fast, although it does not have to be. However, the speed of heating can be an advantage and it is often possible to accomplish in seconds or minutes what could take minutes, hours, and even days with conventional heating methods. The governing parameters are the mass of the material, its specific heat, dielectric properties, geometry, heat loss mechanisms, and coupling efficiency, the power absorbed in the material, and the output power of the microwave or RF heating system. All other things being equal, the speed may be doubled by doubling the output power. The advantages of microwave and dielectric heating includes the following:

- Process speed is increased, as described above.
- Uniform heating may occur throughout the material. Although not always true, often the bulk heating effect does produce uniform heating, avoiding the large temperature gradients that occur in conventional heating systems.
- Efficiency of energy conversion. In this type of heating, the energy couples directly to the material being heated. It is not expended in heating the air, walls of the oven, conveyor or other parts. This can lead to significant energy savings. Also, the energy source is not hot and plant cooling savings may be realized.
- Better and more rapid process control. The instantaneous on-off nature of the heating and the ability to change the degree of heating by controlling the output power of the generator means fast, efficient, and accurate control of heating.
- Floor space requirements are usually less. This is due to the more rapid heating.
- Selective heating may occur. The electromagnetic field generally couples into the solvent, not the substrate. Hence, it is the moisture that is heated and removed, whereas the carrier or substrate is heated primarily by conduction. This also avoids heating of the air, oven walls, conveyor, or other parts.
- Product quality may be improved. Since high surface temperatures are not usually generated, overheating of the surface and case hardening, which are common with conventional heating methods are eliminated. This often leads to less rejected product.
- Desirable chemical and physical effects may result. Many chemicals and physical reactions are promoted by the heat generated by this method, leading to puffing, drying, melting, protein denaturation, starch gelatinization, and the like.

Microwave and RF Drying

The mechanism for drying with microwave and dielectric energy is quite different from that for ordinary drying. In conventional drying, moisture is initially flashed off from the surface and the remaining water diffuses slowly to the surface. Although the potential of energy transfer for heating is the temperature gradient, which results in energy transfer to the interior of the material, the potential for mass transfer is the mass concentration gradient existing between the wet interior and the drier surface. This is often a slow process, diffusion rate limited, which requires high external temperatures to generate the temperature differences required. With internal heat generation, in microwave and RF systems, mass transfer is primarily due to the total pressure gradient established because of the rapid vapor generation within the material [5]. Most of the moisture is vaporized before leaving the sample. If the sample is initially very wet and the pressure inside the sample rises very rapidly, liquid may be removed from the sample under the influence of a total pressure gradient. The higher the initial moisture, the greater is the influence of the pressure gradient on the total mass removal. Thus, there is in effect a sort of "pumping" action, forcing liquid to the surface, often as a vapor. This leads to very rapid drying without the need to overheat the atmosphere and perhaps cause case hardening or other surface overheating phenomena. The advantages of microwave and RF drying are as follows:

- Efficiency: in most cases, the energy couples into the solvent, not the substrate
- Non-destructive: drying can be done at low ambient temperatures; no need to maintain high surface temperatures, leading to lower thermal profiles
- Reduction of migration: solvent often mobilized as a vapor; therefore does not transfer other materials to the surface
- Leveling effects: coupling tends toward the wetter areas
- Speed: drying times can be shortened by 50% or more
- Uniformity of drying: by a combination of more uniform thermal profiles and leveling
- Conveyorized systems: less floor space, reduced handling
- Product improvement in some cases: eliminates case hardening, internal stresses, and other problems

Equipment for Microwave and RF Heating and Drying

The heating mechanisms for microwave and RF heating are similar, but the means of achieving them are somewhat different. The basic components of these systems are a means of generating the high-frequency energy: the generator; and a means of applying it to the work-piece: the applicator. These are described below.

Generators

The basic function of the generator is to convert the alternating current of 50 or 60 Hz to the high frequencies desired. The means of doing this are quite different for dielectric and microwave systems.

RF systems usually employ negative grid triode tubes, although some systems operating in the 50-100 MHz range use beamed power types. RF circuits are usually simple, self-excited oscillators of the Hartley, Colpitts, or tuned plate-tuned grid type. These circuits usually consist of coils and capacitors and/or coaxial lines. The load to be heated may actually form part of the tank circuit capacitance, it may be separately tuned or inductively or capacitively coupled to the oscillator, or it may be a combination of these, being directly connected to the oscillator and also partially tuned to the oscillator circuit. The latter has a distinct advantage in drying applications, in which the electrodes may be partially tuned to the oscillator circuit on the side of the tuning curve, where they will be de-tuned when a dry load with a lower dielectric constant is present. Thus a wet load will cause an increase in electrode voltage, and a dry load will cause a power reduction. In this way there is a self-limiting or leveling effect [5]. Note that energy not developed in the load can be lost in the coupled circuits, causing less efficiency and wasted power.

A negative grid tube is a variable power device that draws current from the supply line only in aufficient quantity to supply circuit losses plus the power consumed in the load. Thus, most dielectric systems control power to the load by varying the RF electrode voltage, often by means of a variable capacitor. In many instances this system is self-limiting and self-regulating, supplying power only as demanded by changes in size or electrical characteristics of the load [5]. (Note that new solid state power supplies based on 50 ø technology are now also available.)

Microwave systems operate on a nominal fixed frequency of 915 or 2,450 MHz, with the frequency controlled by the tube dimensions and geometry. They must also be shielded to prevent excessive radiation of harmonics, as well as for safety. A microwave generator consists of a DC power supply and a tube: either a magnetron or a klystron. These tubes are constant output power devices, and power to the load may be controlled by sensing the load requirements and controlling the input power accordingly, usually by indirectly varying the dc anode voltage. Although magnetrons and klystrons are capable of withstanding a reasonable degree of mismatching, manifested as power reflected back to the tube, precautions must be taken to avoid overheating or in other ways damaging these tubes which is usually accomplished by means of ferrite circulators.

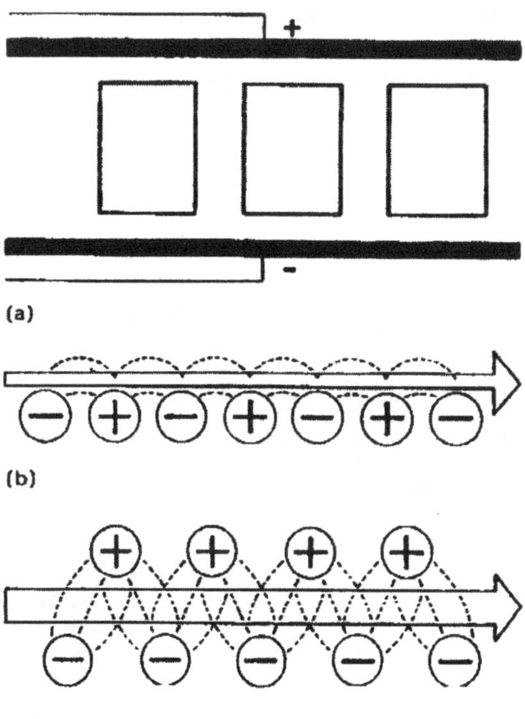

Figure 4. Electrode configurations for RF heating systems. (a) Platen type for bulky objects. (b) Stray field type for thin webs. (c) Staggered type for thick webs or board [2].

Applicators

The means of applying microwave and RF energy to a work piece differ in a very significant manner. Microwave energy is transported through free space, often as random E fields, and must be focused upon the load. On the other hand, dielectric energy is usually applied by means of electrodes, in which the field oscillates through the load, which is placed between the electrodes yielding uniform voltage orientation.

RF Systems. There are many types of electrodes, but they may be described in three basic categories, which are illustrated in Figure 4.

- Platen type: This usually consists of flat plates in pairs, between which the work piece may be held in a batch system or pass on a conveyor belt. Often, a conveyor belt may represent one of the plate electrodes. This is especially useful for bulky objects. A drawback is that as the plates become widely separated the high field strengths required may cause voltage discharge, which can burn the load.
- Stray field type: Here the load, usually in the form of a thin web, passes over the electrodes of alternating polarity. Since the load represents the path of least electrical resistance, the dielectric field passes through it, causing heating.

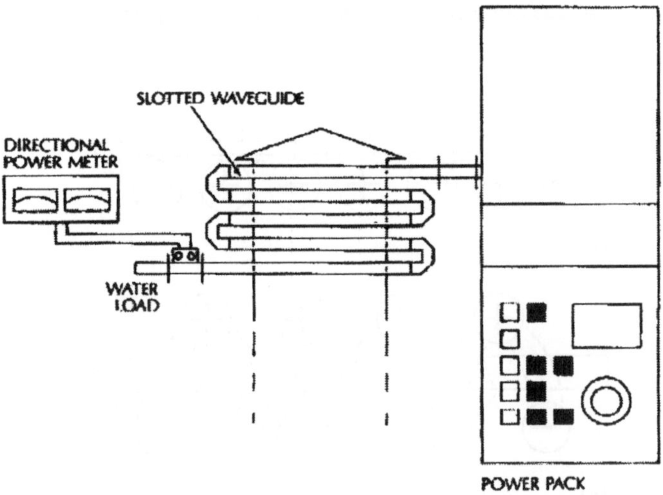

Figure 5. Slotted or serpentine waveguide [2].

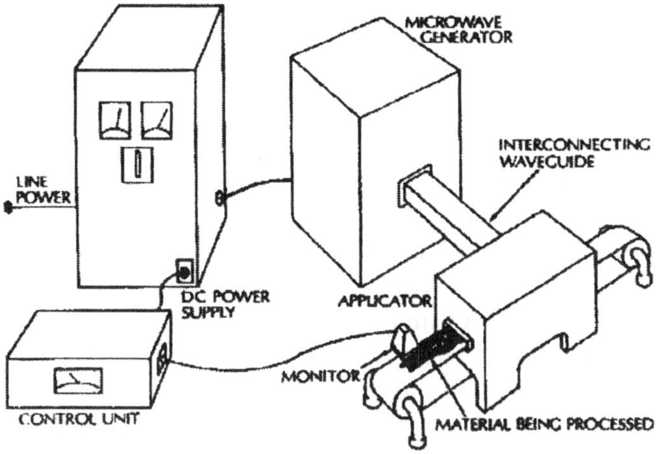

Figure 6. A typical microwave heating system utilizing a conveyorized cavity applicator. A feedback system monitors the heated material and automatically adjusts the output power of the magnetron to control the final material properties [2].

- Staggered type: This is usually for sheet materials and thick webs. The distance between the electrodes is kept to a minimum in order to achieve heating without arcing.
- Other types: Electrodes may also be shaped to conform to the geometry of the load or may be part of the conveyor or hydraulic or pneumatic press.

The entire system is confined within a metal housing to prevent leakage of radiation. At the same time, hot air of controlled temperature, humidity, and/or velocity may be passed through the applicator.

Microwave Systems. After generating the microwave energy it must be transported to the applicator. This is usually accomplished by means of waveguides, although coaxial cable is also useful for lower power. A waveguide is ordinarily a hollow rectangular metallic conduit, usually made of brass or aluminum. Its interior dimensions are carefully chosen to control the nature of the microwave field presented to the applicator. Applicators are of several major types and are always constructed of metal.

Waveguide applicators: Waveguides themselves may be used as applicators. Since the electric field may be maximum in the center of a waveguide, it is possible to pass a material through this intense field to obtain very efficient heating. A good example is the heating of filamentary materials.

Traveling wave applicators: These applicators are also known as slotted, folded, or serpentine waveguides. A slot is cut into the narrow sides of the waveguide, and several waveguides are joined together as shown in Figure 5. A thin sheet material, such as paper or textile, may be passed through the slots. The microwave energy makes several passes through the load, heating it as it travels. These are highly efficient heating systems, although they may cause some side-to-side non-uniformity.

Cavity applicators: This is a large class of applicators, but they are probably the most common type. Home microwave ovens are a typical example. They consist of metal boxes, which may be used for conveyorized systems or in batch operations. The microwave energy may be coupled into this applicator by means of waveguide or coaxial cable through a single port or multiple ports (see Figure 6). There is an industrial system in which over 100 magnetrons are separately introduced into the cavity. In this type of applicator, the load usually represents only a small fraction of the volume of the applicator and is subjected to the microwave field being reflected from the sides of the applicator and passing through it from all sides. This causes a three-dimensional bulk heating effect that is unique and of great use. In order to ensure uniformity of heating, a number of steps may be taken, usually in combination: moving or turning the load in the applicator by means of conveyors or turntables; providing mode stirrers, which often resemble slowly rotating fans and increase the number of modes in the oven, causing reflective scattering of energy; using multiple inputs for the microwave energy; using multiple microwave sources whose slight differences in frequency cause different mode patterns; and choosing the cavity dimensions to support the maximum number of modes, the so-called resonant multi-mode cavity.

Industrial Applications of Microwave and RF Heating

The industrial applications of microwave and dielectric heating are many and varied. In some cases, the application is unique to one form of energy or the other; in other cases, either form may be used. Although it is not possible to give any hard and fast rules for selecting one over the other, there are some guidelines that may be followed [5].

Guidelines for the Selection of Microwaves or RF Heating

- *Size of Load*—If the load is very large or very wide, dielectric heating may be preferred. The depth of penetration is also directly proportional to wavelength, and in dielectric heating this may be measured in meters; in microwaves it is in centimeters. On the other hand, if the piece is small, microwave heating is preferred.
- *Watt Density*—If the watt density requirement is very high, microwaves may be preferred, to avoid arcing and burning of the material. For example, a bulky product consisting of loosely packed particles with a loss factor less than 0.05 favors microwaves.
- *Power*—If the power requirement is high, over 50 kW, economics favor the RF system. However, there are microwave systems of several hundred kilowatts.
- *Geometry*— If the product has an irregular shape with no rectangular cross section, the multimode microwave cavity will provide more uniform heating.
- *System Compatibility*— If the system requires the use of pneumatic or hydraulic presses, metal conveyors, metal dies, or tenter frames, dielectric heating may be the only choice.
- *Self-Regulation*— If the load fluctuates rapidly or goes through drastic changes in dielectric constant or dissipation factor during its heat cycle, partially tuned electrodes and instantly variable power from dielectric heating are advantageous.
- *Self-Limiting*— If the load has a low-frequency loss tangent greater than 1.0, it is more resistive than capacitive, and in many cases this element disappears as the load dries or cures. RF will heat the resistive elements, and

it will be more self-limiting in these areas and will not overheat or overdry. However, if the load has a low-frequency dissipation factor of less than 0.5, then it is mostly capacitive, and reduction of the dielectric constant as it dries or cures will be the predominant change and will permit better self-limiting at microwave frequencies. For example, dielectric heating has very little leveling effect on the moisture content of paper below about 5% because it is self-limiting in that region, but microwave energy could be used to dry the paper to near zero moisture content.

TABLE 3. RF INSTALLATIONS

Application	Number of Installations	Typical Power(kW)
Plastic welding	400	10 - 150
Textile drying	40	50 - 500
Wood drying	10	300 - 500
Wood gluing	400	10 - 500
Ceramic drying	4	100
Food, post baking	10	50 - 200
Food, thawing	20	100
Pre-heating thermoset resins	100	8 - 15
MICROWAVE INSTALLATIONS		
Bacon cooking	50	75 - 400
Core drying	15	50 - 100
Food cooking	6 - 25	100 - 600
Meat tempering	400-500	40 - 150
Medical waste	22	75

TABLE 4. DIELECTRIC PROPERTIES OF VARIOUS MATERIALS

@2450 MHz	ε′	ε″	tan δ
WATER (DISTILLED, 25C)	78.0	12.0	0.15
WATER + 0.5M NaC1 (25 C)	41.9	67.0	1.60
ICE (- 12 C) 3.2	0.003	0.001	0.0009
BEEF (COOKED, 30 C)	30.5	9.6	0.31
HAM (PRECOOKED, 20 C)	42.9	22.8	0.53
POTATO			
RAW (25 C) 57.3	15.7	0.27	
MASHED (30 C)	72.5	24.0	0.33
CORN OIL (25 C)	2.5	0.14	0.06
BUTTER (SALTED, 20 C)	4.6	0.60	0.13
BUTTER (UNSALTED, 20 C)	2.9	0.45	0.16
BREAD DOUGH	22	9	0.41
BREAD	4.6	1.20	0.26
DOUGLAS FIR (20 C 0% MOISTURE)	2.0	0.02	0.01
" " (20 C; 20% MOISTURE)	3.5	0.90	0.26
NATURAL RUBBER (25 C)	2.2	0.01	0.005
NEOPRENE (26 C)	2.8	0.14	0.05
PAPER	2.7	0.20	0.06
POLYETHYLENE	2.3	0.001	0.0004
THERMOSET POLYESTER (20 C)	4.0	0.02	0.01
TEFLON (20 C)	2.1	0.001	0.0003
NYLON (20 C)	2.4	0.02	0.01
SODA LIME GLASS (20 C)	6.0	1.20	0.20
BOROSILICATE GLASS (20 C)	4.3	0.02	0.005
GLASS CERAMIC (20 C)	6.0	0.03	0.01
ALUMINA (20 C)	9.0	0.005	0.001
FUSED QUARTZ (20 C)	4.0	0.001	0.0001
@915 MHz	**ε′**	**ε″**	**tan δ**
WATER (DISTILLED, 25 C)	77.5	1.2	0.02
WATER + 0.5M NaCl (25 C)	69.0	269.0	3.90
BEEF (COOKED,30 C)	35.4	16.0	0.45
HAM (PRECOOKED,20 C)			
POTATO			
RAW (25 C)	65.1	19.6	0.30
BOILED (23 C)	38.0	11.4	0.30
PAPER	2.8	0.2	0.07
POLYETHYLENE	2.26	0.002	0.001

Operational Microwave and RF Industrial Heating Systems

Table 3 gives some examples of current installed microwave and RF systems in the United States. It is by no means complete but indicates the broad range of applications possible [6,7].

Conclusions

While microwave and RF processing have been available for utilization by industry for 50 years their actual use is limited. However, they offer such unique benefits that their applications will continue to grow. When they are properly applied they may make a significant improvement to the profitablity of a manufacturing process in a manner not achieved by any other means.

REFERENCES

1. IEC CISPR Publication 11, Second Ed. (1990)
2. R.F. Schiffmann, Handbook of Industrial Drying, Volume 1, 345-372, Marcel Dekker,Inc., New York, NY (1995).
3. W.R. Tinga, Proceedings of the International Microwave Power Institute Short Course for Users of Microwave Power, 19-29, Manassas, VA (1974).
4. A.C. Metaxas, Transactions of the International Microwave Power Institute, Volume 2, 19-47, Manassas, VA (1974).
5. D.W. Lyons, J.D. Hatcher, and J.E. Sunderland, *J. Heat and Mass Transfer.* 15:897-905 (1972).
6. M. Preston, in Theory and Applications of Microwave Power in Industry, 65-85, International Microwave Power Institute, Manassas, VA (1971).
7. W.R. Paules Jr. and G. Walzer, First World Congress on Microwave Processing, Orlando FL (1997).
8. R.F. Schiffmann, Microwaves: Theory and Application in Materials Processing III, The American Ceramic Society, Westerville, OH (1995).

Fundamental Interaction Mechanisms Between Microwaves and Matter

R.E. Newnham, S.J. Jang, M. Xu and F. Jones

Abstract

A molecular and crystallographic view of the absorption of microwave energy by ceramic materials is presented, with examples drawn from both structural and electronic ceramics, together with the liquid and polymeric materials used in processing them. Among the absorption mechanisms discussed are dipole reorientation, space charge phenomena, ferrimagnetic resonance, and tails from the far infrared. Other important topics include relaxation spectra, field partitioning, and thermal runaway.

Introduction

Many different physical phenomena are involved in the microwave processing of ceramics. Microwaves are electromagnetic waves ranging from 1m to 1mm in wavelength at frequencies from 0.3 to 300 GHz. The interaction with matter takes place through the electric field vector E(V/m) and the magnetic field vector H(A/m) belonging to the microwave. The purpose of this paper is to review the fundamental interaction mechanisms between microwaves and matter.

Electric Loss

When subjected to an electric field, materials polarize creating an electric polarization P(C/m^2) equal to the dipole moment (C-m) per unit volume (m^{-3}).

There are four important polarization mechanisms in solids, and three of them lead to losses in the microwave region: (1) space charges arising from localized electrical conduction, (2) rotating electric dipoles, and (3) ionic polarization associated with far-infrared vibrations.

In discussing these phenomena, it is helpful to describe the losses in terms of the real and the imaginery parts of the dielectric constant K*. The electric field of the microwave is given by

$$\mathbf{E} = \mathbf{E_0} e^{i\omega t} \tag{1}$$

where E_0 is the amplitude of the field, ω is the angular frequency, t the time, and $i = \sqrt{-1}$. The resulting electric flux density is

$$D = D_0 e^{i(\omega t - \delta)} \tag{2}$$

where δ is the phase angle associated with the time lag in polarizing the material. The electric flux density (electric displacement) comes from the applied electric field and the electric polarization:

$$D = \varepsilon_0 E + P = \varepsilon E \tag{3}$$

where ε_0(= 8.85 x 10^{-12} F/m) is the permittivity of free space, and the dielectric constant is the relative permittivity of the material

$$K^* = \frac{\varepsilon}{\varepsilon_o} = \frac{D}{\varepsilon_o E} = \frac{D_o e^{-i\delta}}{\varepsilon_o E} = K' - iK'' \tag{4}$$

The real part of the dielectric constant is in phase with the field

$$K' = \frac{D_o}{\varepsilon_o E} \cos\delta \tag{5}$$

and the imaginary part is out of phase

$$K'' = \frac{D_o}{\varepsilon_o E} \sin\delta \tag{6}$$

The loss factor tan δ is K'' /K'.

As pointed out by Sutton (1989), the power absorbed per unit volume is

$$W = \frac{1}{2} E_o^2 \omega \varepsilon_o K'' = \frac{1}{2} E_o^2 \omega \varepsilon_o K' \tan\delta \tag{7}$$

Microwave absorption therefore increases with field intensity, frequency, loss factor and dielectric constant . It is also important to realize, however, that in microwave processing we are dealing with inhomogeneous materials. Generally there are ceramic particles, organic vehicles, and porosity present in the green ceramic body. In this case we must be aware that electic fields are also inhomogeneous (Meek, 1987). Flux continuity ($D_1 = D_2$) is required across interfaces if both phases are insulators. This means that electric fields are largest in regions of small permittivity.

Conduction Losses

Most ceramics are wide band gap semiconductors with resistances that decrease rapidly with temperature. Sintered alumina is an excellent high voltage insulator with resistivity in excess of 10^{14} Ω-m at room temperature, but it drops to less than 10^{10} Ω-m at 400°C and 10^5 Ω-m at 1000°C. Other ceramic insulators such as steatite and procelain decrease even faster. At 400°C the resistivity of steatite is about 10^8 Ω-m, and porcelain with its mobile alkali ions is only about 10^4 Ω-m.

Metallic behavior is observed in some transition metallic oxides. Rhenium trioxide (ReO_3) is comparable to silver and gold with resistivity less than 10^{-8} Ω-m at room temperature. RuO_2, Ti_2O_3, PdO, and tungsten bronzes such as $Na_{0.8}WO_3$ are also excellent conductors.

The importance of conduction losses in microwave heating is borne out in the experiments of Walkiewicz, Kazonich and McGill (1988). In this study a wide variety of metal powders, inorganic chemicals, and mineral specimens were heated in a microwave oven and the temperatures recorded as a function of time. The results are summarized in Table 1.

TABLE 1. MICROWAVE HEATED MINERALS, CHEMICALS, AND METAL POWDERS. TEMPERATURES WERE RECORDED AFTER A FEW MINUTES HEATING AT POWER LEVELS NEAR 1 KW. (MCGILL, *et al*, 1988) APPROXIMATE RESISTIVITIES ARE GIVEN IN PARENTHESES.

Metal powders (10^{-6} – 10^{-8} Ω-m) Al, Co, Cu, Fe, Mg, Mo Moderate heating to about 400°C	Oxide minerals (10^4 – 10^{14} Ω-m) SiO_2, Al_2O_3, $KAlSi_3O_8$, $CaCO_3$ Very little heating only about 80°C
Alkali Halides (10^4 – 10^5 Ω-m) KCl, KBr, NaCl, NaBr, LiCl Very little heating. Only about 50°C	Sulfide semiconductors (10^{-3} – 10^{-5} Ω-m) FeS_2, PbS, $CuFeS_2$ Easily heated to about 1000°C.
Mixed Valent Oxides (10^{-2} – 10^{-4} Ω-m) Fe_3O_4, CuO, Co_2O_3, NiO Easily heated to about 1000°C	Carbon and Graphite (~10 Ω-m) Easily heated to 1000°C.

It is apparent from these observations that conduction promotes the coupling of microwave energy into the powder. Transition metal oxides and sulfides such as magnetite and pyrite are easily heated, but insulators like alumina and silica are not.

Note, however, that metal particles are not heated as well, despite their high electrical conductivity.

Metals do not heat as well as semimetals and narrow-band semiconductors because electric fields cannot penetrate much below the surface. Skin depth, δ, is defined as the depth at which the electric field drops to 1/e = 0.368 of the surface value. It is related to frequency *f*, permeability μ and conductivity σ by the relation

$$\delta = 1/\sqrt{\pi f \mu \sigma} \tag{8}$$

For copper, $\sigma = 5.8 \times 10^7\ \Omega^{-1}\mathrm{m}^{-1}$, $\mu=\mu_0$, where μ_0 is the permeability of free space, and the skin depth is approximately 1 cm at 60 Hz and less than 1 Ìm at microwave frequencies. Since field penetration is proportional to σ(-1/2), microwave energy heats semimetals and semiconductors better than copper. Note also that the resistance of metals increases with temperature, unlike that of semiconductors which show a rapid decrease. This promotes rapid heating and thermal runaway for the latter, as explained later.

The inhomogeneous nature of green ceramics can be understood using the Maxwell-Wagner model (von Hippel, 1954). For a two-layer system consisting of equal amounts of conducting high-permittivity material (e.g., ferrite) and an insulating low-permittivity material (air or polymer), the relaxation frequency at which absorption occurs is

$$\omega = \sigma/\varepsilon. \tag{9}$$

This lies in the microwave region when the resistivity ρ, expressed in ø-m, is about 10/K, where K is the dielectric constant. Hence, semiconductors and semimetals have about the optimum conductivity for ceramic processing at microwave frequencies.

Oxide Semiconductors

Transition metal oxides like magnetite with mixed cation valences are among the best microwave absorbers with very rapid thermal runaway. In electroceramic technology these mixed valency oxides are used as NTCR (negative temperature coefficient of resistance) thermistors. The electrical resistance decreases exponentially with increasing temperature. Typical of the NTCR thermistor materials, also know as controlled valency semiconductors, are titanium-doped hematite (Fe_2O_3:Ti) and lithium-doped bunsenite (NiO:Li). Reacting hematite with rutile in air yields

$$\left(1-\frac{x}{2}\right)Fe_2O_3 + xTiO_2 \rightarrow Fe^{3+}_{2-2x}Fe^{2+}_xTi^{4+}_xO_3 + \frac{x}{4}O_2 \tag{10}$$

This is an n-type semiconductor in which electrons are transferred between iron atoms of different valence

$$Fe^{3+} + e^- \rightleftarrows Fe^{2+} \tag{11}$$

p-type NTC thermistors are made from nickel oxide doped with lithium.

$$(1\text{-}x)NiO + \frac{x}{2}Li_2O + \frac{x}{4}O_2 \rightarrow Ni^{2+}_{1-2x}Ni^{3+}_xLi^+_xO \tag{12}$$

The whole conduction process involves charge transfer between trivalent and divalent nickel ions.

$$Ni^{2+} + h^+ \rightleftarrows Ni^{3+}. \tag{13}$$

Doped nickel oxide has the rocksalt structure with lithium partially replacing nickel in the cation sites. Ionic radii for Ni^{2+}(0.84Å), Ni^{3+}(0.74Å), and Li^+(0.88Å) all favor octahedral coordination with oxygen.

The resistivity of $Ni_{1-x}Li_xO$ ceramics decreases drastically with increasing lithium content. The restivity at $x=10^{-4}$ is about 1000 Ωm and at x=0.1 it drops to 0.01 Ωm. The color is another indication of increased conductivity. The green color of pure nickel oxide deepens to black with increased doping.

For semiconducting compositions near $Ni_{0.95}Li_{0.05}O$, the band gap is about 0.15eV. The physical origin of this narrow band gap is attributed to the weak attractive forces between the negatively charged portion of the lattice where Li^+ dopant ions are located, and positively charged regions occupied by the compensating Ni^{3+} ions. Charge is neutralized best when these ions are next nearest neighbors, but under the action of electric field or temperature, the electron hole located at the Ni^{3+} drifts away rather easily to other nickel sites. Polarization of the surrounding oxide lattice also contributes to the band gap energy.

Electrical conductivity σ is proportional to the charge carrier concentration n, the charge of each carrier q, and the mobility μ:

$$\sigma = n\, q\, \mu. \tag{14}$$

In thermistors and other materials subject to thermal runaway, the temperature dependence of the conductivity is of great importance. Both n and μ are strongly dependent on temperature. For a semiconductor, the carrier concentration varies exponentially with temperature

$$n \sim \exp(-E/kT) \tag{15}$$

where E is the energy required to liberate charge carriers, k is Boltzman's constant and t is the absolute temperature. The temperature dependence of the mobility depends on its physical origin. For most scattering processes, mobility follows an inverse power law

$$\mu \sim \exp(E'/kT). \tag{16}$$

in which mobility decreases with increasing temperature because of atomic thermal vibrations, but a different temperature dependence is observed for hopping processes. Here the mobility depends on thermal excitation, and increases exponentially with temperature

$$\mu \sim T^{-b} \tag{17}$$

E′ is the energy observed in the hopping process. Summing up, the temperature dependence of the electrical conductivity is

$$\sigma(T) \sim T^{-b}\, e^{-(E+E')/kT} \tag{18}$$

Since exponentials tend to dominate, the electrical resistance of a mixed valency oxide can be described by

$$R = Ae^{B/T}. \tag{19}$$

For the transition metal oxides used as NTC thermistors, R lies in the range 1-10^4 ohms, and B is 2000 - 6000°K. The temperature coefficient α describes the fractional change in resistance with increasing temperature:

$$\alpha = \frac{1\,dR}{R\,dT} = \frac{B}{T^2} \tag{20}$$

At room temperature, the resistance drops by 2-6%/°K for a typical thermistor. Resistance changes typical for thermistors made from transition metal oxides are shown in Figure 1

Thermal Runaway

A typical current-voltage relationship for a transition metal oxide thermistor is shown in Figure 2. This is a so-called static characteristic in which the current is set to constant value and the system is equilibrated. The steady state voltage and temperature are then recorded for each current setting. Runaway takes place if the current is not carefully controlled.

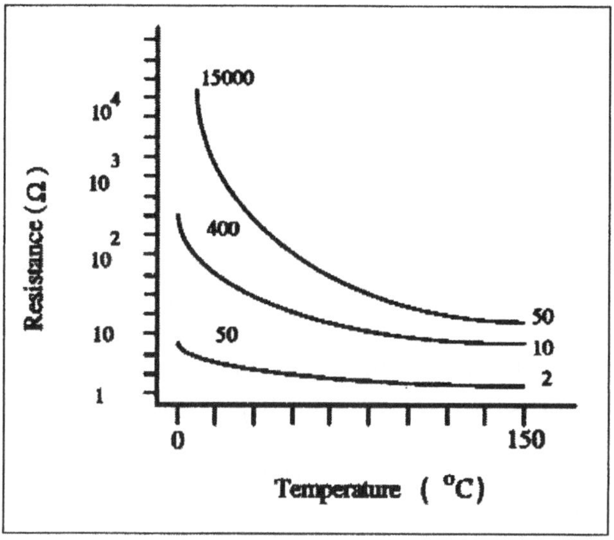

Figure 1. Temperature dependence of typical NTC thermistors.

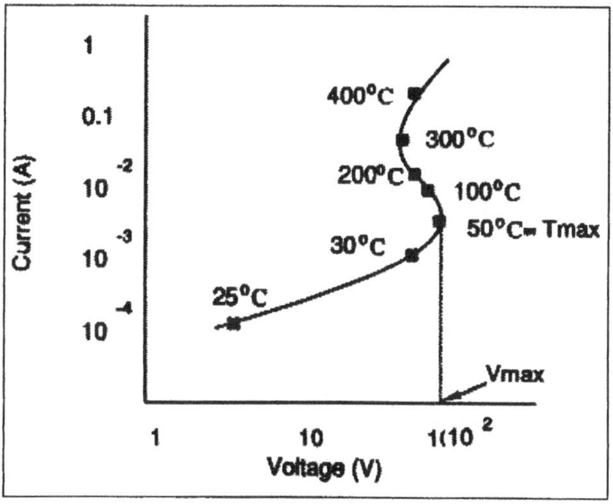

Figure 2. Current-voltage characteristic for a transition metal oxide thermistor. Runaway takes place above T_{MAX} and V_{MAX}.

The relationship between the I-V characteristic (Figure 2) and the R-T relationship (Figure 1) is established through Ohm's Law and the power balance equation

$$R = V/I = Ae^{B/T} \tag{21}$$

$$W = VI = AD(T-T_0). \tag{22}$$

W is the input power (watts), T the specimen temperature, T_0 the temperature of ambient, and D the coefficient in Newton's law of cooling. D is a measure of the thermal coupling to the surroundings per unit surface area of the specimen. When surrounded by air, the specimen is cooled by convective air currents, and D ranges from 5 W/m^2K for free convection to 50 W/m^2K for forced convection (Tummala). The values of D are substantially higher, typically 100 - 10,000 W/m^2K, when in thermal contact to the surroundings takes place through solids or liquids.

Converting Ohm's law and the power balance equation to logarithmic form and summing the two, we obtain an expression relating voltage and temperature:

$$2\ln V = \ln AD + \ln (T-To) + B/T. \tag{23}$$

Thermal runaway takes place when T exceeds T_{MAX} and V exceeds V_{MAX}. To determine these conditions, we take the derivative of the voltage-temperature relation at $T = T_{MAX}$, $V = V_{MAX}$ and

$$2\frac{d}{dT}\ln V = 0 = \frac{1}{T_{MAX} - T_0} - \frac{B}{T_{MAX}^2} \tag{24}$$

$$T_{MAX} = \frac{B}{2} \pm \sqrt{\frac{B^2}{4} - BT_0} \tag{25}$$

For B = 2000 - 4000 K, and T_0 = 30°C, T_{MAX} is in the range 45-85°C, which is not far above room temperature. Therefore many transition metal oxides are on the verge of thermal runaway, as evidenced by the microwave heating experiments. The corresponding values for input current and power are obtained from the equation balancing input and output power using the appropriate value for D.

Microwave Absorption in Silicates

Conduction losses are important in glasses as well. Fused silica has very low tan δ values (<0.001) over wide frequency and temperature ranges. Alkali ions promote dielectric loss in glass. At low frequencies the loss is associated with ion transport through the silicate network giving a space charge contribution to the dielectric permittivity. At 1 MHz the dielectric constant of sodium silicate glasses containing 30 mol% Na_2O. 70 mol% SiO_2 are about twice as large as SiO_2 glass.

Vibration losses become important at higher frequencies. Localized motion of the alkali ion in silicate cages can lead to strong absorption in the microwave range. The undesirable heating of Corning 9609 glass-ceramic cooking ware is an excellent example (MacDowell, 1984). Sodium nepheline ($Na_4Al_4Si_4O_{16}$) is the crystalline phase in this ceramic whose microwave losses are far higher than those of natural nephaline ($KNa_3Al_4Si_4O_{16}$). The glass-ceramic has a tan δ of about 7% at 2.45 GHz compared to 1% for the mineral nepheline. The high loss is attributed to the "rattling" motion of the smaller Na^+ ion occupying the crystallographic site of the larger K^+ ion. Sodium nepheline dishes are unsuitable for use in microwave ovens because of the high heat-up rate. The rattling motion of alkali ions can be pictured as localized conduction, or as a reorientable dipole between the cation and the coordinating anion cage. In this respect it bears a similarity to the double potential model used to describe proton motion in ice and ferroelectric potassium dihydrogen phosphate.

In most glasses, the tan δ values continue to climb with higher frequency. These losses are especially prominent in lead-bearing glasses and can be thought of as tails from far infrared spectra The atomic motion associated with these losses involve motions of heavy ions, or in some cases loosely bonded fragments of the silicate network, against the more tightly bonded portions of the network.

Tails From the Far Infrared

Most people have watched the ghostly "Tales from the Crypt," but few have given much thought to tails from the far infrared. They, too, are rather cryptic in that their origin is very obscure.

In crystals, infrared resonances involve the interaction of electromagnetic waves and the electric dipoles associated with lattice vibrations. The "reststrahlen" in alkali halides come from the excitation of a mode in which alkali ions vibrate out of phase with halogens. The frequency of this peak moves toward the far infrared and microwave region for heavy ions like Cs^+ and I^-. Weaker bonding and heavier masses both contribute to the frequency shift. This idea is utilized in choosing infrared windows for optical systems.

Similar relationships are observed in polymer and composite systems. Vibrational motions in polymers involve large numbers of atoms (and hence large masses) and weak secondary bonding forces between polymer chains. The dielectric properties of polymers at microwave frequencies have been reviewed by A.J. Bur (1985). High frequency dielectric losses arise from various portions of the polymer structure. These include pendulum-like motions of bulky side groups, crankshaft motions of the main chain, "freezing-in" motions accompanying the rubber to glass transformations, and oscillations of the tiny crystallites present in semicrystalline polymers. Generally speaking, the smaller segments of polymer chains resonate at higher frequencies and are the last to be frozen out at low temperatures.

Poly(ethylene oxide) or PEO is a typical crystalline polymer with a melting point of 60°C and Tg near -67°C. The helical chain has a backbone of carbon and oxygen atoms:

```
  |   |       |   |
- C - C - O - C - C - O -
  |   |       |   |
```

PEO has a modest dielectric constant of 4.3 at room temperature but a large absorption factor (tan $\delta \sim 0.1$) in the GHz range. Microbrownian motions of the main chain has been suggested as the cause of this absorption peak. Like glycerine, this peak is strongly dependent on temperature and frequency. At -50°C, the loss peak in PEO has dropped six orders of magnitude to the kilohertz range. (Connor, et al., 1964).

In amorphous polymers it is often difficult to identify the origin of microwave losses. Low density polyethylene $(CH_2)_n$ has twice the microwave loss level of high density polyethylene. (Amrhein, 1972). The tan δ values increase slowly with frequency similar to the behavior of the high Q resonator ceramics. The higher losses of low density polyethylene have been attributed to the amorphous regions where portions of polymer chains are able to oscillate in response to the microwave field.

Dipole Reorientation

The loss spectrum of water illustrates the contributions of conductivity and dipole reorientation to the dielectric constant. Figure shows the classical relaxation spectrum at room temperature. Water molecules have large dipole moments because of their nonlinear molecular geometry. Under an electric field the dipoles reorient easily at room temperature producing a large dielectric constant of nearly 80. At microwave frequencies there is a rapid decrease in the dielectric constant accompanied by a large loss peak which is utilized in cooking foods. The presence of water in meat and vegetables couples the microwave energy into the food. At lower frequencies there is additional loss arising from conduction. The temperature dependence of the dielectric properties at 3GHz and 1MHz are quite different. The microwave loss peak moves to higher frequencies with increasing temperature causing the 3GHz loss to decrease. On the other hand the 1 MHz loss increases with temperature as the water becomes more conducting. It is interesting to note that the dipole reorientation effects in water persist below the freezing point. The dielectric constant of ice is almost as high as water but the relaxation frequency drops rapidly into the kilohertz range and no longer couples strongly to microwaves.

Other polar liquids behave in a manner similar to water. Table 2 lists the microwave properties of several polar and non-polar liquids. Both the real and the imaginary parts of the dielectric constant of polar liquids are orders of magnitude larger than nonpolar liquids. Dipole reorientation is one of the dominant absorption mechanisms in the microwave region.

Many organic liquids show strong dispersion effects. Glycerol ($CH_2OHCHOHCH_2OH$) is a highly viscous polar liquid with a boiling point of 290° C and a melting point of 18° C. Dielectric spectra taken over a wide range of temperatures resemble the curious frequency-dependent curves of lead magnesium niobate (PMN) and other relaxor ferroelectrics. The broad, diffuse melting transition is accompanied by peaks in K′ and K″ which depend strongly on both temperature and measuring frequency. When measured at 100 Hz, K″ peaks at 60° C, and then shifts rapidly to higher temperatures for higher frequencies. At 10 kHz the peak is at -40° C at 10 MHz, 0° C, and at 100 MHz, 25° C. Thus the absorption maximum in glycerol moves rapidly toward the microwave region as the temperature rises (Morgan and Yager, 1940).

It is this "tunability" of energy absorption that makes microwave processing especially interesting. Unlike normal furnaces with broad band heat sources, monochromatic sources such as microwave ovens and high power lasers offer the possibility of shifting the heating from one constituent to another as reaction and densification take place. Exploitation of this time-temperature-frequency feature seems likely.

TABLE 2. MICROWAVE DIELECTRIC PROPERTIES OF LIQUIDS AT 3 GHZ AND 25°C (VONHIPPEL, 1954)

	K	tan δ
Polar Liquids		
Water (H_2O)	77	0.16
Methyl Alcohol (CH_3OH)	24	0.64
Ethyl Alcohol (CH_3CH_2OH)	6.5	0.25
Nonpolar Liquids		
Heptane ($CH_3(CH_2)5CH_3$)	2.0	0.0001
Carbon Tetrachloride (CCl_4)	2.0	0.0004

Microwave Resonance Spectra

Piezoelectric resonances also give rise to loss spectra, sometimes at very high frequencies. For thickness resonance, the fundamental frequency f is

$$f = \frac{1}{2t}\sqrt{\frac{c}{\rho}} \text{ (Hz)} \tag{26}$$

where c is the stiffness (approximately 10^{11} N/m^2 for an oxide), ρ the density (~5000 kg/m^3) and t refers to the specimen thickness in meters. Millimeter thick transducers have resonant frequencies in the MHz range, and extrapolating these results to the microwave range, it is apparent that micron-size piezoelectric particles will resonate at GHz frequencies. Ceramic grain sizes and ferroelectric domain sizes are in the micron range and therefore show electromechanical losses in the microwave range. Experiments on $LiNbO_3$ crystals, bicrystals and ceramics confirm these ideas (Yao Xi thesis).

The microwave losses in $BaTiO_3$ are shown in Figure 3. Note the drastic reduction in loss in specimens placed under DC bias. With domain walls locked in position there is a large decrease in K.

Figure 3 Dielectric permittivity and loss tangent of $Ba(Ti_{0.85}Zr_{0.11}Sn_{0.04})O_3$ ceramics change rapidly in the microwave region. Poling the ceramics inhibit domain wall motion and lowers both K and tan δ. (Poplavko, 1964).

Dimensional resonances of a different sort are employed in the dielectric resonators used in cellular mobile radios, marine satellite communications, and microstripline filters for SHF-TV converters. Dielectric resonance involves electromagnetic waves rather than the acoustic waves utilized in piezoelectric devices. The resonant frequency is controlled by the dielectric constant K (= $\varepsilon/\varepsilon_0$) and relative magnetic permeability $\mu_R = \mu/\mu_0$.

$$f = \frac{c}{\lambda\sqrt{K^{*}\mu_R}} \tag{27}$$

where c is the speed of light in vacuum and λ the wavelength. Resonance occurs when the specimen size corresponds to an integral number of wavelengths. High K dielectrics reduce the wavelengths and miniaturize the microwave filters.

The microwave loss spectra of typical dielectric resonators are shown in Figure 4. Note that high K ceramics have higher losses than do low R materials, and that the losses generally increase at higher frequencies suggesting that the origin of these losses is in the far infrared. tan δ values for dielectric resonators are very small, typically 10^{-3} to 10^{-4}, and are probably associated with low frequency tails of far infrared absorption peaks.

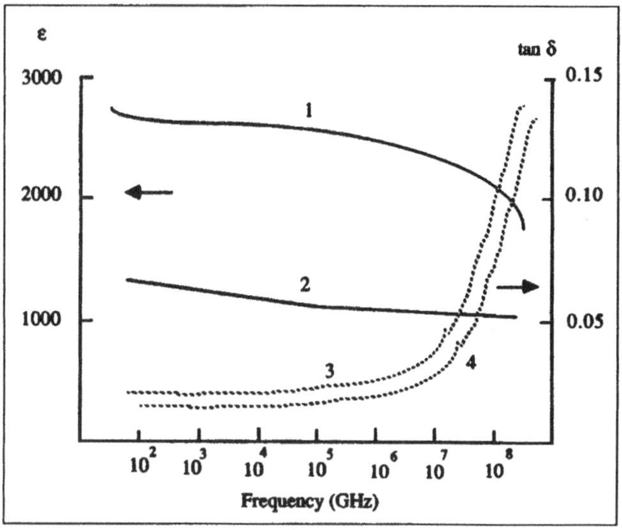

Figure 3. Frequency dependence of the parameters of a polydomain barium titanate crystal in weak fields. (1) Permittivity; (2) permittivity in a bias field of 10 kV/cm; (3) loss-angle tangent; (4) loss-angle tangent in a bias field of 10 kV/cm.

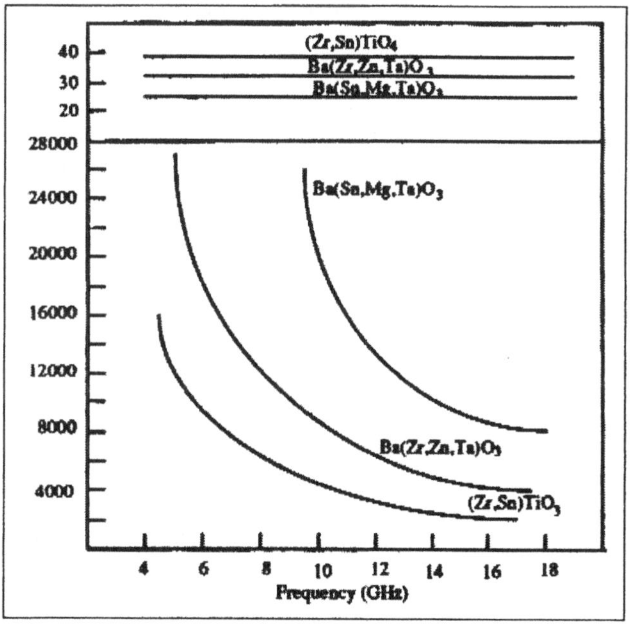

Figure 4. Microwave dielectric constants and loss spectra for ceramics used as dielectric resonators (Wakino,1986).

The lowest frequency infrared mode involves vibrations of the large cation against the oxygen octahedra (Figure 5). The mode softens in high permittivity ferroelectric perovskites and eventually condenses at the Curie temperature to give spontaneous polarization. This helps to explain why high K dielectrics are generally more absorbing than Low K solids in the microwave region. The infrared peaks of dielectric resonator materials have been investigated by Wakino (1986).

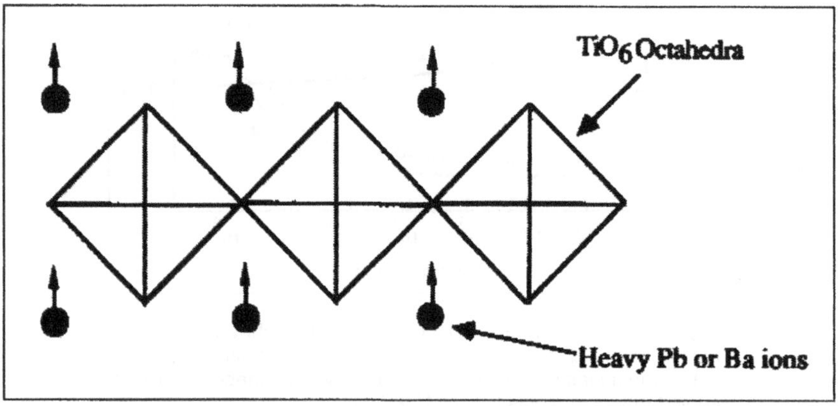

Figure 5. Far infrared vibration mode for oxide perovskites. The vibrational mode of heavy ions of Pb or Ba, and TiO_6 octahedra in the oxide perovskites.

Magnetic Loss

Most of the interactions between microwaves and matter take place through the electric vector of the electromagnetic wave, but the magnetic field vector can also induce substantial microwave losses, especially in magnetic materials with aligned spins. Some of the loss mechanisms in magnetic materials are listed in Table 3.

TABLE 3. LOSS MECHANISMS IN MAGNETIC SOLIDS (CHIKAZUMI, 1964)

Hysteresis loss rising from irreversible domain wall displacements
Eddy current loss from field-induced electric currents
Magnetic aftereffects associated with electron diffusion, usually from Fe^{2+} to Fe^{3+}
Dimensional resonances coming from electromagnetic standing waves
Magnetic resonance from precessional motion of unpaired dipole magnetic moments
Domain wall oscillations

The magnetic field vector of the microwave is represented by

$$\mathbf{H} = \mathbf{H_0 e^{i\omega t}} \tag{28}$$

The resulting magnetic flux density is

$$\mathbf{B} = \mathbf{B_0 e^{i(\omega t-\delta)}} = \mu_0 \mathbf{H} + \mathbf{I} \tag{29}$$

where μ_0 is the permeability of free space (4π x 10^{-7} H/m) and I is the magnetization (magnetic dipole moment per unit volume).

The magnetic permeability μ is analogous to the electric permittivity E.

$$\mu = \frac{B}{H} = \frac{B_0 e^{-i\delta}}{H_0} = \mu' - \mu'', \tag{30}$$

and the magnetic loss factor

$$\tan\delta_m = \frac{\mu''}{\mu'} \tag{31}$$

The relative permeability $\mu_R = \mu/\mu_0$ is the magnetic equivalent of the dielectric constant.

The magnetic permeability of ferrites is strongly dependent on frequency. The resulting losses arise from conduction, ferrimagnetic resonance, and domain wall oscillations.

In magnetic ceramics, much of the microwave loss is associated with ferrimagnetic resonance which is observed in ferrites with the spinel, magnetoplumbite and garnet structures. As illustrated in Figure 6, ferrimagnetic resonance involves precessional motions of the sublattice magnetization vectors about an effective internal magnetic field H_e. The effective field is caused by domain wall resonance or by magnetocrystalline anisotropy. High frequency absorption in the microwave region is found in magnetic materials with large anisotropy coefficients K_1. Soft ferrites such as the manganese ferrites with the spinel structure generally show the highest loss in the MHz region whereas hard ferrites with the magnetoplumbite structure have much higher resonant frequencies up in the GHz range.

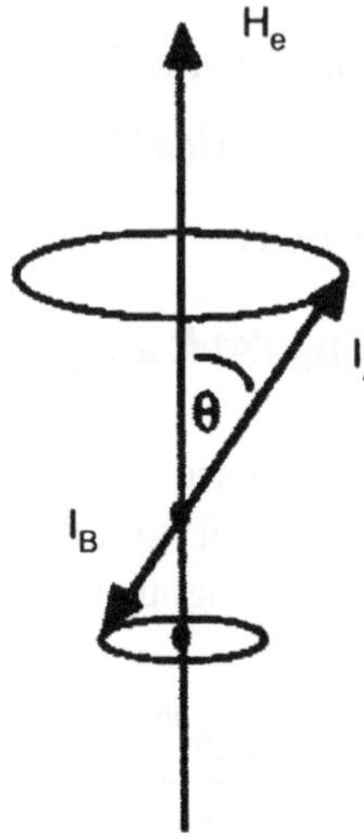

Figure 6. Microwave resonance in ferrimagnetic oxides in which the sublattice magnetizations I_A and I_B precess about an intemal effective field H_e.

Summary

The interactions between microwaves and matter are many and various. Conduction losses from localized space charge effects are responsible for microwave loss in transition metal oxides with mixed valence states. The inhomogeneous nature of green ceramics can be modeled using Maxwell-Wagner theory and thermal runaway follows a process similar to that observed in NTC thermistors.

Alkali ions cause microwave loss in glasses and other silicates, especially when loosely coordinated in a manner allowing considerable rattling room. Oscillating Na ions can be pictured either as localized conduction or as reorientable dipoles. The two polarization mechanisms lose their identity in situations like this.

Organic compounds and polymers are useful in ceramic processing. Dipole relaxation spectra are common in water, alcohol and other polar liquids, leading to intense energy absorption in the microwave regions. Polymers also absorb in this region of the electromagnetic spectrum with smaller segments of the polymer chain resonating at higher frequencies and continuing to lower temperatures.

Localized resonances caused by piezoelectric electromechanical coupling can also cause microwave loss. Domains in ferroelectric ceramics are on the right size scale to promote loss in the GHz region.

Dielectric resonator materials such as $ZrTiO_4$ utilize electromagnetic resonance rather than electromechanical resonance. These are generally low-loss ceramics but the tan ‰ values increase with frequency, suggesting a loss mechanism associated with infrared vibrations. Soft mode condensation in ferroelectrics has a similar origin.

Finally, not all microwave losses take place through the electric vector of the wave. The magnetic field vector can initiate energy transfer through magnetic resonance effects in which the unpaired electron spins precess about internal fields. Anisotropy effects in hard ferrites and oscillating domain walls often create the internal fields.

REFERENCES

1. E.M. Amhein, Ann. N.Y., " High Frequency Microwaves", *Acad. Sci.*, **196** 179 (1972)
2. A.J. Bur, "Dielectric Properties of Polymers at Microwave Frequencies: A Review", *Polymer* **26** (7) 963-977 (1985).
3. S. Chikazumi, Physics of Magnetism. Wiley, New York, (1964).
4. T.M. Conner, et. al., " Microwave Loss in Polymers", *J. Appl. Chem.*, **14,** 74 (1964).
5. Morgan and Yager, " Organic Dielectrics", *Ind. Eng. Chem.* **32,** 1519 (1940).
6. W.H. Sutton, Microwave Processing of Ceramic Materials. *Ceram. Bull.*, **68** (2)376-386(1989).
7. J.F. MacDowell, Microwave Heating of Nepheline Glass-Ceramics, *Ceram. Bull.*, **63** (2) 282-286 (1984).
8. S.L. McGill, J.W. Walkiewicz, and G.A. Smyres, The Effects of Power Level on the Microwave Heating of Selected Chemicals and Minerals in Microwave Processing of Materials, Vol. 124. Mat. Res. Soc., Pittsburgh PA, 1988.
9. T.T. Meek, "Proposed Model for the Sintering of a Dielectric in a Microwave Field." *J. Mater. Sci. Lett.*, **6,** 638-640 (1987).
10. Y.M. Poplavko, "Permittivity Dispersion of Barium Titanate Type Ferroelectric", *Sov. Phys.–Solid State,* **6,** [1] 45-49 (1964).
11. R. Tummala and E. Rymazewski, Microelectronic Packaging Handbook, p.171.
12. A.R. von Hippel (Ed.) Dielectric Materials and Applications M.I.T. Press, Cambridge, MA 1954.
13. K. Wakino, "High Frequency Dielectrics and Their Applications", Proc. ISAF, 1986.
14. Xi Yao, "Dielectric and Piezoelectric Behavior of Lithium Niobate Bicrystals and Polycrystal Ceramics," Ph.D., Thesis, The Pennsylvania State University, (1982).

Microwave Material Interactions and Process Design Modeling

W.R. Tinga

Abstract

Simple but useful heating rate models are developed to evaluate the relationships between energy storage in and energy absorption by dielectric materials at the macroscopic and microstructure level. By including the temperature dependence of a dielectric, some approximate process heating rate models are explored to illustrate the importance of the dielectric's temperature behavior. Practical effects of material boundaries within resonant, traveling wave applicators are discussed and modeled from an engineering point of view. Using these models, different system design alternatives can be analyzed to evaluate their effects on the overall heating process.

Microwave Process Model

Mathematical modeling of parts of the microwave heating process has resulted in useful but often difficult to interpret data [1-7]. To obtain a better intuitive understanding of the relative importance of the many system parameters, this paper addresses the modeling from a physical understanding point of view.

Assuming an arbitrary microwave power source, applicator, and product we can view a typical microwave heating system schematically as shown in Figure 1. Material properties, its shape and that of the applicator determine how energy is distributed in the process material. Power from the source establishes electromagnetic fields which force the material's molecules to store potential energy by taking on new equilibrium positions within the material at the microwave frequency rate. In this process the electromagnetic force fields cause the molecules to dissipate some of the applied kinetic energy as heat. This heat establishes a temperature distribution or field in the material thereby changing its microwave properties, its density and possibly its chemical properties. Heat losses, heat diffusion, density and specific heat changes will modify that temperature field. By far the most influential factors determining energy storage and loss are the effects of temperature and density and we will limit our discussion mainly to temperature. Because of the inherent feedback paths in the process loop, it is not at all clear how any particular heating process will proceed. Stable or unstable heating conditions are both possible depending on the intrinsic feedback properties of the heating process. Precise theoretical calculations to predict the electric and temperature fields are difficult because of the interactive and non-linear nature of the process parameters even though progress is being made in this area [7]. Our approach will be to isolate some of the fundamental quantities of interest and to establish some approximate, but useful engineering models and guidelines to improve our understanding of the relative importance of the variables.

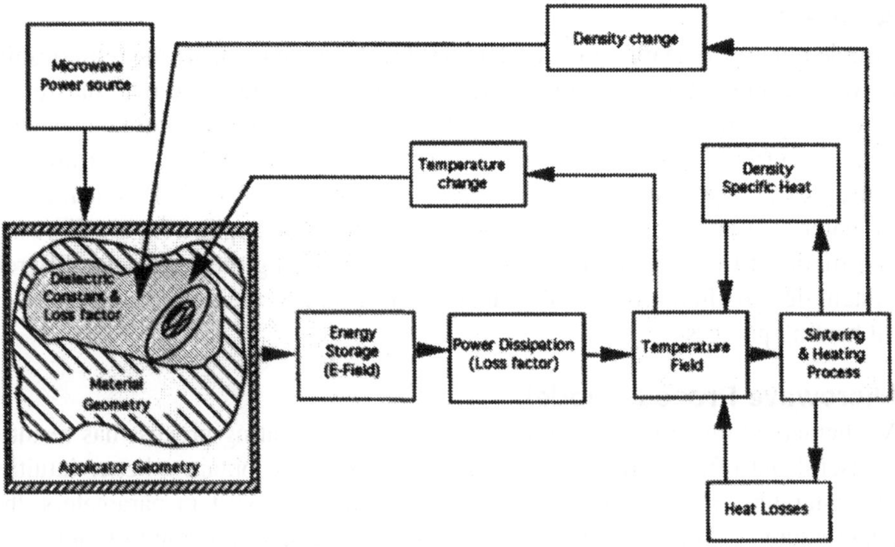

Figure 1. Model of microwave heating process.

First, it is helpful to recognize that the dielectric constant is a measure of a material's ability to store electric energy. The higher the dielectric constant the larger the capacity to store charge or electric energy. Conversely, the higher the dielectric constant, the smaller the internal field in the material in proportion to the externally applied field. All materials, however, exact a price for energy storage. Primarily, energy is extracted by the material from the imposed electric force fields to do the work to re-arrange its charge distribution as demanded by the applied field. We model this work done as the effective dielectric loss factor which automatically accounts for all the internal mechanisms which cause electromagnetic energy dissipation in the material. This description of energy storage and dissipation holds true when we look at a large chunk of material, the macroscopic view, or when we zoom in on, say, a single crystal grain in a ceramic.

It is well-known that the time-average power, in Watts, dissipated in a given material volume, V, due to molecular loss mechanisms, is given for non-magnetic materials by [8]

$$\mathbf{P} = \frac{1}{2}\int_V \mathbf{E} \cdot \mathbf{J}\,dv = \frac{1}{2}\int_V \sigma |\mathbf{E}|^2 dv = \frac{1}{2}\int_V \omega\varepsilon'' |\mathbf{E}|^2 dv \quad (\mathbf{W}) \tag{1}$$

where E = E(x,y,z,f,ε) is the actual peak electric field at the point of interest and; σ and ε'' are the total effective conductivity and loss factor respectively and $\omega = 2\pi f(\text{rad / s})$. Re-writing this integral equation into a differential equation form gives the time-average power density at a point as

$$\frac{dP}{dV} = \pi f \varepsilon'' |\mathbf{E}|^2 \quad (\mathbf{W / m^3}) \tag{2}$$

We emphasize that only the spatially average power via the integral equation can be evaluated experimentally, not this power density at a point. These equations are valid at the macroscopic and microscopic level and give the rate of energy delivery to a material. *If, and only if,* E and ε'' are constant in a given volume, may we write the total power absorbed by volume V as

$$\mathbf{P} = \pi f \varepsilon'' |\mathbf{E}|^2 \mathbf{V} \tag{3}$$

Heating Rate Modeling

If heat conduction is neglected, which is not always warranted, then we can write an approximate heating rate equation, for an isotropic rigid solid, as [9]

$$\frac{dT}{dt} = \frac{q}{C_p \rho} \quad (^\circ\mathbf{K / s}) \tag{4}$$

where C_p is the molar heat capacity in J/kg°K, ρ is the mass density in kg/m^3, T is the temperature in °K, t is time in s . In the case of microwave heating of a material, q=dP/dV represents the internal heat generation at a point in W / m^3. which combined with (2) and (4) yields the heating rate

$$\frac{dT}{dt} = \frac{\pi f \varepsilon''}{C_p \rho} |\mathbf{E}|^2 \quad (^\circ\mathbf{K / s}) \tag{5}$$

This equation is true at the particular point where the value of $\bar{E}$ is determined and when heat losses are negligible. Even though this result is widely used, it hides the intrinsic parameter dependencies on temperature, it does not model heat losses, and $\bar{E}$ is not explicitly known. In high temperature processing this neglect is not warranted. For a uniform plane wave propagating in an infinite, lossless material of relative permittivity, $\varepsilon r'$, input power and field strength are related by [8]

$$|\bar{E}|^2 = \frac{2\eta_0 P_{in}}{A\sqrt{\varepsilon_r'}} \quad (V/m)^2 \tag{6}$$

where A is the cross-sectional area transverse to the direction of propagation, η_0, is the free space wave impedance (η_0,=377 ohms), and P_{in} is the total power in Watts delivered through area, A, into a volume of the material of thickness d. We now assume the material to be lossy, and all of P_{in} to be absorbed in volume Ad. In practice this condition is nearly met when we have a thin slab of lossy material in a plane wave resonator or when we have a slab of low-loss material with very large d. In both cases, $\bar{E}$ remains nearly uniform. Along with these assumptions, we show the parameter's temperature dependence explicitly and using (6) in (5) we obtain

$$\frac{dT}{dt} \cong k_1 \frac{\varepsilon_r''(T)\, P_{in}}{C_p(T)\, \rho(T) \sqrt{\varepsilon_r'(T)}} \quad (°K/s) \tag{7}$$

where

$$k_1 = \frac{2\pi f \eta_0 \varepsilon_0}{A} \quad (1/m^3) \text{ and } \varepsilon_0 = 8.85 \times 10^{-12} \tag{8}$$

It is now quite clear that the microwave process heating rate, for uniform plane wave power delivery, is very temperature dependent. For waveguide propagation, only the constant k_1 changes [10] while for resonant systems a different k_1, containing a temperature dependent Q-factor, is needed [11].

Refining the Heating Rate Model

Since heat losses were not accounted for in the above model, it is not capable of accurately predicting transient or steady state heating rates or temperatures in the material. At the high temperatures needed for sintering, one of the primary heat loss mechanisms is that of radiative heat loss. Adding a radiation loss term [12,13] results in

$$C_p(T)\, \rho(T) \frac{dT}{dt} = k_1 \frac{\varepsilon_r''(T) P_{in}}{\varepsilon_r'(T)} + \frac{\sigma A}{V}\left[e_w(273+T_w)^4 - e(273+T)^4\right] \quad W/m^3 \tag{9}$$

where now σ is the Stefan-Boltzmann constant

$$\sigma = 4.92 \times 10^{-8} kg \cdot cal/m^2 \cdot hr \cdot (°K)^4 = 0.5714 \times 10^{-7} J/s \cdot m^2 \cdot (°K)^4$$

where e_w, T_w and e, T are the wall and sample surface emissivities and temperatures (°C). It is clear that at high enough temperatures, the second radiation term will dominate and cause the heating rate to decrease and the temperature to approach an equilibrium value. To counter this heat loss, insulation can be placed around the material and the equation becomes

$$C_p(T)\,\rho(T)\frac{dT}{dt} = k_1\frac{\varepsilon_r''(T)P_{in}}{\varepsilon_r'(T)} + \frac{\sigma A}{V}\left[e_{ins}(273+T_{ins})^4 - e(273+T)^4\right] W/m^3 \quad (10)$$

where e_{ins}, and T_{ins} are the emissivity and temperature of the inner surface of the insulation surrounding the sample. Heat conduction is still neglected even though this may not be a valid assumption in some cases.

Using this model (eqn's. 9,10) in Mathcad for fully dense mullite and alumina yields the results shown in Figures 2 and 3 for the insulated and non-insulated cases respectively. Actual temperature dependent dielectric data are used. The results are valid under the stated assumptions. For these ceramics, $\varepsilon r'(T)$ has a positive temperature coefficient and has a small decreasing effect on the heating rate, dT / dt. During sintering, both density and heat capacity increase by about a factor of two [11,14] causing the heating rate to change from the data shown. For comparison, the case of mullite, if its loss factor were independent of temperature, is also shown.

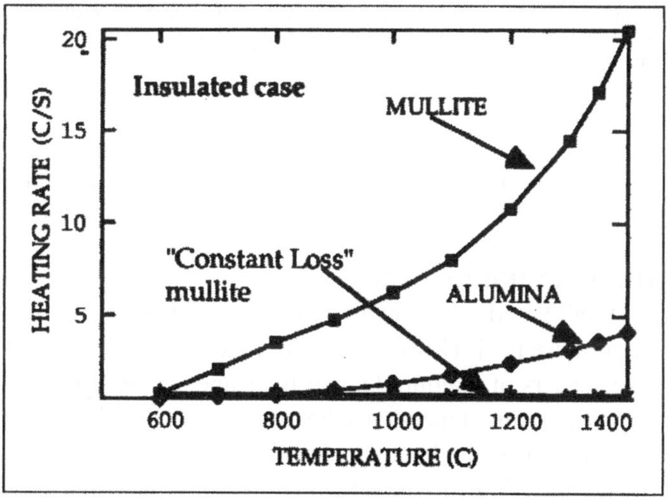

Figure 2. Predicted heating rate versus temperature (eq. 10) for mullite and alumina accounting for $\varepsilon^*(T)$ as given by Xi *et al.* [14] and Beale *et al.* [11]; f = 2.45 GHz, V = 1 x $10^{-6}m^3$, A = 0.00005 m^2, P_{in}= 100W, T_w = T_{ins} = T (insulated), e_{ins} = e = 0.5 $C_p(T)$ = constant = 775 J / kg °K, and $\rho(T)$ = constant = 3980 kg / m^3. Uniform sample heating is assumed.

Note that the heating rate of mullite is four to sixteen times that of alumina at 1400°C. Tests on the very low-density material, zeolite, in our microwave heated dielectrometer [14] have confirmed that heating rates in excess of 700°C are possible for P_{in} = 50W for a sample size similar to that used in the simulation.

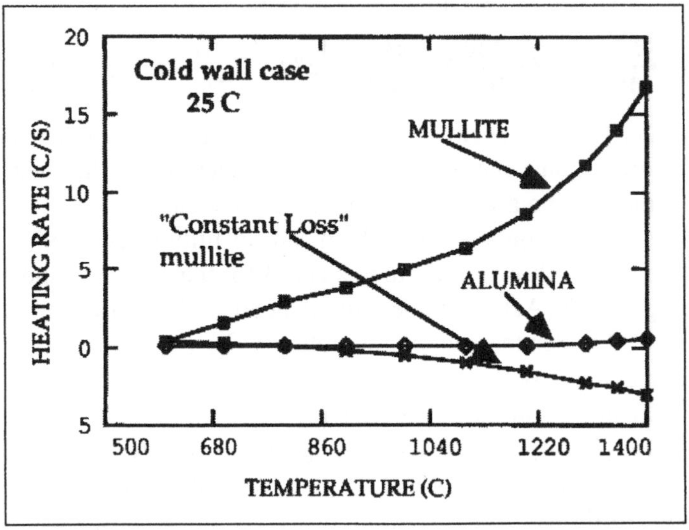

Figure 3. Predicted heating rate versus temperature (eq. 9) for mullite and alumina accounting for ε^* (T) as given by Xi *et al.* [14] and Beale *et al.* [11]; f = 2.45 GHz, V = 1 x $10^{-6}m^3$, P_{in} = 100W, T_w = 25°C (not insulated), A = 0.00005 m^2, e_{ins} = e = 0.5, $C_p(T)$ = constant = 775 J / kg °K.and $\rho(T)$ = constant = 3980 kg / m^3. Uniform sample heating is assumed.

Heat loss results are directly proportional to the sample's surface area, which was deliberately kept small to illustrate the trends shown in the figures.

Relative Parameter Importance

In general, $C_p(T)$, $\rho(T)$, and $\varepsilon r'(T)$ approximately double from room temperature to sintering temperature [14]. However, the loss factor, $\varepsilon r''(T)$, typically increases by two or three orders of magnitude over the same temperature range. Clearly the loss factor's temperature dependence becomes the controlling factor in establishing a material's microwave heating rate. From (9) it is evident that the only parameters which can offset this loss factor effect are the input power and the heat losses. Consequently, process design must concentrate on the control of Pin or heat loss or a modification of the $\varepsilon r''$ (T) behaviour by the mixing of suitable starting materials.

Microstructural View

Some debate on the effect of preferential grain boundary heating is found in the literature. Johnson [15] argues that the average grain to boundary temperature differential must be miniscule if such preferential heating takes place. Yet, Meek [16] presents experimental evidence that grain boundaries of microwave sintered samples are amorphous which suggests that grain boundary melting took place. One possibility is that grain boundary impurities depress the boundary's melting point causing the amorphous phase. However, conclusive evidence of what really happens at the grain boundary during microwave heating, is still lacking. Consequently, we consider it useful to present a summary of why large differential heating rates can be expected on the microstructure scale.

From dielectric mixture theory we know that, at the microstructural level, the E-field can vary significantly across grains, voids and grain boundaries. Meek [17], among others [13,18], showed that the heating rate in a low-density unsintered, part of a ceramic mixture is much greater than that in the sintered or dense parts of the mixture. Solving the heating rate equations at the microstructural level is inherently a complex task. Using approximate engineering models, however, we can draw some useful conclusions about differential heating rates at the microstructural level.

Consider the microscopic, but highly schematic, view of a dielectric depicted in Figure 4. Isolating a confocal ellipsoidal inclusion in a 3-component mixture, as shown, allows an estimate to be made of the ratio of the shell's electric field, representative of a grain boundary, to the field in the center inclusion. By using (2) we can write the power absorption ratio as

$$\frac{\Delta P_{abs2}}{\Delta P_{abs_3}} = \frac{\varepsilon_2''}{\varepsilon_3''}\left[\frac{|\overline{E}_2|}{|\overline{E}_3|}\right]^2 \tag{11}$$

Assuming very thin spherical shells or grain boundaries, gives a field ratio of 1.5 [13]. If a shell to grain loss factor ratio of 10 is assumed and estimating, for a grain of alumina with a glassy grain boundary $\varepsilon r'$=l0, $\varepsilon r'$=4 we get a shell to grain power absorption ratio of 22.5. Besides, the ratio of the local field to the average field in the bulk material may reach values of 10 or larger. Neglecting heat conduction, which assumption Johnson [15] challenges, it can be shown [13] that

$$\frac{dT_2}{dt} = \frac{\varepsilon_2''}{\varepsilon_3''}\left(\frac{\varepsilon_3'}{\varepsilon_2'}\right)^2 \frac{\rho_3}{\rho_2}\frac{C_{P_3}}{C_{P_2}}\frac{dT_3}{dt} \tag{12}$$

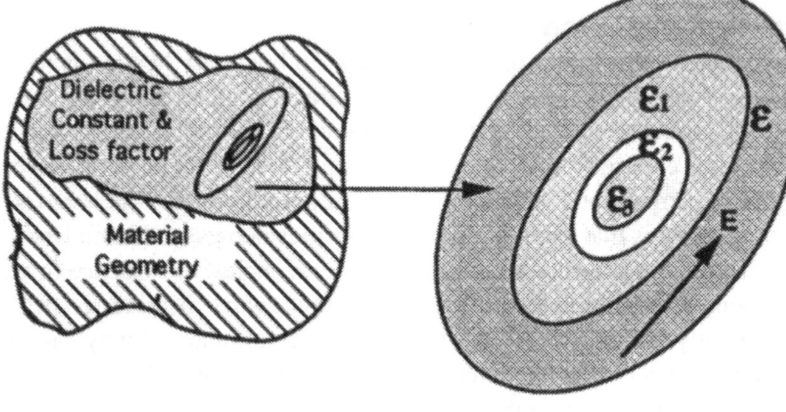

Figure 4. Isolated grain and grain boundary in a dielectric.

If we assume a reasonable grain to grain boundary density ratio of 2 and equal heat capacities, we have $(dT_2/dt)/(dT_3/dt) = 45$. For disk-shaped grains, the grain boundary to grain heating ratio could increase to 180. Observe that, since $\sqrt{\varepsilon} \propto \rho$ [19] the shell's grain boundary heating rate is proportional to the square of the density ratio. Whatever the cause of the enhanced microwave sintering, very large differential heating rates at the microstructural level are physically possible. At present, the implications of such high rates are not at all clear.

Practical Effect of Applicator and Material Boundaries

Recall from our process model in Figure 1 that the boundaries of a process material and the applicator geometry also affect the resulting fields in the material and hence the heating performance. In all cases, travelling waves are delivered by the microwave generator to the applicator. If the applicator is a dominant mode travelling wave structure containing the process material, the electric field and the resulting power density decay exponentially in the direction of travel with a subsequent decay in the temperature profile. However if the material's dimension in the direction of wave travel is shorter than 2-5 field penetration depths, reflected waves at the downstream boundary travel in the backward direction and cause an exponential power deposition in the backward direction. Any combination of forward and backward waves cause standing waves. Their adjacent stationary peaks and valleys are spaced by $\lambda_0/4\sqrt{\varepsilon'_r} \propto \rho$ Where λ_0, is the free space wavelength.

In case our applicator is a resonant structure, forward and backward travelling waves are combined to form characteristic standing waves (mode patterns) within the material and the applicator. Often this results in considerable focusing of the fields in particular locations within the resonator. Resulting energy patterns depend mainly on the material properties, the applied frequency, the applicator's size and geometry and are, to first order, determined by TEM or plane wave behaviour [20]. Power penetration depth in a material contained in a multimode cavity is nearly independent of the excited cavity modes if no large interface reflections occur for cases when $\varepsilon r' \gg 1$. Moreover, this penetration depth calculated for plane waves incident at large oblique angles is different by only about 5% from that calculated for the normal incidence case. In short, in the absence of large internal reflections and with $\varepsilon r' \gg 1$, first-order engineering design calculations can be made using the plane wave propagation model even in multimode resonators. Field focusing can be used to advantage for sintering small samples but causes severe problems in the uniformity of heating in practically sized samples, at least at the commercially used microwave heating frequencies.

There are a few different ways to counteract the non-uniformity caused by standing waves. Increase the frequency by an order of magnitude, as done by Oak Ridge researchers [21], or somehow create field boundary conditions which change with time. For example, moving a wall in a multimode resonator through a half wavelength will cause the physical peak and valley locations to interchange [22]. Of course, a similar effect can be obtained by moving the material through the applicator or moving the field patterns around by means of a mode stirrer. At present, the first method is not commercially economical because of the cost of, for example, a 28 GHz source. The last two schemes are practical in two dimensions. Maslowski et al [3], discovered that the introduction of a transverse inhomogeneity in a TEM guiding structure, e.g., an insulating layer, can greatly improve the longitudinal heating uniformity since that inhomogeneity excites other field components shifted in space phase.

Finally we briefly discuss leaky and surface wave structures which may be useful applicators even for high temperature designs. One good starting reference is the work of A.A. Oliner reviewed by Itoh [23].

Leaky waves, which are non-uniform plane waves, cause power deposition in a material, adjacent to the leaky applicator surface, which decay exponentially with distance into the material. The decay constant is a function, first of all, of the applicator structure. In general, leaky wave structures are more difficult to analyze but can produce very high intensity E-fields, comparable to those in high Q resonators [24]. Corrugated or reactance modulated surfaces produce surface waves which travel along the surface with very low attenuation. Again the fields decay, perpendicularly away from the surface, in an exponential fashion. Nonetheless, such structures largely eliminate standing wave effects and can possibly be used to advantage in high temperature processing. Leaky wave applicators can, under limited loading conditions overcome the penetration depth limit of normal field incidence [25].

One of the more successful approaches to distribute energy more uniformly in large multimode resonators or microwave tunnels, has been the use of slotted waveguides, or leaky guides, coupled to one or more sides of the resonator. These slowly leak, or radiate, a calculable fraction of their available power into the multimode resonator. Judiciously spaced slots distribute the available power to different oven regions, thereby counteracting the non-uniformity caused by the resonator's standing waves and the product which loads the resonator.

Conclusion

First and second order microwave heating rate models show that the temperature dependent loss factor will be a major determining factor in predicting temperatures in a material above 500 °C. Power input and heat-loss parameters are the only control variables available to the designer of a high temperature microwave heating process. At the microstructural level, very high localized heating rates are possible, the implications of which are, at present, not well understood. Applicator design options include slow-wave and leaky-wave designs although such applicators have, as yet, not been used for high temperature processing work.

ACKNOWLEDGMENT

Financial support for this work from the National Science and Engineering Research Council of Canada is gratefully acknowledged.

REFERENCES

1. W.J.L. Jansen and E.J. Michelis, "Computer Modeling of Dielectric Heating", High Frequency Processing and Heating conference, N.V. KEMA, Arnhem, The Netherlands, 1989.
2. D.M. Van Dommelen and P.F. Stefens, "Temperature Distribution in High Frequency Heated Dielectrics", J. of Microwave Power and Electromagnetic Energy, pp. 121-126, 1987.
3. R. Maslowski, F. Vermeulen and S. Chute, "Promoting Axial Uniformity in Radio-Frequency Heating of a Lossy Material in a Waveguide by Introduction of an Insulating Gap", J. of Microwave Power and Electromagnetic Energy, Vol 27(1), pp. 33-37, 1992.
4. D.G. Watters, M.E. Brodwin, G.A. Kriegsman, "Dynamic Temperature Profiles for a Uniformly Illuminated Planar Surface" Mater. Res. Soc. Symp. Proc. (24), pp. 129-134, 1988.
5. Ceramic Transactions, *Microwaves: Theory and Application In Materials Processing*, 21, edited by David E. Clark, Frank D. Gac and Willard H. Sutton The American Ceramic Society Inc., 1991.
6. *Microwave Processing of Materials III*. edited by William B. Snyder, Jr., Willard H. Sutton,

Magdy Iskander and P. Lynn Johnson, Mater. Res. Soc. Symp. Proc. 269, San Francisco, CA, 1991.

7. Gregory A Kriegsmann, "Thermal Runaway in Microwave Heated Ceramics: A One-Dimensional Model", J. Appl. Phys. 71(4), pp. 1960-1966, 1992.
8. David M. Pozar, *Microwave Engineering,* Addison-Wesley, p.36-37, 40, 1990.
9. The Essentials of Heat Transfer I, Research and Education Association, New York, 1987.
10. G. Roussy and J.M. Thiebaut, "Calorimetric Microwave Loss Measurements of Granular Materials, J. of Microwave Power, Vol 19(3), pp. 181-186, 1984.
11. Guy O. Beale and Francisco J. Arteaga, "Automatic Control to Prevent Thermal Runaway During Microwave Joining of Ceramics", Mater. Res. Soc. Proc. Vol 269, pp. 265-270, 1992.
12. D.Palaith, R.Silberglitt, C.C.M. Wu, R. Kleiner, E.L. Libelo, "Microwave Joining of Ceramics", Mater. Res. Soc. Symp. Proc. (24), pp. 255-266, 1988.
13. W.R. Tinga, "Mixture Laws and Microwave-Material Interactions" *Progress in Electromagnetics Research. PIER 6, Dielectric Properties of Heterogeneous Materials,* Editor A. Priou, pp. 1-40, 1992.
14. W. Xi and W.R. Tinga, "A High Temperature Microwave Dielectrometer", Ceramic Transactions, Microwaves: Theory and Application In Materials Processing, 21, edited by David E. Clark, Frank D. Gac and Willard H. Sutton The American Ceramic Society Inc., pp 215-224, 1991.
15. D. Lynn Johnson, "Microwave Heating of Grain Boundaries in Ceramics", J. Am. Ceram. soc. 74(4), pp. 849-850, 1991.
16. T.T. Meek, X. Zhang and M. Rader, "An Analysis of the Individual Grain Structure of an Oxide Heated Using Microwave Radiation and Heated Conventionally", Ceramic Transactions, *Microwaves: Theory and Application In Materials Processing,* 21, edited by David E. Clark, Frank D. Gac and Willard H. Sutton The American Ceramic Society Inc., 1991.
17. T.T. Meek, "Proposed Model For the Sintering of a Dielectric in a Microwave Field", J. Mat. Science Letters (6), pp. 638-640, 1987.
18. W.R. Tinga, W.A.G . Voss, D.F. Blossey,, "Generalized Approach to Multiphase Dielectric Mixture Theory", J. of Appl. Phys., Vol. 44(9), pp. 3897-3902, 1973.
19. S.O. Nelson, "Method for Determining Dielectric Properties of Solids from Measurements on Pulverized Materials", 1987 IEEE MTT-S Int. Microwave Symp. Digest, Vol. 1 pp. 461-463, 1987.
20. P.O. Risman and T. Ohlsson, "2450 Mhz Microwave Heating Distributions in Food Material Slabs and Cylinders", High Frequency Processing and Heating conference, N.V. KEMA, Arnhem, The Netherlands, 1989.
21. H.D. Kimrey, M.A. Janney, and R.L. Beatty, "Microwave Sintering of Zirconia-Toughened Alumina Composites", Mater. Res. Soc. Proc. Vol 189, 1990.
22. J. Chang, and M.E. Brodwin, " A Novel Dynamic High Order Multimode Microwave Applicator", International Microwave Power Symposium (IMPI), 1992.
23. T. Itoh, "Perspectives on Guided Wave Phenomena", IEEE, MTTS Digest, pp 133-136,1988.

24. J.E. Gerling, "Striping Apparatus For Highways", U.S. Patent # 3,472,200, Oct.14, 1969.

25. H. Guo, Y. Carmel, W.R. Lou, L. Chan, J. Rodgers, D.K. Abe, A. Bromborsky, W. Destler and V. L. Granatstein, " A Novel Highly Accurate Synthetic Technique For Determination of the Dispersive Characteristics in Periodic Slow Wave Structures", IEEE Trans. Microwave Theory Tech., vol. MTT40, p.2086-2094, 1992.

Modeling of Multi-Frequency Microwave Sintering of ZnO Ceramic

A. Birman, B. Levush, Y. Carmel, D. Gershon, D. Dadon, L.P. Martin and M. Rosen

Abstract

To gain better insight of microwave heating of ceramic materials and to optimize this process, we have developed a code modeling the heat propagation in the ceramic. The model takes into account the variation of the heat conduction coefficient, specific heat and permittivity with temperature. In addition, the heat conduction coefficient and permittivity dependence on the porosity is modeled using Maxwell Gamet mixing theory (MGT). The code was used to model experiments of ZnO microwave heating. These computations prove that the model used in the code is capable of reproducing the salient features of the heating process The code was modified to allow modeling of simultaneous microwave heating at different frequencies. In particular, heating of ZnO sample with two microwave sources at widely separated frequencies (2.45GHz and 28GHz) showed that it is possible to keep a homogeneous temperature profile across the sample throughout the sintering process.

Introduction

Use of microwaves for ceramic sintering has the potential of producing high quality ceramic bodies of complex shape using less energy. This process however is extremely complex and optimizing it requires a large number of trial and error experiments. Therefore, without proper computational modeling further progress is hindered [1]. We have developed a one dimensional heat transport code to model heat propagation and microwave energy absorption during the sintering process Microwave absorption is governed by the dielectric properties of the ceramic material which depend on the temperature, porosity and frequency. Specific heat and heat conduction coefficients depend on the temperature and heat conduction further depends on the porosity All this is taken into account in our model. The code was first tested against the analytical results of Kriegsmann [2] and then used to model sintering of ZnO ceramic. The numerical results were found to be in good agreement with the data taken in our laboratory during ZnO sintering. The model was further expanded to allow multi frequency processing. In particular, it is shown that with two microwave sources (2.45GHz and 28GHz) applied simultaneously it is possible to achieve a homogeneous temperature profile across the sample throughout the sintering process.

Theoretical Model

To obtain the temperature distribution in the sample at each moment one has to solve the heat transport equation with the appropriate boundary conditions. For the ID case this equation is given by:

$$c_p(T)\frac{dT}{dt} = \frac{d}{\rho(T)dx}[\kappa(T,\rho)\frac{dT}{dx}] + \frac{P(T,\rho,f,t)}{\rho(T)} \qquad (1)$$

where c_p is the heat-capacity, κ is the heat-conduction, ρ is the density and P is the power absorbed in the ceramic. The boundary conditions are either (1) no thermal flux, (2) prescribed temperature temporal profile, (3) heat loss by radiation and convection. For microwave sintering the power P absorbed per unit volume is given by:

$$P = 2\pi f \varepsilon_0 \varepsilon'_r \tan\delta \,|E|^2$$

where f is the frequency of the microwave source, ε_0 is the permittivity of free space, ε_r is the relative dielectric constant, tan‰ is the loss tangent and | E | is the local electric field. Strictly speaking the spatial distribution of the electric field should be found by solving the field equations in the ceramic. However, in this work the electric field is assumed to decay within the sample according to:

$$E = E_0 \exp(\frac{-x}{d}) \qquad (2)$$

where d is the skin depth and E_0 is the electric field in the unloaded cavity, which is proportional to the microwave power temporal profile g(t). The skin depth is further related to the frequency of the microwave source and to the loss tangent through:

$$d(T,\rho,f) = \frac{0.0675}{f(GHz)\sqrt{\varepsilon'_r}\sqrt{\sqrt{1+\tan^2\delta}-1}} \; [m] \qquad (3)$$

Modeling of Microwave Sintering of ZnO Ceramic

To perform the computation knowledge of the dependence on temperature, porosity and frequency of all the relevant quantities is required. The specific heat temperature dependence on the temperature in the range relevant to the sintering process is shown in Fig. 1. The data was taken from Ref [3] and fitted to a curve of the form $a + bT + cT^2$ ($a = 0.140$, $b = 1.77 \times 10^{-5}$, $c = 2360$). The temperature dependence of the heat conduction coefficient [3] [4], is shown in Figure 2.

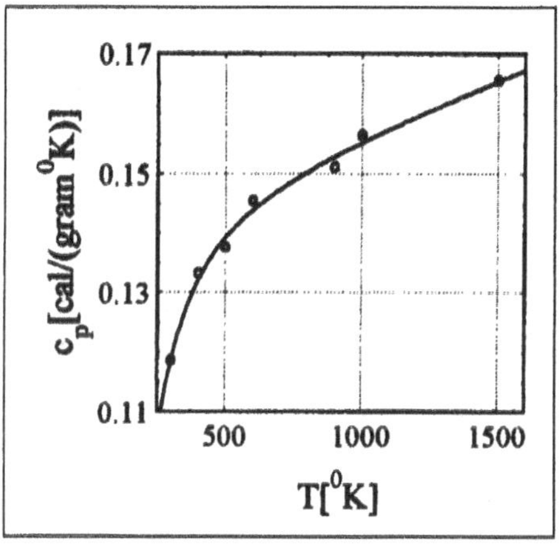

Figure 1. Specific heat of ZnO versus temperature.

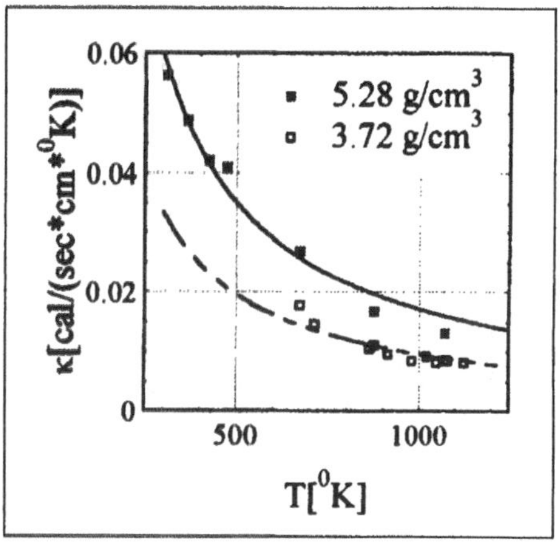

Figure 2. Heat conduction of ZnO versus temperature

The two sets of experimental data correspond to two densities 5.28 g/cm³ (95% of the theoretical density) and 3.72 g/cm³ (67% of the theoretical density). Both sets of data were fitted to a curve of the form $a/(b+T)$. For the high density we found $a = 16.8$ and $b = -18.5$ and for the low density we found $a = 9\ 4$ and $b =-18.5$. According to Maxwell Gamet mixing theory [5] heat conductivity of porous materials κ_m, is related to that of the full dense material through:

$$\kappa_m = \kappa_c \frac{(d-1)\upsilon_c}{d-\upsilon}$$

where κ_c is the heat conduction coefficient for fully dense ZnO ceramic, υ_c is the ratio of the density to that of theoretical density and d is either 2 or 3 for a 2D or 3D mixing model correspondingly. For $\upsilon_c = 0.95$ and $\upsilon_c = 0.67$ the ratio of the corresponding heat conduction coefficients is predicted to be 0.620 for $d = 3$ and 0.557 for $d = 2$. Note, that the ratio found from the fitted data is 0.559 which is in agreement with MGT

The real and imaginary parts of the dielectric constant of fully sintered ZnO of theoretical density were measured at 2.2GHz [6] and are shown in Figure 3 and Figure 4 .

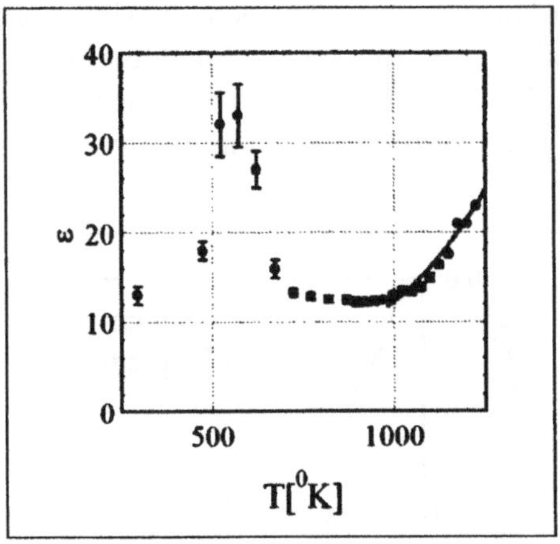

Figure 3. Relative real part of the dielectric constant of ZnO vs. temperature

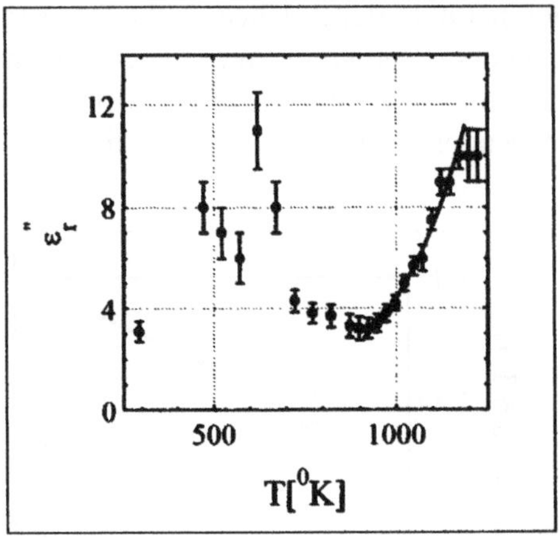

Figure 4 Relative imaginary part of the dielectric constant of ZnO vs. temp.

At temperatures above 900°K we fitted the data to an exponential curve $ae^{(bT)}$ We found b = 2.82×10^{-3} for ε and b = 4.77×10^{-3} for ε". The exponential dependence on the temperature of ε is corroborated by earlier data [7] where the exponent is found to be 2.65×10^{-3} (note that this data was taken at a low frequency of 1kHz). For the sample sintered in our laboratory the complex permittivity at room temperature of a fully sintered sample ε_{c0} was measured and found to be $\varepsilon_{c0}' = 19 \pm 2$ $\varepsilon_{c0}'' = 9 \pm 2$ (which is somewhat different from the data obtained in [6]). For our simulations we used the following:

$$\varepsilon_c' = \varepsilon_{c0}' \text{ for } T < 900°K \qquad \varepsilon_c' = \varepsilon_{c0}' \exp[0.00282*(T-900) \qquad \text{for } T > 900°K$$

$$\varepsilon_c'' = \varepsilon_{c0}'' \text{ for } T < 900°K \qquad \varepsilon_c'' = \varepsilon_{c0}'' \exp[0.00477*(T-900) \qquad \text{for } T > 900°K$$

For the relative dielectric constant the dependence on the porosity we used MGT [5], namely:

$$\varepsilon_m = \varepsilon_c \frac{2\upsilon_c\varepsilon_c + (3 - 2\upsilon_c)}{\upsilon_c + (3-\upsilon_c)\varepsilon_c}$$

where ε_c is the complex permittivity for fully dense ZnO. The permittivity at room temperature was also measured over the frequency range 0.2 GHz to

20 GHz [8] and the data fitted to a linear logarithmic curve:

$$\varepsilon_{c0}' = -11.5 \log_{10} f(\text{GHz}) + 23.5 \quad \text{and} \quad \varepsilon_{c0}' = -5 \log_{10} f(\text{GHz}) + 11.1$$

During the sintering process the density of the sample is continuously changing thus modifying the thermal and electrical properties throughout the sample. Hence, the heat transport equation should be supplemented by an equation relating the density change to the local temperature and the heating rate. In this work however the density was assumed to be a single valued function of the temperature. The density as function of the temperature was measured [8] and preliminary results are shown in Fig. 5. For temperatures above 650°C and below 1100°C a third order polynomial was fitted to the data. For temperatures above 1100°C we assumed the density to be linearly extrapolated to full density. Note that densification does not start before 650°C. In solving the heat transport equation eq. (1), we have used the above fitted data for the density change with the temperature, supplemented with the condition that density is only increasing (solidification is irreversible).

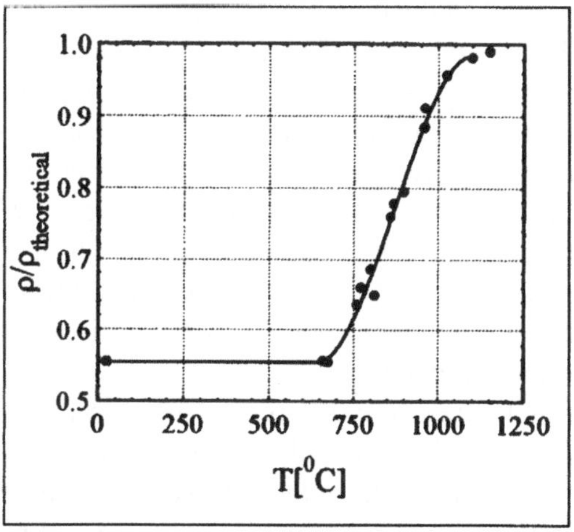

Figure 5. Relative density vs. temperature.

Simulations and Experimental Results

Our simulation of the experiment set-up is shown schematically below.

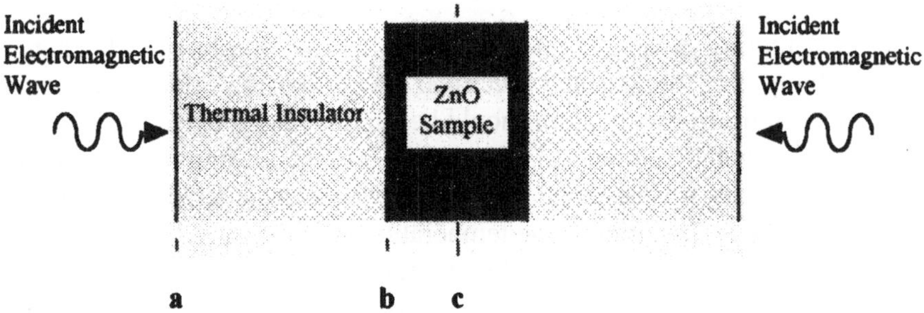

The cylindrical ZnO sample of height 1.65cm amd diameter 3.3cm is represented by a 3.3cm thick slab sandwiched between 4cm thick slabs of low density porous ZnO ($\rho/\rho_{theoretical}$= 0.12) representing the thermal insulation (not undergoing sintering). Plane a is kept at room temperature and the simulation is symmetric with respect to plane c. The electric field is assumed to decay within the sample (from b to c) according to eq. (2). To compute the skin depth (eq. (3)) the value of tan δ at b is used In the experiment the input power g(t) had a step-like shape as shown in Fig. 6 and the surface and core temperatures were measured with two type K thermocouples. For this input power, eq. (1) was solved yielding the temperature distribution in the sample as function of time and, in particular, the temperature at the core (plane c) and at the surface (plane b). The measured and computed core temperatures are shown in Fig. 7, while the difference between the core and the surface temperature is shown in Fig. 8. The electric field E_0 was first estimated from the measured g(t) and then adjusted so that the maximum calculated core temperature (at 28800 sec) is approximately equal to the maximum measured temperature.

We note that for the temperature difference further refinements may be needed to improve the modeling. This includes such issues as (1) more accurate spatial distribution of the electric field in the sample (2) improved modeling of the densification process (3) 2D computations (2D modeling of microwave sintering of Alumina had been reported in [9]).

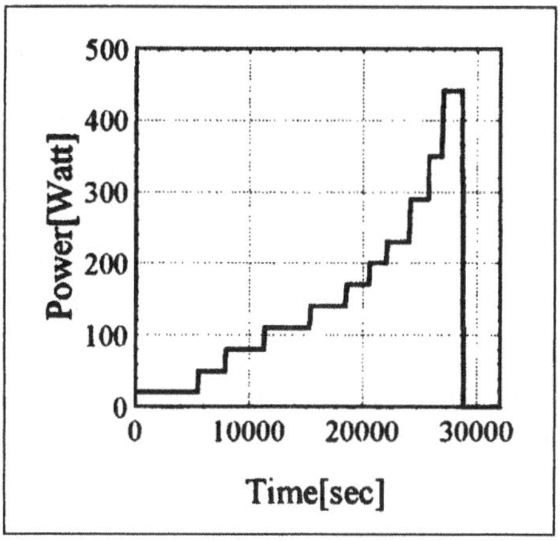

Figure 6. Microwave input power temporal profile.

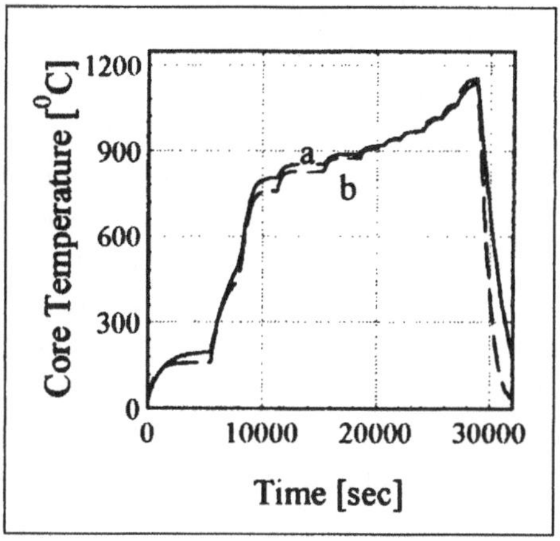

Figure 7. Core temperature: (a) measured (b) computed (dashed line).

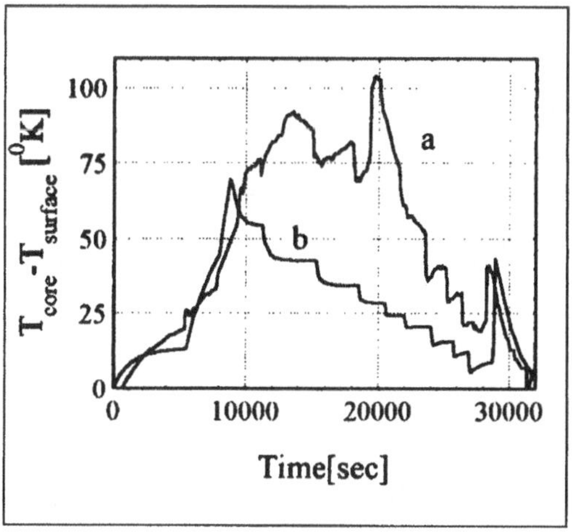

Figure 8. T_{core}-$T_{surface}$ vs. time (a) measured and (b) computed.

Application of Two Microwave Frequencies in the Sintering Process

One of the advantages of microwave sintering is higher heating rates. However, this might lead to nonuniform temperature distribution during processing This nonuniformity may be fixed by applying microwave heating at two distinct frequencies simultaneously. Having shown that our code is capable of predicting behavioral trends of the sintering process we set forth to simulate an accelerated heating process. For example, a ZnO sample is heated for 300sec at a constant rate of about 200°C per minute. The results of these numerical experiments are shown in Fig. 9. The curves in this graph are the difference between the core and surface temperature versus time for three cases: (a) only a 2.45GHz microwave source is applied, (b) only a 28GHz source is applied, (c) both sources are applied.

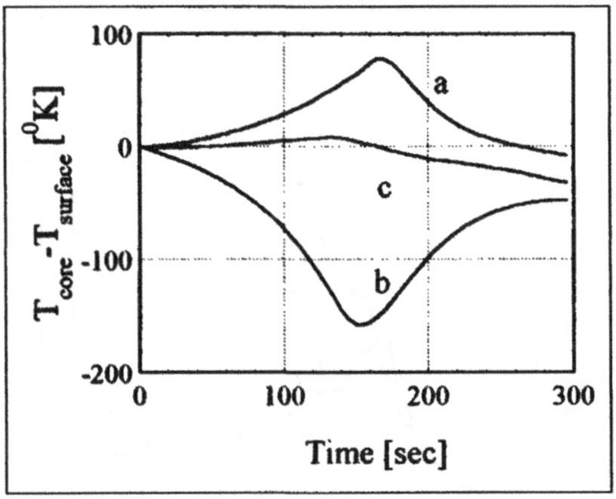

Figure 9. Calculated T_{core}-$T_{surface}$ vs. time: (a) 2.45 GHz, (b) 28 GHz, (c) 2.45 GHz and 28 GHz

For the low frequency and low temperatures the skin depth is 2.41cm which is larger than the sample's size, which therefore is heated almost uniformly. However, the conductive cooling of the surface causes the core temperature to exceed that of the surface. With the increase of the temperature, the skin depth along with the heat conduction decreases, thus increasing the temperature difference. For the high frequency the initial skin depth is 0.3cm which is smaller than the sample size. Therefore most of the power absorbed is close to the surface and the core is heated mainly by conduction, which causes the core's temperature to be lower. At higher temperatures the heat conduction increases due to the densification of the ceramic. Thus, the temperature gradient decreases in both cases. Using the two frequencies simultaneously, the temperature difference is reduced. In this example, the same temporal power profile was assumed for both frequencies, however, the temperature gradient can be further decreased by using different temporal power profiles.

Conclusion

Any attempt to simulate the sintering process requires detailed knowledge of the thermal and electrical physical parameters as function of the temperature, the porosity and the frequency applied, as well as the electrical field spatial distubution in the sample. This work is an attempt to model the sintering process of ZnO ceramic. The physical parameters dependence on ρ, T and f was in part based on measurements done in our laboratory and partly compiled from the literature. We found qualitative agreement with the experiment's results. We show that in the case of high rate sintering it is possible to achieve a history of uniform temperature distribution in the sample by applying two frequencies simultaneously.

ACKNOWLEDGMENT

Work supported by U.S. Department of Energy Division of Advanced Energy Projects. We are grateful to J. Binner and T. Cross for allowing us to use their facilities at the University of Nottingham In particular, many thanks to J. Binner for making it possible to have D Gershon perform the measurements. We would also like to thank V. Semenov for helpful discussions and comments.

REFERENCES

1. M. F. Iskander, "Computer modeling and numerical simulation of microwave heating systems" MRS Bulletin 18 pp 3036 (1993), and references therein.
2. G. A Knegsmann, "Thermal runaway in microwave heated ceramics" I. Appl. Phys 71 (4) pp 19601966 (1992).
3. Y.S. Touloukian, R.W. Powell, C.Y. Ho, P.G. Klemens, "Thermophysical Properties of Matter" The TPRC Data Series (Plenum NewYork 1970).
4. M.P. Shascalski, "Akustichekie Kristally" Nauka, 1982.
5. W. D. Kingery, H.K. Bowen, D.R. Uhlmann, "Introduction to Ceramic" John Wiley & Sons, Inc, New York 1975.
6. D. Gershon, J. Binner, T. Cross, and N. Greenacre (Private communication).
7. E F Tokarev, I.B. Kobyakov, I.P. Kuzmina, A.N. Lobachev, G.S. Pado, "Dielectric and Piezoelectric Properties of Zincite in the 4.2-800K temperature Range" Sov. Phys Solid State 17, No 4 (1975).
8. L.P. Martin, D. Dadon, D. Gershon, B. Levush, Y Carmel and M. Rosen, "Characterization of microwave processed ZnO" Paper presented in this conference.
9. J.R Thomas, J.D. Katz and R.D. Blake, "Temperature distribution in microwave sintering of Alumina cylinders" Mat. Res. Symp Proc 347 p 311 (1994).

Equipment Safety for Microwave and Radio Frequency Processing of Ceramics

J.F. Gerling

Abstract

The use microwave technology is increasingly being investigated for existing and new materials processes. Particular advantages and benefits have been identified for applications in ceramics. As progress is made towards commercializing these applications, those who are unfamiliar with the processing equipment will need to be aware of the associated safety issues. This paper reviews the current literature on related health effects, regulatory safety standards, equipment designs practices and guidelines for safe equipment use.

Introduction

Whenever microwave or radio frequency (RF) technology is considered for a particular industrial process, one of the first concerns to be raised is "Is it safe?" Quite often the answer is "Yes, depending on how you use it." Admittedly, such a response is hardly comforting to a key decision maker who is new to or unfamiliar with microwave processing. Without a thorough understanding of the issues regarding the safe use of microwave technology, a cautious and slightly skeptical manager might well be reluctant to invest in a new process having unforeseen and potentially costly consequences.

An understanding of a few basic topics relating to the health and safety issues involved in microwave and RF (hereinafter the term "microwave" includes RF frequencies) processing provides confidence that it can be as safe as most conventional heating technologies. The most important topics relate to (a) known health effects from working with microwave equipment, (b) established regulatory standards and guidelines for safety equipment use, (c) design practices of industrial equipment that ensure adherence to safety standards, and (d) recommended practices and procedures for safe equipment operation.

The following sections present an overview of these topics, highlighting certain key fundamentals that will help provide a general understanding and working knowledge for ceramists involved in materials research and process development. Certain topics, particularly equipment design, require a much greater degree of study to fully understand all aspects of importance. As specific applications progress towards commercialization, ceramists will prudently seek the advice and involvement of experts in these fields to ensure a successful and safe implementation of their technology.

Health Effects Related to Microwave Equipment

Public concern over the health and safety issues relating to human exposure to electromagnetic fields has increased dramatically in recent years. Much of this concern is due to the rapid proliferation of cell phone use worldwide, even though electromagnetic fields are emitted by numerous other natural and man-made sources.[1] As a result, numerous studies have been and are being conducted to determine the health effects and risks. Findings covering a broad range of frequencies have been published.

Of particular interest are studies involving the "non-ionizing" microwave frequencies, generally defined as ranging from 3 kHz to 300 GHz, whose energy levels are an extremely small fraction of that required to ionize tissue and disrupt cellular DNA.[2] This is in contrast to Gamma and X-ray frequencies having energy levels sufficient to ionize (displace electrons within the atomic structure of) exposed materials. The primary effect in biological materials of microwave radiation exposure is thermal[3], thus presenting a potential health risk due to overheating. However, much of the public concern has been focused on whether any athermal effects exist and pose a health risk.

Cancer Related Effects of Microwave Exposure

Numerous recent epidemiological studies on the relationships between various cancers and exposure to electromagnetic fields have been reviewed by Elwood.[4] Sources of emissions studied included radio and television transmitters, cell phones, radar and occupational environments, while cancer types studied included brain, lung, testicular, lymphatic and hematopoietic cancers, adult and childhood leukemia and Hodgkin's disease. Not all studies gave details of the level of exposure, but those that did indicate exposure levels were within established regulatory or otherwise recommended guidelines where such guidelines exist.

The results of these studies are inconsistent at best and show no statistically significant evidence of a link between incidences of cancer and exposure to electromagnetic fields at microwave frequencies. In most cases the incidence rate showed no significant increase in risk from casual exposure. For example, some studies on broadcast transmitters indicated a marginally increased risk amongst residents near one such emission source but no increased risk near other similar sources. One study on occupational exposure actually showed a decrease in mortality rate that was attributed to a "healthy worker effect".

Non-Cancer Related Effects

Clinical and epidemiological studies of effects on cataracts, sexual function and fertility, spontaneous abortion and birth defects, neurological and cardiovascular disorders, and other non-cancer epidemiological effects have been conducted.[1] As in the cancer related studies, the results are largely inconsistent and, in some cases, subject to scrutiny due to confounding by other causes. For example, inconsistent findings between studies on cell phone use conducted in different countries were suspected to be partly due to cultural influences.

Where conclusive evidence of an effect exists, it has mostly been shown to be due to a thermal response of the tissue. An interesting example is the auditory response from high intensity pulse modulation of electromagnetic fields, whereby thermal expansion in soft tissues in the head is conducted to the ear.[5] This annoying effect, however, is not a health risk provided the heat absorption is not sufficient to cause tissue damage. On the other hand, the eyes and testes are considered particularly vulnerable to excessive heating due to their comparative lack of blood flow as a heat dissipation mechanism.

While the above mentioned studies have failed to provide convincing evidence of a strong relationship between exposure and cancer, it should be noted that one cannot conclude from these results that other possible hazards do not exist. Furthermore, most studies generally cover a relatively short time frame, thus no conclusions can be made as to the likelihood or risks of any long term effects.[6]

Thermal Hazards to Microwave Exposure

Thermal injury is a time-temperature phenomenon whereby the rate of tissue cell protein destruction exceeds its rate of self-repair for an amount of time sufficient to terminate cell metabolism.[7] The rate of protein destruction increases with temperature, thus decreasing the time required for thermal injury as illustrated in Figure 1.

The rate at which a given volume of tissue is heated by electromagnetic energy varies according to frequency and power density. Higher frequencies have a shorter penetration depth and thus will dissipate heat in a more concentrated volume nearer the skin surface. Similarly, a focused or "contact" source of power concentrates its energy within a smaller volume, resulting in a greater rate of local temperature rise. Therefore, a lower power level is required to cause burns by high frequency and/or focused exposures.[8] As examples, (a) a 20 Watt laser can cause

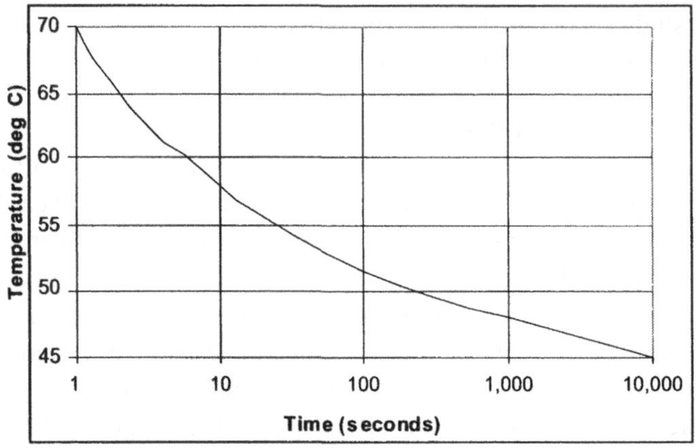

Figure 1. Threshold temperature vs. time at temperature for skin burns.

a burn within a few seconds whereas a 150 Watt light bulb can make a person feel comfortably warm for hours, and (b) a given power level at RF frequencies will cause a lower rate of temperature rise at a given location in a human body than at microwave frequencies due to the difference in penetration depth.

Cardiac Pacemakers

Many electronic devices can be subject to faulty operation if not properly shielded to prevent radio frequency interference (RFI). While some very early pacemakers were designed without RFI shielding, most or all pacemakers manufactured since the mid-1970's include such shielding. A series of studies conducted to determined the maximum threshold of interference for safe operation indicated that newer models could withstand levels well above 1 mW/cm^2.[9] As a result, an editorial was published by the American Medical Association stating that the pacemaker interference issue "does not at this time constitute an important clinical problem."[10]

High Voltage

Although the health effects of electromagnetic exposure seem to generate the most public interest and concern, the hazards associated with high voltage are of equal importance to those working with industrial microwave equipment. Almost all microwave generators contain high voltage circuitry which can be lethal while the equipment is in operation and, in some cases, during non-operation depending on the equipment design. Most microwave generators contain circuitry that stores a high voltage electric charge in one or more capacitors. Although most do, not all designs provide for automatic safe discharge of stored energy, thus creating a potential hazard to service personnel.

Injury or death may result when the human body becomes part of an active electrical circuit having a current capable of overstimulating the nervous system or damaging internal organs.[11] The extent of injury due to exposure to high electrical energy depends on the type (AC or DC) and magnitude of electrical current, the path in which the current flows through the body, and duration of current flow. Direct contact with electrical energy, such as due to a high voltage arc to the body, often results in burns to the skin and internal tissue.

A 50 or 60 Hertz (Hz) alternating current (AC) of 20 milliamps (mA) flowing through the chest area for an extended period can cause death due to respiratory paralysis, whereas a current of 100 mA can cause ventricular fibrillation.[12] Under dry conditions, the body's resistance is approximately 100,000 ohms, while wet or broken skin can reduce the resistance to 1,000 ohms. Thus, exposure to common 120 Volts AC mains voltage under dry conditions will produce a current of only 1.2 mA which is barely perceptible, but under wet conditions the resulting 120 mA current can cause death due to ventricular fibrillation. High voltage energy can further reduce the body's overall resistance to 500 ohms by breaking down the skin layer. This results in extremely high current flow which can cause cardiac arrest and internal organ damage.

The human body can tolerate up to five times the level of direct current (DC) than AC.[13] Typical power supplies used in microwave ovens generate approximately –4,000 volts, which can result in a body current of 8 amps. Industrial microwave generators typically generate much higher voltages. Thus, even with the higher tolerance threshold, exposure to these voltages can quickly cause death due to cardiac arrest.

Regulatory Standards for Safety

Electromagnetic Exposure

An important parameter used in establishing guidelines for reducing the risk of injury due to exposure to electromagnetic field is the "specific absorption rate" (SAR) which is usually expressed in units of watts per kilogram (W/kg).[14] Numerous studies have determined the minimum SAR at which a risk of thermal injury exists, and various international government agencies have adopted standards to limit exposure such that the maximum safe SAR is not exceeded.[1,3,8]

Most European countries have adopted guidelines established by the International Committee on Non-Ionizing Radiation Protection (ICNIRP) for maximum safe occupational and general public exposure levels for whole-body average SAR and localized SAR.[15] These guidelines give rise to reference levels for plane wave power densities as given in Table 1 for frequency ranges of interest.

The International Electrotechnical Commission (IEC) has specified a standard for industrial microwave heating equipment which defines maximum exposure levels for equipment operating under "normal" and "abnormal" conditions.[16] Under this standard, which is applicable to equipment operating in the frequency range from 300 MHz to 300 GHz, the power density is measured at least 5 cm from any accessible location on the equipment and limited to 5 mW/cm^2 during "normal" operation and 10 mW/cm^2 during "abnormal" operation.

In the USA, the Federal Communications Commission (FCC) has adopted standards based on recommendations developed by the American National Standards Institute (ANSI) and Institute of Electrical and Electronics Engineers

TABLE 1. ICNIRP RECOMMENDED EXPOSURE REFERENCE LEVELS FOR PLANE WAVE POWER DENSITIES (mW/cm^2).

Frequency Range	Occupational	General Public
10-400 MHz	1	0.2
400-2000 MHz	f/400	f/2000
2-300 GHz	5	1

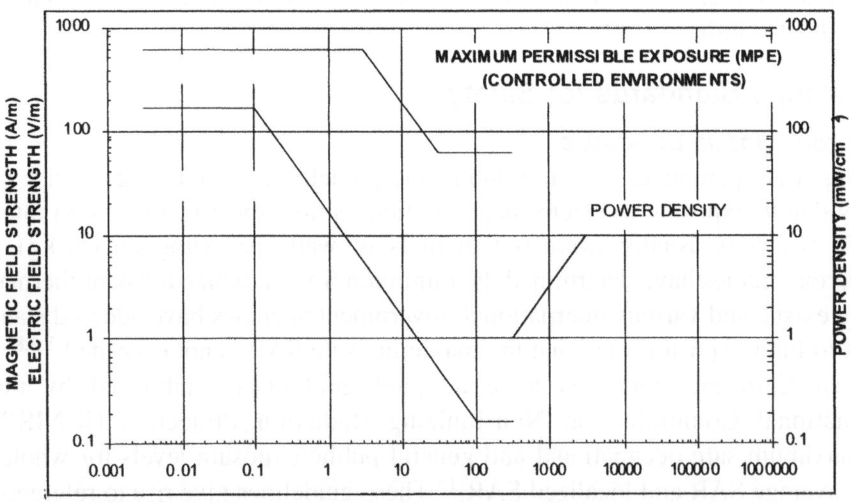

Figure 2. ANSI/IEEE recommended maximum permissible exposure (MPE) for controlled environments per standard C95.1-1999.

(IEEE) as defined under IEEE standard C95.1.[17] This standard defines the maximum permissible exposure (MPE) averaged over a time period for controlled and uncontrolled environments as summarized in Figures 2 and 3. Table 2 highlights MPE values for specific industrial, scientific and medical (ISM) frequencies commonly used in materials processing applications.

The Occupational Safety and Health Administration (OSHA) has established a regulation applicable to frequencies from 10 MHz to 100 GHz whereby exposure is limited to no more than 10 mW/cm^2 measured at 5 cm from the emission source and averaged over a 0.1 hour period.[18] OSHA further establishes regulations for hazard labeling of equipment.

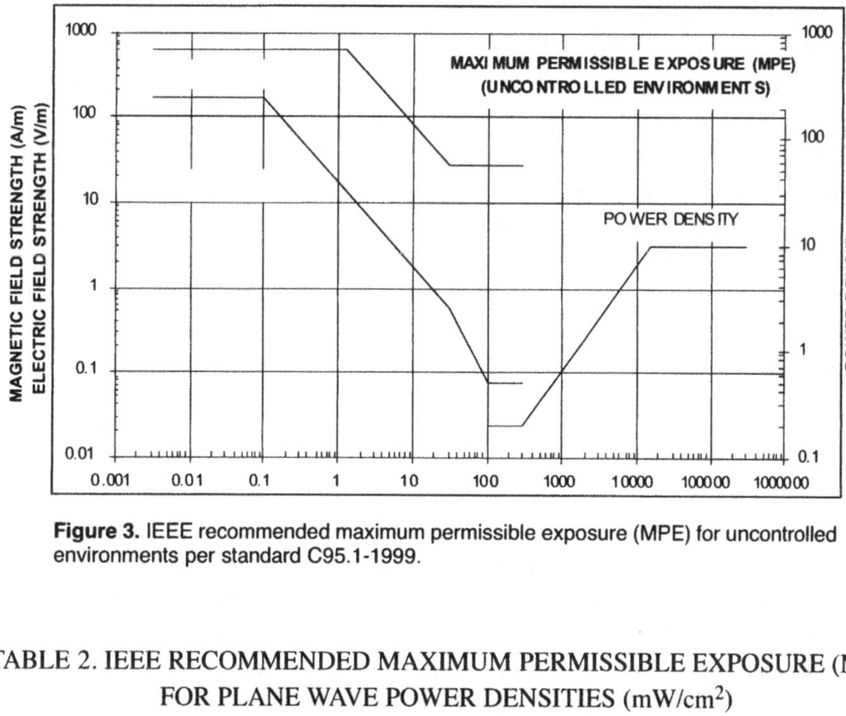

Figure 3. IEEE recommended maximum permissible exposure (MPE) for uncontrolled environments per standard C95.1-1999.

TABLE 2. IEEE RECOMMENDED MAXIMUM PERMISSIBLE EXPOSURE (MPE) FOR PLANE WAVE POWER DENSITIES (mW/cm^2) AT ISM MICROWAVE FREQUENCIES.

Frequency	Controlled Environments	Uncontrolled Environments
915/896 MHz	3	0.6
2.45 GHz	8.2	1.6
5.8 GHz	19	3.9

High-Voltage and Equipment Safety

Safety guidelines have been established for voltages used with and generated by microwave equipment. Although the National Electrical Code (NEC) defines high voltage as greater than 600 Volts AC[19], various safety requirements for electrical wiring apply to all voltages. These requirements, applicable in the United States, include circuit overload and ground fault protection devices, wire sizes and materials, conduits and shielding, enclosures and workspace clearances. Specific requirements apply to industrial machinery which includes most microwave processing equipment. Similar standards and requirements have been established internationally by British Standards (BS), European Committee for Electrotechnical Standardization (CENELEC), International Electrotechnical Commission (IEC), and Japanese Industrial Standards (JIS).

Microwave generators, power supplies and other components that make up industrial processing systems are governed under various standards established for product safety. Underwriters Laboratory (UL) and American National Standards Institute (ANSI) standards generally apply to products used in the United States, compliance to which is evaluated and certified by a Nationally Recognized Testing Laboratory (NRTL). Similarly, products used internationally must also comply with safety standards established within the countries where used, including those by Standards Council of Canada (SCC), the European Union "Low Voltage"[20] and Machinery[21] directives (CE Mark) and, to a limited extent, the Chinese Compulsory Certification (CCC Mark). Certain industry groups also have specific requirements for equipment, such as the SEMI S2 Environmental, Health, and Safety Guideline for Semiconductor Manufacturing Equipment.[22]

Equipment Design for Safety

Microwave processing systems generally consist of three main elements: a microwave power generator, power delivery (waveguide) components and a process cavity (applicator). Systems can be as small and simple as a typical residential microwave oven or as large and complex as a industrial food processing conveyor. Many different types of generators and waveguide components are available and can be configured in a variety of ways depending on the process characteristics.[23] However, they all perform essentially the same function: to safely and reliably generate and deliver microwave power to the process cavity.

The process cavity itself can also be configured in many ways, but its primary function is to effectively "couple" the delivered microwave energy to the material being heated in such a manner as to meet the requirements of the heating process. Process cavities are generally designed for either "batch" or "continuous" material processing, while some are designed for semi-continuous flow.

Many common design practices for safety apply to all types of commercial and industrial equipment used in materials processing as defined in the standards mentioned above. In addition to these commonalities, microwave processing systems utilize several features that are uniquely necessary to ensure reliable and safe operation. Most pertain to suppression of emissions, or "microwave leakage," while others may be specific to certain materials processes.

Microwave Leakage Suppression

Once high power microwave energy is generated, it must be contained almost entirely within the processing system in order to comply with regulatory standards for safety and interference. The term "Faraday cage" (British physicist Michael Faraday, 1791-1867) has often been used to describe a microwave cavity in that it is an enclosure capable of blocking the entry or exit of electromagnetic waves. The ideal Faraday cage is a metallic enclosure with no holes, broken seams or openings of any kind. For all practical purposes, the term also applies to microwave generators and waveguide components. But for obvious reasons, no practical microwave processing system can be a completely continuous metallic enclosure as it must, as a minimum, allow access to its interior for entry and exit of the material to be heated. In most cases it must also have seams or openings suitable for assembly, viewing, ventilation, sensor insertion and other process related necessities.

Material Entry/Exit for Batch Processing. The most common type of opening in a microwave heating system is the door of a household microwave oven. One can easily see that the door does not make complete contact with the oven cavity, leaving the casual observer to wonder what is preventing the microwave energy from radiating out of the oven. The door is designed with a nondissipative "reactive" quarterwave choke along the opening perimeter that effectively blocks the transmission of energy at a specific frequency.[24] Most industrial systems having batch cavities utilize such door seals, although some employ a direct contact seal using a woven metal mesh gasket or springy beryllium copper fingerstock. In either case, the door seal design is robust and capable of withstanding repeated openings without degradation in performance throughout the life of the equipment.

Material Entry/Exit for Continuous Processing. Continuous flow process systems require openings that allow for material entry and exit on a continuous basis, such as for conveyorized and/or extruded product. Depending on the size and geometry of these openings as required for the process material, the means for suppressing microwave leakage is either reactive, as in the case described above for door seals, or dissipative whereby microwave energy is actively absorbed within the structure of the opening, or a combination of both.[25] In either case, microwave energy is attenuated at some rate as it passes through the opening. The rate of attenuation through these openings depends on the opening size and attenuation method used. Thus, the opening typically has some length in order to attenuate the energy to sufficiently safe regulatory levels, with larger openings generally having a lower attenuation rate and requiring greater length. The term "tunnel" is often applied to openings for continuous flow processing.

Cavity Ventilation and Viewing Ports. Almost all microwave ovens have a window in the cavity door through which the cooking process can be observed. Close inspection of this window reveals the presence of a screen made of tiny perforations in thin metal. Each hole in the screen acts as a high-pass filter that effectively blocks the transmission of electromagnetic energy at microwave frequencies yet allows transmission of visible light. Other openings in the oven cavity, such as for ventilation and illumination, are perforations in the cavity wall that function in a similar manner.

As microwave energy propagates through a waveguide or other restricted path, it is attenuated exponentially as a function of wavelength and the size of the path according to

$$P_1 = P_0 e^{-2ax} \tag{1}$$

where P_1 and P_0 are the field strengths at the path exit and entrance, x is the path length and a is the rate of attenuation in dB/unit length as energy propagates along the path.[26] Figure 4 illustrates the rates of attenuation through a circular waveguide as functions of the inside diameter for various ISM frequencies. The diameter at which the attenuation rate approaches zero is the "cut-off" diameter for the respective frequency, and waveguides smaller than this dimension are said to be below or beyond cut-off. These curves are used to determine the minimum screen thickness, or hole "length," that provides the necessary total attenuation for a given perforation hole diameter and incident field strength.

For example, suppose the highest internal field strength at a cavity opening is determined to be 1 kW/cm^2 and the regulatory limit to be met outside the cavity is 1 mW/cm^2. Using equation 1, the minimum attenuation for a 0.020″ thick screen is 345 dB/inch. Figure 4 then yields a maximum perforation hole diameter of roughly 0.08″.

Cut-off Tubes. The waveguide below cut-off principle is also commonly used for larger unobstructed openings, referred to as "cut-off tubes," such as for optical thermometry, fluid flow, etc. However, the curves in Figure 4 apply only to unfilled tubes. When filled with a dielectric material, the effective inside diameter is increased by a factor equal to the square root of the material's dielectric constant. This lowers the effective attenuation rate and increases the minimum tube length, although it will be counteracted somewhat for high dielectric loss materials that absorb energy propagating through the tube. The analysis is further complicated when the tube is filled only partially or with multiple materials of various dielectric properties, such as when flowing water in Teflon tubing through the cut-off tube.

Feed-thru Openings: Inserting a metallic object, such as a thermocouple probe or wire, through a cavity opening can be extremely dangerous by causing excessive microwave leakage. Unless it is in intimate contact with the opening wall, the metal object in conjunction with the metallic opening wall will have a transmission line effect, thus allowing microwave energy to propagate freely through the opening and radiate from its end. Materials having a high dielectric constant can also create a similar effect when inserted through a cavity opening, such as passing a small column of water passing through a comparatively large cut-off tube. Electrical wires passing thru a cavity opening require special techniques such as capacitive filtering to prevent leakage and other forms electrical interference in the external circuitry. Given the complexities involved with proper feed-thru design, professional advice and assistance should be sought whenever installing feed-thru devices in microwave cavities.

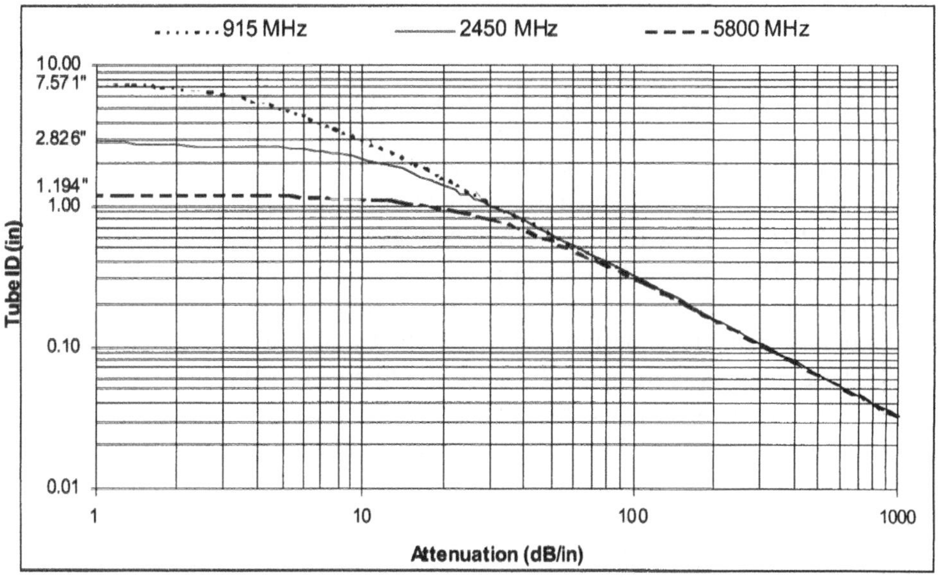

Figure 4. Attenuation curves for circular waveguide at three ISM frequencies.

Safety Interlock Devices

Interlock devices are typically defined as mechanical and/or electrical components or sub-systems that prohibit the operation or functioning of other components or sub-systems if certain conditions are not satisfied. In the case of microwave processing systems, this usually applies to the devices that prohibit the operation of the microwave generator if a condition exists that might cause excessive microwave emissions.

Cavity Access. Regulatory standards require interlock devices at all locations where the microwave cavity can be opened without the use of tools. Of course, the most common access point is the cavity door, although many industrial systems have other features that require periodic opening for maintenance or other such purposes. Interlock devices prevent the microwave generator from operating if the opening is not closed, and they are typically designed such that a failure of the interlock device leaves the system in a safe non-operative mode. US government standards applicable to residential and commercial microwave ovens require redundant interlock devices that are concealed and inaccessible, at least one device of which having a means of fail-safe monitoring.[27] IEC standards also require redundant and concealed interlock devices at all cavity access locations, as well as interlocks for the presence and/or flow of material through the cavity if necessary to maintain emissions within the guidelines.[16]

Waveguide Flanges. An often neglected location for microwave leakage due to inadvertent carelessness during equipment maintenance is a waveguide flange connection. Many systems are designed to allow easy disconnection of a waveguide component for routine maintenance, some even without the use of tools. In such cases an interlocking device is required to ensure proper re-assembly before bringing the system back on-line. In fact, literal interpretation of the IEC standards would require interlock devices at all waveguide flange connections, although this practice is rarely found or deemed necessary in industrial equipment.

High Voltage. As described earlier, microwave generators contain high voltage circuitry that can be lethal when contacted inadvertently during operation. For this reason, interlock devices on enclosure access covers are typical of most contemporary equipment designs and required by almost all regulatory standards. These safeguards are generally sufficient to protect service personnel from exposure to stored high voltage energy hazards. Most circuit designs provide for safe electrical discharge within the time required to remove safety covers and gain access the internal circuitry.

Although not required by regulatory standards, many equipment designs also incorporate interlock wiring into electrical connectors associated with inter-modular high voltage circuitry, thus disabling the high voltage circuits whenever these connectors are not properly mated. However, even when the equipment is designed with such protection, quickly disconnecting an electrical cable and exposing its contacts may not allow sufficient time for stored energy to be discharged. Thus, caution must be exercised ANY time an electrical cable is disconnected.

Warning Labels and Visual Indicators

Many regulatory standards require various labels and indicators alerting the equipment operator of a potentially hazardous condition.[3,16,20,22,27] Warning labels are required for both electrical shock and electromagnetic radiations hazards and must indicate briefly (a) the nature of the hazard, (b) the potential consequences of exposure to the hazard, and (c) necessary actions to prevent exposure. Special labels are required for cavity doors and other openings to warn against allowing foreign objects to interfere with the door seals.

Visual indicators are required to alert nearby personnel that the equipment is in operation. These indicators must be plainly visible during normal operating conditions or where personnel might typically be present.

System Architecture and Process Safety

Fire and explosion prevention is particularly challenging as it combines the volatile characteristics of certain materials processes with the unique propensity for ignition in a microwave system.[28] Inert gas purging is commonly used to remove oxygen from the cavity atmosphere, and regular cleaning of interior surfaces helps to reduce the sources of fuel. But these practices may not be practical for or applicable to all processes. The system developer must have a thorough understanding of both the materials process and microwave technology. Recognizing that such hazards often are not completely avoidable, the design must minimize the hazard potential and effectively deal with a hazard event.

There are many other aspects of the overall system design that are considered when minimizing the risks and hazards associated with microwave processing, including walk-in cavity access, high temperatures, automated mechanisms and moving parts.[29] A thorough review of these considerations is well beyond the scope of this paper, so materials process developers are encouraged to seek the advice of equipment experts having experience in similar processes.

Guidelines for safe equipment operation

Common sense plays an important role in the safe operation of any product regardless of the magnitude of its associated hazards. Apart from that, a few basic practices and habits should be institutionalized when operating microwave equipment.

Safety Training. Basic first aid treatment and emergency rescue procedures should be a part of a formal safety training program. Many hazards, particularly those associated with high voltage, can be prevented from becoming lethal if immediate rescue action is taken. For example, electrocutions resulting in ventricular fibrillation do not necessarily cause immediate death[11], thus emergency cardiopulmonary resuscitation (CPR) can be administered to potentially save the life of the victim.

Equipment Training. All personnel engaged in operating and/or maintaining industrial microwave equipment should receive proper training to be aware of the hazards associated with their respective tasks. As a minimum, all product user manuals and other documentation provided by the equipment manufacturer should be read and thoroughly understood before attempting operation or service. Safety training should emphasis awareness and recognition of both microwave radiation and high voltage hazards and appropriate responses when identified. In addition, service personnel should be knowledgeable and competent with regard to safe troubleshooting and repair of high voltage circuitry.

Buddy System. Never operate or perform maintenance on industrial microwave equipment while alone. In the event of hazard exposure, the victim may be incapacitated and need immediate medical attention for survival.

Leakage Detection. Every user of an industrial microwave heating system should have on hand or ready access to a high quality microwave leakage detector, also referred to as a microwave survey meter. Leakage should be checked on a periodic basis and always after any maintenance has been performed on the equipment. Upon detecting excess emissions, the equipment should be immediately shut down and the cause of emissions investigated and corrected before restarting. Also, the leakage detector should be calibrated at least annually to ensure proper operation.

Periodic Maintenance. Many hazards result from inadequate or improper equipment maintenance practices. A formal program of periodic inspection and maintenance should be implemented that addresses all potential hazards and hazard sources. The program should include not only the microwave equipment itself but also any other tools or equipment required for the maintenance tasks. The most basic tasks to be included are (a) to inspect for microwave leakage, (b) clean the cavity interior and all internal waveguide surfaces, and (c) replace damaged or broken electrical connectors even if still functional.

Conclusion

While there are indeed many hazards associated with industrial microwave and RF processing equipment, the risks associated with them are no greater than with conventional equipment when proper safeguards have been implemented. Microwave heating technology has been a vital component of industrial production since the

1940's and will continue to be applied as new materials processes are developed. Newcomers to this technology can develop a sense of security with its use by learning and understanding the basic concepts about health effects, regulatory standards, design safeguards and operational guidelines presented here.

REFERENCES

1. "Health Effects from Radiofrequency Electromagnetic Fields," National Radiological Protection Board, UK, Volume 14 No. 2, 2003.
2. "Effects of Electric and Magnetic Fields," *Report 7 of the Council on Scientific Affairs (I-94)*, American Medical Association, USA, 1994.
3. "Non-Ionizing Radiation: Radiofrequency and Microwave Radiation," Occupational Safety & Health Administration, USA, 2003.
4. J.M. Elwood, "Epidemiological Studies of Cancers and Exposures to Radio Frequencies," pg. 507-514 in *Microwave and Radio Frequency Applications,* American Ceramic Society, 2003.
5. "Mobile Phones and Health," Report of the Independent Expert Group on Mobile Phones (IEGMP), Chilton, UK, 2000.
6. "Establishing a Dialog on Risks From Electromagnetic Fields," World Health Organization, Geneva, Switzerland, 2002.
7. A.R. Moritz and F.C. Henriques, "Studies in Thermal Injury II: The Relative Importance of Time and Surface Temperature in the Causation of Cutaneous Burns," *American Journal of Pathology* Vol. 23, pg. 695-720, 1947.
8. J.M. Osepchuk, "Radiofrequency Safety Issues in Industrial Heating Systems," pg. 125-137 in *Microwaves: Theory and Applications in Materials Processing,* Ceramic Transactions, American Ceramic Society, Vol. 21, 1991.
9. J.M. Osepchuk, "Debunking a Mythical Hazard", *Microwave World,* International Microwave Power Institute, Vol. 2 No. 6, 1981.
10. N.P. Smyth, et al, "The pacemaker patient and the electromagnetic environment," *Journal of the American Medical Association,* Vol. 227 No. 12, pg. 1412, 1974.
11. V. Casini, "Overview of Electrical Hazards," *Worker Deaths by Electrocution,* National Institute for Occupational Safety and Heath, 1998.
12. C.F. Dalziel and W.R. Lee, "Re-Evaluation of Lethal Electric Currents," IEEE Transactions on Industry and General Applications, Vol. IGA-4 No. 5, pg 467-476, 1968.
13. Ibid.
14. R.F Cleveland and J.L. Ulchek, "Questions and Answers about Biological Effects and Potential Hazards of Radiofrequency Electromagnetic Fields," FCC OET Bulletin 56, Fourth Edition, 1999.
15. "Guidelines for Limiting Exposure to Electric, Magnetic, and Electromagnetic Fields (up to 300 GHz)," International Commission on Non-Ionizing Radiation Protection, Germany, 1997.
16. "Specifications for Safety in Industrial Microwave Heating Systems," *Safety in Electroheat Installations,* Publication 60519-6, International Electrotechnical Commission, Geneva, Switzerland, 2002.

17. "IEEE Standard for Safety Levels with Respect to Human Exposure to Radio Frequency Electromagnetic Fields, 3 kHz to 300 GHz," Institute of Electrical and Electronics Engineers, New York, USA, 1999.

18. "Nonionizing Radiation," Section 1910.97, *Code of Federal Regulations,* Title 29: Labor, Part 1910, Subpart G, 1996.

19. "National Electrical Code, 2002 Edition," NFPA 70, National Fire Protection Association, Quincy, MA, 2001.

20. "Council Directive of 19 February 1973 on the harmonization of the laws of Member States relating to electrical equipment designed for use within certain voltage limits," 73/23/EEC, Office for Official Publications of the European Communities, Luxembourg, OJ L 77, 1973.

21. "Directive 98/37/EC of the European Parliament and of the Council of 22 June 1998 on the approximation of the laws of the Member States relating to machinery," Office for Official Publications of the European Communities, Luxembourg, OJ L 207, 1998.

22. "Environmental, Health, and Safety Guideline for Semiconductor Manufacturing Equipment," S2-0200E, SEMI, California, USA, 2002.

23. J.F. Gerling, "Waveguide Components and Configurations for Optimal Performance in Microwave Heating Systems," pg. 559-566 in *Microwaves: Theory and Applications in Materials Processing V,* Ceramic Transactions, American Ceramic Society, Vol. 111, 2001.

24. A.C. Metaxas and R.J. Meredith, "Multimode Oven Applicators," pg. 147 in *Industrial Microwave Heating,* Peter Peregrinus Ltd., London, UK, 1983.

25. R. Meredith, "Choking (attenuation) tunnels for continuous-flow applicators," pg. 229 in *Engineers' Handbook of Industrial Microwave Heating,* Institute of Electrical Engineers, London, UK, 1998.

26. G.L. Ragan, "Microwave Transmission Circuits," *Radiation Laboratory Series,* Massachusetts Institute of Technology, McGraw-Hill, New Work, USA, Vol. 9, pg. 644, 1948.

27. "Performance Standards for Microwave and Radio Frequency Emitting Products," *Code of Federal Regulations,* Section 1030.10: Microwave Ovens, Title 21, Vol. 8, 2003.

28. R.F. Schiffmann, "Fires in Microwave and RF Heating Systems: Causes and Prevention," pg. 255-362 in *Microwaves: Theory and Applications in Materials Processing V,* Ceramic Transactions, American Ceramic Society, Vol. 111, 2001.

29. R.F. Schiffmann, "Designing Industrial Microwave Heating Systems for Safe Operation: Batch Ovens," pg. 269-279 in *Microwave and Radio Frequency Applications,* American Ceramic Society, 2003.

Fundamentals and Application of Dielectric Heating Technologies for Materials Processing: A Review

J.W. Cresko

Abstract

If any field of scientific study can be characterized as multidisciplinary, it is the study of microwave (MW) and radio-frequency (RF) dielectric heating. For more than a half-century, physicists, engineers, chemists, materials scientists, biologists, food scientists and medical doctors—among others—have sought to exploit the interaction of thesenon-ionizing electromagnetic (EM) waves and materials.

The study of MW and RF heating (commonly referred to as dielectric heating since dielectric materials have strongly bound polar molecules that are energized by these EM waves) continues to be a very dynamic field, attracting researchers and industrialists from a wide range of disciplines, but it is also relatively small. Many of the key persons responsible for early technological advances in MW and RF heating equipment and applications are at or near retirement. Persons new to the field may find it difficult to gain a global view of the process of developing MW and RF applications. Recent advances have been made in computer modeling, equipment design and technology that can help novice and experienced persons alike to better evaluate MW and RF for materials processing. It is the intent of this paper to examine fundamental concepts important from the process development/materials engineering perspective, and to review the state-of-the-art of dielectric materials processing in context with its history.

Background

Industrial dielectric heating is a small subset of the utilization of electromagnetic (EM) energy. Fundamental discoveries in electromagnetics can be traced to giants such as Faraday (discovery of EM induction in 1831), Maxwell (developed EM theory in 1864) and Hertz (experimental verification of EM wave propagation in 1885); these discoveries were followed closely by applications by great inventors such as Morse (telegraph in 1844), Bell (telephone in 1876) and Marconi (wireless telegraphy in 1894).[1] These beginnings have led to today's enormous array of devices (i.e., receivers, amplifiers, vacuum tubes, crystal detectors, antennas, magnetrons, klystrons and gyrotrons, etc.) and applications (i.e., communications, radar, GPS, atmospheric measurements, remote sensing, diathermy, materials measurements and heating[2]). Hertz had noted the ability of high-frequency energy to produce heat, as did d'Arsonval who in 1895 directed the first high-frequency heat therapy unit at the Hôtel-Dieu Hospital in Paris, France. In the 1940s, interest in the use of EM heating for commercial and industrial applications expanded after advances made in MW technology during WWII. Much of the MR and RF technology was developed for food applications (i.e. tempering of frozen foods, precooking of bacon and chicken, pasta drying, finish drying of potato chips and biscuits).[3] Nonfood applications were investigated and developed: RF for adhesives curing, textile drying and wood drying in the 1940s; MW for urethane foam heating, rubber curing and foundry sand core drying, and RF for PVC plastic welding in the 1960s; MW and RF generated plasmas for diamond and diamond-like coatings and MW power supplies for ultraviolet curing equipment in the 1980s; and MW for powdered metal sintering and accelerated chemistry in the 1990s.

The advantage of *volumetric heating* was realized early on, with MW and RF heating and drying applications deployed for their abilities to overcome slow conduction/convection heating rates of poor thermally conducting materials. However, industrial MW and RF heating equipment is more expensive to acquire than the traditional conduction/convection equipment they are to replace. And if the economics of the process are not well understood, the operating costs can be higher as well. Indeed, MW water heaters do not exist because it is far more expensive to use microwaves for the purpose of simply increasing the temperature of water. For example, consider an electric immersion heater versus a microwave system for the purpose of heating water on a 1 kW basis.

The immersion heater requires minimal engineering design: the sheath material and watt density ratings must be compatible with the liquid, and installation is simple—typically by way of flange mount or screw plug into an existing vessel. For heating water, a watt density of 15 W/cm^2 is typical, requiring only about 67 cm^2 of surface area per kW at a cost of approximately $100/kW. A factor of safety is designed into such applications to account for decreased heat transfer due to fouling, but maintenance and replacement are simple and inexpensive.

Contrast this with an attempt to use a MW system to heat water. In this case engineering costs are high, stemming from variations of heating efficiency based on water chemistry as well as the safe physical installation of such a system, requiring custom MW applicator and system design. Variations from the design basis (e.g., process variations of water chemistry) could significantly reduce the heating efficiency. Since MW systems utilize a separate power supply that powers the magnetron, waveguides to transfer energy to the process and protection systems to prevent damage of the magnetron tube (which has a finite life) have maintenance costs that are much higher. The cost basis of installed MW systems is typically $1,000 per kW, an order of magnitude greater than that of an immersion heater. Hence, for the simple task of raising water from one temperature to another, the use of MW is economically infeasible.

It is likely that many of the early commercial and industrial MW and RF successes were serendipitous in that resultant product benefits (i.e., improved product quality, ability to process heat sensitive materials, new products) evolved from attempts to improve process efficiency. Failures may be due to technological barriers, especially during scale-up (from laboratory to full-scale production), but historically the inappropriate application of dielectric heating technologies can often be traced to an insufficient analysis of the economics.[4]

Another difficulty encountered when developing MW and RF processes is the complexity of constructing a comprehensive understanding of the physics of the process and system. The physics of EM wave/material interaction is complex, and is compounded by the coupled heat and mass transfer as well as the materials physics and chemistry. Further, because the material to be processed (the load) becomes an integral part of the overall system, the equipment design—especially the applicator design—is far more critical than in traditional heating processes. However, when this technology is applied appropriately, significant efficiency advantages can be realized, and in many cases EM energy becomes the enabling technology.

Harnessing the Wave/Material Interaction

The propagation of an EM wave within a material is a function of the material's electrical properties, which are most conveniently described by complex permittivity, permeability, wavenumber and impedance. As the wave enters a dielectric material, there is some attenuation due to molecular interactions—for instance polar molecules will tend to align themselves with the electric field. The incident wave is described by the wave equation, where[5]

$$\nabla^2\mathbf{E} - \mu\varepsilon\frac{\partial^2\mathbf{E}}{\partial t^2} = 0 \qquad \text{and} \qquad \mu\varepsilon = \frac{k^2}{\omega^2} \tag{1,2}$$

As such, the wavenumber is defined:

$$k = \frac{\omega}{v} = w\sqrt{\mu\varepsilon} = \frac{2}{\lambda}\pi \tag{3}$$

where λ is the wavelength, ω the angular frequency = $2\pi f$ and v the wavespeed (v = speed of light in a vacuum). The ratio of the electric to the magnetic field components is defined as the wave impedance η, and is measured in ohms:

$$\eta = \sqrt{\frac{\mu}{\varepsilon}} = \frac{\mathbf{E}}{\mathbf{H}} = \frac{V/m}{A/m} = 377\ \Omega \text{ in free space} \tag{4}$$

Attenuation may be caused by a number of interaction mechanisms of the incident wave within the media, in which the bulk effect may be described by the complex wavenumber k and impedance η:

$$k = k' - jk'' \qquad \text{and} \qquad \eta = \eta' - j\eta'' \tag{5, 6}$$

Since k is complex, permittivity and permeability are also complex:

$$\varepsilon = \varepsilon' - j\varepsilon'' \qquad \text{and} \qquad \mu = \mu' - j\mu'' \tag{7, 8}$$

The permittivity and permeability are often expressed as the ratio with respect to free space:

$$\varepsilon = \varepsilon' = \frac{\varepsilon}{\varepsilon_o} \qquad \text{and} \qquad \mu = \mu' = \frac{\mu}{\mu_o} \tag{9, 10}$$

where $\varepsilon_o = 8.854 \times 10^{-12}$ F/m and $\mu_o = 4\pi \times 10^{-7}$ H/m. In the dielectric heating literature, one often sees values of the dielectric constant reported as κ rather than ε. Further, the relative permittivity should technically be written as $\varepsilon_r = \varepsilon_r' + \varepsilon_r''$; however, the subscript r is not commonly shown in the literature, and the expression for the complex relative permittivity is $\varepsilon_r = \varepsilon' + \varepsilon''$. As such, the real part of the relative permittivity ε' is equivalent to κ' the ratio of the capacitances of a material to vacuum:

$$\varepsilon' = \kappa' = \frac{C}{C_o} \tag{11}$$

The real part of the permittivity ε' describes how much of the incident electric field is stored in the material (e.g., the capacitance), and is greater than one for most solids at room temperature. The imaginary part of the permittivity (ε'') is typically referred to as the loss factor and describes how much of the incident field is dissipated. The loss factor (ε'') is greater than zero and is not the same as the loss tangent (although sometimes in the literature it may not be clear which value is being presented). The loss tangent is the ratio of energy lost to energy stored, and mathematically is the vector sum of the real and imaginary components at 90° out of phase:

$$\tan\delta = \frac{\varepsilon'}{\varepsilon''} \tag{12}$$

An example of a material that is essentially transparent to EM (at 25 °C and 2.45 GHz) is Teflon (polytetrafluoroethylene—PTFE), with typically reported values of $\varepsilon' = 2.04$–2.08 and $\varepsilon'' = 0.00031$–0.0005, which vary little from the values of a vacuum: $\varepsilon' = 1$ and $\varepsilon'' = 0$. Teflon or similar materials are often used as containers, vessels, etc. in MW or RF applications. Contrast this with a ceramic such as silicon nitride, with typical reported values of $\varepsilon' = 7.0$–8.0 and $\varepsilon'' = 0.1$–0.3 (at 25 °C and 2.45 GHz).[6]

Permeability is analogous to permittivity, but describes the interaction of the magnetic field with a material. The real part of the permeability μ' is the ratio of the inductance of a material to the inductance of a vacuum:

$$\mu' = \frac{L}{L_0} \tag{13}$$

Many materials are nonmagnetic, especially organics, but ferrous materials such as iron have appreciable permeability. It can be seen from Eqs. (1) and (2) that both the permittivity and permeability affect the interaction of the EM wave with a material. Indeed, such materials are used as susceptors to improve the heating response of materials in MW and RF heating.

The complex permittivity and permeability yield four parameters that fundamentally describe how a material will respond to an EM wave. If the reflection of an incident wave is not too high, accurate values of the effective permittivity and permeability (and the related wavenumber and impedance) can be determined. If the material is conductive, the reflection is high and dielectric heating is usually impractical, and alternate types of heating (e.g., EM induction)[7] should be investigated. Although the wavenumber and impedance do not give different information from the permittivity and permeability, they group together information that is useful to the materials engineer. For instance, the design of a low-reflection surface will involve:

- Matching the interface to free space so that the incident energy is effectively coupled, i.e., matching the effective material impedance η to the impedance of free space = 377 Ω (refer to Eq. (4), and
- Attenuating the coupled energy within the material, where the imaginary part of the wavenumber gives insight into absorption (refer to Eq. (3).

When a susceptor is introduced into a matrix, the complex values of the effective permittivity, permeability, impedance and wavenumber are altered. There are a number of methods for predicting the effective dielectric properties of certain types of composite materials. Models such as those of Maxwell Garnett,[8] Bruggeman[9] and Looyenga[10] are effective medium theories (EMT) that can predict an insulating matrix (host) of volume fraction v_{ins} loaded with a given volume fraction v_{cond} conducting inclusions, but these models are limited owing to the dependence on a variety of variables including v_{cond}/v_{ins}, inclusion size, inclusion dispersion, inclusion size distribution, inclusion geometry and conductivity. For example, Maxwell Garnett can break down at high loading of conductive inclusion levels where percolation[11,12] can occur and a discontinuous jump of electrical properties occur; empirical data show a jump in conductivity at critical loading of v_{cond}, dependent upon inclusion, from >10 Ω-cm to <1 Ω·cm.[13]

Designers of materials with specific electrical properties requirements approach the problem by engineering the complex permittivity/permeability. Applications for composites with specifically engineered electrical properties include electromagnetic interference (EMI) shielding materials,[14] conductive adhesives,[15] radar absorbing materials (RAM),[16] and integrated electronic packaging to protect sensitive electronic equipment as well as attain strict compliance with EMI/RFI emission levels.[17] This materials engineering approach is useful for heating applications in other areas, for example, food products;[18,19] materials otherwise nonresponsive to EM energy can be modified by dielectrically responsive inclusions.

Measuring the Dielectric Properties of Materials

Values of the dielectric properties of select materials can be found in the literature, but may be reported only at one or two particular frequencies. Further, the composition of the reported material may not be accurately known, or how the dielectric properties may change during processing (i.e., as a function of temperature). Any comprehensive evaluation of MW and RF heating technology will require accurate values of the material's dielectric properties.

Electromagnetic energy, at very low power levels, can be used to probe the properties of materials. In the case of MW measurements, transmission and reflection (S_{11} and S_{21}) information is acquired, typically by means of a vector network analyzer (VNA) and test fixtures designed for the particular frequency (i.e., a resonant cavity fixture).[20] The transmission and reflection data are reduced to the complex expressions of the effective permittivity, permeability, wavenumber and impedance via numerical techniques.[21,22] Various methods of measuring the dielectric properties across a frequency range are possible, including transmission line techniques (waveguide and coaxial fixture),[23] coaxial probe[24] and free-space techniques.[25,26]

Loss Mechanisms and Heating of Materials by EM Energy

When an EM wave impinges upon a material, some amount of the energy is *reflected* from the interface (the higher the impedance mismatch, the more energy is reflected) or *transmitted* into the material. When that surface is highly conductive (the conductivity σ is σ_m, e.g., the conductivity of a metal) at the given frequency, the energy is principally reflected back from the surface and performs no useful heating. Materials that are very transparent may allow most of the transmitted energy to pass through the material unimpeded, but lossy/conductive dielectrics can convert energy not reflected into heat energy. The interaction of this energy within the material results in the development of a current density described by Ampere's law[5]:

$$\nabla \times \mathbf{H} = \alpha \mathbf{E} + \frac{\delta \mathbf{D}}{\delta t} \tag{14a}$$

where $\mathbf{D} = \varepsilon \mathbf{E}$, and considering that the impinging wave is a plane wave $\mathbf{E} = \mathbf{E}e^{j\omega t}$

$$\nabla \times \mathbf{H} = \mathbf{J} = \alpha \mathbf{E} + j\omega\varepsilon \mathbf{E} \tag{14b}$$

where the current density $\mathbf{J}$ is the sum of the combined conduction loss $\sigma\mathbf{E}$ (where σ is the conductivity of the medium) and the displacement loss. The combined losses of a material will determine how and to what extent a material will absorb energy of a particular frequency. However, this simplified expression does not differentiate the different contributions to the losses from different mechanisms.

The physical/chemical mechanisms by which molecules respond to EM are frequency dependent. A material may have several mechanisms that contribute to its effective permittivity; each of these will exhibit a characteristic resonant frequency. The mechanisms for losses at RF frequencies are generally due to ionic conductivity, and its contribution to the loss factor (ε'') falls off as the frequency increases (the ionic conduction contribution to ε'') is proportional to $1/\omega$). In the MW region, losses are generally due to rotation of induced or permanent dipoles that manifest themselves as a *resonant peak* of ε'' as ε' is *relaxing*. Electronic and atomic polarization can also contribute to the dielectric properties, but the actual resonance response is observed at much higher frequencies (i.e., in the infrared and ultraviolet ranges); for many dry solids electronic and atomic polarization are the dominant dielectric loss mechanisms. Some insight into factors contributing to losses can be ascertained by reconsidering Eq. (14b) with the complex expression for permittivity (Eq. (7)):

$$\mathbf{J} = (\alpha + j\omega\varepsilon_o\varepsilon' + \omega\varepsilon_o\varepsilon'')\,\mathbf{E} \tag{14c}$$

which, upon rearranging gives:

$$\mathbf{J} = j\left\{\varepsilon' - j\left[\frac{\sigma}{\omega\varepsilon_o} + \varepsilon''\right]\right\}\mathbf{E} \tag{14d}$$

where the relative effective loss factor shows the contributions from conductivity and dipolar relaxation:

$$\varepsilon''_{\text{eff}} = \left[\frac{\sigma}{\omega\varepsilon_o} + \varepsilon''\right] \tag{15a}$$

The unique loss mechanisms of specific materials that make up ε'' can include contributions from dipolar (d), electronic (e), atomic (a) and Maxwell–Wagner (MW) effects, such that an expanded expression for the effective loss factor consists of:[27]

$$\varepsilon''_{\text{eff}} = \frac{\sigma}{\omega\varepsilon_o} + \varepsilon''_{\text{d}} + \varepsilon''_{\text{e}} + \varepsilon''_{\text{a}} + \varepsilon''_{MW} \tag{15b}$$

At sufficiently high levels of EM power, the interaction of the applied field and a material can induce heating by way of these mechanisms. For materials that are very lossy, the wave's velocity and wavelength are significantly decreased (note Eq. (3)) and most of this energy can be converted to heat. Conveniently, the power density P (W/m^3) inside a material can be described in terms of the angular frequency (ω), the loss factor (ε'') and the peak applied electric field ($\hat{E}$):[28]

$$P = \frac{1}{2}\omega\varepsilon''_{\text{eff}}\,\varepsilon_o\hat{E}^2 \tag{16}$$

When described in terms of the average electric field inside of the dielectric (P_d) (based upon the RMS voltage), the average power density is:[29]

$$P_d = \omega\varepsilon''_{\text{eff}}\,\varepsilon_o E^2_{RMS} \tag{17}$$

If there are also magnetic losses within the material (from ferrous iron powder dispersed in the dielectric matrix, for example), an effective permeability (analogous to the effective permittivity) also contributes to the average power density:

$$P_d = \omega\varepsilon''_{\text{eff}}\,\varepsilon_o E^2_{RMS} + \omega\mu''_{\text{eff}}\,\mu_o H^2_{RMS} \tag{18}$$

The average power density does not describe the variation in power density within a material. EM energy entering a lossy material will give up its energy to the material. As a result, the electric field and the power decay. The skin depth (δ) is defined as the depth where the electric field is $1/e$ (about 37%) of its incident value; the power penetration depth (D_p) is defined as the depth where the power density is $1/e$ (about 37%) of its incident value. Note that $D_p = \delta/2$ and the power at the skin depth is only 14% the incident power. This has important implications in the choice of dielectric heating, for example, when considering wavelength with respect to the geometry of the material to be heated.

When assessing the feasibility of a dielectric heating process, it is convenient to empirically assess the effectiveness by measuring the heating rate (°C/s) for a given specific power (W/g). A *measure of efficiency* (MOE) can be calculated by dividing the *heating rates by specific power:*

$$\text{MOE} = \text{HR/SP} \quad (\text{g}\cdot{}^\circ\text{C}/(\text{W}\cdot\text{s})) \tag{19}$$

The MOE can provide a direct comparison of the heating efficiency between various EM heating techniques, but it is important to realize that the effect of applicator efficiency is implicit in this calculation. As such, the MOE does not, for example, necessarily compare MW vs. RF; it does compare MW of a certain applicator design vs. RF of a certain applicator design.

As an example, consider preliminary testing to compare RF vs. MW to heat an emulsion of poly(vinyl chloride) (PVC) in water.[30] The RF power was applied to the product in a capacitive applicator arrangement (SRF-1) as well as a Strayfield applicator arrangement (SRF-2) (applicator design is discussed below). Samples were heated by MW in a Cober LBM 1.2A laboratory microwave research oven (Cober Electronics Norwalk, CT, 203-855-8755). The static tests are summarized in Table 1.

TABLE 1. SUMMARY OF STATIC RF AND MW TESTS

Test configuration*	Specific power (W/g)	Heating rate (°C/s)	Weight loss (%)	MOE (g·°C/W·s)
SRF–1	7.4–7.7	0.5–1.2	1.0–11.0	118×10^{-3}
SRF–2	4.5–17.9	0.6–2.5	0.4–0.7	71×10^{-3}
SMW–1	4.5–11.2	0.5–1.4	0.2–1.4	132×10^{-3}

Test Configuration: SRF–1 = Static RF–1 → 'Parallel plate' applicator; SRF–2 = Static RF–2 → "Strayfield" applicator; SMW–1 = Static MW–1 → "Multi-mode" cavity.

The applicator design is a critical component of the overall system design, but it is an iterative process (Figure 1) where all conceptualization, feasibility testing, design, engineering and scale-up should be considered within the context of the overall process.

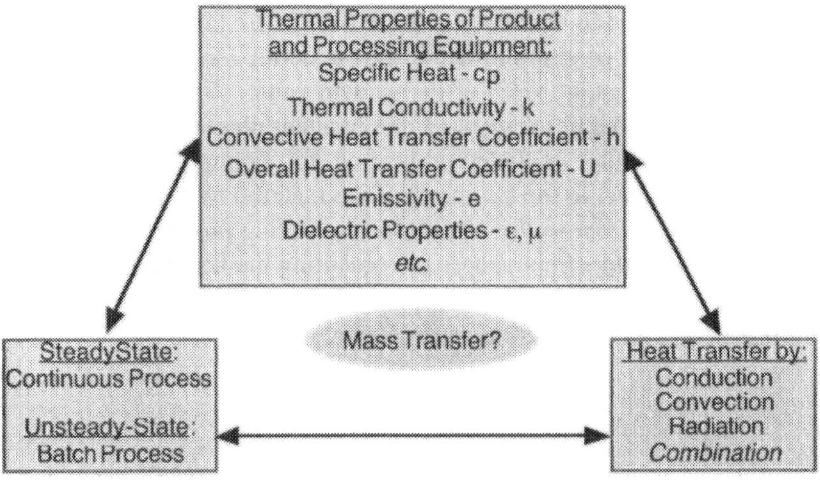

Figure 1. Dielectric heating process development is an iterative process

In recent years, commercial EM simulation programs have been adapted from communications applications to MW heating applications. No software package is able to thoroughly model all aspects of the process, but simulations of the EM and thermal processes allow a parametric study that can speed the overall process development.

Recent Trends in RF Equipment Technology

RF dielectric heating occurs in low-conductivity materials when stimulated by electromagnetic radiation in the frequency range of approximately 10 to 300 MHz. In industrial heating applications this is typically accomplished through applicators that apply the energy by one of three basic mechanisms pictured in the schematics of Figures 2, 3 and 4. Figure 2 shows a capacitive setup where the **E**-field develops perpendicular to the parallel plates, effective for processing relatively thick materials. Figure 3 shows a Strayfield setup where the **E**-field develops parallel to a sequence of alternating electrodes, typically used for processing relatively thin materials. Figure 4 shows a staggered Strayfield setup where the **E**-field develops components both parallel and perpendicular to the material, appropriate for materials of intermediate thickness.

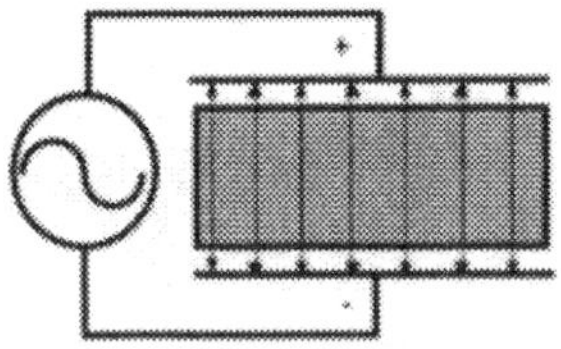

Figure 2. Parallel plate mechanism

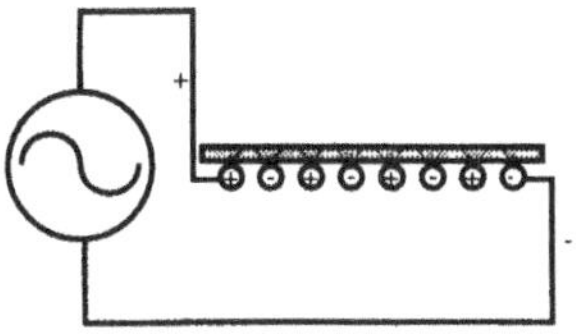

Figure 3. Strayfield mechanism

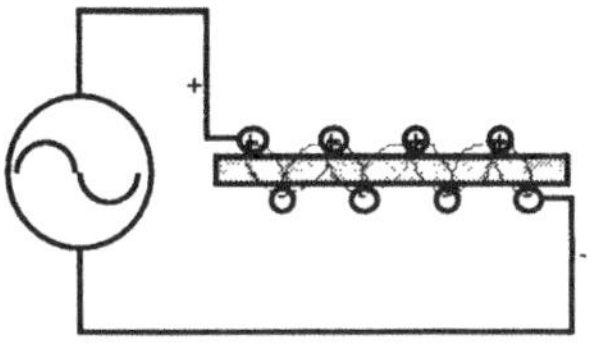

Figure 4. Staggered strayfield mechanism

These arrangements are typical in industry and are effective at heating essentially solid, planar materials. But what if the material is free-flowing, such as polymer resin pellets or a granular food product? The electrodes can be arranged in a vertical configuration and converted to a cylindrical arrangement such that the material can flow through the annular space bounded by the electrodes (Figure 5).

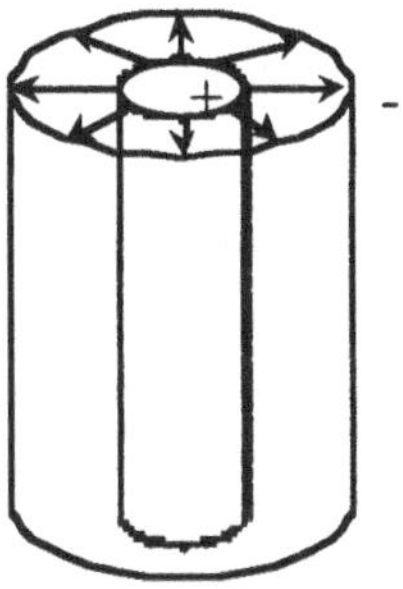

Figure 5. Vertical RF electrode.

The standard power supply of RF equipment for more than 50 years has been the free-running oscillator. Recently, the use of 50 Ω technology has become an option. The 50 Ω solid state RF generator technology (Advanced Energy, Inc., Voorhees, NJ, http://www.advanced-energy.com; Sairem S.A., Lyon, France, http://www.sairem.com) is preferred for control and precision power delivery.[32,33] The equipment applies RF energy of absolutely fixed frequency through an automatic impedance matching network (see Figure 6), which matches the constant impedance output of the RF power supply (50 Ω) to the impedance of the overall load (applicator and product).

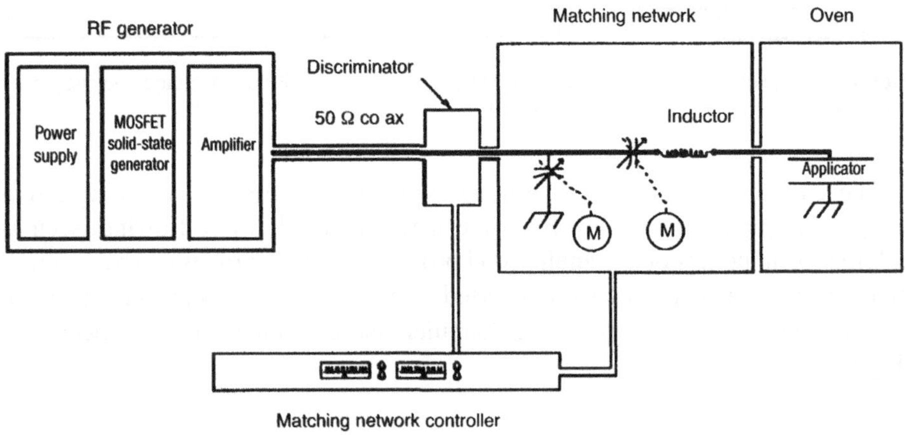

Figure 6. Schematic of a 50 Ω RF heating system.

Changes to the overall load characteristics can commonly occur during start-up or during material changes. The use of 50 Ω technology gives significant feedback control advantages, since the automatic matching network adjusts for load impedance variations. The system adjusts to ensure a constant forward power into the product by monitoring and actively controlling the reflected power to zero. Also, since frequency does not drift as with older free-running oscillator systems, there is no difficulty complying with the allowable industrial, scientific and medical (ISM) bands.

Contemporary Applications

The use of MW and RF dielectric heating technology has traditionally been viewed as a tool for improving a process, as a means to cost savings through improved efficiency. Some typical applications are outlined in Table 2. While cost savings by way of process improvement is certainly justified, it is likely the future advancement of this technology will be based upon the use of MW or RF as an enabling technology.

The term enabling technology infers a value added to a product that is not readily achievable by other means. This has been a principal focus of many interdisciplinary researchers in recent years, resulting in advances that not only address process efficiency, but product value as well. Recent research areas are indicated in Table 3. [34,35,36]

TABLE 2. TRADITIONAL INDUSTRIAL APPLICATIONS OF MW AND RF TECHNOLOGIES

Radio frequency	Microwave
Drying (textiles, paper, wood)	Tempering (meats, vegetables, fruits)
Plastics welding	Curing rubber
Plastic preform heating	Vacuum drying pharmaceuticals
Post baking crackers/biscuits	Ceramics sintering
Resins & adhesives curing	Bacon cooking
Plasma generation	Plasma generation

Conclusions

MW and RF technologies have a significant history and promising future. Interdisciplinary research is key to MW & RF applications from a process development and materials engineering perspective, establishing these technologies as not only important techniques to make process efficiency improvements, but in many cases as enabling technologies.

TABLE 3. AREAS OF CONTEMPORARY MW AND RF RESEARCH

- Microwave processing of ceramics
  - Joining and welding
  - Binder burnout and sol-gel processing
  - Sintering, annealing and ablation
- Microwave and radio frequency processing of polymers
  - Composites curing
  - Polymerization acceleration
  - Solid phase polymerization
  - MW processing of heat sensitive polymer emulsions
- Microwave synthesis and reactions
  - Organic synthesis—accelerated reactions and new materials
  - Catalysis
  - Inorganic reactions—hard coatings, oxides and complexes
  - Synthetic diamond production
- Microwave processing of conductive materials
  - Curing of carbon fiber reinforced plastics (CFRPs)
  - Sintering of powdered metals
  - Metals and conductive materials processing
- Biomedical research and applications
  - RF ablation and MW hyperthermia of cancerous tumors
  - MW imaging of cancerous breast tissue
- Waste remediation and environmental applications
  - In-situ RF treatment of contaminated soils
  - Medical and bio-medical waste sterilization by MW
  - Hazardous waste processing by MW
  - MW and RF solvent extraction and separation techniques
- Semiconductor Manufacture
  - MW and RF semiconductor material processing

REFERENCES

1. Sobol, H. and K. Tomiyasu, "Milestones of Microwaves," *IEEE Trans. Microwave Theory Tech.,* Vol. 50, No.3, pp. 594-611, (March 2002).
2. Osepchuck, J.M., "A History of Microwave Heating Applications," *IEEE Trans. Microwave Theory Tech.,* Vol. MTT-32, pp. 1200-1223, (Sept. 1984).
3. Decareau, R.V., "Microwaves in the Food Processing Industry," Academic Press, Inc., Orlando, FL, (1985). ISBN: 0-12-208430-6
4. Studt, T., "The Ten Rules of Technology Transfer," *R&D Magazine,* Vol. 46, No. 2, (Feb., 2004).
5. Metaxas, A.C., "Foundations of Electroheat," John Wiley & Sons, Inc. Chichester, UK, (1996). ISBN: 0-471-95644-9
6. Harris, N.H., et al, "Dielectric Properties of Ceramics at Microwave Frequencies," *Microwaves: Theory and Application in Materials Processing,* Ceramic Transactions, Vol. 21, pp. 235-242, (1991). ISBN: 0-944904-43-2
7. Cresko, J.W. and Roberts, P.L., "Curing Carbon Fiber/Epoxy Composites by Inductively Coupling RF Energy," *Proceedings of the 36th Microwave Power Symposium,* pp. 45-48, San Francisco, CA April 18-20, 2001.
8. Maxwell Garnett, J.C., *Philos. Trans. R. Soc.* **203**, 385, (1904); 205, 237, (1906).
9. Bruggeman, D.A.G., *Ann. Phys. (Leipzig)* **24**, 636, (1935).
10. Looyenga, *Physica,* **31**, 401, (1965).
11. Bhattacharya, S.K., "Metal Filled Polymers: Properties and Applications," Marcel Dekker Inc., New York, (1986).
12. Kusy, R. P., and Turner, D. T., *Nature,* **229** (2) 58, (1971).
13. Helme, B.G.M., "Microwave Dielectric Properties of Elastomers Containing Particulate Metal Fillers," *Mat. Res. Soc. Proc.* Vol. 189, pp 517-522, (1991).
14. Bigg, D.M., Stutz, D.E., *Polym. Compos.* (USA), **4** (1), pp 40-60, (1983).
15. Gooch, J.W., Daher, J.H., "Elastomeric Conductive Sealants Protect Aircraft from Corrosion," *Adhesives Age,* pp. 17-20, (May, 1990).
16. Embury, J.F., Manuel, H.L., Kagan, J.A., "Measurement and Prediction of Complex Permittivity and Permeability of Particulate Coatings in the Centimeter and Millimeter Wave Regions," *Proc. of Int'l. Symp. Multiple Scattering of Waves in Random Media and Random Rough Surfaces,* pp 473-477, editors Varadan & Varadan, The Pennsylvania State University, (1985).
17. Meyers, J., "Metal Filled Resins Sparked by EMI Shielding Needs," *Modern Plastics,* pp. 20-21, (August, 1994).
18. Shukla, T.P. and Anantheswaran, R.C., "Ingredient Interactions and Product Development for Microwave Heating," *Handbook of Microwave Technology for Food Applications* (Edited by Datta, A.K. and Anantheswaran, R.C.), pp. 355-389, Marcel Dekker, (2001).
19. Shukla, T.P., "Microwaveable Food Ingredients," *Cereal Foods World,* 43 (10) 770, (1998).
20. ASTM, *Standard Test Methods for Complex Permittivity (Dielectric Constant)*, Designation D 2520-86, Method B (Resonant Cavities), American Society for Testing and Materials.
21. Altshuler, H. M., "Handbook of Microwave Measurements," editors M. Sucher & J. Fox,

Wiley & Sons, New York, (1963).

22. Nicholson, A.M. and Ross, G.F., "Measurement of the Intrinsic Properties of Materials by Time Domain Techniques," *IEEE vol. IM-TC,* (January, 1990).

23. Baker-Jarvis, J., Vanzura, E.J. and Kissick, W.A., "Improved Technique for Determining Complex Permittivity with the Transmission/Reflection Method," *IEEE Transactions MMT,* **38** (8), p. 1096, (August 1990).

24. Blackham, D., "Use of Network Analyzer and Coaxial Probe to Determine Complex Permittivity," *Mat. Res. Soc. Proc.* Vol. 189, pp 123-128, (1991). ISBN: 1-55899-214-6

25. Ghodgoankar, D.K., et al, "Free Space Measurement of Complex Permittivity and Complex Permeability of Magnetic Materials at Microwave Frequencies," *IEEE Transactions on Instrumentation and Measurement,* Vol. 39 (2), (1990).

26. Ghodgoankar, D.K., et al, "Free Space Method for Measurement of Dielectric Constants and Loss Tangents at Microwave Frequencies," *IEEE Transactions on Instrumentation and Measurement,* Vol. 38 (3), (1989).

27. Metaxas, A.C. and Meredith, R.J., "Industrial Microwave Heating," *IEE Power Engineering Series 4,* Peter Peregrinus Ltd., London, (1993). ISBN: 0-906048-89-3

28. Roussy and Pearce J.A. "Foundations and Industrial Applications of Microwaves and Radio Frequency Fields," John Wiley & Sons, Inc. Chichester, UK, 1995. ISBN: 0-471-93849-1

29. Metaxas, A.C., "Heating with Electromagnetic Fields—A Unified Approach," *Teaching Monograph 4 of the Electroproduction Education Series,* The Electricity Council, London – EC 5055/9.89

30. Cresko, J.W., "Innovative Application of Dielectric Heating Technologies to Concurrently Reduce Energy Usage and Emissions in Polymer Manufacturing Processes," *Proceedings of the 3rd World Congress on Microwave and Radio Frequency Processing,* Sydney, Australia September 22-26, 2002.

31. Yakovlev, V.V., "Comparative Analysis of Contemporary Electromagnetic Software for Microwave Power Industry," in *Microwave and Radio Frequency Applications,* Ceramic Transactions, Vol. 111, pp. 551-558, Edited Folz, D.C., et al., (2003). ISBN: 1-57498-158-7

32. Mavretic, A., Gergin, E., and Stach, J. "Materials Processing by RF Power," pp. 561-567 *Proceedings of the 7th International Conference on Microwave and High Frequency Heating,* Valencia,. Spain (Sept. 13-17, 1999). ISBN: 84-7721-781-5.

33. Mavretic and Allen, J., "Integration of Fifty Ohm Technology for Materials Processing," pp. 267-273, *Proceedings of the 6th International Conference on MW & HF Heating,* Fermo, Italy, (Sept. 9-13, 1997).

34. "Microwaves: Theory and Application in Materials Processing IV" (Ceramic Transactions Vol. 80 - *Proceedings from the 1st World Congress on MW Processing),* Edited by Clark, D.E., Sutton, W.H. and Lewis, D.A., (1997). ISBN: 1-57498-025-4

35. "Microwaves: Theory and Application in Materials Processing V" (Ceramic Transactions Vol. 111 - *Proceedings from the 2nd World Congress on MW & RF Processing),* Edited by Clark, D.E., Binner, J.G.P. and Lewis, D.A., (2001). ISBN: 1-57498-103-X

36. "Microwave and Radio Frequency Applications (Ceramic Transactions Vol. 111 - *Proceedings from the 3rd World Congress on MW & RF Applications),* Edited by Folz, D.C., Booske, J.H., Clark, D.E. and Gerling, J.F., (2003). ISBN: 1-57498-158-7

Putting It Into Practice

Microwave Drying of Electrical Porcelain: A Feasibility Study

W.A. Hendrix and T. Martin

The technical feasibility of using microwave energy in combination with convection heating to enhance drying rates for electrical porcelain bodies has been experimentally investigated. In preliminary studies, drying times for solid and cavity core bodies were reduced by 80-90% using various levels of microwave energy along with a range of convection conditions. Predictions based on this work indicate the possibility of further reduction in drying times through optimization of microwave and convection parameters. Furthermore, implementation of this new drying process would greatly impact the porcelain electrical insulator manufacturing process, reducing energy consumption, in-process inventory, factory space, and, very possibly, labor requirements and product losses.

Introduction

One type of production process for electrical porcelain insulators requires drying of extruded cylindrical bodies in two stages: first, to a point where the body can be shaped, and then to a very low moisture content before glazing and firing. Drying times during the first stage, when the body is above the so-called critical moisture content, are controlled by the need to minimize stress relief cracking. Heat and mass transfer limitations control the drying rate, below the critical moisture content, in the latter part of the first stage and during the second stage of drying.

Conventional drying processes used for electrical porcelain bodies involve a trade-off between time, energy usage, and product quality. These bodies can be dried relatively defect-free at ambient conditions, but may require very long time periods depending on size, e.g., as much as several weeks. The total drying time can be reduced through the use of a thermal dryer, but only to a point where energy consumption is not excessive or unacceptable quality results.

The drying time and energy requirement for conventional ceramic drying depend greatly on two rate-limiting factors. Heat is added to the body through its surface, typically by convection, and then must diffuse into the body by conduction to effect water removal. The rate of surface heat transfer is limited not only by the current convection dryer technology and economics, but also by the need to minimize unwanted surface changes. The rate of conduction depends on the thermal properties of the body, which vary as moisture content decreases. This behavior tends to increase drying times and energy requirements, especially during the latter portion of the drying cycle when mass diffusion rates are low.

Combining microwave with convection seeks to overcome the heat transfer deficiencies of conventional drying. Since the microwaves heat the water in the body directly, the internal and surface heating can be uncoupled. This circumstance allows the surface heating to be minimized and results in more uniform moisture and temperature gradients. The net effect anticipated is an increase in drying rates without problems caused by uneven volume changes and case hardening (overdrying of the surface, which reduces its permeability to diffusion of water). Figure 1 illustrates schematically the impact of a drying process that combines microwave with convention.

Based on the attributes of microwave heating, its application to industrial drying processes could yield substantial productivity and quality benefits. For example, specific benefits sought in the manufacture of porcelain electrical insulators include:

- Reduced process time. The use of microwave heating has the potential to significantly reduce the drying time (Fig. 1).

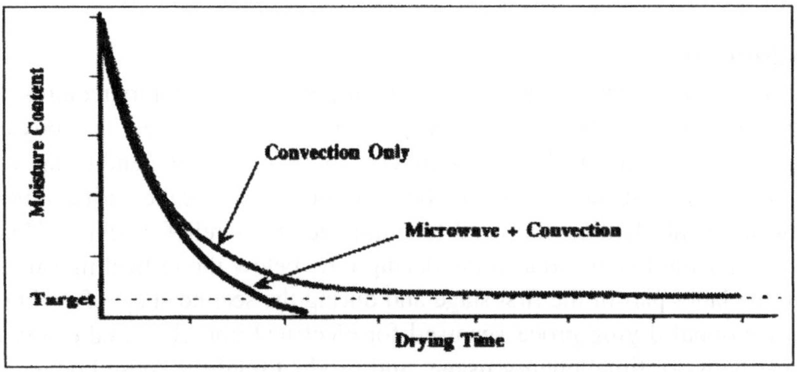

Figure 1. Impact of microwave heating on drying behavior.

- Reduced in-process inventory. Reduction in drying time translates directly into a lower in-process inventory of ceramic bodies needed to satisfy production demands.
- Reduced floor space. Less factory space is a consequence of smaller in-process inventory.

- Reduced energy consumption. The use of microwave heating results in lower convective heating requirements and shorter run times.
- Increased production capacity. Reductions in processing time, in-process inventory, and space requirements can be used to increase manufacturing capacity within the same facility.
- Improved product quality. The more precise control afforded by microwave heating provides the potential for improvement in product quality and yield.

This paper describes the results of a test program performed to evaluate the technical feasibility of using combined convection and microwave heating to dry electrical porcelain bodies. The following discussion provides a description of the testing and experimental results observed along with the preliminary conclusions and plans for future investigations.

Description of Testing

The scope of the testing involved laboratory investigation of the microwave-assisted drying of porcelain test bodies for moisture contents ranging from those experienced just after extrusion to those experienced just before glazing. This approach allowed evaluation of the drying characteristics above the critical moisture content, where stress relief cracking is an important concern, and below this point, where heat and mass transfer limitations primarily control the drying rate.

Two different types of extruded electrical porcelain bodies were used in the test program, as shown in Fig. 2. All of the test bodies were cylindrical in shape with approximate dimensions of 17.2 cm in diameter and 25.4 cm in length. One type of test body was solid and weighed about 11.4 kg (referred to as a solid post body). The other test body was similar, but contained a 5.1 cm diameter blind hole extending about two-thirds of the way down its longitudinal axis (referred to as a cavity core post body). The weight of the cavity core samples was slightly less than the solid samples at just under 10.9 kg. Both of these test body types are representative of smaller production sizes for porcelain electrical insulators. The conventional drying time is around 150 h for the solid body and around 130 h for the cavity core body.

The drying trials were performed in a 915-mHz batch microwave oven capable of continuously varying microwave power in a range of 0-6 kW. Even microwave field distribution is provided by a waveguide splitter that introduces energy into the cavity at two locations, 180° out of phase. An electrically driven mode stirrer and carousel are also contained in the cavity to assist in promoting even distribution of the microwave heating.

Figure 2. Solid and cavity core test bodies.

The microwave oven has been modified to allow for simultaneous or sequential convective heating with humidity control. Sensible heating of the convection air up to 120°C is accomplished using a bank of four individually controlled 5-kW electric heating elements. The air temperature can be controlled to within ±3°C at steady-state conditions. The convection air may flow upward or downward in the microwave cavity with impingement velocities over 15 m/s.

Moisture may be added to the oven environment for humidity control using a free-standing 9.1-kg/h, 6-kW electric boiler equipped with an accumulator tank. The humidity is controlled manually based on readings from a hygrometer mounted in the oven cavity.

The major control variables measured during testing were net microwave power input and convection conditions (temperature, volume flow rate, and relative humidity). Microwave power level was observed using a HewlettPackard Model 432A bolometer with application time controlled manually using an electronic timer. Convection conditions (air temperature, volume flow rate, and relative humidity) were monitored using hand-held instruments or from instrumentation in the control panel of the oven.

The major response variables measured during the testing consisted of surface temperature and moisture levels, internal temperatures and moisture levels, moisture loss rate, sample quality, and origin of sample failure.

Surface and internal temperatures were taken using a hand-held digital thermocouple equipped with the appropriate probe The moisture loss from a test sample was tracked by weighing it on a laboratory balance. A cross check of test body moisture content was performed by sampling it and using an Ohaus Model

MB200 moisture determination balance. In the case of multiple samples, moisture content was determined from weight change after drying in a laboratory oven. Sample quality and origin of sample failure were assessed visually with photographic documentation employed where appropriate.

Test Results

The experimental investigation addressed thc effects of a range of microwave and convection conditions on the drying behavior of the test samples. Measurements were made of relevant process parameters to establish temperature and moisture profiles, drying rates and time, and impact on quality. The first set of trials established the handling and treatment procedures and response to various microwave and convection conditions using the solid test bodies. Next, a series of tests parametrically evaluated combined microwave and convection drying of these same bodies. A third group of trials involved parametric evaluation of the small cavity core test bodies. A total of 52 individual experimental tests were conducted.

Drying of the test bodies at moisture levels above the critical moisture content was found to be particularly challenging. A series of more than 20 trials was required to establish suitable conditions for drying without stress relief cracking. Figure 3 provides an example of a failure experienced during the test program.

It was found that once below the critical moisture content, where body shrinkage is essentially complete, greater microwave power could be used to increase drying rates. However, even this technique has its limits. Microwave power was increased on one sample to a point where the sample exploded due to excessive internal pressure.

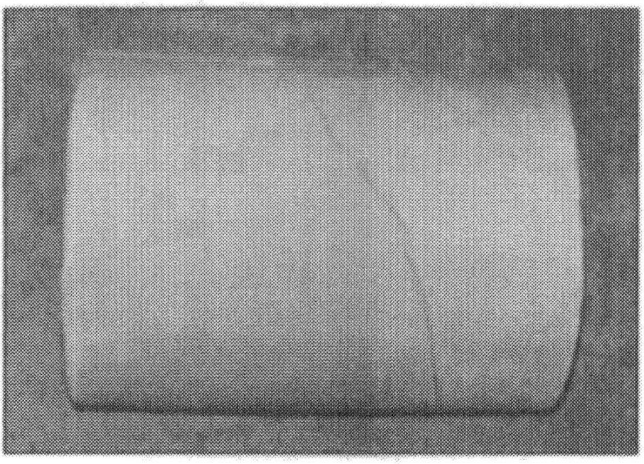

Figure 3. Test sample failure.

Figure 4 presents the dimensionless drying curves for the small solid and cavity core bodies during the shrinkage portion of the drying cycle. Note that the cavity core body exhibits slightly faster drying behavior due to its greater surface-to-volume ratio. Other than this minor difference, the drying behavior of the two test body types is essentially the same, i.e., either may be considered representative of the drying behavior.

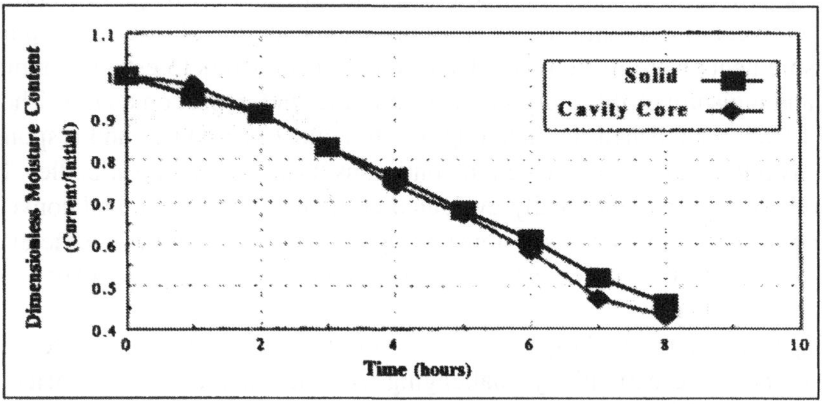

Figure 4. Dimensionless drying curves during shrinkage portion of the drying cycle.

Figure 5 shows the drying rate curves for two small cavity core bodies. The curves are very similar to each other and characteristic of those observed for conventional drying. The drying rate increases initially to a constant rate period above the critical moisture content, then exhibits falling rate behavior thereafter.

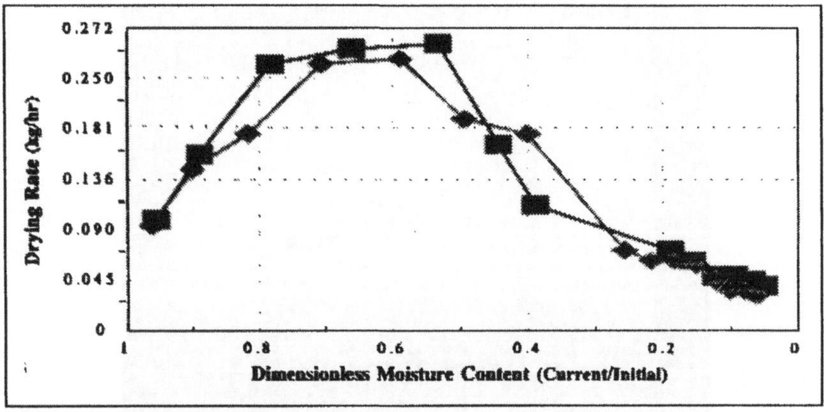

Figure 5. Drying rate curves for cavity core bodies.

Figure 6 shows the drying curves corresponding to the drying rate curves shown in Fig. 5. The total drying time is observed to be on the order of 15 h, which represents an order of magnitude improvement over the conventional drying cycle. Furthermore, the drying time appears to be quite repeatable with equal amounts associated with shrinkage and nonshrinkage portions of the cycle.

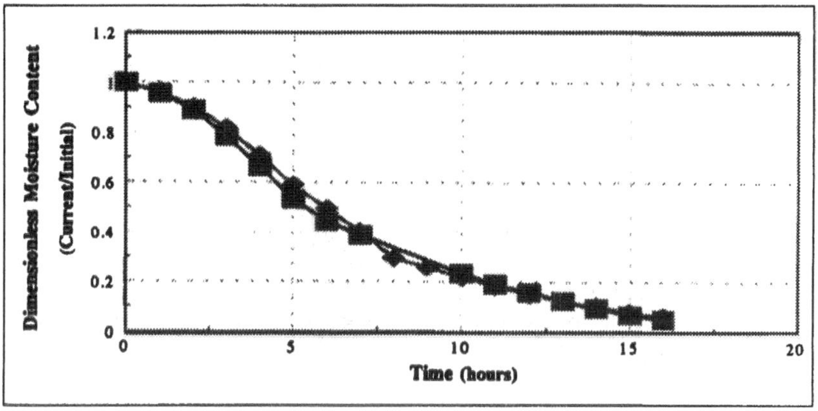

Figure 6. Drying curves for cavity core bodies.

Conclusions and Plans for Future Work

Extensive laboratory testing has demonstrated the technical feasibility of combined convection and microwave drying of electrical porcelain insulator bodies. This testing resulted in a reduction in drying times of 80-90% over those required using conventional equipment. This success has been achieved by careful control of process conditions in order to avoid stress relief cracking associated with uneven shrinkage of the bodies. Although preliminary at this point in time, the drying results have been applied to a typical insulator manufacturing process to yield the following estimated cost reductions:

- 75% annual savings in energy costs under the assumption that the drying process is completely fired by natural gas, utilizing an engine-generator set to produce all of the electricity and convection heat required.
- 90% one-time savings in inventory costs because of reduced dryer cycle times.
- 90% annual savings in factory space costs due to reduction of in-process inventory.

- 50% annual savings potential in labor costs due to reduced manual handling and automation.
- 50% annual savings potential in reduced product losses because of better dryer control.

Laboratory testing activities are continuing in order to further optimize the combined convection and microwave drying conditions and to extend them to the drying of much larger bodies. Successful completion of this work would be more representative of an actual insulator manufacturing operation and is expected to lead to on-site pilot-plant testing followed by a full-scale combined microwave and convection dryer installation.

ACKNOWLEDGMENTS

The authors gratefully acknowledge the Lapp Insulator Company and the Rochester Gas and Electric Corporation for their support of the experimental work described here. Mr. Dennis Moraski and Mr. John Keder, both of microGas Corporation, are also recognized for their involvement in the testing and very helpful discussions of the test results.

Microwave Energy Versus Convected Hot Air for Rapidly Drying Ceramic Tile

D.A. Earl, D.E. Clark and R.L. Schulz

Abstract

The purpose of this study was to determine if microwave energy could provide advantages over the conventional hot air method currently used for rapidly drying ceramic tile. Tiles consisting of a typical fast-fire body formula were dried to 0.5% moisture using a 2.45 GHz, 950 W microwave oven and a natural gas-fired roller dryer. Statistical methods were employed to develop equations for predicting microwave energy consumption, tile % moisture and surface temperature given drying time, tile volume and % relative humidity. Microwave drying was found to require 36% less energy than hot air drying. Moisture was removed and surface temperature elevated at faster rates using microwave energy.

Introduction

Ceramic tiles are dried to less than 1.0% moisture before firing to increase green strength (to reduce breakage) and to avoid tile explosions due to rapid body expansion during the initial stages of firing. Current drying methods involve natural gas combustion and hot air convection.

Only about 44% of the energy produced in a conventional tile dryer goes towards heating tiles and evaporating water. The remaining energy is consumed by the dryer structure, refractories and exhausted air [1]. There are also drying speed limitations with prevailing systems. Conduction heating produces thermal gradients which can cause differential shrinkage cracks if tiles are heated too rapidly.

Microwave energy is not currently used to dry tile, but has been applied previously to drying and firing many other types of ceramics [2-4]. The objective of this study was to determine if microwave energy would provide any economic or techmcal advantages over the current methods used to dry ceramic tiles.

Experimental

The tile body selected for the drying tests consisted of a typical fast-fire wall tile formula, as shown in Table 1.

TABLE 1. TILE BODY FORMULA USED IN DRYING EXPERIMENTS.

Weight %	Material	Composition [5]
35	Illitic Clay	$Al_{2-x}Mg_xK_{1-x-y}(Si_{1.5-y}Al_{0.5+y}O_5)_2(OH)_2$
20	Wallostonite	$CaSiO_3$
25	New York Talc	$3MgO{\cdot}4SiO_2{\cdot}H_2O$
20	Prophyllite	$Al_2O_3{\cdot}4SiO_2{\cdot}H_2O$

These materials were wet milled to produce a 7.0% moisture mix, then pressed into 5.75 cm diameter by 0.70 cm thick tile discs using a Carver lab press. Tiles were dried using a conventional natural gas-fired horizontal roller dryer or a 2.45 GHz multimode oven with a glass turntable. The microwave was rate at 950 W and the unit delivered approximately 500 W heating power to the sample (~50% efficient).

A Scotchtrak IR-1600 portable infrared heat tracer, a Hanna portable HI-8064 thermohygrometer and a Metler digital balance were used to measure tile surface temperature, microwave cavity % relative humidity and tile weight loss, respectively, following each drying trial. To avoid water vapor condensation between the tiles and glass turntable, tiles were elevated using strips of Ultra-Poly G-24MMW plastic.

Initially, tiles exploded during microwave heating due to the rapid heating rate. To control the tile heating rate, 500 cm^3 of water was introduced into the microwave cavity as a susceptor material. Water was selected because its dielectric loss factor tends to decrease non-linearly with temperature [6]. This phenomenon results in a decrease in the amount of input power absorbed by the water during a drying cycle. Thus, power for heating the tiles gradually increases as the water temperature increases. This condition better simulates typical time/temperature curves used to dry tiles in natural gas dryers and eliminated the tile explosions.

Figure 1 shows the microwave drying set-up. A 3 ft diameter, 1725 rpm Dayton fan was used to blow air into a series of 2 mm diameter holes located on the inside wall opposite a side port. The fan was operated during some experiments to separate the effects of humidity on the dependent variables.

The natural gas roller dryer was set at a typical production peak temperature of 175°C for the drying experiments. Higher temperatures caused the tile discs to crack.

Twenty-seven microwave drying tests were performed. Drying time, tile volume in the microwave and % relative humidity were varied while tile % moisture and surface temperature were measured following each experiment. Repeat tests were also conducted to confirm test method validity. Statistical methods were used to quantify variable relationships.

Also, modulus of rupture, thermogravimetric analysis (TGA), dilatometry and average fluorine (F) evolved were compared for tiles dried with each process to less than 1.0% moisture.

In order to obtain a useful comparison between microwave and hot air drying methods, the energy consumed during both processes was estimated. The most energy efficient natural gas-fired roller dryer found during this study consumed 2411 J of energy per cm^3 of tile. Determination of energy consumed during each microwave drying test was accomplished by:

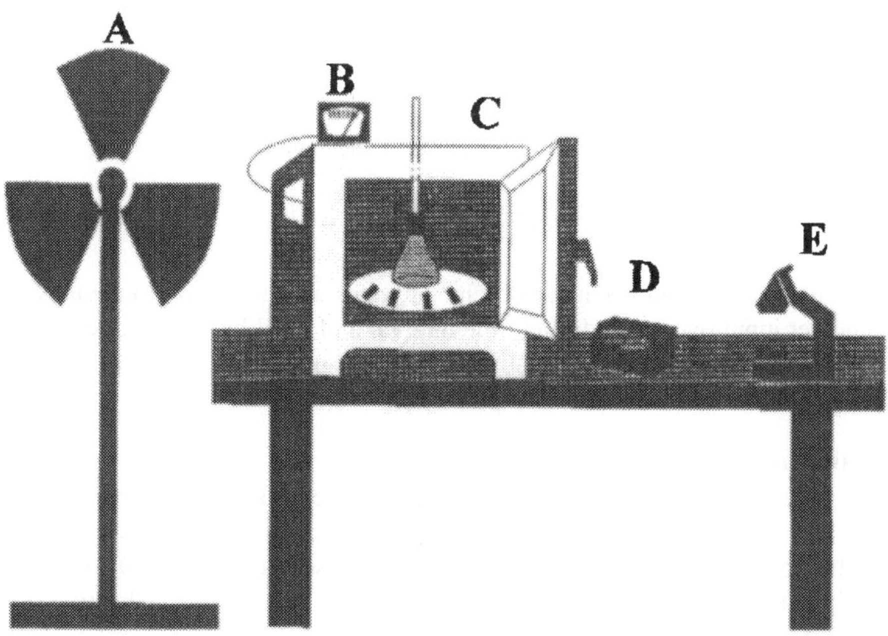

Figure 1. Microwave drying setup: A = Dayton fan used to reduce humidity in the microwave; B = humidity gauge; C = microwave cavity including beaker with 500 cc of water, rubber stopper, glass tube and plastic strips; D = digital balance; and E = infrared heat sensor for temperature measurement

1. Energy input (E_i) = 950 W microwave power input x heating time interval.
2. Water calorimetric tests were conducted to determine the relationship between average power absorbed (P_{av}) by the 500 cm^3 of water during each microwave trial and heating time. The P_{av} was calculated using [7]:

$$P_{av} = \rho C_p \, dT/dt \, V \tag{1}$$

where:

ρ and C_p	=	density and specific heat of water, respectively
T	=	water temperature (°C)
T	=	microwave heating time (s)
V	=	volume of water = 500cm3

3. A least squares regression method [9] was used to fit an equation to descabe P_{av} in terms of heating time.
4. The energy consumed by the 500 cm^3 of water (E_2) during each trial was estimated by

$$E_2 = \int P_{av} dt \tag{2}$$

5. The energy used for drying tiles per unit tile volume is

$$(E_1 - E_2)/(\text{tile volume, cm}^3) = \text{J/cm}^3 \tag{3}$$

This method is valid because 500 cm^3 of water was used only to create variable power input over each drying time interval and would not be necessary in a microwave oven where power input could be programmed to produce specific heating versus time data.

Results and Discussion

Microwave calorimetric tests revealed an inverse relationship between Pav absorbed by water and water temperature. This correlates to the variation in dielectric loss factor of water versus temperature observed by Datta [9]. The relationship between Pav absorbed by the water (watts) and microwave heating time (seconds) was found to be

$$P_{av} = 346.71 + 3108.83/t - 12.45\sqrt{t} \tag{4}$$

The regression fit was very good with an R^2 (coefficient of determination) value of 0.996 and an estimated accuracy of ± 7 watts.

By varying the amount of water and microwave heating time, it was determined that the calculated P_{av} leveled off at a maximum of 416 W. This would make the microwave oven only 44% efficient (water vapor and wall losses not considered). Generally, home microwave ovens are ≈ 50% efficient [1]. The below average efficiency may be due to the age of the unit (7 years).

The microwave experimental relationships quantified using statistical methods were:

$$E/V = 594.78 + 6.43t^2 - 22970.40/V - 42.25\sqrt{V} + 42{,}147.53t/V \quad (5)$$

$$\%M = 2.19 + 0.011V - 0.043t^2 + 3.6/t - 96.48/H \quad (6)$$

$$T = 58.61 + 26.26\sqrt{t} - 3.6 \times 10^{-6} V^3 + 6.6 \times 10^{-6} H^3 - 0.206V/t \quad (7)$$

$$d\%M/t = -0.086t - 3.601/t \quad (8)$$

$$dT/dt = 0.205V/t^2 + 13.131\sqrt{t} \quad (9)$$

E/V = energy consumed to dry tiles per unit tile volume (energy density, J/cm^3)
V = tile volume (cm^3)
t = drying time (minutes)
H = % relative humidity in microwave cavity
%M = tile % moisture
T = tile surface temperature (°C)
d%M/dt = rate of moisture removal
dT/dt = rate of tile surface temperature increase.

Equations 5 through 9 were used to adjust microwave drying dependent variables to conditions of 65% relative humidity which was the level measured in the hot air roller dryer. It was not possible to obtain this specific humidity in the microwave due to the limited control of air flow. Therefore the equations were used to extrapolate data in order to make parallel drying method comparisons. Figure 2 shows that it required 31% less energy per tile volume (1665 J/cm^3) to dry tiles to less than 1.0% moisture in 10 minutes when compared to hot air drying (2411 J/cm^3). Also, a 30% reduction in drying time was achieved by increasing the microwave energy per tile volume from 1665 J/cm^3 to 2604 J/cm^3. Tile discs cracked when attempts were made to increase the hot air drying energy. It appears that microwave volumetric heating allows for more energy input and faster drying without cracking the tiles. The surface temperature of microwave dried tiles was elevated faster than hot air dried tiles with similar energy inputs as shown in Figure 3.

Figure 4 demonstrates that the rate of moisture removal in the early drying stages was much higher during microwave processing. This may be due to the fact that overall volumetric heating is a much faster process than conduction heating of a ceramic due to its low thermal conductivity [10]. Tile volume and drying time in equations 5 and 6 were iterated using a computer to find the minimum microwave energy per unit tile volume necessary to dry tiles to 0.5% moisture. Figure 5 shows

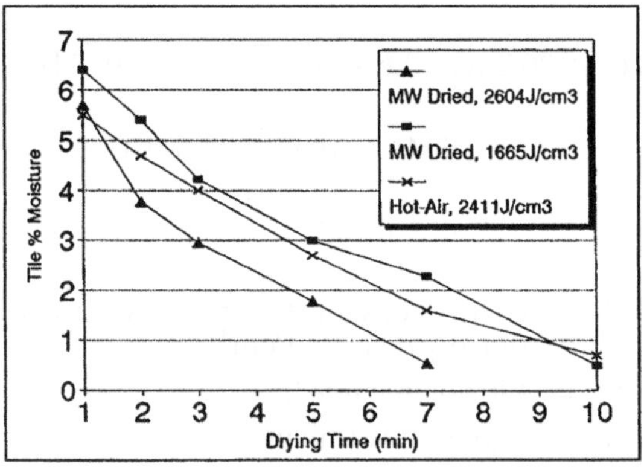

Figure 2. Tile % moisture versus drying time at 65% relative humidity.

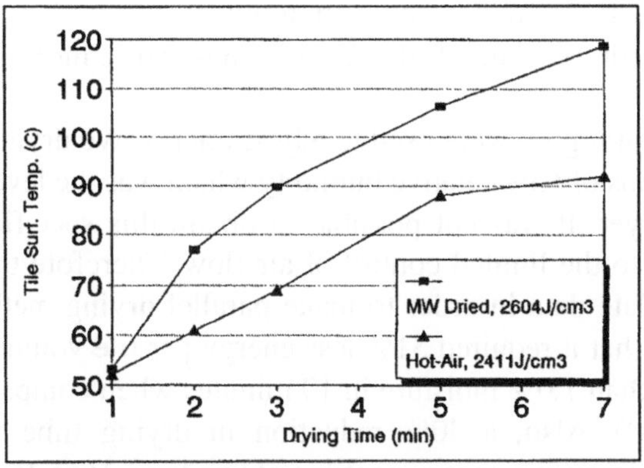

Figure 3. Tile surface temperature versus drying time at 65% relative humidity.

that a minimum input energy per unit tile volume of 1543 J/cm^3 was required to dry 492 cm^3 of tile to 0.5% moisture. The estimated drying time was 11.8 minutes. The 492 cm^3 tile volume was very close to the actual interior microwave cavity cross-sectional area. Therefore, the microwave dried most efficiently when loaded with a single layer of tiles. Energy consumption savings for this case when compared to hot air drying was 36%.

Modulus of rupture, TGA and dilatometric analysis of dried discs yielded no significant difference between microwave and hot air processed samples.

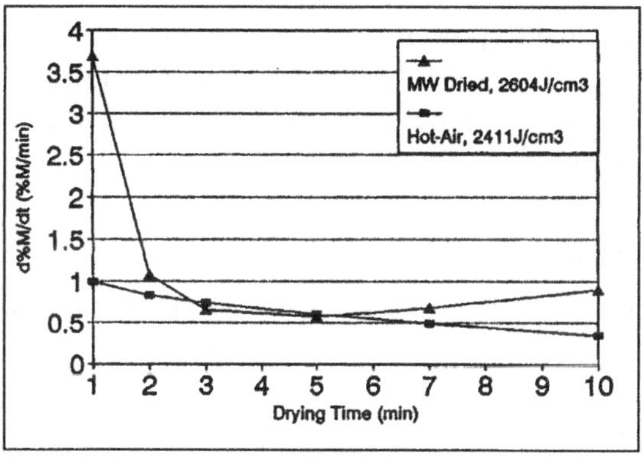

Figure 4. Rate of moisture removal versus drying time at 65% relative humidity.

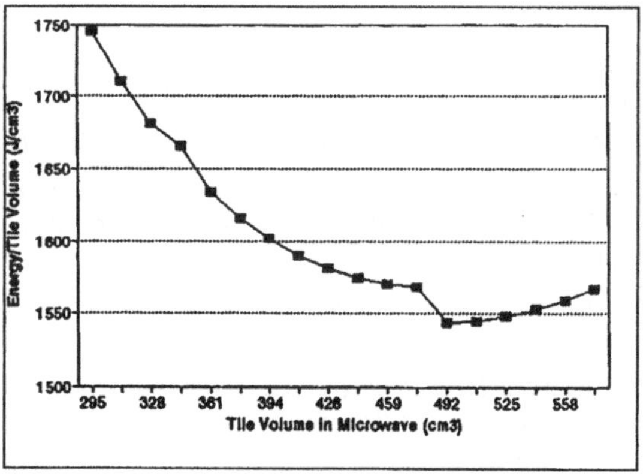

Figure 5. Energy density to dry tiles of 0.5% moisture versus tile volume during microwave drying at 65% relative humidity.

An analysis of fluorine evolved from two tiles during both drying methods showed that the microwave processed samples evolved 33% less fluorine, on average. These results were statistically inconclusive due to the small sample size and therefore, should be re-tested.

Summary

The replacement of hot air methods for drying ceramic tile with microwave energy has the potential to reduce energy costs. This study found 36% less energy was required. Tile moisture was removed and surface temperature was elevated at faster rates with microwave energy when compared to hot air. Moisture content and surface temperature of tiles dried with microwave energy were found to be non-linearly related to power input, tile volume, heating time and % relative humidity. More research should be conducted to confirm that less fluorine is evolved during microwave processing and to determine if this translates to a total reduction after drying and firing.

ACKNOWLEDGMENTS

The authors thank Huntington/Pacific Ceramics, Inc. and Florida Tile Industries, Inc. for support during this work.

REFERENCES

1. G. Nassetti and G. Timellini, *Infrared Radiation and Microwaves in the Drying Process for the Production of Ceramics,* Italian Ceramic Center Int. Pub., Bologna, Italy (1990).
2. D.E. Clark, *Microwave Processing of Materials,* Workshop given for local industry at the University of Florida, Gainesville, FL (1989).
3. G. Fisher, *Ceramic Industry,* 121, pp. 40-42 (1983).
4. L. Sheppard, *Amer.Cer.Soc. Bull.,* 67, pp. 1556-1561 (1988).
5. R. Grimshaw, *The Chemistry and Physics of Clays,* 4, pp. 97-124 (1971).
6. W.H. Sutton in, *Microwaves: Theory and Application in Materials Processing II,* Ceramic Transactions, Vol. 36 (D.E. Clark, W.R. Tinga and J.R. Laia, Jr., eds), American Ceramic Society, Westerville, OH, pp. 10 (1993).
7. A. Metaxas and R. Meridith, *Industrial Microwave Heating,* Peter Peregrinus, London, UK (1983).
8. R. Bowles and S. Haller, *Problem Solving with Experimental Design, Multiple Correlation, Multiple Property Optimization and Partitioning Variability,* SSI Management Consultants, Inc., Cleveland, OH (1990).
9. A. Datta, *Chem. Eng.,* 86[6], pp. 47-51 (1990).
10. *Chemical Engineering Handbook* 5th Ed. (R. Perry and C. Chilton, eds), 10-6, McGraw-Hill, NewYork, NY (1973).

Microwave Sintering Kilogram Batches of Silicon Nitride

P.S. Apte and W.D. MacDonald

Abstract

Microwave heating has been shown to be a very useful process for sintering ceramic materials because it helps overcome several constraints originating from their thermal properties and from furnace design. Thus volumetric heating helps avoid the effects of thermal conductivity, shadowing and convective heat transfer. One of the major limitations is ceramic materials susceptibility to non-uniform heating and thermal runaway which limits the volume of material that can be processed. The challenge is to heat up commercially viable batches of material in a controlled fashion without local overheating. The work done at Sherritt Inc. has been focused on the scale up of the heating process. It has been shown that the uniform hot zone volume can be extended to over 3 litres. The experimental work used silicon nitride as the ceramic material and the uniformity of the heating process was assessed by evaluating the properties of the sintered pieces. The results show that it is possible to sinter over 1800 grams of silicon nitride based samples in about two hours using less than 5 kWhrs/kg.

Introduction

Microwave processing for sintering ceramic materials was proven to be feasible three decades ago through the work done at General Motors when several spark plugs were sintered simultaneously [1]. This work also showed that the process was economically advantageous in terms of the energy used for sintering. There have been other reports dealing with the energy efficiency of microwave heating which have shown that compared to conventional electric furnaces, microwave furnaces could be up to an order of magnitude lower in terms of energy consumption [2]. This work also showed that the energy efficiency can be obtained only as the batch size was increased. The critical factor in obtaining a larger batch size is the ability to heat up a larger volume of material uniformly. This is a particularly difficult thing to achieve since microwave furnaces are very susceptible to thermal runaway if there is any non uniformity [3]. It is essential that thermal runaway be consistently eliminated for any successful commercial application. Katz [4] gave a general review of microwave sintering of ceramics that summarizes work up to 1992.

There have been several studies dealing with the sintering of silicon nitride. The most notable is the work done on sintering, annealing and reaction bonding at Oak Ride National Laboratory by Janney, Kimrey, Tiegs, Holcombe and others [5,6,7]. Their work has addressed several fundamental issues regarding the sintering of silicon nitride. Physical properties of sintered materials were addressed in this work but only to evaluate the efficiency of microwave heating for larger volumes.

Katz [8] and his coworkers at Los Alamos National Laboratory have reported extensive work on sintering aluminum oxide. Our work was restricted to silicon nitride because here a batch operation, controlled atmosphere and electric heating are features common to both microwave and conventional sintering.

The research and development work at Sherritt Inc. has been directed towards scale up to a "reasonable" size by commercial standards. In this context, reasonable is defined as a batch size that could process the same volume of material as a typical commercial batch furnace over a 24 hour period. Based on discussions with furnace makers this size was determined to be between 1.5 to 2 kg (for silicon nitride) to be processed in about two hours. In terms of a hot zone, this corresponds to a cylindrical zone about 15 cm (6 in.) diameter by 15 cm long.

The heating of ceramic bodies by incident electromagnetic radiation is shown schematically in Fig. 1. In conventional heating the incident flux is in the visible part of the spectrum, penetrates ceramic bodies to a very thin skin depth (a few micrometers) and the interior of the ceramic body heats up mainly by conduction. Those pieces that do not 'see' the radiation directly have to depend on radiation from other ceramics and hence they heat up at different rates than those that do 'see' the radiation. Sintering cycles are generally slow because they have to account for these non-uniform heating conditions.

At a frequency of 2.45 GHz microwaves penetrate most ceramic materials to depths of several centimeters and for many of the commercially produced ceramic bodies this exceeds the dimensions of the pieces. Microwave energy is absorbed throughout the ceramic, based on its dielectric properties and volumetric heating is obtained. The surface of the ceramic bodies is generally slightly cooler because of heat loss from the surface by radiation. The most important factors in determining the sintering process are insulation and the prevention of thermal runaway. If a hot spot is formed, it tends to absorb more microwave energy than its surroundings causing thermal runaway. Microwave heating offers the following advantages:

- The heating can be very rapid,
- Shadowing is not a factor,
- The sintered body has a fine grained microstructure, and
- Surface degradation is minimized.

One of the main features of the technology is the use of a powder bed in which the pieces to be sintered are placed. The use of a powder bed for sintering silicon nitride is common practice. In this case, the bed composition has to be altered to make it suitable for use in a microwave field. The powder bed performs several useful tasks in the sintering cycle such as

- Initiate heating from room temperature when silicon nitride itself does not couple well with microwaves,
- Insulates the ceramic pieces once they couple and start heating,
- Enables a switch over from heating the powder bed to heating the ceramic pieces based on differences in bulk density,
- Provides control over the atmosphere, and
- Acts like 'kiln furniture' holding the samples apart.

The experimental work at Sherritt started by developing a process for sintering about 25 grams of silicon nitride, representing a heating volume of about 50 mL. Successive designs and developments progressed from 100 to 1800 gram batches, representing a volume of about 3 litres for the largest batch. Further scale up based on this technology is only dependent on the availability of sufficient microwave power. The uniformity of the heating process was assessed by the uniformity of the properties and the absolute values of the properties, which pro-

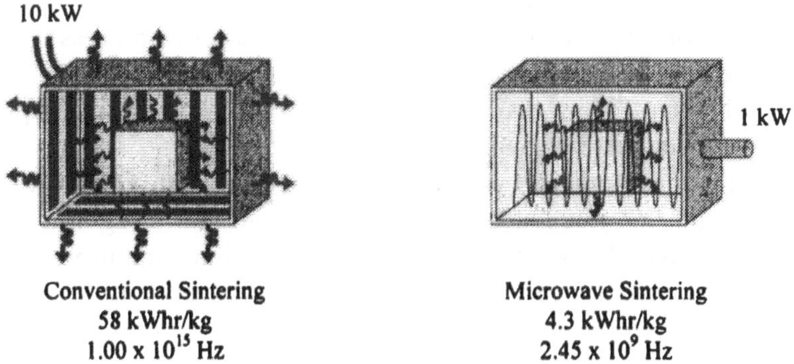

Figure 1. Schematic representation of heating by microwave and visible radiation. The penetration or skin depth of microwaves is orders of magnitude greater than for light. Most of the energy used in conventional high temperature furnaces is lost to the surroundings.

vide an indication of the advantages of microwave heating. The data reported in this paper addresses some of the processing parameters such as the energy utilization, the effect of cycle time, powder preparation and the properties of the sintered samples.

Experimental

The process described in this paper has been developed for pressureless sintering of silicon nitride. The sintering aids consisted of alumina and yttria. The silicon nitride powder was grade SN-E- 10[1], the alumina was grade A-16[2], and the yttrium oxide was Grade C fine[3].

The apparatus used for the experimental work consisted of a 5 kW microwave power supply[4] fed into a three port circulator. Forward and reflected power are measured and reported on the control panel. The power is fed to the applicator through a triple stub tuner which is used as a matching network and enables better utilization of the power. A three-port circulator and a dummy water load ensure that the reflected power reaching the magnetron is small. The microwave power is fed into an applicator consisting of metallic container of a suitable size and shape for the number of pieces to be sintered. It is possible to drill small holes, ~ 5/8"dia., in the applicator to flow in gases or to insert a temperature probe provided no conducting material is inserted through these holes. Microwave power is controlled by a computer which also gathers data from the different temperature probes, the forward and reflected power meters and microwave leak detectors placed around the applicators. The leak detectors are independently linked to the power supply such that if a leak greater than 0.5 milliwatts per square centimetre is detected at any time through the run the power is shut off.

The temperature inside the hot zone is measured by an iridium tipped sapphire probe[5]. The surface temperature of the heating casket is monitored by pyrometry and calibrated with the hot zone temperature. Once a set of process conditions has been developed for a particular product, only the surface temperature is monitored in production. The process is controlled by the power input curve provided the temperature attained falls within a well-defined band.

Silicon nitride, alumina and yttria powders, other process additives, binders, lubricants and other organic aides are mixed in a ball mill using hexane as the milling medium and alumina balls to facilitate the milling and mixing process. The powders are mixed for 24 hours and the solvent is removed using a rotary evaporator. The dried powder is compacted in a hydraulic press using tungsten carbide tooling at a pressure of 275 MPa (40,000 PSI), and subsequently cold isostatically pressed at 350 MPa (55,000 PSI). The pressed compacts range from 12 to 25 cm in diameter, 5 to 12 cm thick and weigh 3 to 35 g.

The pressed compacts are placed in an alumina crucible and filled with a powder bed. The alumina crucible is insulated by a refractory thermal blanket. The assembly is placed in an applicator large enough to accommodate the number of pieces to be sintered and the power is applied. The actual power cycle varies because the maximum power required to sinter depends on the number of pieces to be sintered as shown in Table I.

The samples are removed from the insulation blanket and their density is measured to evaluate whether the heating had been adequate and uniform. This cycle is repeated by making appropriate changes to the power rate and the maximum in the power cycle depending on sample density and surface quality. When the power is too high, the parts are "overcooked" as evidenced by silicon droplets accumulated on the part surface. Once a satisfactory density and density distribution is obtained, the cycle is repeated to ensure reproducibility. The data from the power measurements is then analyzed to obtain a sintering program for the particular applicator design and load.

Results and Discussion

Process Trials

Over the last three years, the scale up of the microwave sintering process has progressed from sintering a few samples at a time to over 500 samples simultaneously. To accomplish this, over 750 sintering runs have been made. A selection of the results from representative runs from each time period in the development of the process is shown in Table 1.

It can be seen that the efficiency of microwave power utilization and the density improves as the mass of material processed increases. The recent power consumption is consistent with the results of Patterson et al [2] who calculated

TABLE 1: SUMMARY OF SCALE UP RESULTS ON MICROWAVE SINTERING OF SILICON NITRIDE.

No.	No. of Parts	Sample Mass (grams)	Hot Zone Volume (litres)	Max. Power (Watts)	Time (min.)	Specific Energy* (kWhr/kg)	Density (g/cm^3)	Density Range (±g/cm^3)
1	5	25	0.05	900	70	50.0	3.10	0.10
2	10	50	0.20	900	90	25.0	3.16	0.04
3	49	412	0.75	2400	100	7.9	3.21	0.03
4	104	333	0.90	2250	120	8.7	3.25	0.01
5	560	1790	3.00	5000	120	3.6	3.26	0.007

*Microwave net energy absorbed, as obtained from input minus reflected power.

the net power consumption for alumina to be 3.8 kWhr/kg of alumina. The maximum power required for sintering the 1.8 kg batch was 5 kW, which is the limit of our power supply.

The maximum density obtained represents over 99% of the theoretical density and the variation is less than 0.007 g/cm^3 as shown in Fig. 2. which is comparable to the accuracy of measurement by Archimedes principle. Further densification can be attained by hot isostatic pressing the samples at 1800° C in a nitrogen atmosphere at 206 MPa. However, the maximum density increase was never greater than 0.03 g/cm^3. In earlier trials, major deviations in density were observed when the alumina crucible cracked and the protective atmosphere was lost.

Product Evaluation

As part of our scale up process, product quality standards have been developed for each stage, from incoming powder inspection to final quality assurance before shipping. For the rest of this paper, we focus on the recent results of the microwave sintering stage. For each run following sintering, the parts are visually inspected for distortion, cracks, and surface damage. Sintered densities are obtained from preselected samples, including standards, in certain locations within the hot zone. The weight change for these samples are measured as well as the linear shrinkage. At least one sample from each run is polished for metallographic examination and porosity evaluation as per ASTM B276. Hardness measurements are made using Rockwell "A" and Vickers at a 10 kg load. Typical property data is shown in Table 2.

The data shows that the sintered material has excellent physical properties and a fine grained structure expected from volumetric heating of the ceramics. The range in Vickers hardness values reflect the dependence of that measurement on the applied load. At low loads. ~ 1kg, the material gives high hardness values with considerable scatter. At higher loads, the hardness and the scatter decrease, as shown in Fig. 3, indicating that a more reliable value is obtained. That this is

TABLE 2: PHYSICAL PROPERTIES OF MICROWAVE SINTERED SILICON NITRIDE

Density	3.26 g/cm^3
Hardness	94.1 ± 0.1 Rockwell A
	16.7 – 17.5 GPa Vickers
Toughness	7.5 MPa m$^{1/2}$
Phase Structure	>95%ß · Si_3N_4
Grain Size	Acicular, approximately 0.5 x 5 μm

not a particular characteristic of microwave silicon nitride is indicated by the similar trend for a competitive material.

A fractograph of a sintered sample, shown in Fig. 4. reveals a very fine grain size and an aspect ratio of about 10 on the acicular grains. The lack of surface degradation is shown in Fig. 5 on a CNGA-432 cutting tool insert blank. The Vickers hardness variation from surface to center was ±0.5 GPa. Only a minimum of grinding is required to bring the sample within the cutting tool class G tolerance of ± 0.001″. Hardness measurements made on the as sintered surface are typically only 0.2 points lower in Rockwell A hardness than on a ground and polished internal surface. Since it is possible to make sintered silicon nitride with a very thin degraded zone, it appears feasible to make near net shape ceramics by microwave sintering.

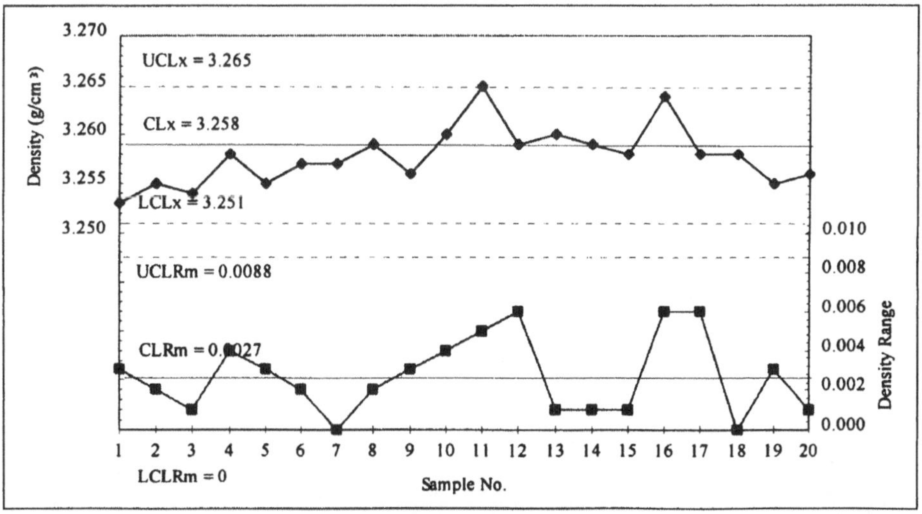

Figure 2. Variation in sintered density on a typical run as shown by an x-Rm chart. Note that the density variation is less than 0.2%, and the process is well under statistical control.

Conclusion

It is possible to scale up microwave sintering to a heating volume of 3 litres. Further extension of the uniform hot zone is feasible with a larger power supply. The sintered product obtained has uniformly high density and excellent physical. Microwave sintering does provide a fine grained product with reduced surface damaged layer as would be expected from a rapid, volumetric heating process.

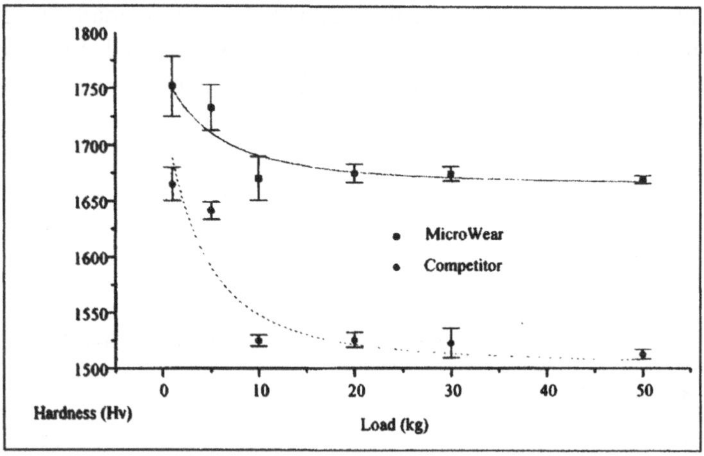

Figure 3. Vickers macrohardness versus indent load on microwave sintered silicon nitride and a competitive conventionally sintered silicon nitride. The error bars, which are plus/minus one standard deviation, decrease with load.

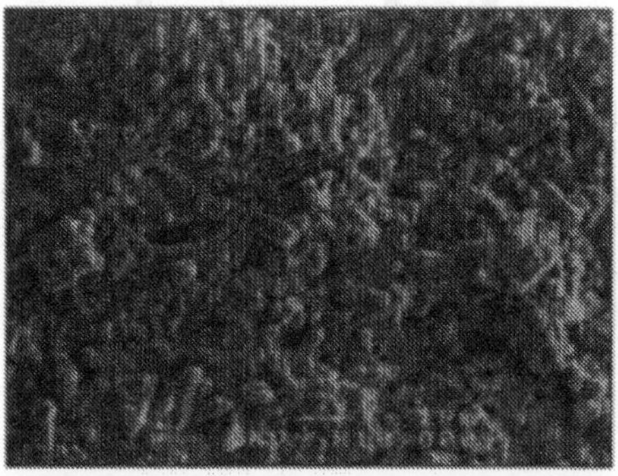

Figure 4. SEM fractograph of a microwave sintered wear part. The ß-phase needles are embedded in glassy matrix that has a grain size on the order of 0.2 to 0.4 μm. Original 10,000X.

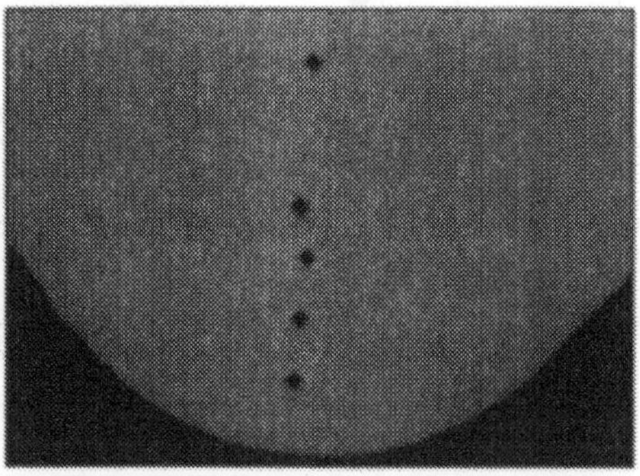

Figure 5. Polished section through the cutting tip of a CNGA-432 style insert blank. The closest microhardness indent is approximately 0.005″ from the surface and gave a reading of 17.4 GPa. Original 100X.

ACKNOWLEDGMENTS

The authors would like to thank the technologists and co-p students involved in this project for their help. The authors also appreciate the permission of Sherritt, Inc. to publish this work.

REFERENCES

1. Schubring, N.W., "Microwave Sintering of Alumina Spark Plug Insulators", *Am. Ceram. Soc.*, Electron. Div., Grosinger, NY (1983)
2. Patterson, M.C.L., Kimber, R.M., Apte, P.S., and Roy, R., "Batch Process for Microwave Sintering of Si_3N_4" in Microwave Processing of Materials, Mater. Res. Soc. Symp. Proc. (Snyder, W.B. et al eds)., 25772 (1991)
3. Johnson, D.L, Skamser, D.J., and Spotz, M.S., "Temperature Gradients in Microwave Processing: Boon and Bane", Ceramic Transactions, 36: Microwaves: Theory and Application in Material Processing 11 (Clark, D.E., Tinga, W.R. and Laia, J.R. eds) 133-46 (1993)
4. Katz, J.D., "Microwave Sintering of Ceramics", *Annul Rev. Mater. Soc.* 22, 153-70 (1992)
5. Holcombe et al., "Method for Heat Treating and Sintering Metal Oxides with Microwave Radiation" U.S. Patent #4,880,578 (1989)
6. Tiegs, T.N., Kiggans, J.O. and Kimrey, H.D., "Microwave Processing of Silicon Nitride", Microwave Processing of Materials 11, Mater. Res. Soc. Symp. Proc., 189 (Snyder, W.B. *et al.* eds) 267-72 (1991)
7. Janney, M.A. and Kimrey, H.D., "Microwave Sintering of Alumina at Powder Science, II, B, Ceramic Transactions, 1, 919-24 (1988)
8. Katz, J., and Blake, R.D., *Am. Ceram. Soc. Bull.* 70, 1304-08 (1991)

Sintering of Traditional Ceramics by Microwaves (84GHz and 2.45GHz)

**S. Takayama, M. Mizuno, S. Obata,
T. Shimada, K. Satake, M. Sato, T. Mutoh,
T. Shimotuma, S. Ito, K. Ida, T. Inoue,
K. Esaki, O. Motojima and M. Fujiwara**

Abstract

The application of the microwave heating to the traditional ceramics is reported here. The traditional ceramics is a mixture of several kinds of compounds and produces a liquid phase at the sintering temperature. This makes the sintering process harder to understand than the pure fine ceramics. We tried the microwave sintering of traditional ceramics and analyzed the results in comparison to conventional heating. Tableware could be sintered by microwaves as well as by conventional heating. Microwave uses cesser energy. Based on our results traditional kilns are being converted to microwave kilns.

Introduction

The heating power by microwave is expressed as

$$P \sim 2\,\pi\,E^2\,\varepsilon\,\tan\,\delta, \tag{1}$$

where E is electric field by microwave, ε is dielectric constant, and ε tan δ is the dielectric loss [1]. The microwave power is absorbed uniformly in the body of the sintered material (workpiece), assuming the temperature and electric field are uniform all over the body. Microwave operating at 84 GHz 2 kW (100 kW peak, 2% duty operation) were fed into a stainless steel chamber (650 mm diameter x 1100 mm). The 50x50 mm samples and cups and saucers were places into the cavity.

Low density pure alumina walls covered them for thermal insulation. The heating rate by microwave sintering was as high as 20°C/minute, while one by conventional sintering was about 2°C/minute.

The liquid phase is produced in the ceramics when the temperature exceed about 900°C. It absorbs microwave power higher than the particles. As the volume of liquid phase occupies about half of entire weight and the particle diameters are small, the liquid phase and the solid phase are almost equal temperature by conductive heat transfer. The density, bending stress and SEM photographs show that the microwave sintered traditional ceramics are almost equivalent or a little better than that of sintered by conventional furnace. Similar sintering results were obtained in the 2.45 GHz 3 kW experiments. The uniform temperature distribution in the workpieces by volumetric heating with microwave can help the stress free rapid sintering. The real tableware (cups, bells and dolls) with many kinds of glaze could be sintered as fine as conventional heating. A pilot plant is under construction for this particular use in Japan.

Experimental

Samples Preparation

Kaolin-feldspar-alumina (Al, A2) and kaolin-feldspar-quartz system (Sl, S2) were used as typical materials. Kaolin-feldspar-alumina bodies were prepared with clay 7.0 mass%, kaolin 31.4 mass%, feldspar 31.5 mass%, and alumina 30.1 mass%. Al and A2 used two different particle sizes of alumina. Sl and S2 were prepared with clay 8.0 mass%, kaolin 36 mass%, feldspar 36 mass%, and quartz 20 mass%, and clay 6.0 mass%, kaolin 27 mass%, feldspar 27 mass%, and quartz 40 mass%, respectively. The samples were calcined at 850 degreeC before the microwave sintering.

Sintering Procedure

The uniform heating can be realized, if the electromagnetic field is uniform all over the body of workpiece and the surface is thermally shielded well. The microwave reflect many times on the metallic chamber wall. If the characteristic dimensions of the chamber is a hundred times larger than the wavelength, electric field is considered to be uniform in it. If the wave length is of the order of the dimensions of the applicator, such as 2.45 GHz, special devices will reduce the non-uniformity of electric field produced by the standing waves.

The key to microwave volumetric heating is how to protect the surface from the heat transfer. The radiative loss increases in proportion to 4th power of temperature. The convective and conductive losses are proportional to the temperature gradients between the surface of the work and thermal insulator. This surface cooling produces temperature gradient in the body of the work.

It must be noted to ensure the successful microwave heating that the inner thermal insulator will be heated to equal temperature to the workpieces by microwave.

84 GHz System

Figure 1 shows the schematic of the arrangement for heating system. The microwaves were injected to the chamber from 88.9 mm corrugated wave guide with TEM00 Gaussian beam. The chamber works as a non-resonant cavity.

The test pieces were surrounded with low density pure alumina walls. It is thin enough for the microwave to penetrate the wall into the works. The dielectric loss of the insulator is estimated equivalent or lower than that of test pieces at 84 GHz. The temperature difference between the test pieces and inside wall of the insulator are small to ignore the cooling by infrared radiation from the surface of test pieces during the heating. The gyrotron is designed to generate 100-500 kW 84 GHz microwave. It was operated at 100 kW with pulse modulation in which microwave generated in 10 to 30 msec in each second. The average microwave power was controlled from 1 to 3 kW.

2.45 GHz System

Microwave from two magnetrons, which generate 1.5 kW at 2.45 GHz, were injected from two ports into the chamber. The beat of waves by small differences of frequencies of magnetron eased non-uniformity by the standing mode in the chamber. In addition, microwave circulator stirred the mode in the chamber.

The double layer thermal insulator was used. The inner thermal insulator made of mullite was installed between the workpieces and outer insulator made of pure alumina. The microwave loss decreases with frequency in the alumina insulator. The inner surface is colder than the workpieces. The microwave absorption is almost equivalent to the work in the mullite insulator. The temperature of it rises almost equal to the work. This double layer blanket realize quasi-adiabatic conditions for the works during the sintering.

Heat Curve

Figure 2 shows the heating curves of the microwave heating at 84 GHz and the conventional firing. The temperature rise time was almost 10 times faster than conventional firing. The cooling phase was not controlled. The microwave power is controlled to keep the rise time up to about 920°C, then the injected microwave power was reduced.

For microwave heating at 2.45 GHz, the heating curve of the work and the inner thermal insulator made of the mullite at 2.45 GHz. The inner thermal insulator became higher than the workpieces under about 900°C. The heating curve of the workpiece and inner wall crosses over, when the liquid phase is produced in the work at 900°C or higher.

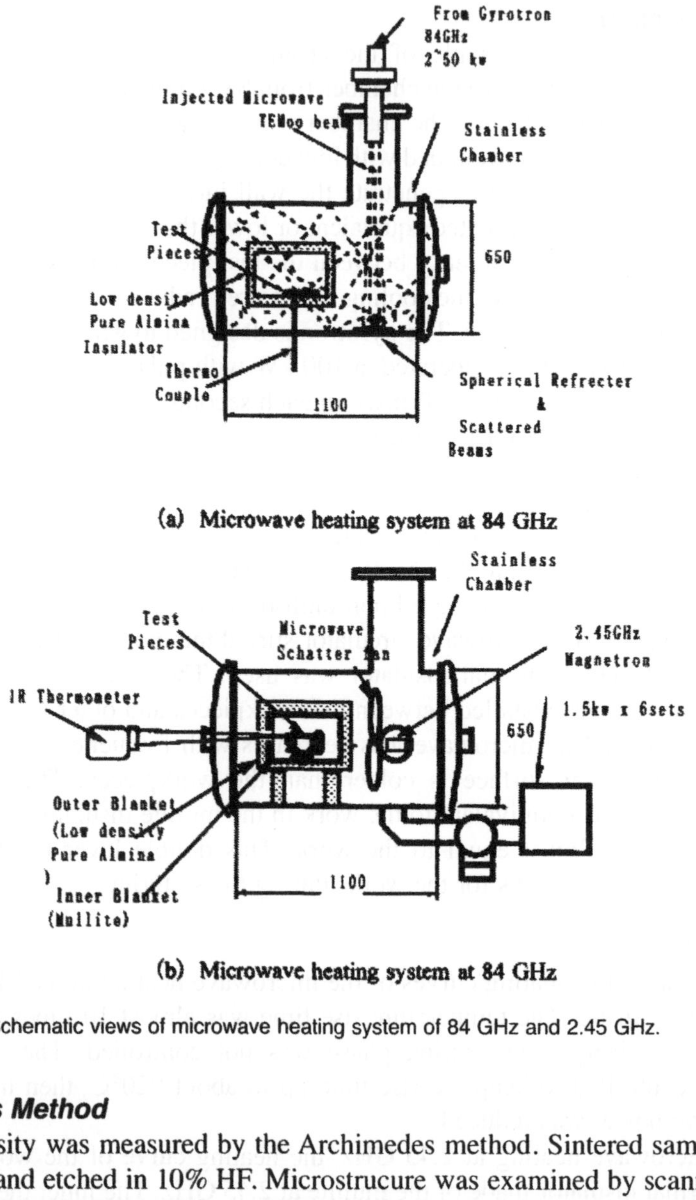

Figure 1. Schematic views of microwave heating system of 84 GHz and 2.45 GHz.

Analysis Method

Bulk density was measured by the Archimedes method. Sintered samples were polished and etched in 10% HF. Microstrucure was examined by scanning electron microscopy (SEM). The bending strength measurements carried out with an inner span of 30 mm and crosshead speed of 0.5 mm/min at room temperature by the three-points method and averaged in three to five measurements.

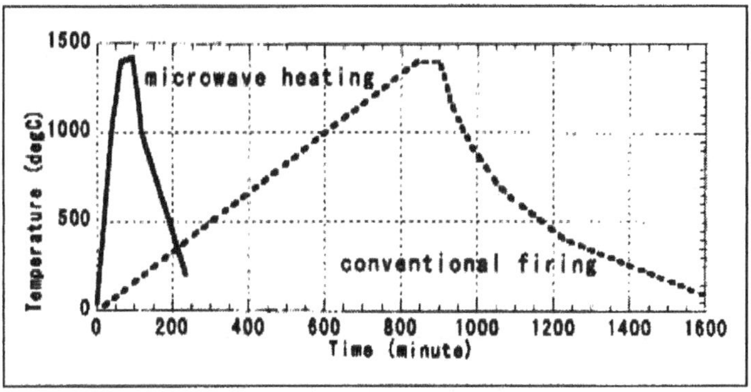

Figure 2. The heat curve of the microwave heating and the conventional firing.

Results

Analysis of Sintered Samples by Using the Microwave

Tables I and II show the apparent porosity, bulk density and bending strength of kaolin-feldspar-alumina bodies and kaolin-feldspar-quartz bodies, which were sintered by using 84 GHz and 2.45 GHz microwave. As a result, the apparent porosity, bulk density and bending strength show that the microwave sintered traditional ceramics are almost equivalent or a little better than that of sintered by conventional furnace.

Compared with the samples sintered by a conventional fumace as shown Table I, almost equivalent sintering results were obtained by the 2.45 GHz 3 kW. The ceramics with glaze could also be sintered as fine as conventional firing.

TABLE I. POROSITY, BULK DENSITY AND BENDING STRENGTH OF KAOLIN–FELDSPAR–ALUMINA BODIES

	Temperature (°C)	No	Apparent porosity (%)	Bulk density (g/cm³)	Bending strength (MPa)
Microwave sintering	1220	A1	0 03	2.81	221
at 84 GHz	(0.5 hr)	A2	0.03	2.81	137
Microwave sintering	1220	A1	0 03	2.83	
at 2.45 GHz	(0.5 hr)	A2	0.01	2.82	
Conventional sintering	1220	A1	0.04	2.75	197
	(1.0 hr)	A2	0.03	2.80	139

TABLE 2. POROSITY, BULK DENSITY AND BENDING STRENGTH OF KAOLIN–FELDSPAR–QUARTZ BODIES

	Temperature (°C)	No	Apparent porosity (%)	Bulk density (g/cm^3)	Bending strength (MPa)
Microwave sintering	1150	S1	0.03	2 45	144
(0.5 hr)	S2	1.20	2.32	140	
Conventional sintering	1150	S1	0 04	2.44	163
(1.0 hr)	S2	3.52	2.25	151	

The liquid phase is made with the feldspar when the temperature of samples became greater than up to 900°C. Figure 3 shows the difference of each bulk density with respect to the liquid phase. The difference is large between the microwave sintered bodies and the conventional sintered bodies that include a little amount of feldspar. This can be explained that the temperature of the liquid phase became higher than the rest of solid component in the body and that the viscosity of liquid phase becomes low and hastens the sintering.

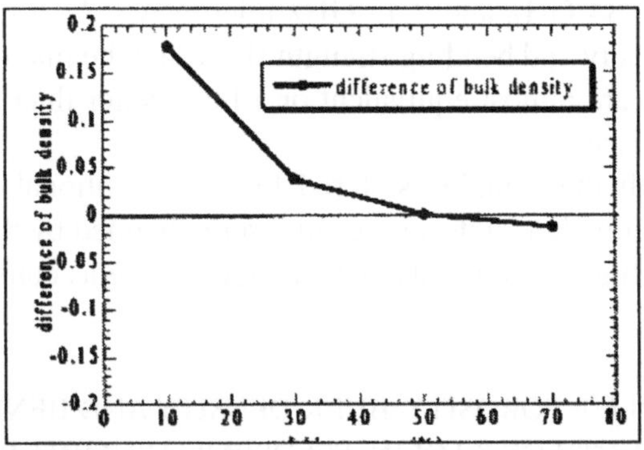

Figure 3. The difference of bulk density between the microwave sintered and the conventional sintered with respect to the amount of feldspar.

Analysis of Microstructure by Using SEM

The microwave sintered samples at 84 GHz and 2.45 GHz were observed by SEM. The microstructures were similar to conventional sintered samples and the difference between the surface position and the inside position can not be found[2].

Discussion

Uniform Heating in the Rapid Sintering

Conventional sintering is progressed by the thermal conduction from surface to inside in traditional ceramics. The large scale crack happens by temperature gradient across the sample. In microwave volumetric heating, the temperature gradient will be small, if the thermal insulator is designed properly. The skin depth of microwave are given by

$$D = \frac{3.22 \times 10^7}{f\sqrt{\varepsilon}\tan\delta} \qquad (2)$$

in which f is microwave frequency, Â is dielectric constant and tan δ is called the loss tangent. For example, the dielectric loss and the half power depth of alumina, quartz and mullite are 0.009, 0.0004, 0.00054 and 4500 mm, 67800 mm, 69450 mm at 2.45 GHz, respectively [1,3]. The skin depth is much longer than the thickness, *i.e.* 5 mm, of the sample. It ensures uniform heating all over the volume of the sample.

Rapid Sintering Process by Microwave Heating

The traditional ceramics is progressed densification on account of the viscous flow in the liquid phase. The sintering processes are greatly influenced by the quantity and the viscosity in the liquid phase. The amount of liquid phase and the viscosity by the liquid phase are decided by composition of raw materials and the temperature. The liquid phase has larger dielectric loss.

The kaolin-feldspar-quartz bodies are composed of feldspar, quartz and metakaolin made of the amorphousness by the dewatering from kaolin mineral. The kaolin-feldspar-alumina bodies consist of metakaolin, feldspar and alumina. These bodies contain many pores at the initial stage of sintering. The feldspar becomes the liquid phase at the final stage of sintering and the liquid phase changes to a glassy phase during cooling to room temperature. Corundum changes from the alumina which is kept at the particle size without reactions with others at the final stage. These bodies are found to contain a glassy phase more than 50 mass% by means of quantitative analysis of X-ray diffraction. This microstructure of bodies contains mullite less than 1 μm in size. The glassy phase forms the matrix around the crystal phase in the traditional ceramics.

In the microwave heating, the caloric power in the bodies is proportionate to the dielectric loss. The microwave sintering is greatly influenced by the selective heating of the liquid phase.

Therefore, we will discuss the relationships of temperature between the liquid phase and the solid state in the traditional ceramics in the following.

Initial Stages of Sintering

If each composition has a different temperature, thermal stresses are produced. These thermal stresses are relaxed by texture of matrix and the thermal conduction. Although, the micro cracks may be formed in the bodies, they will be filled with the liquid phase during the final stage of sintering. Therefore, rapid microwave sintering is able to sinter as uniformly as similar conventional sintering.

Final Stages of Sintering

The temperature of the liquid phase is rapidly raised by the selective heating. The liquid phase wets the particle when the temperature is more than 900°C. The temperature difference between the solid and liquid phase are equibalance by thermal conduction. The thermal conductivities of mullite and corundum are about 4 w/m·k and 6 w/m·k at 1200°C, respectively. This discussion is supported by the similarity of characteristics between the microwave sintering and the conventional sintering .

The average temperature of samples is greatly influenced in the liquid phase of microwave heating. If a sample contains much liquid phase, the average temperature of the sample becomes similar to the temperature of the liquid phase. As shown Figure 3, we confirmed the possibility of the low temperature sintering when there is little liquid phase in the traditional ceramics.

The liquid phase sintering of alumina was observed by Ishizaki et. al. by using the 4 wt% additives which produced the liquid phase at sintering. The density of alumina reached up to 95% by microwave sintering at 200°C lower temperature than the conventional sintering.

We consider that the alumina are sintered by the solution and reprecipitation of alumina, and the alumina are sintered at the lower temperature than conventional sintering because the selective heating to the liquid phase hastens the solution and reprecipitation. Therefore, we need to discuss the sintering process, such as the viscus flow or solution and reprecipitation, at the liquid phase sintering of the ceramics by microwave heating.

We compared the rapid microwave heating to the conventional firing in traditional ceramics and discussed influences in the selective heating by microwave. The great influence of selective heating to the liquid phase appears even if the liquid phase is little produced in the sintering process. Then, we expect that the microwave sintering process appears different process from conventional sintering.

Industrial Applications

On the basis of these basic experiments and analysis, the sintering of real tablewares (cups, bells and dolls) and industrial components (electrical insulators, etc), it must be noticed that the products can be sintered with many kinds of glazes. Oxidation and deoxidation can be controlled by the atmosphere in the thermal insulator. Single stage processing including drying and calcination can be applied to the sintering of electrical insulators and other industrial components. Figure 4 shows the microwave sintered cups at 2.45 GHz.

Figure 4. The microwave sintered cups, bell and doll at 2.45 GHz.

Conclusion

The heating rate was as high as 20°C/minute to reach to the sintered temperature of 1230°C in less than an hour, while 2 to 3°C/minute is normal in the conventional firing. The energy is estimated to be at most 1/10 of conventional system to sinter a batch of workpieces. We consider that the viscous flow in the liquid phase, which appeared at higher temperature of 920°C, plays an important role on microwave sintering.

The experiments were conducted at the frequency of 84 GHz and the lower frequency of 2.45 GHz. The 2.45 GHz system can sinter the traditional ceramics as well as higher frequency.

Uniform heating can be realized by beating and scattering the microwaves in the applicator with optimized blanket in 2.45 GHz.

The cost of 2.45 GHz magnetrons is much less than that of gyrotrons. The system will be competitive to the conventional kilns. More over, as the traditional ceramics are almost transparent to the 2.45 GHz microwave for long skin depth, it sinters the materials with large volume or with poor heat conductivity in shorter time at higher temperature rise rate than in the conventional firing.

We propose to apply 2.45 GHz sintering to the traditional ceramics industry. A pilot oven has been constructed in Toki-City Japan.

More conveniently, existing conventional kilns will be converted to microwave oven just by entering a stainless steel box and optimized blanket and by applying 10-50 kW 2.45 GHz microwave, the productivity will increase and enviromnental load will be reduced to 1/10[4].

ACKNOWLEDGMENTS

The authors gratefully acknowledge the help of Miss Hisako Kuroishi and Yoshikazu Hasegawa (Gifu Prefectural Institute for Ceramics Research and Technology) and Dr. Byron Peterson (National Institute for Fusion Science). They gratefully thank the support of Yoshio Nikawa and Takayuki Nakamura (Kokushikan University).

REFERENCES

1. C. Shibata, "Industrial Microwave Power Engineering", 1984, Denkishoin corporation.
2. M. Mizuno, *et. al.*, "Analysis of Microwave Sintered Porcelain", 2000, in Proc. of this conference.
3. S. Takayama, *et al.* "Measurements of Dielectric Loss and Microwave Absorption in Traditional Ceramic", 2000, in Proc. of this conference.
4. Masato, *et al.* "Design of 2.45 GHz Microwave Furnace for Traditional Ceramics Sintering", 2000, in Proc. of this conference.

Microwave Plasma Sintering of Alumina

M.P. Sweeney and D.L. Johnson

Abstract

A tunable single mode (TM_{012}) microwave cavity was designed to initiate and sustain a plasma for sintering ceramic materials. Almost 100% of the incoming microwave energy could be coupled to the plasma under optimal tuning conditions. An optical fiber thermometer was used to continuously monitor the temperature of alumina specimens during plasma sintering. The sintering temperature and resulting density of the specimens were dependent on gas content at constant net power absorption.

Introduction

The history of plasma sintering dates back to 1968, when Bennett et.al. first reported developing a 2.45 GHz microwave-induced plasma heat source for sintering ceramics.[1] Alumina compacts were found to sinter more rapidly in a plasma and had higher densities, greater strengths and finer grain sizes than conventionally sintered specimens. Evidence of enhanced densification in a plasma also was observed.[2] Specimens that were partially sintered in a discharge showed a negligible increase in densification upon further heating in a conventional furnace at the same temperature. However, densification would continue when the same specimen was reimmersed in the plasma, and its final density would be equivalent to that of specimens that were plasma sintered continuously for the same amount of time. These results suggest that a plasma sintering environment may be more than a simple heat source.

Extensive work in the field of plasma sintering has been conducted for over ten years at Northwestern University. Microwave applicators and RF and DC hollow cathode plasma devices have been used to generate plasma discharges. Rapid densification rates and anomalous heating characteristics were reported [3-5] Knowlton first showed that the final densities of RF plasma sintered specimens were strongly related to the amount of dopant gas present during sintering, whether directly added to the plasma or injected into the plasma by desorption from the surface of the ceramic powder particles.[6] Chen further quantified these results and reported that outgassed alumina specimens did not show any significant amount of shrinkage in a 100% argon plasma, but could be heated to sintering temperatures when small amounts of polyatomic gases such as N_2, O_2, H_2 and H_2O were added to the argon.[7] The net power consumption of the plasma could not be accurately determined for these experiments.

The present work was undertaken to further investigate the effects of plasma composition on microwave induced plasma heating of alumina. A tunable single mode (TM_{012}) resonant cavity that had on-line power monitoring and tuning capabilities was designed and implemented for the microwave plasma sintering studies. It has been noticed that a microwave discharge takes on a shape related to the filled pattern of its electromagnetic mode. [8] The TM_{012} mode was chosen because the electric field lines run parallel to the coaxial plasma tube.[9]

Experimental

The single mode (TM_{012}) tunable cylindrical resonant cavity used as the microwave plasma smtering apparatus was designed in accordance with the descriptions published by Asmussen.[8] A schematic representation is shown in Figure 1. The adjustable movable plate short of the cavity served two purposes: tuning the TM_{012} mode and improving impedance matching with the plasma. A coaxial launch probe assembly was used to transmit 2.45 GHz microwave power into the cavity. Both the insertion depth of the probe into the cavity and the position of its short circuit could be individually adjusted to optimize tuning of the system. The 40 mm I.D. fused quartz plasma tube was cooled by compressed air and the outer brass walls were water cooled. A microwave power directional detector was used to continuously monitor power absorption into the cavity.

The alumina specimen configurations used in this study were 4 mm diameter x 150 mm rods and thimble shaped specimens, 8 mm ID, 11 mm OD and 10 mm high. All specimens (30 m^2/g alumina with 3% polyvinyl butyral binder) were isostatically pressed and pre-sintered at 650°C in air to burn out the binder. Both types of specimens were held stationary in the cavity, with the rods extending along most of the length of the cavity and the thimbles rested inverted in the center of the plasma on the end of a single crystal sapphire tube . An optical fiber thermometer (OFT) lightpipe sensor was inserted into the interior of the thimble to provide temperature readings.

The sintering runs for all specimens began by subjecting them to a low power argon plasma to drive off adsorbed gases from the powder particle surfaces without causing sintering. Gas composition and flow rates were adjusted and the power and pressure were brought up to their working levels to initiate sintering.

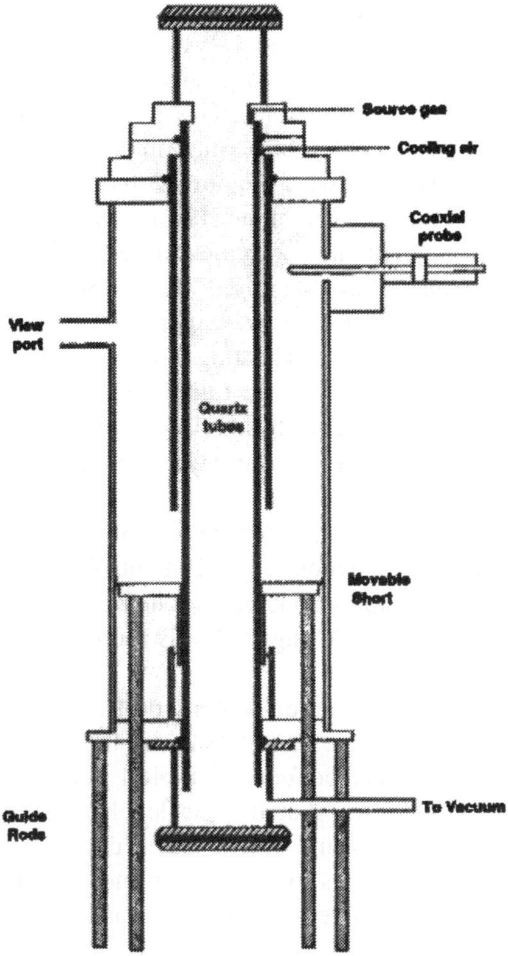

Figure 1. Schematic cross section of the TM_{012} plasma sintering cavity.

The alumina rods were plasma sintered with varying amounts of nitrogen or oxygen in argon. The sintering conditions, which were constant for all runs, were 475 W net power absorbed, 14 kPa total pressure, 10 ml/min STP total gas flow rate, and 5 minute soak. The diameters of the rod specimens were measured before and after sintering as a function of position in the cavity. Using the OFT, the same time/temperature profile was reproduced under different plasma gas conditions for the thimble experiments. Here the flow rate was fixed at 2.5 ml/min STP at a pressure of 15 kPa. The final densities of the specimens were determined by Archimedes' method.

Results and Discussion

Initial experiments showed that under optimal tuning conditions essentially 100% of the forward power could be coupled to the plasma. Certain plasma operating conditions resulted in the melting of quartz and alumina after tuning.

The normalized final shrinkage of plasma sintered alumina rods as a function of position in the cavity is shown in Figures 2 and 3. Rods sintered in a 100% argon plasma showed no evidence of densification with the same net power that caused shrinkage with diatomic molecules present. As the overall content of diatomic gas in the source gas was raised, the shrinkage of the rods increased. The length of the plasma hot zone was approximately constant regardless of diatomic gas content. The enhanced heating effects may be attributed to the additional heating modes of diatomic gases (atomic recombination). The maximum temperature (maximum shrinkage) position was located in line with the coaxial probe, rather than at the center of the cavity as expected for the undistorted TM_{012} mode.

Figures 4 and 5 show the relative density and normalized shrinkage of the sintered rods as a function of diatomic gas content. At low percentages, oxygen provided greater densities and shrinkage rates than nitrogen for the same volume content and same net power. At higher levels the densification effects of both gases converged.

Figures 6 and 7 show the time-temperature profiles of thimble specimens sintered in a variety of plasmas. A summary of the operating conditions and resulting thimble densities can be found in Table 1. As expected, more net power was required to bring the lower diatomic gas levels to the same temperatures as for the higher quantities. Plasma source gases doped with nitrogen required greater overall net power levels to reach the same temperature as those doped with oxygen. A significant increase in density was seen with the 50% nitrogen sintered thimbles.

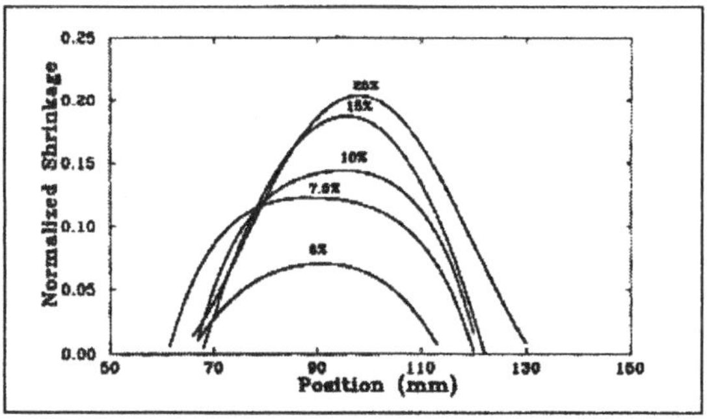

Figure 2. Normalized diametral shrinkage as a function of distance from the top of the cavity. Indicated are volume percent levels of oxygen in an argon carrier gas.

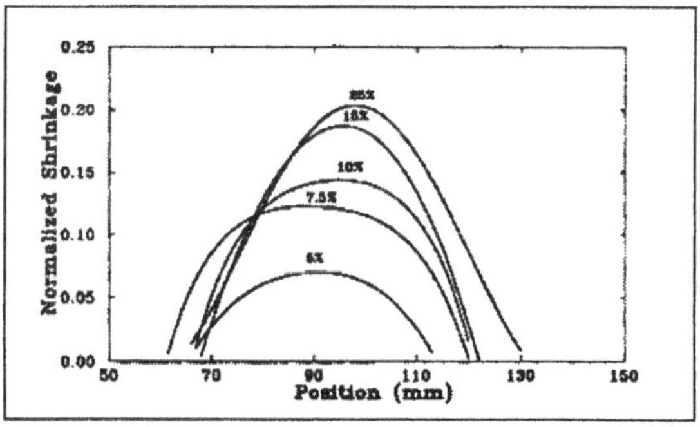

Figure 3. Normalized diametral shrinkage of plasma sintered alumina rods as a function of distance from the top of the cavity. Indicated are volume percent levels of nitrogen in an argon carrier gas.

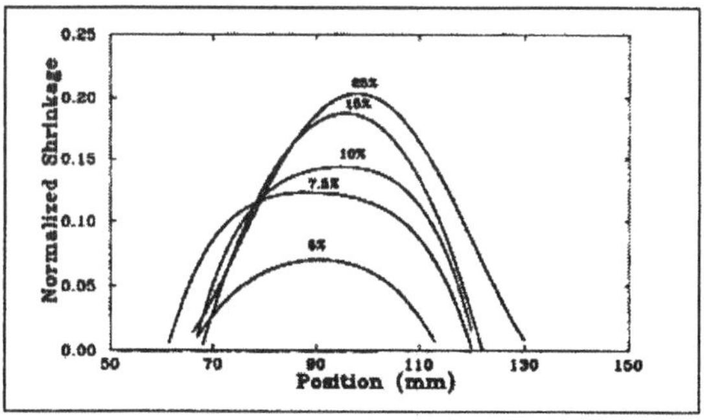

Figure 4. Maximum density of alumina rods sintered in the plasma as a function of diatomic gas content. Data of Figures 2 and 3.

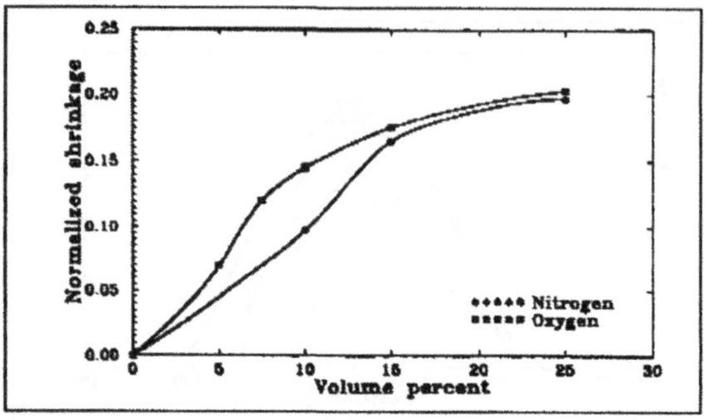

Figure 5. Normalized maximum diametral shrinkage of plasma sintered alumina rods as a function of diatomic gas content.

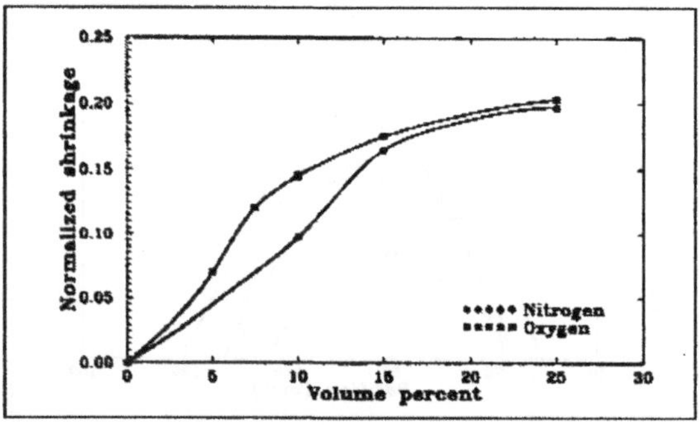

Figure 6. Indicated temperature-time profiles of alumina thimble specimens sintered under different oxygen gas contents. Label letters refer to Table 1. Low power was applied first to remove adsorbed gases from the specimen.

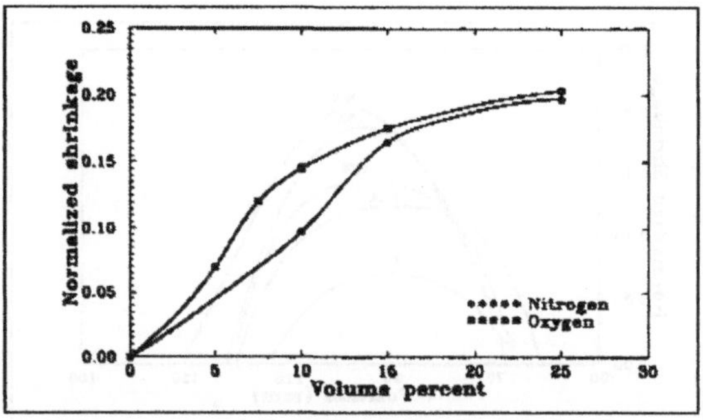

Figure 7. Indicated temperature-time profiles of thimble specimens sintered under different nitrogen gas contents. Label letters refer to Table 1.

TABLE 1. TIME–TEMPERATURE THIMBLE RUNS

Run ID	Volume percent of Additive Gas	Net Power (watts)	Relative Density
a	50% O_2	600	79.1%
b	10% O_2	880	84.2%
c	50% N_2	870	94.9%
d	50% N_2	820	93.8%
e	10% N_2	1180	83.8%

After the present work was completed, Hansen observed that the OFT reads a temperature that is biased to significantly lower values because of translucency of the thimble specimens at the wavelength of radiation sampled by the OFT.[10] While the temperatures shown in Figures 6 and 7 are substantially below the true temperatures, it is reasonable to suppose that the error is approximately the same for all specimens since they do not differ greatly in density (and therefore translucency).

Conclusion

The single mode (TM_{012}) tunable resonant cavity was shown to be a viable means for microwave-induced plasma sintering. Advantages of this system include realtime monitoring of the net power absorbed by the plasma and tuning capabilities to optimize power coupling. Sintering studies with alumina showed that the previously recognized enhanced sintering characteristics in diatomic gases were not a function of power absorbed by the plasma, but rather depended upon the chemical composition of the plasma.

ACKNOWLEDGMENTS

This work was supported by the National Science Foundation Grant DMR8719077, and made use of Central Facilities supported by the National Science Foundation through the Northwestern University Materials Research Center, Grants Nos. DMR 85-20280 and 88-21571. The contributions of Prof. M.E. Brodwin, Department of Electrical Engineering and Computer Science, particularly the design of the launch probe, are gratefully acknowledged.

REFERENCES

1. C.E.G. Bennett, N.A. McKinnon, and L.S. Williams, "Sintering in Gas Discharges," *Nature,* **217,** 1287-88 (1968).
2. C.E.G. Bennett and N.A. McKinnon, "Glow Discharge Sintering of Alumina," in *Kinetics of Reactions in Ionic Systems,* Edited by T.J. Gray and V.D. Frechette, Plenum Press, New York, 1969; pp. 408-12.

3. Joung Soo Kim and D. Lynn Johnson, "Plasma Sintering of Alumina," *Am. Ceram. Soc. Bull.*, **62** [5] 620-22 (1983).

4. D.L. Johnson, V.A. Kramb, and D.C. Lynch, "Plasma Sintering of Ceramics," in *Emergent Process Methods for High-Technology Ceramics.* Vol. 17, Edited by R.F. Davis, H. Palmour III, and R.L. Porter, Plenum Publ. Corp., 1984; pp. 207-211.

5. E.L. Kemer and D.L. Johnson, "Microwave Plasma Sintering of Alumina," *Am. Ceram. Soc. Bull.,* **64** [8] 1132-36 (1985).

6. J.M. Knowlton, "Radio Frequency Plasma Sintering of Alpha-Alumina Doped with Various Oxides," MS Thesis, Northwestern University, 1985.

7. M.Y. Chen and D.Lynn Johnson, "Effects of Additive Gases on Radio Frequency Plasma Sintering of Alumina," *J. Mat. Sci.,* accepted for publication.

8. J. Asmussen, Jr., R. Mallavarpu, J.R. Hamann, and H.C. Park, "The Design of a Microwave Plasma Cavity," *Proceedings of the IEEE* **1,** 109-117 (1974).

9. J. Asmussen and R. Garard, "Precision Microwave Applicators and Systems for Plasma and Materials Processing," in *Materials Research Society Symposium,* Vol. 124, (Materials Research Society, 1988), pp. 347-352.

10. J.D. Hansen, private communication, 1991.

Microwave Processing of Ceramics, Composites and Metallic Materials

D. Agrawal, J. Cheng, Y. Fang, and R. Roy

Abstract

The versatility of the application of microwave energy has been long recognized in a variety of areas, such as communication, rubber vulcanization, paper/textile/wood drying, food processing, pharmaceutical and biomedical fields. Over the last two decades, microwave energy is being increasingly applied in the area of materials (ceramics, composites and metals) processing for various purposes. The increased interest in microwaves for these areas is due to its certain advantages over conventional heating, such as the reduction in cycle time, energy savings, product quality improvement and eco-friendliness. For over a decade, the Penn State's microwave research center has been working on a wide variety of materials, including oxide and non-oxide ceramics, electroceramics, hard metal composites (WC/Co), transparent ceramics, nanostructured ceramics and metallic materials. Remarkable achievements have been made in the sintering and syntheses of these materials using microwaves. Recent developments have shown that all common metals (Cu, Fe, Ni, Ti, Al, Co, Ag, Au), refractory metals (W, Re, Mo), their alloys and steel compositions have been sintered successfully using microwaves, in many cases, to full density very rapidly with improved properties. The selective heating feature of microwave heating has been utilized for metal–metal brazing and joining in a few minutes.

Introduction and Background

The application of microwave energy to process various kinds of materials in an efficient, economic and effective manner is emerging as an innovative technology with great commercial potential and many advantages. In the area of processing of ceramics and metals, it has been a kind of sintering and synthesis revolution in terms of extraordinary enhancements in materials diffusion and reaction kinetics. Though microwave energy has been in use for over 50 years in a variety of applications, including communications, food processing, rubber vulcanization, paper/textile/wood drying, pharmaceutical and biomedical fields, it has been relatively recently that microwaves have been useful in materials processing, especially synthesis and sintering of ceramics. The very first application of microwave energy was in communications and this area still dominates microwave applications: almost all communications, in space and mobile/cordless phone technology are based on microwave frequencies.

Recent developments and innovations made in microwave processing in the last decade have attracted worldwide attention in academia and industry. The most prominent advances made in the past few years include synthesis and sintering of traditional and advanced ceramics in < 30 minutes, sintering of tungsten carbide- (WC-) based composites, fabrication of transparent ceramics, processing of electroceramics, fabrication of nanostructured ceramics, sintering and melting of metals, effect of microwave **E** and **H** fields on the processing of various kinds of materials, and design of continuous microwave systems enabling the commercialization of the laboratory scale technology for various important materials. This paper will review some of these developments.

Five comprehensive reviews by Clark and Sutton [1], Schiffmann [2], Katz [3] Sutton [4,5] are more than adequate to give the interested readers a broad picture of the status of microwave processing research until 1996 and provide a historical perspective of the application of microwave energy to materials processing. The decade of 1980 to 1990 witnessed widespread interest and much research activity producing very interesting and exciting results by many researchers all over the world, but most of these results have been confined to the laboratory scale. New innovations have been reported since the early 1990s. These innovations opened up new avenues of research triggering a renewed activity all over the world by various groups, especially in Europe, Japan and China.

Microwave Processing

Microwave processing of materials, which includes heating and sintering, is fundamentally different from conventional processing which involves radiant/resistance and/or convection heating followed by transfer of thermal energy via conduction to the inside of the work-piece through thermal conductivity. For many applications, it is rather a slow process and takes considerable time to achieve thermal equilibrium. One can heat any material in a conventional furnace. In the case of microwave heating, the absorption of microwave energy is followed by volumetric heating involving a conversion of electromagnetic energy into thermal energy. In this process, the role of thermal conductivity is minimized, and heating is instantaneous and rapid and is a function of the material being processed. The heat is generated internally within the material instead of originating from external sources. In general, microwave heating provides a very rapid method for energy conversion as compared to conventional methods where energy transfer heats the material. Due to volumetric heating and the ability to heat the material uniformly from the inside-out, microwave processing offers unique advantages, especially in food and rubber processing and in drying operations.

Microwaves are a small part of the electromagnetic spectrum with wavelengths ranging from 1 mm to 1 m in free space and frequencies between 300 GHz and 300 MHz, respectively. Based on microwave–matter interaction, most materials can be divided into three categories: opaque, transparent and absorbers. It is well-recognized that bulk metals are opaque to microwaves and are good reflectors at room temperature; this property is used in radar detection. However, as will be shown in subsequent sections, metals in powder form are very good absorbers of microwaves and are heated very effectively. Further, bulk metals, if pre-heated to moderate temperatures, also become good microwave absorbers. Most other materials either are transparent or absorb microwaves to varying degrees at ambient temperature. The degree of microwave absorption and, consequently, the heating profile, changes dramatically with the rise in temperature. Microwave heating is material dependent, so only those materials that couple in with the microwave field will heat. The rate of heating will depend upon their degree of absorption, which is a function of various factors, including the dielectric loss, grain size, frequency and electrical conductivity. Selective heating of matter can be beneficial for certain applications, especially for metal–metal joining and brazing.

Microwave heating has significant advantages over conventional heating [6-9] in materials processing, such as substantial energy savings (in some cases, as much as 90% energy savings have been reported [10]), very rapid heating rates, considerably reduced cycle time, product quality improvement and eco-friendliness. It has been generally observed that microwave-sintered products possess finer microstructures and other unique features leading to considerable improvement in the mechanical properties and, hence, exhibit better overall product performance. This feature is now becoming the major driving force attracting many ceramic and metal industries to make serious efforts to commercialize the technology for their products.

Recent Developments

Microwave energy for ceramic processing experienced a big push in the eighties. Ceramic processes where microwaves had been applied include process control, drying of ceramic sanitarywares, calcination, decomposition of gaseous species by microwave plasma and sintering of oxide ceramics by microwave plasma. Some of these areas have been commercially developed. These applications involve the use of microwaves at low temperatures (<500°C) and, therefore, could easily be scaled-up. However, high-temperature processing successes achieved with microwaves in the laboratory could not be so easily translated to the industry due to many hurdles encountered while scaling-up. Recently, there have been reports indicating that some success has been achieved in commercializing microwave sintering of tungsten carbide-based cutting tools [11] and alumina-based products [10]. It is hoped that, in the near future, microwave processes will be successfully developed for commercialization of many specialty materials.

In the case of most ceramics, microwave heating is mainly characterized by the dielectric loss of the material. However, there are other factors that significantly contribute to microwave heating, such as ionic conductivity, degree of porosity, particle size, electrical conductivity and magnetic coupling. The exact mechanisms for microwave heating and sintering have not yet been very well explained. The interaction between microwaves and matter takes place through the electric field vector and the magnetic field vector of the electromagnetic field and involves polarization and conduction processes. Classically, there are various absorption mechanisms identified in microwave–matter interaction. Some of them are dipole reorientation and conduction of space and ionic charge, which are primarily found in insulators or dielectric materials. The vast majority of papers dealing with microwave heating of solids ascribe the heating to energy loss mechanisms of the electric vector. Very recently, experimental findings have demonstrated that magnetic losses also play an important role in microwave heating/sintering of bulk materials for a wide range of conductor (metals) and semiconductor materials [12,13].

In 1994, Cherradi *et al.* [14] published a paper in which they showed that the magnetic field made substantial contributions to the heating of alumina (at high temperature), semi-insulators and metallic copper. In their experiments, they had always heated the sample in both **H** and **E** fields simultaneously which caused a complicated interplay of the different absorption and conduction mechanisms. At Penn State, with very careful design of a single mode cavity, we separated E and H fields at 2.45 GHz, and then heated various materials in more or less pure electric and magnetic fields separately. In general, the conductive samples, such as metal powders and WC, could be much more efficiently heated in the magnetic field. On the contrary, the pure ceramic samples (insulators), such as Al_2O_3 and ZnO, showed much higher heating rates in the pure electric field. The structural state of the materials plays an important role in the microwave–materials interaction. For example, the powder compact copper sample absorbed a significant amount of microwave energy, but the solid bulk sample did not under the same conditions. From the data generated in this study and from the general theory of energy loss in various materials when placed in a microwave field, it is no longer possible to ignore the effect of the magnetic component, especially for conductor and semiconductor materials. The contributions to the magnetic loss mechanism can be hysteresis, eddy currents, magnetic resonance and domain wall oscillations [12].

Microwave Sintering of Important Ceramics

Al_2O_3. Alumina is the most common ceramic and has been widely used in microwave-sintering research by many scientists working in the field. Because of its highly refractory nature, it is very difficult to sinter to full densification unless suitable sintering aids or some special processing techniques are adopted. Many people have successfully sintered alumina to high densification using microwaves [4]. In general, the sintering of alumina using microwave processing can be achieved much faster and at a lower temperature than with a conventional process. Alumina samples microwave sintered at 1400°C with no hold time were 98% dense [15]. In conventional heating, it requires at least 1600°C and 2 hours of soaking time to achieve the same degree of densification and with substantial grain growth. Using microwaves, nearly full density was achieved at about 200°C less than the conventional temperature, as shown in Figure 1. This work now has been extended to fabricate some commercial alumina products with substantial improvement in the quality of the sintered product. For example, sol–gel-prepared alumina grit was sintered to full density using a continuous microwave process [16]. The alumina grit powder developed by Carborundum Universal Madras (India) is a kind of agglomerated granule with an average particle size of 0.8-1.0 mm. Microwave sintering of this material at 1500°C for 15 minutes provided a density of 3.96 g/cm^3, which is very close to theoretical density. Table 1 compares the property data with conventionally prepared alumina grit. It is obvious that microwave-sintered alumina grit possesses much higher abrasion index and hardness values. The abrasion index of this grit was 94 and Vicker's hardness 2317 kg/mm^2. In another study [17], large objects of alumina with diameters of 1 to 10 cm and lengths up to 2 m were also prepared using a continuous microwave system. These parts of 98% density at 1400°C were very uniformly and homogeneously produced.

ZrO_2. Fine-grained zirconia ceramics were microwave sintered at 1360°C for 2 min in a multimode, 2.45 GHz system. The sintered density was about 97.8% and average grain size was 0.25 μm. In another study of phase transition in Ca-stabilized ZrO_2 using microwave heating, almost complete transformation from tetragonal to cubic phase was obtained at 1600°C in 10 min; conventionally it takes several hours (~10 hours) to achieve the same degree of transformation (unpublished data).

TABLE 1. COMPARATIVE PROPERTIES OF MICROWAVE AND CONVENTIONALLY SINTERED ALUMINA GRIT.

	Sintering conditions	Microwave	Conventional
Density g/cm^3	1450°C/15 min	3.70	3.92
	1400°C/45 min	3.94	3.96
	1500°C/15 min	3.90	3.89
Abrasion Index	1450°C/15 min	95	68
	1400°C/45 min	100	65
	1500°C/15 min	94	94
Vickers Micro Hardness (kg/mm^2)	1450°C/15 min	2205	732
	1400°C/45 min	2388	1026
	1500°C/15 min	2317	1885

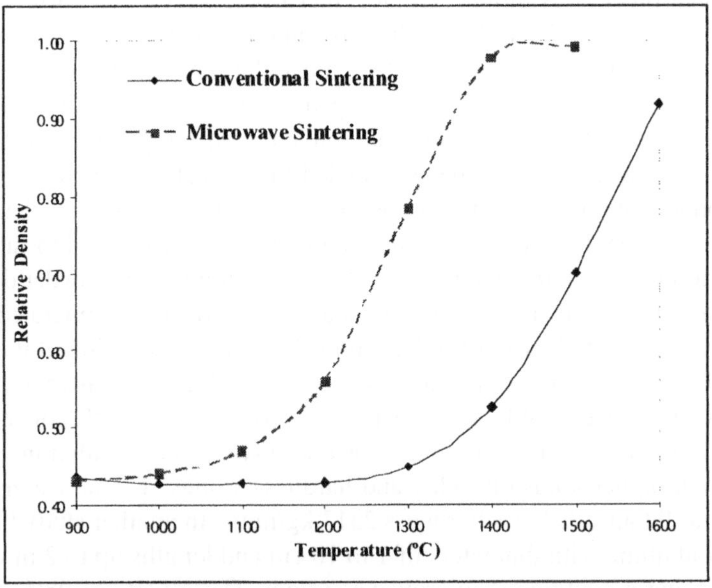

Figure 1. Sintered density vs. temperature plots for microwave and conventionally sintered alumina with zero soak time.

PZT: $PbZr_{0.52}Ti_{0.48}O_3$: $PbZr_{0.52}Ti_{0.48}O_3$. PZT is a common ferroelectric material with the perovskite crystal structure. It is conventionally fabricated at temperatures over 1200°C for several hours. One serious problem associated with PZT fabrication is the high loss of PbO due to lengthy sintering times. This is a serious environmental issue. In the microwave-assisted process, single-phase

PZT was obtained at temperatures as low as 600°C [18] by using TiO_{2-x}. The use of non-stoichiometric TiO_{2-x} enhances microwave absorption and increases the reaction kinetics substantially as is evidenced from the following examples. The use of nonstoichiometric precursors also leads to different reaction pathways for the formation of PZT. The rate of formation of PZT with microwave synthesis is controlled by a mechanism other than simple thermal diffusion. In another study, PZT samples were microwave sintered at temperatures 150° less than the conventional processing [19] resulting in finer grain size and minimal PbO loss. Table 2 shows a comparison of PbO loss in conventional and microwave processes for different PZT compositions.

***$BaTiO_3$* (BT).** Virtually all solid state reactions for the synthesis of materials in a conventional process are conducted under isothermal conditions; i.e., two or more phases involved are at the same temperature. However, using microwaves for materials synthesis involving two or more phases with different microwave absorptions may cause a situation known as an "anisothermal" state in which, on the microscopic level, one phase has a widely different temperature than the other phase. This may be the key to understanding the "microwave effect" that causes dramatic enhancements in reaction and material diffusion rates. In the synthesis of PZT (as shown above) and $BaTiO_3$, we have adopted the anisothermal approach to cause reactions to occur very quickly, producing the desired phase in a few minutes. For the synthesis of BT, we used $BaCO_3$ (a poor microwave absorber) and TiO_{2-x} (an excellent microwave absorber). The reaction of the mixture of these two phases in a microwave field occurs radically different from the conventional isothermal heating situation. Table 3 lists the sequence of phases formed as a function of temperature and time in conventional and microwave settings. In the microwave case, at 250°C with no soak time, hexagonal $BaTiO_3$ appears; at 900°C in 5 min nearly pure tetragonal phase $BaTiO_3$ is formed. On the other hand, the conventional process even at 1300°C for a 1 hour soak time does not produce any XRD-detectable $BaTiO_3$ phase [20].

TABLE 2. PBO LOSS DATA DURING THE SINTERING OF PZT BY VARIOUS PROCESSES.

Open crucible method/PZT type	MEGAC*	TRS 200*	TRS 200B*
Conventional firing (10°/min)	3%		
Fast conventional firing (25°/min)	1.35%	5.7%	13.3%
MW process, 2.45 GHz (25°/min)	0	<0.5%	0.1-0.5%
MW process, 30 GHz (30–50°/min)	0	2.3%	0.2%

*Commercial PZTs from different vendors.

TABLE 3. SYNTHESIS OF BATIO$_3$ USING BACO$_3$ AND TIO$_{2-x}$ AS PRECURSORS.

Temp./ soak time	$BaCO_3$ (%)	TiO_2 (%)	Ba_2TiO_4 (%)	$BaTiO_3$ Hexagonal (%)	$BaTiO_3$ Tetragonal (%)
Conventional heating					
900°C/2 min	57	43	0	0	0
950°C/2 min	67	33	0	0	0
900°C/1 hour	62.5	21	16.5	0	0
1100°C/1 hour	36	38	26	0	0
1200°C/1 hour	29	37	34	0	0
1300°C/1 hour	0	0	100	0	0
Microwave heating					
250°C/0 min	45	52	0	3	0
400°C/0 min	28	26	0	42	4
500°C/1 min	13	8	0	54	25
600°C/5 min	11	12	0	30	47
700°C/5 min	7	9	0	14	70
900°C/5 min	0	1	0	0	99

$Ba(Mg_{1/3}Ta_{2/3})O_3$/$Ba(Zn_{1/3}Ta_{2/3})O_3$. $Ba(Mg_{1/3}Ta_{2/3})O_3$ (BMT) and $Ba(Zn_{1/3}Ta_{2/3})O_3$ (BZT) with perovskite crystal structure are good dielectric materials for microwave resonators because of their high quality factors (Q) and moderate dielectric constants. These remarkable materials are perhaps the most refractory oxides (melting point >3000°C). Therefore, very high temperatures (>1600°C) are required to sinter them in a conventional furnace. Often, to obtain BMT/BZT ceramics with high densities, sintering aids such as Mn and Sn are used, but the sintering aids also degrade the dielectric properties. We have synthesized and sintered BMT/BZT single-phase materials using reduced-oxide precursors and microwave processing [21, 22] without the addition of sintering aids. The use of reduced Ta_2O_{5-x} enhanced the reaction kinetics and produced a single-phase material at much lower temperatures (1300°C/20 min) with higher densification than normally obtained by conventional processes. Microwave-processed BMT samples exhibited densities as high as 97% of theoretical when heated at 1600°C for 30 min. The average grain size in microwave-sintered BMT was about 1 μm, in contrast to 3 μm in the conventionally sintered material. BZT samples with an average grain size < 5 μm were sintered at 1400°C for 5 min to full densification using the microwave process.

NZP Materials. $NaZr_2(PO_4)_3$ is a parent composition of a very large family of materials now known as NZP and characterized by their low thermal expansion behavior [23]. NZP has a special open framework structure with corner-linked

ZrO_6 octahedra and PO_4 tetrahedra. The covalent network skeleton NZP is remarkably stable with respect to temperature and chemical substitution. Compounds based on the NZP structure have found applications as superionic conductors, candidates for ceramic nuclear waste forms and catalysts for NO_x reduction, and they are suitable for low-thermal-expansion applications. Conventional solid state reaction processes to synthesize and sinter NZP compounds require long cycle times and often do not result in high densification or single-phase materials. A novel microwave-assisted single-step procedure for the synthesis of a number of important members of the NZP family in less than 20 min has been developed [24]. This was made possible by identifying the high microwave absorption of sodium di-hydrogen phosphate monohydrate (NHPM) and using this as one of the reactants. The microwave products exhibited high phase purity and good crystallinity. The microwave-assisted synthesis was found to be simple and fast and minimized the loss of volatile species. Interestingly, the combination of microwaves and nonstoichiometric precursors (such as TiO_{2-x} and partially stabilized zirconia in place of regular zirconia) for the preparation of NZP was found to reduce the reaction temperatures and enhance the reaction rates even further. Specific compounds of $NaZr_2(PO_4)_3$, $NaSn_2(PO_4)_3$ and $NaTi_2(PO_4)_3$ were synthesized at temperatures as low as 600°C in 15 min. A microwave-assisted procedure for the rapid sintering of NZP materials has also been demonstrated, and more than 97% densification has been achieved at temperatures much lower than those required for conventional processing.

ZnO Varistor Ceramics

Zinc oxide (ZnO) varistors are electronic ceramic devices possessing highly nonlinear current–voltage characteristics that enable them to be used as voltage surge suppressors. Their typical electrical behavior is controlled by their microstructure (grain size and grain boundary chemistry) and the composition. The important role of the sintering procedure in developing the microstructure necessary to obtain desired nominal voltage (V_{nom}), energy handling capability and clamping performance of the zinc oxide varistors is well-recognized. We have microwave-sintered various types of zinc oxide varistors under different processing conditions. Microwave sintering of ZnO varistor samples showed significant reduction in the cycle time and substantial improvements in the electrical properties. Microwave-sintered samples exhibited better densities, finer grain size,and more uniform microstructures relative to conventionally processed samples (Table 4 and Figure 2) [25].

Electrical characterization of microwave-sintered samples showed higher-voltage and better clamping properties as compared to conventional sintering. A typical *V–I* curve obtained for V275LA4-type radials (type of varistors) is shown in Figure 3. Higher volts/mm response in microwaved samples signifies that (i) smaller devices with similar electrical properties can be used, and (ii) less material can be used to develop similar devices than those that can be obtained through conventional processes.

Multilayer varistors (MLV) based on ZnO as the passive ceramic material and platinum as inner electrodes [25] were also processed successfully using microwave technology. Total cycle time for microwave sintering of MLVs was only 3 hours as compared to 30 hours in a conventional process where considerable time is spent during the binder burnout stage. In contrast, there is no need for a binder burnout stage in the microwave process. A typical microstructure of the multi-layer varistors sintered in a batch microwave oven is shown in Figure 4. The platinum inner electrodes were intact and continuous. This signifies that metal electrodes can be sintered (processed) using microwaves when cofired with a ceramic. Another aspect of this research is that this multilayer composite

TABLE 4. TYPICAL DENSITY and GRAIN SIZE DATA FOR ZnO RADIALS OF TYPE V275LA4.

Sintering condition	Density (g/cm^3 (%TD))	Grain size (μm)
Conv.1250°C-2h	5.58 (98.4)	10
MW1100°C-15min	5.58 (98.4)	6.0
MW1200°C-15min	5.59 (98.6)	7.5
MW1300°C-15min	5.60 (98.8)	9.5

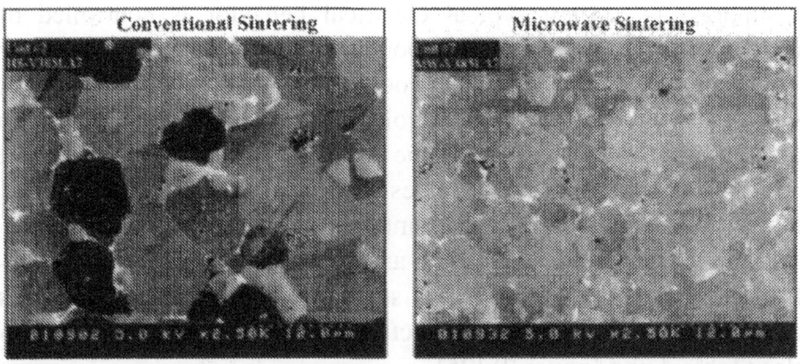

Figure 2. Typical microstructures of conventional and microwave-sintered ZnO radials showing more uniform and finer microstructure of microwave-sintered part than the conventional product.

structure has much better electrical properties than the conventionally processed structure. Data presented in Table 5 indicates that the microwave process yields higher volts/mm parts and gives better clamp performance. Leakage characteristics (IR) and other electrical properties relevant to varistors (peak pulse, energy, life test) have been measured, and the data exhibited a similar trend.

Multilayer Capacitors (MLCs)

Multilayer ceramic capacitors (MLCCs) are used in almost all areas of electronics as an important ceramic component, comprising a large industry that produces over 10^{12} components per year. There is a constant increase in the demand in functionality of electronic devices and requirements for miniaturization, high capacitive volumetric efficiency, high reliability and low production cost. In designing capacitors, dopants are added to a basic $BaTiO_3$ powder to control temperature coefficient of capacitance and magnitude of dielectric permittivity and to limit dielectric loss, maximize insulation resistance and degradation resistance.

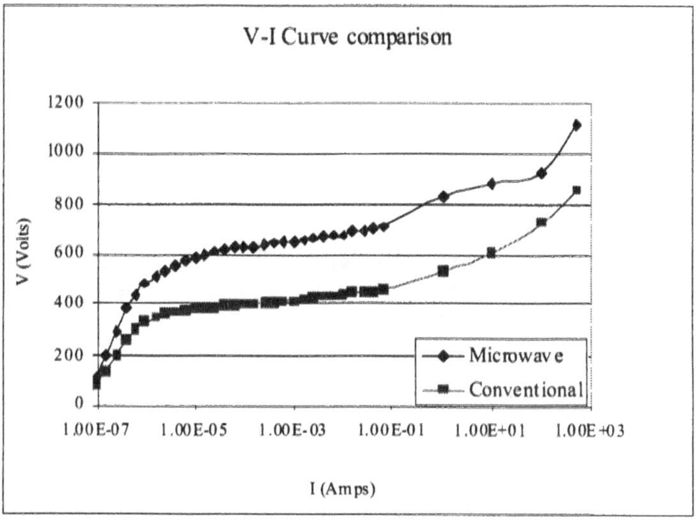

Figure 3. A comparison of *V–I* curves between microwaved- and conventionally sintered ZnO radials.

Over 80% of modern multilayer capacitors involve the cofiring of $BaTiO_3$ dielectric formulations with Ni inner electrodes. In order to cofire a dielectric with Ni, firings have to be conducted at a low p_{O_2}. Typical firings involve temperatures of ~1260–1300°C and atmospheres with $p_{O_2} \approx 10^{-10}$– 10^{-12} atms.

Microwave sintering of Ni-electrode MLCCs was conducted in an intermediate reducing atmosphere of $p_{O_2} \approx 10^{-6}$ atm. It was found that, in the temperature range around 1250°C, the MLCC chips were sintered well, resulting in dense and uniform parts without any delaminations or cracks [26]. Representative SEM micrographs of the fracture surface and free surface of the microwave-sintered

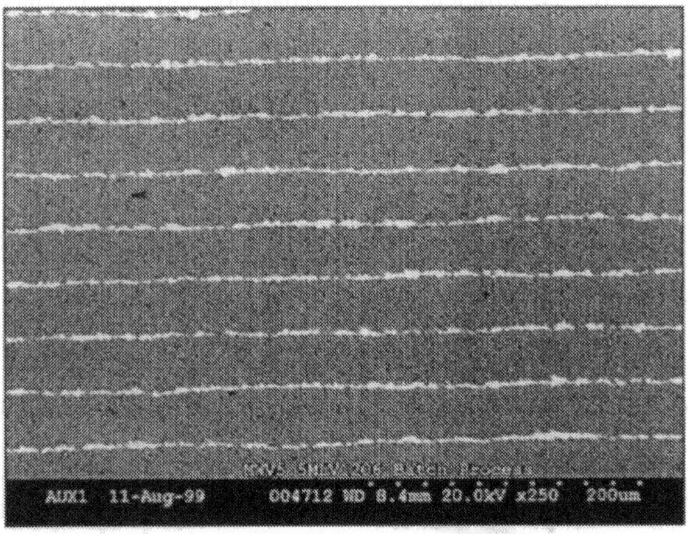

Figure 4. A typical micrograph of a microwave-sintered multilayer varistor.

TABLE 5. ELECTRICAL PROPERTIES OF MICROWAVE- AND CONVENTIONALLY SINTERED MLVS.

V_{nom}	Variance (V_{nom})	Clamp ratio	IR
	Microwave-sintered V5.5mla1206		
19.22	2.45	1.539	0.892
	Conventionally sintered V5.5mla1206		
8.23	2.98	1.927	0.920

Ni-electrode MLCC chips (cross section) are shown in Fig. 5 [27]. Highly dense microstructures were achieved by microwave sintering at about 1250°C. The internal electrodes were found to be continuous. Since the total processing time for microwave sintering was only about 10% of that for conventional sintering, the microstructures suggest that the densification kinetics of the MLCCs were substantially enhanced in the microwave sintering. The average grain size of the microwave-sintered MLCC matrix was 0.5–0.6 mm, similar to that of the conventionally sintered sample. The dielectric properties of the microwave-sintered MLCCs were comparable to the standard products sintered by the conventional process. Compared to the conventional process, microwave sintering conducted in a dry and static atmosphere, with heating rate one order of magnitude higher, heating time one order of magnitude shorter and sintering temperature 100°C lower than the conventional process, produced MLCC parts of similar quality and saved about 90% in processing time.

Transparent Ceramics

Transparency is a valuable optical property of materials. The nature of the material, including grain size, density, crystal structure, porosity and the grain boundary phase, is the main factor that influences the degree of transparency. Glasses are optically isotropic and have no grain boundaries and, therefore, possess excellent transparency. Most noncubic ceramics are anisotropic and polycrystalline. The grain boundaries in a ceramic strongly scatter light. Therefore, to convert a noncubic ceramic having grains larger than the wavelength of light into a transparent

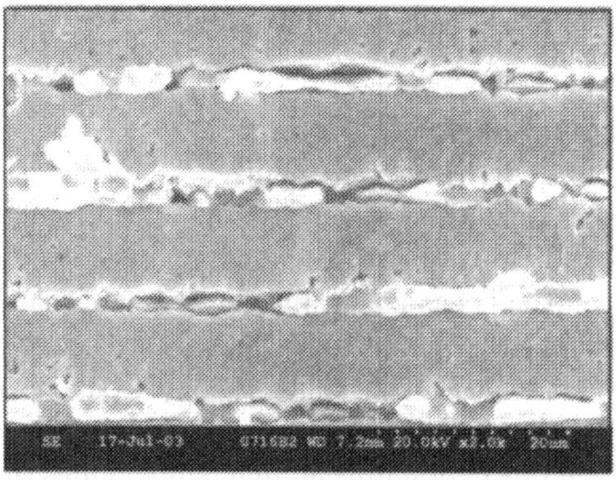

Figure 5. A typical microstructure of microwave-sintered MLCC showing continuity in Ni electrodes and very uniform and homogeneous sintering of $BaTiO_3$ ceramic layers. (This is how conventionally sintered electrodes look and show continuity in electrical conductivity.)

ceramic, one must have very low grain boundary volume and no inter- or intragranular porosity. However, if the grain size is smaller than the wavelength of light (0.4–0.7 μm), the light can transmit through the ceramic. Cubic ceramic materials, such as spinels and AlON, can be made into transparent ceramics even if the grain size is larger than the wavelength of light. To achieve transparency in a ceramic, one must control the grain growth, eliminate porosity and achieve complete densification. The conventional methods to fabricate fully dense and reasonably transparent ceramics involve high temperatures, lengthy sintering conditions and various complex processing steps, which make the processing of transparent ceramics very difficult and expensive. However, the microwave method has been successfully used to fabricate transparent ceramics due to its ability to minimize the grain growth and produce a fully dense ceramic in a very short period of time without utilizing high-pressure conditions [28]. Hydroxyapatite was fully sintered to form a transparent ceramic at 1100°C in 10 min by microwave processing [29]. Densification was shown to be critically dependent on the starting materials. Transparent ceramics of spinel and alumina were also fabricated [30, 31]. Fully dense alumina [32] and spinel ceramics using high purity and submicron sized powders were developed with a reasonable degree of transparencies on laboratory-typesmall samples at 1700°C sintered for 15 min in the microwave system. Fully transparent AlON ceramics were also made using a multimode microwave system at 1800°C [33]. Translucent ceramics of AlN, a well-known high-thermal-conductivity material, were also produced at 1900°C in 60 min using a microwave oven [34]. Recently, a MgO ceramic has been fabricated into a translucent form using nanometer-sized starting powders and LiF as a sintering aid [35]. Figure 6 shows some of the microwave-processed transparent and translucent ceramics of hydroxyapatite, alumina, AlN and ALON.

Microwave Processing of Nanomaterials

Nanostructured materials have been recognized to exhibit superior properties over standard materials due to the presence of nanometer range microstructures in the finished product. However, since the nanopowders are extremely active, and tend to agglomerate and coarsen during sintering at elevated temperatures, it has been a great challenge to sinter the nanophase materials to high densification and still retain the nanostructure in the sintered body. Conventional sintering usually requires holding the compacted bodies at the sintering temperature for several hours. The long processing time at high temperatures often results in substantial coarsening and grain growth. Since microwave processing requires much shorter soak times for sintering, thereby minimizing grain growth, it is possible to fabricate nanostructured materials using microwaves. We have used

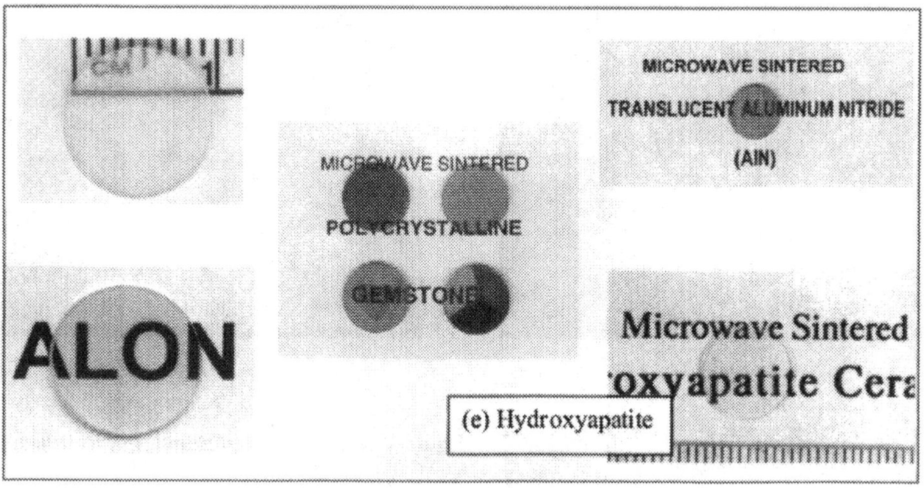

Figure 6. Various kinds of transparent and translucent ceramics fabricated in microwave: (a) pure alumina, (b) doped alumina, (c) AlN, (d) ALON and (e) hydroxyapatite.

nanophase MgO [35], TiO_2 and Cu metal powders to sinter these materials and produce very fine microstructures.

MgO was sintered at 1150–1400°C for 15 to 75 min. TiO_2 and TiO_{2-x} were sintered at 1150°C for 20 min. Nano-Cu was sintered at 868–920°C for 5–20 min. All samples were sintered in a 2.45 GHz multimode cavity. The ramp rate for these sintering experiments was 25-50°C/min. MgO ceramics were fabricated by using high-purity MgO nanopowder containing 0.2–4% LiF as a fugitive additive. LiF helped in improving the density to about 97% at 1150°C in 15 min. Figure 7(a) shows a typical microstructure of an MgO microwave-sintered nanoceramic. Dense TiO_2 ceramics were obtained by microwave sintering of nanophase TiO_2 sintered at 1150°C for 20 min. (Fig. 7(b). Nanophase Cu was microwave sintered at 920°C for 20 minutes to give about 83% density and very limited grain growth (Fig. 7(c)) with an average grain size of about 80 nm.

WC–Co-Based Ceramic–Metal Composites

WC-Co-based composites, due to their unique combination of hardness, toughness and strength, are universally used for cutting tools, machining of wear resistant metals, grinding, mining, geothermal as well as oil– and gas– drilling operations. Conventional methods for sintering WC with Co as a binder phase involve high temperatures (up to 1500°C) and lengthy sintering cycles (~24 hours) in order to achieve high densification. Such conditions favor undesirable

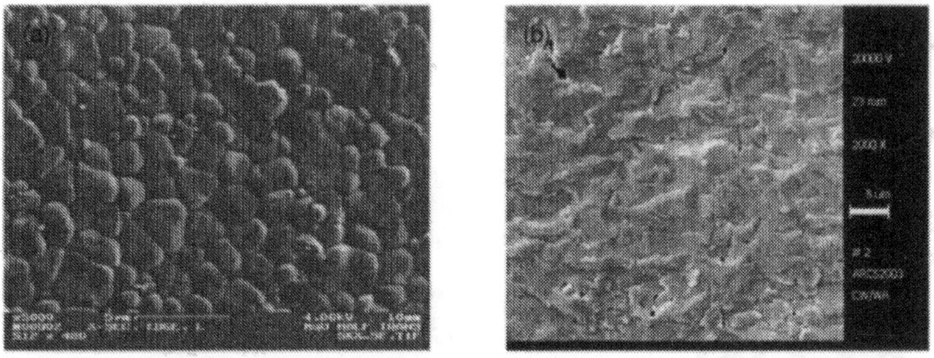

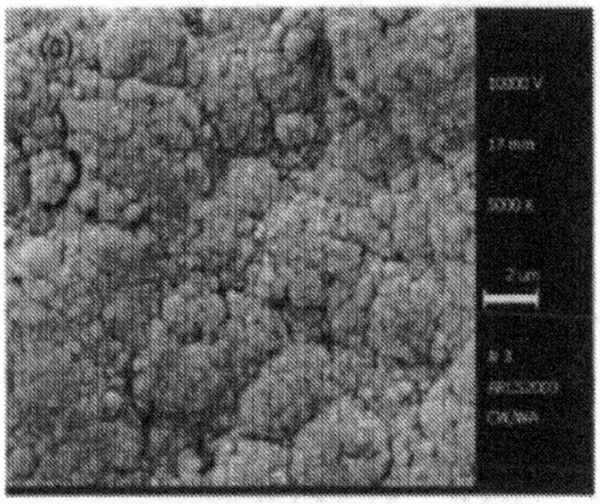

Figure 7. Typical microstructures of microwave-sintered nanophase powders of (a) MgO, (b) TiO_2 and (c) Cu metal.

WC grain growth in the presence of Co melts. Consequently, the mechanical strength and hardness of the tools are diminished. Often, additives such as titanium carbide (TiC), vanadium carbide (VC) and tantalum carbide (TaC), are used to suppress grain growth, but unfortunately such additives, deleteriously affect the mechanical properties of the tools and add substantially to the overall cost. Since microwave heating requires very little time to obtain nearly full densification, grain growth is relatively suppressed and finer microstructures can be obtained.

In 1991, J. P. Cheng in a Ph.D. thesis [36] first showed that WC–Co composites could be sintered in a microwave field. Gerdes and Willer-Porada [37] also reported the sintering of similar WC objects from normal-sized powders. They followed a reactive sintering route using a mixture of pure W, C and Co instead of normal sintering. Using a newly designed microwave apparatus, we were able to fully sinter WC commercial green bodies containing 12% and 6% Co [38], and we observed that microwave-processed WC–Co bodies (sintered at 1250–1320°C for only 10-30 minutes) exhibited better mechanical properties than the conventional parts, fine and uniform microstructure with little grain growth, and nearly full density without adding any grain-growth inhibitors [11, 39-40]. The microstructural examination of microwave-sintered WC-Co samples, in general, exhibited smaller average grain size than the conventionally sintered sample. Microwave-sintered parts have shown significant property improvements without varying the component materials, and without the addition of grain growth inhibitors. Because of this, WC parts produced by the microwave-sintering process exhibit an unprecedented improvement in abrasion resistance (15–30% better), erosion resistance (22% better) and corrosion resistance in 15% HNO_3 (20 times better) without any noticeable loss in hardness or fracture toughness. These improvements in the properties are believed to be due to the fine microstructure, uniform Co distribution and pure Co phase at the grain boundaries in microwave-sintered samples [41]. Figure 8 illustrates some commercial WC–Co parts fabricated successfully using microwave technology. Now several companies are commercially exploiting this technology.

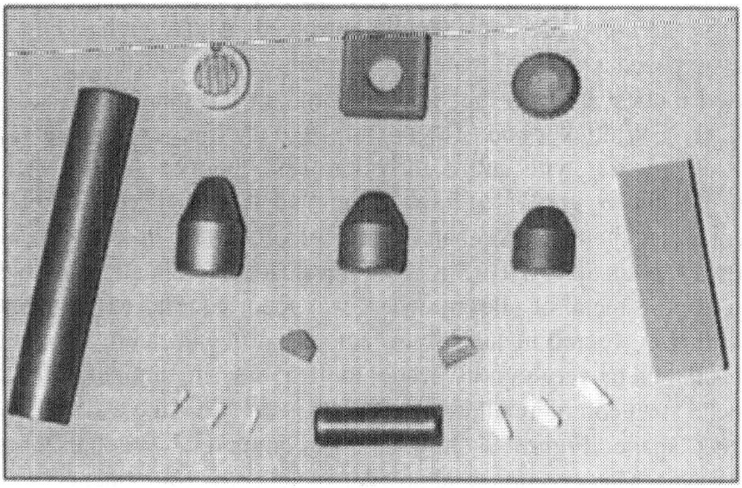

Figure 8. Microwave-sintered WC–Co parts of various shapes and sizes.

Microwave Processing (Sintering, Brazing and Melting) of Metallic Materials

Microwave processing of materials mostly has been confined primarily to ceramics, semi-metals, inorganic, and polymeric materials. There are few detailed reports on microwave processing of metallic materials. The main reason for this was due to the misconception that all metals reflect microwaves and/or cause plasma formation, and hence cannot be heated in a microwave field. However, the researchers did not realize that this observation is valid only for bulk metals at room temperature and not for powdered metals and/or for bulk metals at elevated temperatures. Now it has been proved that metallic materials in powder form do absorb microwaves, and if bulk metals are preheated to a temperature up to 400°C, they also start coupling with the microwave field and are heated very rapidly, so much so that they can be melted.

At 2.45 GHz, it is observed that the skin depth in the bulk metals is very low (of the order of few microns), and, hence, very little penetration of microwaves takes place. However, in metal powders (<100 μm), rapid heating in the material as a whole can occur. Work on microwave interaction with metallic powders was performed by Nishitani [43] who reported that, by adding few percent of electrically conducting powders such as aluminum, the heating rates of the refractory ceramics are considerably enhanced. Walkiewicz *et al.* [43] likewise simply exposed a range of materials, including six metals to a 2.4 GHz field, and reported modest heating (but not sintering) in the range from 120°C (Mg) to 768°C (Fe). Whittaker and Mingos (44) used the high exothermic reaction rates of metal powders with sulfur for microwave-induced synthesis of metal sulphides. Sheinberg *et al.* [45] heated Cu powders coated with CuO to 650°C, but did not sinter them. Narsimhan *et al.* [46] succeeded in heating Fe alloys in a microwave oven up to 370°C in 30 minutes. However, in all these studies, no sintering of pure metal or alloy powders was reported. Our experiments [13] have shown that powdered metals are in fact very efficient microwave absorbers and as a result can be very rapidly heated to their sintering temperatures. Some powder metal steel commercial parts of FC208 and FN208 have also been sintered to near net shape. Figure 9 shows some commercial products sintered using microwaves. Many commercial powder-metal (PM) components of various alloy compositions, including Fe and steel, Cu, Al, Ni, Mo, Co, W, WC, and Sn, have also been sintered with microwaves producing essentially fully dense bodies.

Microwave sintering of PM green bodies comprising various metals, steels and metal alloys produced highly sintered bodies in a very short period of time [47]. Typically, the total cycle time was about 90 min with sintering temperature ranges of between 1100–1300°C and soak times of 5 to 60 min. The mechanical

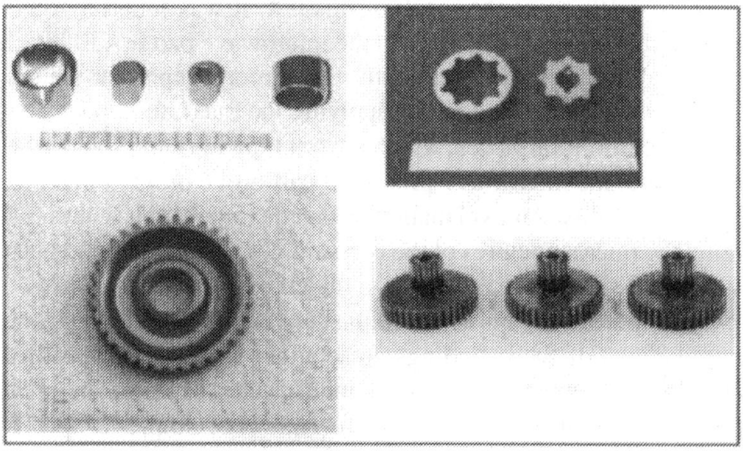

Figure 9. Microwave-sintered commercial metals.

properties, such as the modulus of rupture (MOR) and hardness, of microwave-processed samples were much higher than the conventional samples. As an example, copper steel (MPIF FC-0208 composition) was successfully sintered by the microwave technique to obtain good sintered density, hardness, flexural strength and near net dimensions, thus yielding equivalent or even sometimes superior mechanical properties over those samples produced by conventional sintering [47]. In this material, a Rockwell B hardness (HRB) as high as 82 ± 2 was obtained for samples sintered at 1260°C for 5 min soaking in a flowing forming gas atmosphere. The maximum flexural strength of 1077 ± 10 MPa was obtained for microwave-sintered samples at 1140°C for 20 min.

An examination of the microstructures and porosity distributions of the conventional and microwave-sintered samples reveals that microwaved samples had more uniform microstructures than the conventional samples in which the core had more pores than the surface. In some microwave-processed samples, the cores were slightly denser than the surface. This is typical of most ceramics sintered in a microwave, indicating that heat transport is from the inside out and that the interior of the material may be hotter than the surface. However, using our modified microwave-sintering system, we were able to obtain sintered samples with uniformly dense core and edges. An important distinction in the microstructures of conventional and microwave-sintered samples was that the pores in the microwave-sintered samples had more rounded edges than those of the conventional sample. It is commonly known that sintered products exhibit an increasing ductility when the pore shape is more spherical. This was proved by conducting a standard test for measuring ductility and toughness of hollow cylindrical samples of FC208. Figure 10 shows the result of this test. It was found that the conventional part failed at a load of 320 lbs and the microwave part at 430 lbs, indicating an increase of about 30% in the strength. But a more important feature was the manner in which the parts failed after applying the maximum load. The conventional part broke into four pieces, which is very typical of the standard PM parts. On the other hand, the microwave-processed part broke into two flat pieces, indicating a higher ductility. An explanation of this distinguishing feature was sought by analyzing the microstructures of the two samples. It was revealed that the pores in the microwave-sintered samples had rounded edges (Fig. 10(c)) in contrast to the sharp-edged porosity (Fig. 10(d)) in the conventionally sintered samples.

The implications of these findings are obvious in the field of powder-metal technology. Metal powders are used in industry for diversity of products and applications. The challenging demands for new and improved processes and materials of high integrity for advanced engineering applications require innovation and newer technologies. Finer microstructures and near theoretical densities in special PM components are still elusive and challenging. Increasing cost is also a concern of the industry. Researchers are looking for newer technologies

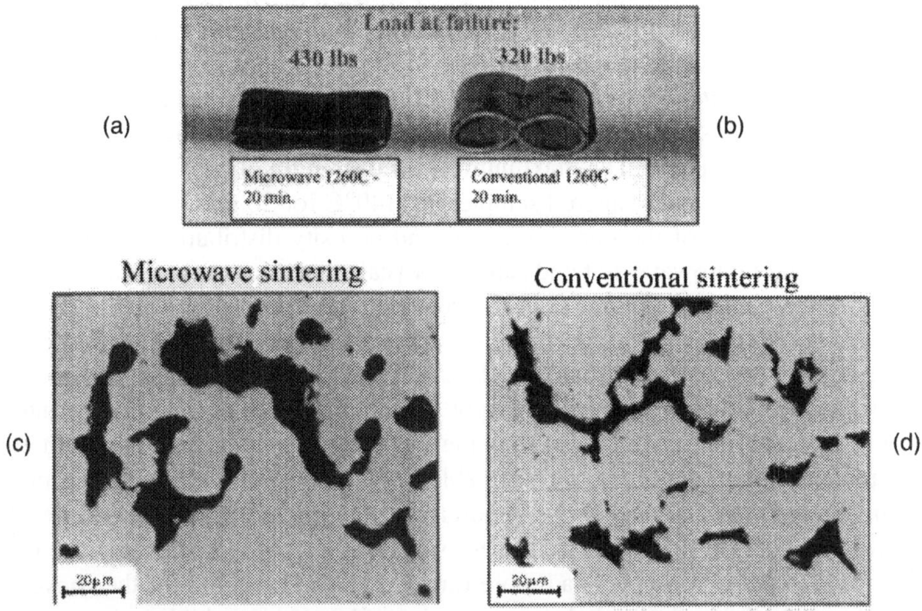

Figure 10. Ductility test results performed on FC-208 steel sample: (a) microwave-sintered and (b) conventionally sintered. Pore shape in (c) microwave sample and (d) conventionally sintered sample.

and processes to meet these demands. The developments as reported in this paper using microwave processing may offer a new method to meet these demands of producing better microstructures and properties in powder-metal products.

As mentioned earlier, microwaves selectively heat powder metals and reflect from bulk metals at room temperature. This feature has been exploited to braze and join bulk metals using powdered-metal/alloy braze materials. We have joined steels, W bulk metals and also brazed super alloys. An example is shown in Figure 11 in which regular steel and cast-iron parts have been joined in a microwave field in 2–3 min using braze powder. The joint is almost perfect, as indicated by the microstructural examination of the sample. We have also repaired and brazed super-alloy-based turbine blades. This work can also be extended to join metal to a ceramic, and also developing ceramic coating on steels or metal coatings on ceramics.

Summary

In the last 5–10 years, many significant developments and advances have taken place in the field of microwave processing of materials. Traditional and advanced ceramics and composites have been fabricated using microwave technology with potential of huge savings in time and energy, with substantial improvements in the properties. It has been demonstrated that, in the case of WC–Co processing,

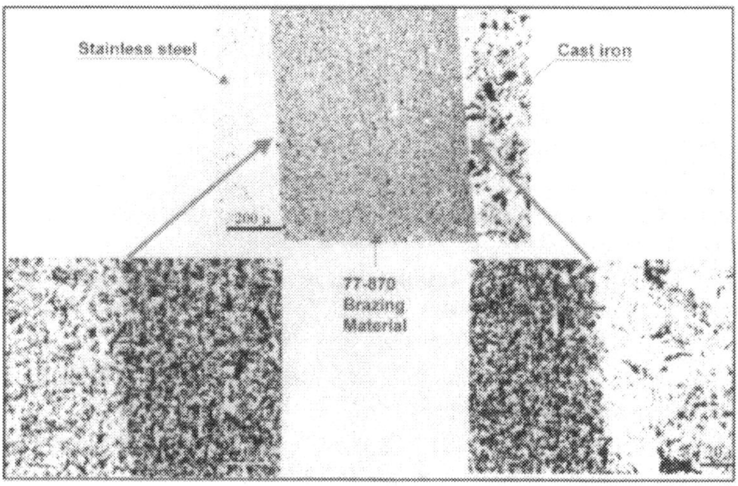

Figure 11. Microwave joining of stainless-steel and cast-iron using braze powder.

microwaves can reduce the cycle time to about one-tenth the cycle time required by conventional means, and produces product with improved abrasion, erosion and corrosion resistance. Refractory oxides, such as alumina and zirconia, have been sintered using microwaves to near full density at much lower temperatures and in dramatically shorter sintering times than normally required in a conventional method. The fabrication of transparent ceramics, including alumina, spinel, mullite, ALN and AlON, has been achieved in a single-step process at ambient pressure. The most recent significant development in microwave processing has been the sintering, brazing, joining and melting of metals.

REFERENCES

1. D. Clark, W.H. Sutton, "Microwave processing of materials", *Annu Rev Mater Sci* 26, 299-331 (1996).
2. R.F. Schiffman, "Commercializing microwave systems: Paths to success or failure," *Ceramic Transactions,* 59, 7-17 (1995).
3. J.D. Katz, "Microwave sintering of ceramics", *Annu Rev Mater Sci* 1992, 22, 153-70 (1992).
4. W. Sutton, "Microwave processing of ceramic materials", Am Ceram Soc Bull 68 376-86 (1989).
5. W. Sutton, "Microwave processing of ceramics - an overview", *Mat Res Soc Symp Proc* 269 3-19 (1992).
6. *Microwave Processing of Materials II,* edited by Snyder Jr WB, Sutton WH, Iskander MF, Johnson DL, Mats Res Soc Proc Pittsburgh USA Vol. 189 (1990)
7. *Microwave Processing of Materials III,* edited by Beatty RL, Sutton WH, Iskander MF, Matls Res Soc Proc Pittsburgh USA Vol. 269 (1992)
8. *Microwaves: Theory and Application in Materials Processing II,* edited by Clark DE, Tinga WR, Laia Jr JR, Ceram Trans Vol. 36 (1993)
9. D.K. Agrawal, "Microwave Processing of Ceramics", *Current Opinion in Solid State & Materials Science,* 1998, 3:480-485
10. "Recent Development of Microwave Kilns for Industries in Japan," M. Sato, T. Mutoh, T. Shimotuma, K. Ida, O. Motojima et al, in *Proc. 3rd World Congress on Microwave and Radio Frequency Applications* eds. 281.189 (2003) ACS Publ.
11. P.D. Gigl, E. Breval , J. Cheng, D.K. Agrawal, R. Roy, "Structure properties of microwave sintered cemented tungsten carbide materials", Presented at First World Congress on Microwave Processing, January 5–9 1997, Orlando , Florida.
12. J. Cheng, R. Roy and D. Agrawal, "Microwave heating behaviors of different materials in electric and magnetic fields," *Mat. Res. Innovat.,* 5, 170-177 (2002).
13. R. Roy, D. Agrawal, J. Cheng, and S. Gedevanishvilli, "Full sintering of powdered-metal bodies in a microwave field," *Nature,* 399, 668-670 (1999)
14. A. Cherradi, G. Desgardin, J. Provost, and B. Raveau, "Electric & magnetic field contributions to the microwave sintering of ceramics," *Elelctroceramics IV,* Vol. II, (eds. Wasner, R., Hoffmann, S., Bonnenberg, D., & Hoffmann, C.) RWTN, Aachen, 1219-1224 (1994).

15. Kristen H. Brosnan, G.L. Messing and D.K. Agrawal, "Microwave Sintering of Alumina at 2.45 GHZ," *J. Am. Ceram. Soc.* 86[8] 1307-12 (2003).

16. J. Cheng, Y. Fang, D. Agrawal, R. Roy, and P.S. Jayan, "Continuous microwave sintering of alumina abrasive grit," *J. Mats. Proc. Tech.*, 108 26-29 (2000).

17. J. Cheng, Y. Fang, D. K. Agrawal, and R. Roy, "Densification of large size alumina objects by continuous microwave sintering," *Ceramic Transactions,* Vol 59: Microwave: Theory and Applications in Materials Processing III, Eds. D.E. Clark, D.C. Folz, S.J. Oda, and R. Silberglitt, pp 457-463, ACS publ. (1995).

18. B. Vaidhyanathan, A.P. Singh, D.K. Agrawal, T.R. Shrout and R. Roy, "Microwave Effects in Lead Zirconium Titanate Synthesis: Enhanced Kinetics and Changes Mechanism," *J. Am. Ceram. Soc.* 84(6) 1197-202 (2001).

19. S. Rhee, D. Agrawal, T. Shrout, and M. Thumm, "Investigation of High Microwave Frequency (2.45 GHz, 30 GHz) Sintering for Pb-based Ferroelectrics and Microscale Functional Devices," *Ferroelectrics,* 261 15-20 (2001).

20. M.D. Mathis, "Microwave synthesis using multicomponent and multiphasic systems {Ph.D. Thesis)," University Park, The Pennsylvania State University, 1997.

21. B. Vaidhyanathan, D.K. Agrawal, T.R. Shrout and Y. Fang, "Microwave Synthesis and Sintering of $Ba(Mg_{1/3}Ta_{2/3})O_3$" *Mats Lett* 42, 207-211 (2000).

22. Y. Fang, D. Agrawal, W. Hackenberher, T.R. Shrout, "Enhanced Sintering of High-Q $Ba(Zn_{1/3}Ta_{2/3})O_3$ Ceramics," *J. Am. Ceram. Soc.* (Submitted).

23. E. Breval and D. K. Agrawal, "Thermal Expansion Characteristics of NZP, $NaZr_2P_3O_{12}$-Type Materials: A Review," *Br. Ceram. Trans.*, 94 [1] 27-32 (1995).

24. B. Vaidhyanathan, D.K. Agrawal and R. Roy, "Microwave synthesis and sintering of [NZP] compounds," *J. Amer. Cera. Soc.*, 87[5] 834-839, (2004).

25. Production of passive devices, D. Agrawal, R. Raghavendra, B. Vaidhyanathan, US Patent #6, 399,012, June 4, 2002.

26. Y. Fang, D. Agrawal, M. Lanagan, T. Shrout, C. Randall, M. Randall, A. Henderson, "Ceramic Materials and Multilayer Electronic Devices," in *Ceramic Materials and Multilayer Devices,* Cer. Trans. Vol 150 Amer. Cer. Soc. Publ. Westerville Ohia USA pp359-366 (2004).

27. Y. Fang, M. Lanagan, D. Agrawal, G.Y. Yang, C. Randall, T. R. Shrout, A. Henderson, M. Randall, A. Tajuddin, "An Investigation Demonstrating the Feasibility of Microwave Sintering of Base-Metal-Electrode Multilayer Capacitors," *J. Amer. Ceram. Soc.* (In review)

28. J. Cheng, D.K. Agrawal, R. Roy, " Fabricating transparent ceramics by microwave sintering," *Am Ceram Soc Bull* 79[9], 71-74(2000).

29. Y. Fang, D.K. Agrawal, D.M. Roy,R. Roy, "Fabrication of Transparent Hydroxyapatite Ceramics by Microwave Processing", *Mater Lett* 23 147–51 (1995).

30. Y. Fang, D.K. Agrawal, D.M. Roy, R. Roy, "Transparent mullite ceramics from diphasic aerogels by microwave and conventional processing," Mat Lett 28 11-15 (1996).

31. T. Saji, "Method for manufacturing artificial sintered gemstone," Japan Patent No. 7-187760 (1995)

32. J. Cheng, D. Agrawal, Y. Zhang and R. Roy, "Microwave Sintering of Transparent Alumina," *Mats. Lett.*, 56, 587-592 (2002)

33. J. Cheng, D. Agrawal, Y. Zhang and R. Roy, "Microwave Reactive Sintering to Fully Transparent Aluminum Oxynitride (AlON) Ceramics," *J. Mat. Sci. Lett* 20, 77-79 (2001).

34. J. Cheng, D. Agrawal, Y. Zhang and R. Roy, "Development of translucent aluminum nitride (AlN) using microwave sintering process," *J. Electroceramics,* 9, 67-71 (2002).

35. Y. Fang, D. Agrawal, Ganesh Skandan and M. Jain, "Fabrication of Translucent MgO Ceramics using Nanopowders," *Mat. Lett.* 58, 551-554 (2004).

36. J. Cheng, "Study on microwave sintering technique of ceramics materials." Ph.D. thesis 1991, Wuhan University of Technology, China

37. T. Gerdes, M. Willert-Porada, "Microwave sintering of metal-ceramic and ceramic-ceramic composites", *Mat Res Soc Symp Proc* 347 531-537 (1994).

38. J. Cheng, D.K. Agrawal, S. Komarneni, M. Mathis, R. Roy, "Microwave processing of WC-Co composites and ferroic titanates", *Mat Res Innova* 1 44-52 (1997).

39. R. Roy, D.K. Agrawal, J. Cheng, "An improved process and apparatus for the preparation of particulate or solid parts," U.S. Patent # 6,004,505 (Dec. 21, 1999).

40. D.K. Agrawal, J. Cheng, R. Roy, P. Seegopaul, "Microwave sintering of nanocomposite tungsten carbide-cobalt hard materials," Presented at First World Congress on Microwave Processing, January 5–9 1997, Orlando, Florida

41. D. Agrawal, A.J. Papworth, J. Cheng, H. Jain and D.B. Williams, "Microstructural Examination by TEM of WC/Co Composites Prepared by Conventional and Microwave Processes," *Proc. 15th Intl. Plansee Seminar 2001,* Eds. G. Kneringer, P. Rodhammer and H. Wildner, Plansee Holding AG, Reutte Vol 2, 677-684 (2001).

42. T. Nishitani, "Method for Sintering Refractories and an Apparatus therefor," U.S. Patent # 4,147,911 (1979).

43. J.W. Walkiewicz, G. Kazonich and S.L. McGill, "Microwave Heating Characteristics of Selected Minerals and Compounds", *Minerals and Metallurgical Processing,* 1988, vol.5, p.39.

44. Whittaker, A. G., Mingos, D. M., "Microwave-assisted solid-state reactions involving metal powders," *J. Chem. Soc. Dalton Trans.* 2073-2079 (1995).

45. Sheinberg, Meek & Blake, "Microwaving of Normally Opaque and Semi-opaque Substances" U.S. Patent # 4,942,278 (July 17, 1990).

46. Narsimhan, Arvidsson, Rutz, & Porter, " Methods and Apparatus for Heating Metal Powders" U.S. Patent # 5,397,530 (March 4, 1995).

47. R.M. Anklekar, D.K. Agrawal, and R. Roy, "Microwave Sintering and Mechanical Properties of P/M Steel", *Powder Metal.* Vol. 44[4], 355-362 (2001).

Microwave Plasma Synthesis of Nanoparticles: Application of Microwave to Produce New Materials

D.V. Szabó, D. Vollath and W. Arnold

Abstract

Nanoparticles and coated nanoparticles with particle sizes in the range from 5 to 10 nm and narrow particle size distribution are produced by a microwave plasma process. This plasma process uses a gas phase reaction, performed in a quartz glass tube passing a microwave cavity. The microwave components necessary for this process, such as magnetrons, wave-guides, directional couplers, tri-stub-tuners, or isolators are commercially obtainable. Two standard frequencies. 0.915 and 2.45 GHz, are available, using the standard type R9 and R26 wave guides. The relatively small dimensions of these wave-guides limit the diameter of the reaction tube in the intersection with the wave-guide. Therefore, a special TE_{11}: cavity with cut-off tubes, characterized by high microwave efficiency, has been developed. A specially developed power variator makes it possible to reduce the microwave power below the minimum power of the magnetron. Careful adjusting and tuning of the equipment is necessary.

Chemical reactions, thwarted under normal conditions by kinetics, can be performed in the microwave plasma. For this process volatile compounds such as chlorides, carbonyls or metallorganics are used as precursors, and mixtures of Ar/O_2, N_2H_2, Ar/H_2S or Ar are used as reaction gas. The precursors are evaporated and introduced into the reaction tube just in front of the plasma zone Examples of particles produced with this process are Al_2O_3, ZrO_2, TiO_2, Fe_2O_3, ZrN, BN, WS_2, MoS_2, and $ZrSe_2$. The particles acquire electrical charges in the plasma and, therefore, repel each other after leaving the plasma zone This makes it possible to use cascaded plasma sources for the production of ceramic coated, double ceramic coated particles, or polymer coated nanoparticles of the core/shell type. Such particles are very interesting as starting material for nanocomposites because the special "nanoproperties" as e.g. superparamagnetism, due to the small particle size, can be achieved for technical applications.

Introduction

Nanomaterials generally are defined as materials with particle sizes below 100 nm. Additionally they exhibit special physical properties related to the small dimension of the particles [1] such as superparamagnetism. These interesting properties mostly are properties of single, isolated particles. If nanoparticles interact together, as in a consolidated part for technical application, they will behave like a large particle. Due to interactions, the special particle size dependent properties are lost. Additionally, it will be nearly impossible to sinter such particles without grain growth. This leads to the loss of the special properties. The solution of these problems is to reduce the interaction between nanoparticles by keeping them "on distance". This can be obtained by using coated nanoparticles as the starting material for nanocomposites. Therefore a very versatile process, the microwave plasma process, has been developed [2,3]. With this process it is possible to produce oxides, nitrides, sulfides, or selenides in one reaction step, as well as ceramic or polymer coated nanoparticles and metal decorated nanoparticles with narrow particle size distribution in a two-step process.

Fundamentals of th Microwave Plasma Process

The microwave plasma process uses gas phase reactions. It draws its advantages from the fact that reactive molecules are partly ionized or dissociated in a plasma. This is a highly activated state and, as a consequence, thermodynamically possible but under normal conditions kinetically thwarted reactions can be performed at reduced temperature. The energy transfer E to a charged particle with a mass m in an oscillating electrical field with the frequency f is:

$$E = \frac{1}{m - f^2} \tag{1}$$

As the mass of electrons is much lower than the mass of ions or radicals, significantly more energy is transferred to electrons than to the latter ones. Thus, the "temperature" of the free electrons is much higher than the "temperature" of the ions [4]. The higher the frequency, the lower the energy transfer, thus lower temperatures can be achieved with higher frequencies. In a plasma, the collision frequency z with uncharged molecules and atoms has to be taken into account. This leads to the modified relation:

$$E = \frac{1}{m} \cdot \frac{z}{f^2 - z^2} \tag{2}$$

As the collision frequency increases with increasing gas pressure, the energy transfer is a function of the gas pressure. The energy transfer increases with the gas pressure for $z < f$, exhibits a maximum at $f = z$, and decreases for $z >$ Vnom$f$.

Two standard industrial frequencies are available: the 2.45 GHz and 0.915 GHz. Due to the significantly different energy content of the different species a non-equilibrium plasma is obtained with relatively low overall temperatures in the system. The temperatures (measured directly behind the plasma zone) can be adjusted in the range of 100 to 500°C in the case of the 2.45 GHz equipment and in the range of 500 to 800°C for the 0.915 GHz equipment. These temperatures are low compared to conventional gas phase reactors. The pressure in the system is adjusted in the range from 10 to 50 mbar. These experimental conditions lead to residence time of the particles in the plasma zone in the range of milliseconds.

As a unique feature of this process the particles formed in the plasma carry electric charges of the same sign. Therefore they repel each other impeding particle growth by agglomeration. As a consequence the particle size distribution of the powders is very narrow. Because of the electric charges it is possible to coat the particles individually with a second phase in a cascaded process.

Microwave Equipment

Magnetrons (2.45 and 0.915 GHz), wave guides (R26, R9) and some other passive components such as a tri-stub-tuner, isolator, or directional couplers are commercially available. For the synthesis of nanoparticles a special microwave cavity, using a rotating TE_{11} mode has been developed [5,6]. This cavity has the advantage of easy plasma ignition and high microwave efficiency in a broad range of gas pressure, as shown in figure 1. The length of the reaction zone can be adjusted to a multiple of the half wavelength of the microwaves. Pinching of the plasma is not observed. The diameter of the reaction tube is selected up to 80 mm in the case of 0.915 GHz magnetrons, using a special absorber with Bragg structure for effective prevention of microwave leakage. This special design enables the production of about 0.5 mole/h in a laboratory system. With this cavity and an optimized design of the Bragg structure it is possible to use reaction tubes up to 150 mm in diameter, which is interesting for an industrial up scaling.

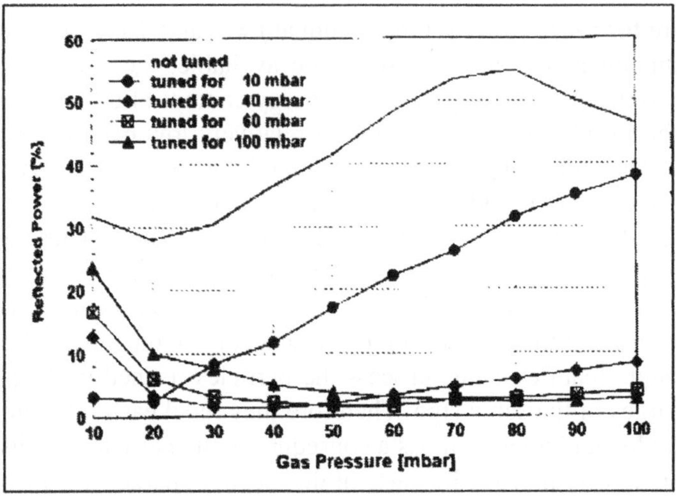

Figure 1. Reflected power of a TE_{11} cavity, tuned for various gas pressures.

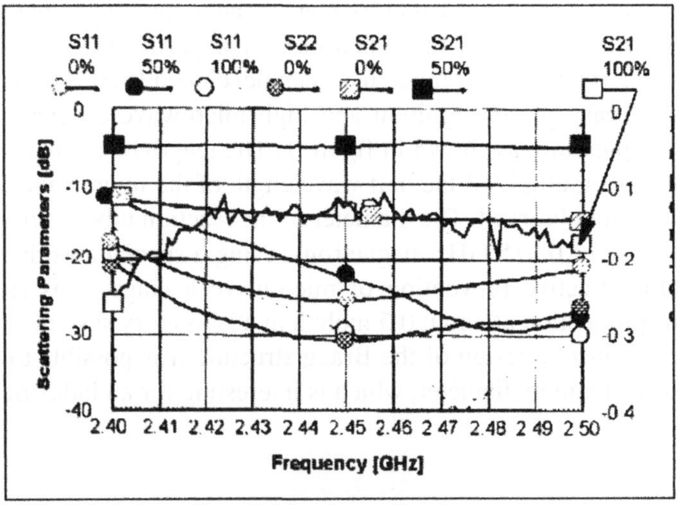

Figure 2. Scattering parameters of a 2.45 GHz power variator. S_{11} and S_{22} denote reflected, S_{21} denotes transmitted power. 0, 50 and 100% mean flap closed, half and fully opened.

A power variator is used to reduce the microwave power below the minimum power of the magnetron to reduce the temperature. In principle, this component is a wave-guide with a flap that can be opened or closed continuously. Figure 2 shows the scattering parameters in dB, characterizing a 2.45 GHz power variator. It can be seen that the reflections for all positions of the flap are below 1% (left scale). A water load absorbs the reflected power. If the flap is open, a transmission of 97% is reached (right scale, highlighted by arrow), if the flap is half closed, the transmission is reduced to 37% of the microwave power. With a completely closed flap the transmission is below 5%. The theoretical value of 0% is not reached, since the flap is contacted mechanically and not galvanically.

Synthesis of Nanoparticles and Coated Nanoparticles

The plasma is generated in a reaction tube made of quartz glass, passing through the microwave cavity. Precursors like chlorides, carbonyls or metal-organic compounds are evaporated outside of the reaction tube and introduced into the system directly in front of the plasma zones. As reaction gas mixtures of $Ar/20vol\%O_2$, $N_2/4vol\%H_2$, or $Ar/1vol\%H_2S$ are used, depending on the particles to be produced.

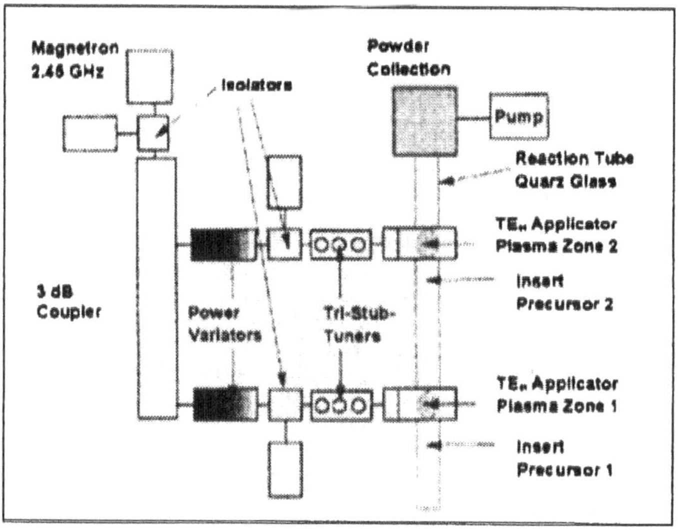

Figure 3. Schematic experimental set-up for the synthesis of ceramic-coated nanoparticles.

Taking advantage of the electrical charges of the as-produced nanoparticles and the relatively low temperatures it is possible to coat the individual particles in a second step. This coating either is a layer of a ceramic [7] or a polymer [8]. The particles formed in step 1 act as nuclei for the condensation of particles formed in step 2. For the polymer coating the experimental arrangement is modified. A monomer is evaporated and introduced directly after the plasma zone. It condenses on the nanoparticles and polymerizes by the temperature in the system and UV radiation from the plasma.

The schematic experimental set-up for the synthesis of ceramic coated nanoparticles is shown in figure 3. The two microwave branches are adjusted independently. using the power variator and the tri-stub-tuners. Isolators in each branch thwart a mutual influence of the microwave adjustment. Particle collection is performed by thermophoresis.

Morphology of Coated Nanoparticles

Figures 4 and 5 show electron micrographs of ceramic coated and polymer coated nanoparticles, respectively [9,10] The cores (dark) and the coating of the individual nanoparticles can clearly be seen. Core and coating must not have a mutual solubility, as this is the case for HfO_2 and Al_2O_3 (figure 4). The lattice fringes of the HfO_2 particles are the visible ones The coating with amorphous Al_2O_3 is about 0 5 nm thick The γ-Fe_2O_3 particles shown in figure 5 are 5–6 nm in diameter, and the polymer coating is 2–3 nm thick. This corresponds to approximately 80 vol%. Therefore it can bee seen easily.

Solids of these powders are prepared by cold pressing and sintering at 1200°C for 15 min in the case of ceramic/ceramic nanoparticles or by "hot-pressing" of ceramic/polymer nanoparticles in the temperature range from 70 to 140°C at 0.7 MPa. Figures 6 and 7 show the typical structures of consolidated parts made of nanocomposites [11,12].

Starting powder for the material shown in figure 6 were ZrO_2 particles. 6 - 8 nm in diameter, coated with 2 nm thick Al_2O_3. After sintering the ZrO_2 cores exhibit a particle size around 10 nm. This means that nearly no grain growth is observed in the ZrO_2 and therefore potentially interesting physical properties of the core are preserved in a densified body.

In the case of the "hot-pressed" Fe_2O_3/polymer nanocomposite (figure 7), made of powder containing 50 vol% polymer, grain growth of the Fe_2O_3 is not observed. It is important to point out that the particle size distribution is very narrow. Striations visible in the micrograph are artifacts from the sample preparation technique.

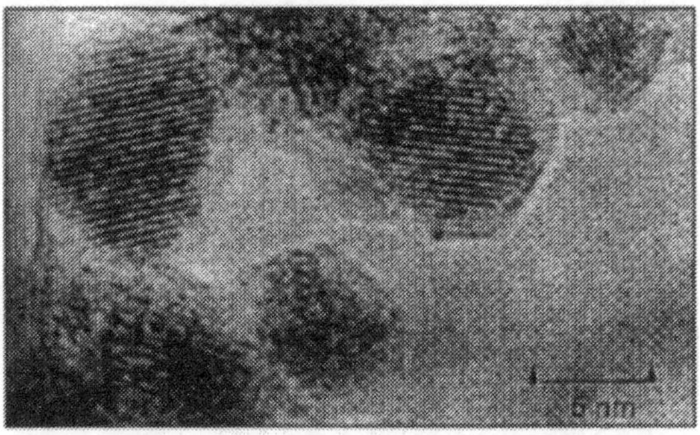

Figure 4. High-resolution electron micrograph of individual HfO_2 nanoparticles coated with amorphous Al_2O_3.

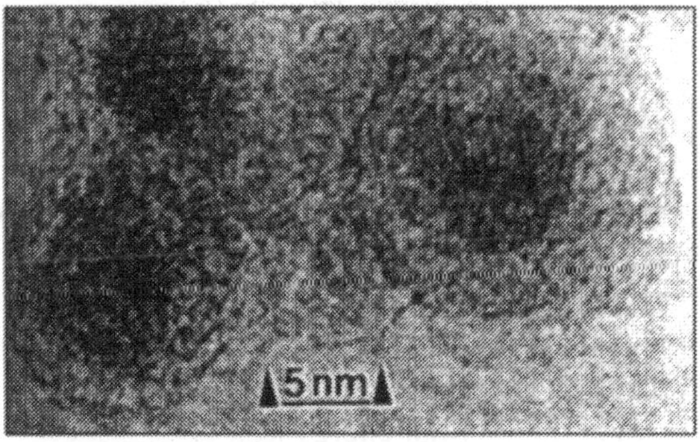

Figure 5. High-resolution electron micrograph of polymer-coated γ-Fe_2O_3 nanoparticles.

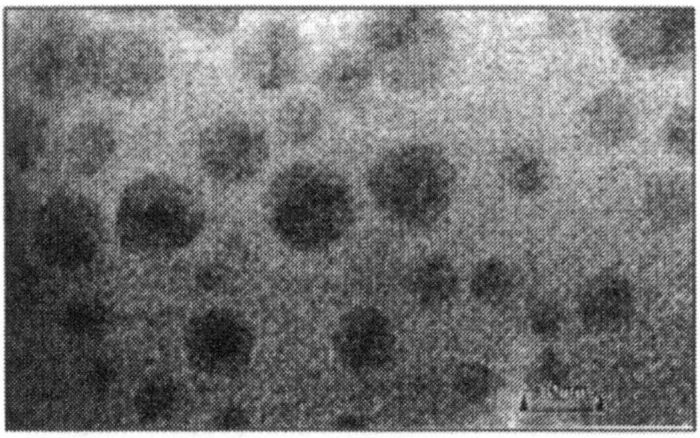

Figure 6. Sintered zirconia–alumina nanocomposite. The zirconia particles remain separated by the surrounding alumina.

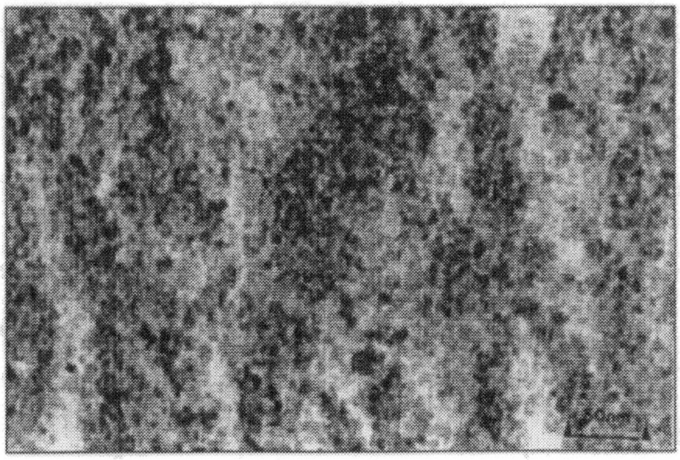

Figure 7. Electron micrograph of a 20 nm thin section of γ-Fe_2O_3/ polymer nanocomposite.

Properties of Coated Nanoparticles

For technical application, one of the very interesting particle size depending physical properties of nanomaterials is superparamagnetism. This means that the vector of magnetization is fluctuating thermally and the material is free of hysteresis in the magnetization curve. A particle is superparamagnetic if the energy of magnetic anisotropy Kv (K: constant of magnetic anisotropy, v: volume of the particle) is smaller than the thermal energy kT:

$$Kv < kT \tag{3}$$

This typical "nano-property" can only be obtained in a technical part by using coated nanoparticles. Due to the coating the dipole-dipole interaction between the magnetic particles is reduced. Figure 8 shows the static magnetization curves of uncoated and ZrO_2-coated y-Fe_2O_3 [11]. It can easily be seen that the uncoated material exhibits a hysteresis, whereas the coated material is free of any hysteresis.

The ultimate proof of superparamagnetism is the Mössbauer-spectroscopy. Normal ferrimagnetic material shows a sextet of lines, whereas superparamagnetic material is characterized by the quadrupole split doublet. Figure 9 compares the Mössbauer spectra of a partly superparamagnetic non-coated (left) and a superparamagnetic polymer coated nanoscaled ferrite (right) [12]. The coated material exhibits no sextet; therefore it is superparamagnetic.

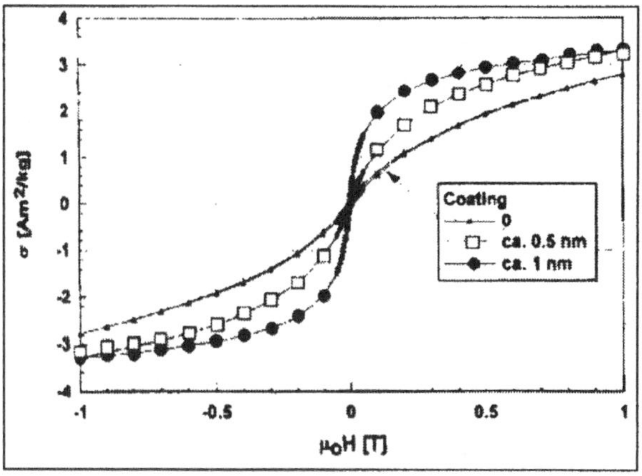

Figure 8. Comparison of magnetization curves of non-coated and different zirconia-coated γ-Fe_2O_3 nanoparticles at 300 K.

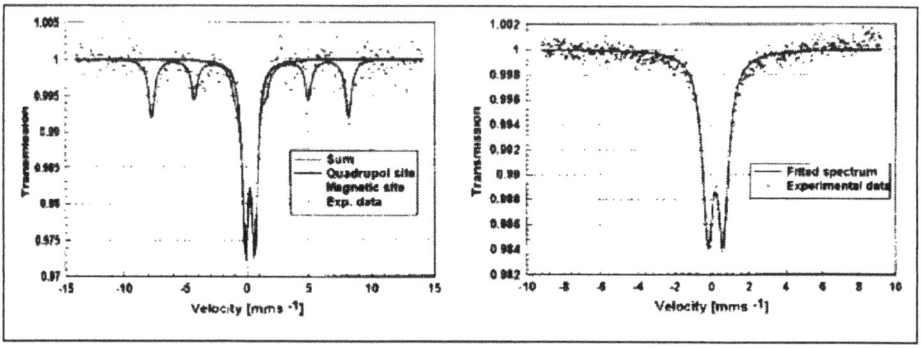

Figure 9. Mössbauer spectra of a partly superparamagnetic γ-Fe_2O_3 nanopowder (left) and a superparamagnetic γ-Fe_2O_3/polymer nanocomposite.

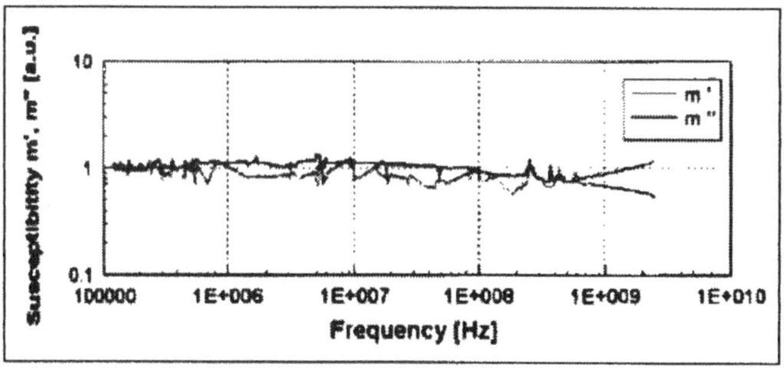

Figure 10. Real and imaginary part of the susceptibility of a $MnFe_2O_4$/PMMA nanocomposite as a function of frequency.

A very important application of superparamagnetic γ-Fe_2O_3 in medicine is the use as drug carrier, for contrast enhancement in NMR (nuclear magnetic resonance) or in local hyperthermia for cancer treatment.

The property essential for technical application in telecommunication is the susceptibility. especially in the high frequency range. Preliminary results of the frequency dependence of the complex susceptibility are shown in figure 10 for a ferrite/polymer nanocomposite. It can be seen clearly that no resonances are observed up to the GHz range.

Summary

The microwave plasma synthesis is well suited to obtain nanosized ceramic and coated particles with particle diameters from 5–10 nm and narrow particle size distributions. Using TE_{11} cavities, the process has been scaled up for laboratory equipment to mole batches. Further up scaling is feasible. It is possible to obtain bulk solids made from coated nanoparticles with special "nano-properties", as superparamagnetism. These superparamagnetic nanocomposites have a high potential of application in the medical and in communication industries.

REFERENCES

1. H. Gleiter, "Nanocrystalline Materials", *Prog. Mater. Sci.,* **33** 223-315 (1989).
2. D. Vollath, K.E. Sickafus, "Synthesis of Nanosized Ceramic Oxide Powders by Microwave Plasma Reactions", *Nanostructured Materials,* **1** 427-437 (1992).
3. D. Vollath, K. E. Sickafus, "Synthesis of Nanosized Ceramic Nitride Powders by Microwave Supported Plasma Reactions", *Nanostructured Materials,* **2** 451-456 (1992).
4. A.D. McDonald, *Microwave Breakdown in Gases,* John Wiley & Sons, New York, 1966.
5. D. Vollath, A. Mobius, "Design and Application of a TE11 Cavity for Microwave Plasma Processing", lst World Congress on Microwave Processing, Florida, 1997.
6. A. Möbius, M. Mühleisen, German Patent application DEGM 29512436.9 (1995).
7. D. Vollath, D.V. Szabó, "Nanocoated Particles: A Special Type of Ceramic Powder", *Nanostructured Materials,* **4** 927-938 (1994).
8. D. Vollath, D.V. Szabó, B. Seith, German Patent 196 38 601 2-43 (1997).
9. D. Vollath, D.V. Szabó, "Nanocomposites - New Functional Materials on the Edge of Industrial Application", *FZK-Nachrichten* **31** 197-205 (1999), in German.
10. D. Vollath, D.V. Szabó, J. Fuchs, "Synthesis and Properties of Ceramic-Polymer Composites". *Nanostructured Materials,* **12** 443-438 (1999).
11. D. Vollath, D.V. Szabó, J. Hauâelt, "Synthesis and Properties of Ceramic Nanoparticles and Nanocomposites", *J Europ. Ceram. Soc.,* **17** 1317-1327 (1997).
12. D. Vollath, D.V. Szabó, J. Fuchs, "Polymer Coated Nanoparticulate Ferrite Powders: Properties and Application", *Mat. Res. Soc. Symp. Proc.* **577** 443-448 (1999).

Hybrid Microwave Firing of Heavy Clay Products

G.V.A. Tayler, M. Anderson and M. Hamlyn

Abstract

Thermal analyses and dielectric property measurements of numerous heavy clay bodies provided data for the selection of optimum body compositions. The addition of carbon increased the complex permittivity of the bodies tested, in such a way that it improved the efficiency of microwave coupling in the firing process. The microwave autogenous firing (MWAF) mechanism was identified.

The key variables that influenced this microwave heating process were examined. The test firing methodology used varying techniques of forward power application. The surface area of the test pieces, the application of preheated secondary air supply to the cavity and the application of modulated-pulsed microwave power were found to influence temperature differential across the section of the test pieces examined. In light of these research findings, the paper proposes some of the design features of a continuous and intermittent, hybrid microwave process for the firing of heavy-clay products.

Introduction

The MWAF process, as applied to the firing of heavy-clay products, has been found to be an effective mechanism for exploiting the efficiencies of the volumetric heating effect of microwave energy. It was found that controlled applications of microwave energy to carbon enriched heavy clay bodies, resulted in minimal amounts of power being required to achieve effective microwave heating of the load. The combination of microwave energy with conventional gas firing has been found to be an effective way of speeding up the firing process, as well as a means of reducing total energy consumption (1).

Figure 1 illustrates the reactions that take place during the heating of a typical heavy-clay material. Varying amounts of energy are required at the different stages of the heating cycle to complete the reactions.

Experimental microwave firings have shown that it is feasible to design and develop a hybrid firing system utilizing the autogenous combustion mechanism that is initiated by a microwave energy source.

Autogenous combustion can be described as the heat transfer process developed from the ignition and combustion of the carbon additive within the heavy-clay body (2). This technique of kiln firing is typically used in clamp kiln firing systems for clay brick. More recently, tunnel kilns have been designed and built with burner/fan circulation zones that recirculate the sensible heat developed from the autogenous combustion in the preheat and furnace zones, thereby

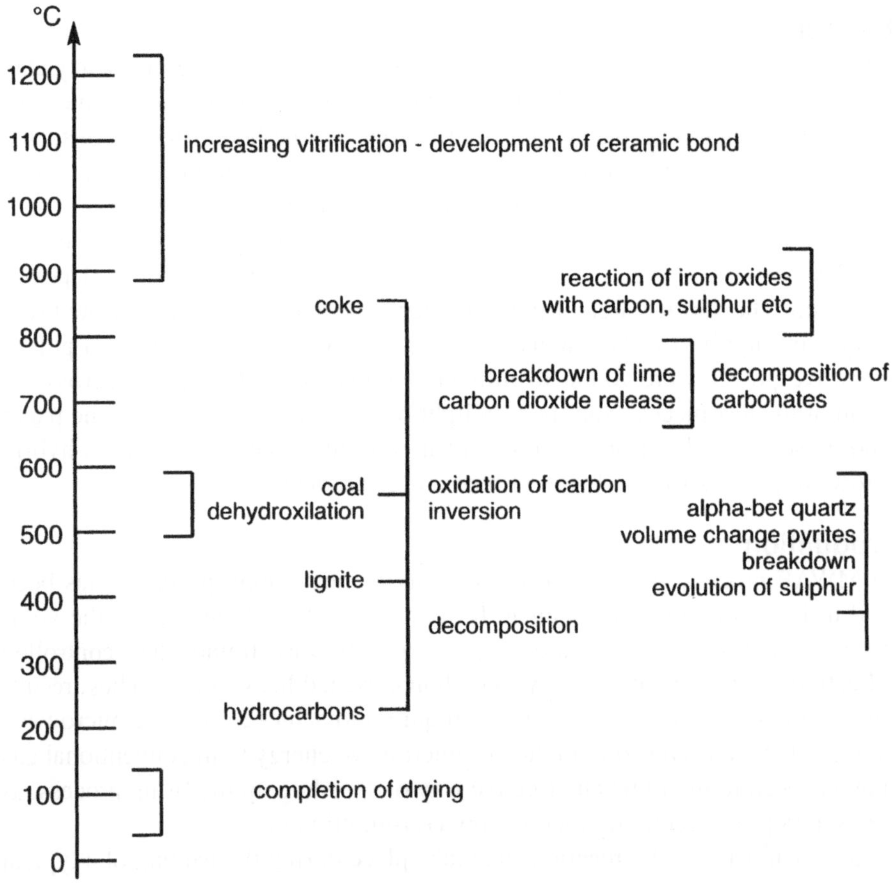

Figure 1. Reactions during heating of a typical heavy-clay mineral.

reducing gas usage. Figure 2 illustrates the typical heating cycle of a heavy-clay tunnel kiln firing process. The temperature cycle is required to follow differing heating rates, to facilitate the effective completion of the different reaction stages illustrated in Figure 1. As a result of this, the firing cycles are lengthy and sluggish, due to the poor thermal conductivity of the clay body, as well as the poor efficiency of the convection and the radiation heat transfer mechanisms

The introduction of the more efficient volumetric heating (3) of microwave energy, at selected portions of the heating cycle, provides the opportunity to reduce the overall cycle time and improve the specific energy usage.

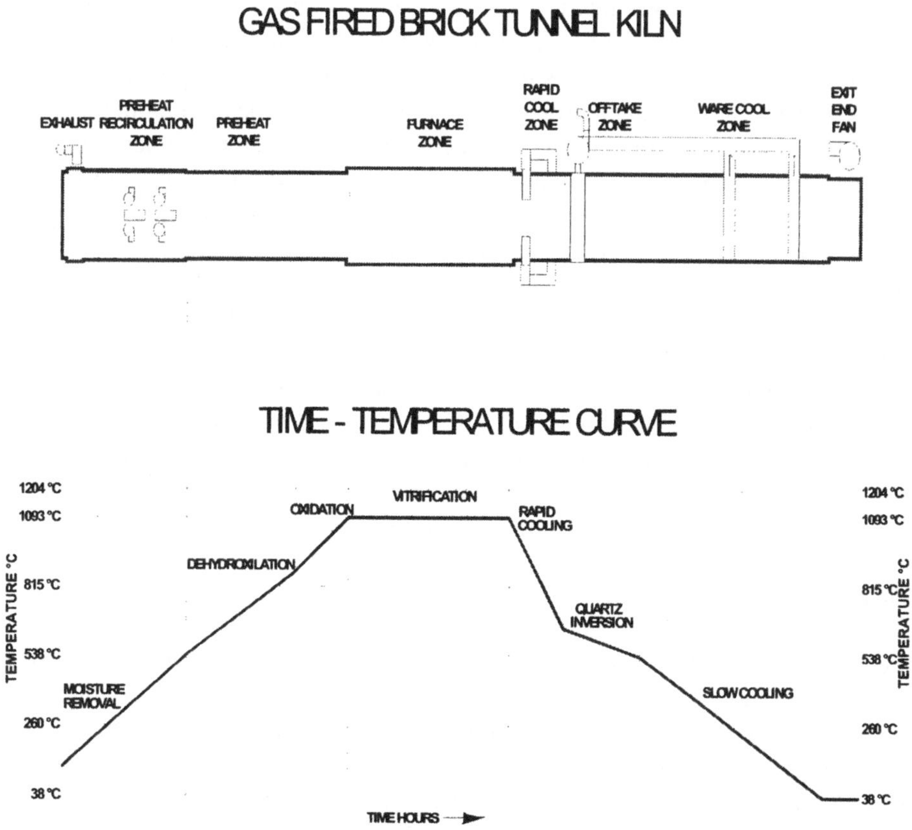

Figure 2. Layout and heating cycle of a typical gas-fired heavy-clay tunnel kiln.

Research efforts have also shown that with this technology, the microwave energy can be applied as a minor proportion of the total energy used in the system.

The hybrid gas/microwave firing technology, using the autogenous combustion microwave firing (MWAF) mechanism, could be developed and adapted for continuously fired brick kilns as well as intermittently fired kilns.

Experimental

Research on the MWAF mechanism was conducted at the TRI laboratories of ESCOM, South Africa, in collaboration with Staffordshire University, UK. The hybrid firing system, used Kanthal resistance elements in combination with a 6 kW, 2450 MHz, microwave system. An insulated stainless-steel-lined cavity was used. The cavity was lined on the sides and bottom with the Kanthal resistance elements rated at 12 kW. The wave guide was fitted with a circulator and stub tuner. A quartz window was installed in the wave guide to prevent heat and moisture from causing magnetron over-heating and possible arching in the waveguide. The control equipment was designed to measure sample temperature using optical fiber thermometry. Forward power and reflected power were monitored to develop an understanding of the firing behavior of the materials during the autogenous firing phase of the cycle. The method of application of the microwave power was adjusted and optimized over numerous firings, in order to achieve the best possible heat transfer mechanism. During the microwave stages of the cycle, the heat transfer mechanism was enhanced greatly by the provision of a convective heat dissipation technique. The measurements taken during the many test firings and the observations of the fired results enabled the researcher to draw some important conclusions on the key variables of this hybrid microwave heating process. The research findings provided guidelines for the design of a continuous or intermittent firing hybrid microwave process.

Clay Bodies

Typical heavy-clay bodies were studied. They consisted primarily of a kaolinitic fireclay combined with silica sand, pulverized fly ash (PFA) additive and carbon fly ash (CFA) additive. The CFA, a byproduct from an industrial processing industry, contained approximately 40% fixed carbon and served as a "doping" agent to increase the complex permittivity of the bodies (4). Figure 3 illustrates how the carbon additions increased complex permittivity of the body. This increased carbon content, resulted in more efficient coupling (5) with the microwave field. The control of this key variable facilitated the development of the MWAF process.

Test firings

The objective of the test program was to use controlled microwave energy applications to ignite (6) carbon within the body. This autogenous combustion of the carbon was sustained by pulsed microwave power applications during the firing cycle. Network analyzer measurements were taken of both an empty and loaded cavity. Adjustments were made to optimize the field distribution within the cavity. The loaded cavity, containing a sample mass of approximately 1.5 kg, indicated good microwave power absorption with minimal reflected power. Increased sample-hack setting density further improved field distribution in the load.

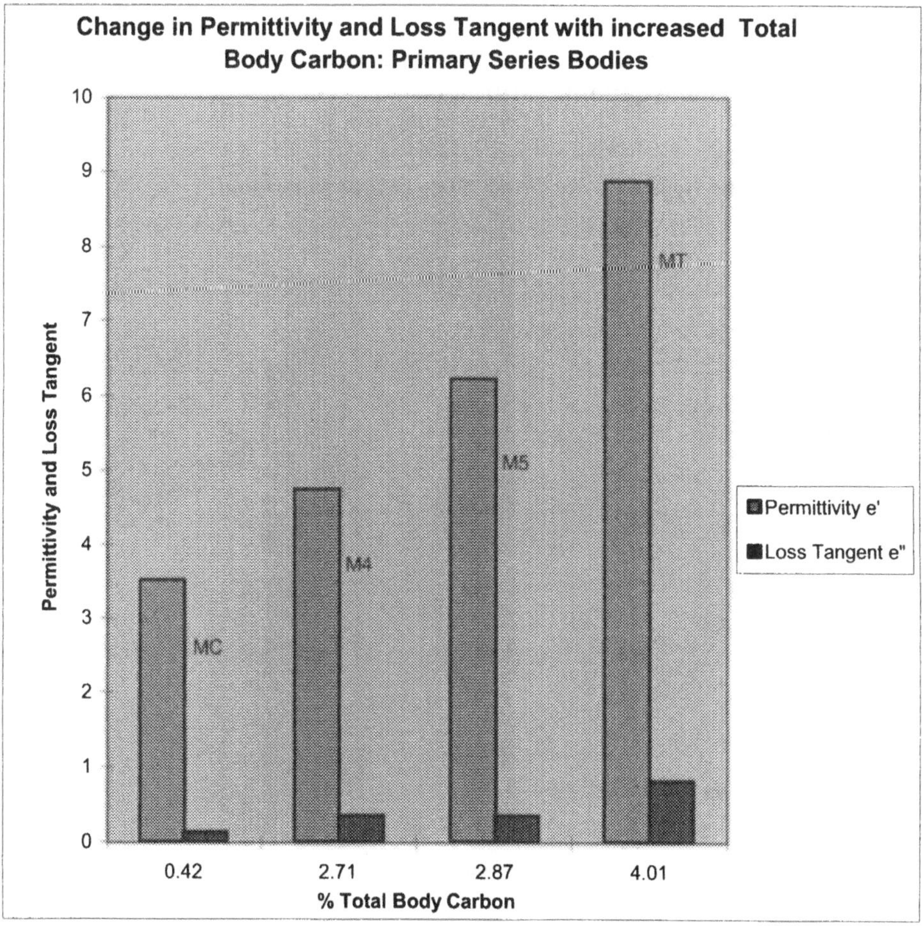

Figure 3. Increased complex permittivity with increased carbon content.

The preheat cycle (from 20°C to 600°C) was programmed to finish over a time cycle at a predetermined ramp rate. The transition from the radiant heat, preheat cycle, to the microwave autogenous firing cycle was operated manually. The energy usage during the preheat cycle was recorded using an energy data tracker. The microwave-firing cycle was programmed to be either a "full forward power cycle," after the 600°C dwell was completed or a "modulated-pulsed forward power cycle" using the automatic control loop facility available in the MICROSOFT LAB VIEW software used with the test furnace. Figure 4 illustrates a firing cycle using the modulated-pulsed method of applying the forward microwave power. Firings were carried out in a "single high" sample setting pattern. Results of each firing were analyzed. A heat dissipation technique was achieved by the introduction of air preheated to 500°C. This technique simulated the conditions of an industrial furnace (7). The best results were obtained with perforated sample pieces that had a greater surface area and produced better carbon burnout with more uniform vitrification.

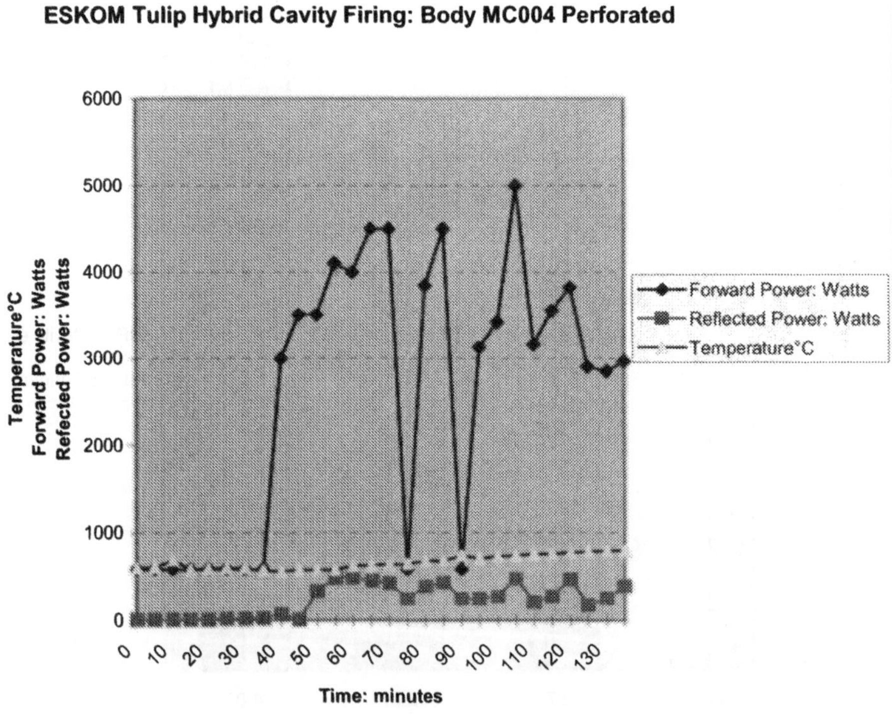

Figure 4. Firing cycle using the pulsed-modulated forward power application.

An optimum carbon content of 2% was found to produce the most effective carbon burnout, resulting in a uniformly vitrified body. This carbon content also produced efficient microwave coupling and, hence, facilitated effective autogenous combustion. These findings provided the guidelines for a proposed hybrid MWAF system.

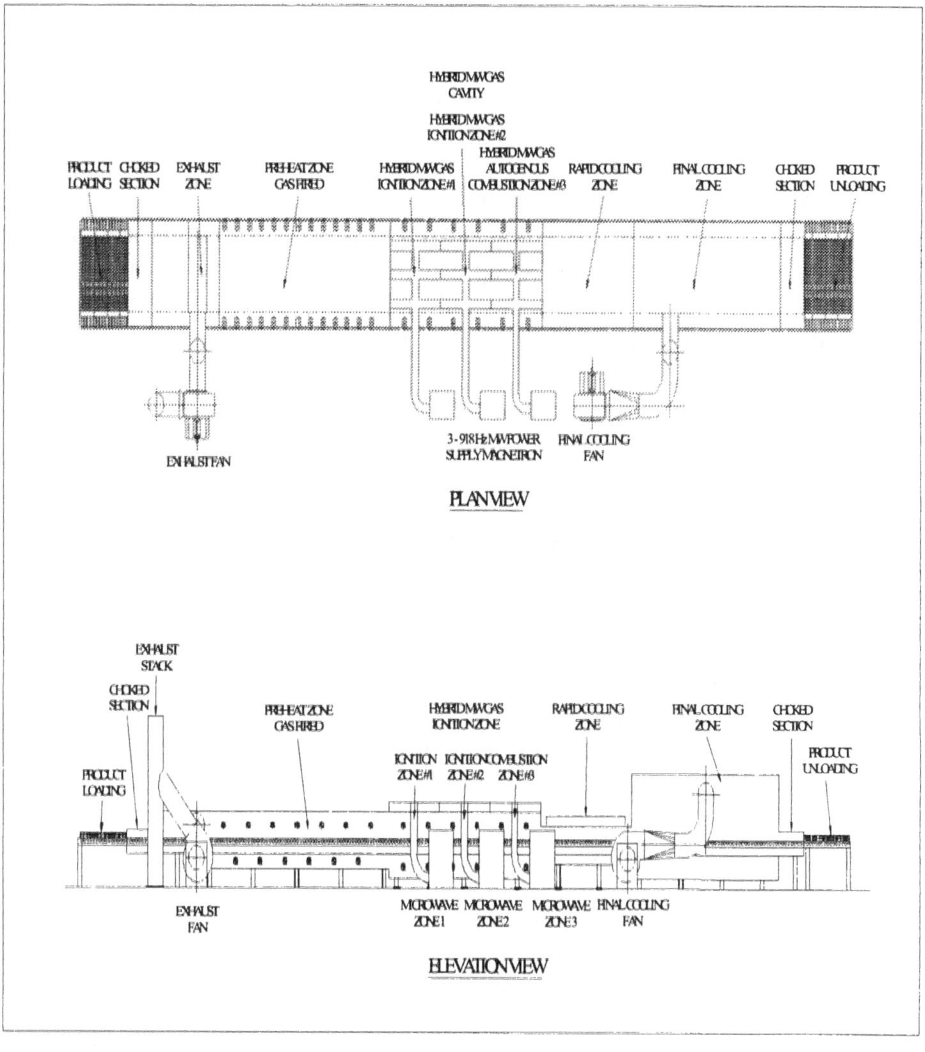

Figure 5. Proposed continuously firing, hybrid MWAF roller-bat kiln.

Proposed Continuous Hybrid MWAF Process

Applying these research findings, the design of a continuously firing, rollerbat kiln is proposed and illustrated in Figure 5. The counterflow of the preheated gasses i.e. the gasses that flow in the opposite direction to the ware passing through the kiln, is an important feature of the hybrid MWAF process. Controlled volumes of this preheated air, at a suitably controlled pressure, in combination with the recirculation of gasses from high velocity gas burners in the preheat and firing zones, ensure that the ware is uniformly preheated before entering the microwave ignition and autogenous combustion zones. The 915 MHz microwave ignition and autogenous combustion zones are hybrid gas/microwave zones. The choice of 915 MHz was made due to better coupling efficiency and depth penetration achievable with this frequency. Pulsed application of microwave power, in these three separate zones, facilitates the autogenous combustion of the carbon in the body and an efficient and rapid heat transfer mechanism. Conventional thermocouple and infrared temperature measurement techniques are used to control gas firing settings and the forward power microwave levels in such a way that the programmed set-point temperatures in the zones are maintained automatically. The heat dissipation effect of the counter flow gasses counteracts the reversed-thermal gradient effects (8) developed by the microwave-heating systems. This important feature prevents thermal runaway from developing, a problem seen in the microwave-only test firings.

Proposed Intermittent Hybrid MWAF Process

The proposed hybrid MWAF intermittent brick kiln consists of an insulated firebrick– (IFB-) lined steel shell, constructed as a 13.8 m diameter circular structure, which supports a domed crown, manufactured from IFB masonry or Z-block refractory fiber. The entire outer surface of the IFB lining is sealed, using a Teflon bonded, stainless-steel foil layer, to render the structure completely gas tight. Firing is carried out in 50 h, using 6 large, high-velocity, pulse-fired gas burners. The kiln pressure is automatically controlled.

In the proposed hybrid intermittent kiln, the microwave energy is launched into the large cavity, using 915 MHz generator–magnetron systems. Four waveguide entry points are positioned symmetrically at the base of the crown. This system is designed to introduce the microwave energy after the dehydroxylation stage of the firing cycle. The pulsed, forward microwave power follows the programmed temperature set point schedule, to achieve the following conditions:

- Maximum uniformity of heat dissipation within the load.
- Minimal "reversed thermal gradient" effects.
- Minimized thermal runaway conditions.

- Maximum degree of carbon burnout.
- Reduced firing cycle time.

For an effective, continuous or intermittent MWAF firing process, there are key operational controls of the process.

- The speed of the product traveling through the unit (continuous firing).
- The composition of the material i.e. total carbon content.
- The level and method of forward power application in the ignition and autogenous combustion stages of the cycle.
- The volume and pressure of combustion air supplied to the cavity.

Conclusions

The work has indicated that the use of a hybrid microwave-firing system to initiate autogenous combustion is highly feasible. The addition of a suitable carbon component improves the complex permittivity of the material being fired. The use of a controlled preheated secondary air is the key to the development of a heating process that effectively minimizes thermal runaway. The effective control the conventional gas firing in combination with the modulated-pulsed microwave portions in the overall energy balance enables the more expensive microwave portion of the total energy requirement to be minimized.

The numerous test firings and the analyses of the results yielded the following.

- The use of perforated test pieces produced improved results due to better circulation and greater surface area, which facilitated a more effective gas diffusion process.
- Body carbon contents, which exceeded 2%, displayed efficient microwave coupling, but detracted from the final properties of the fired body due to the "carbon-trap" phenomenon.
- The best carbon burnout and vitrification results were achieved using an air supply preheated to 500°C and introduced into the cavity through a perforated base on which the samples were mounted. This preheated air supply produced improved heat dissipation and reduced the "reversed thermal gradient" effect.
- A modulated-pulsed forward power technique resulted in better heat distribution within the load.
- An excessive supply of air to the cavity resulted in a higher microwave power demand. An optimum combination of air and modulated-pulsed microwave power is required.

An industrial scale MWAF system, utilizing hybrid gas/microwave firing, would need to adhere to most of the above design requirements to be effective. The research has shown that, using this hybrid microwave/gas technology (9), firing cycles can be reduced by as much as 50%, with resultant energy reductions of as much as 35%.

REFERENCES

1. Bond S.M, "Microwave-Assisted Processing from Laboratory to Production"; Paper presented at the 2nd World Microwave Congress, Florida, (2000).
2. Tayler G.V.A., "The use of Microwave Energy to Initiate Autogenous Combustion for the Firing of Heavy Clay Products"; Ph.D. thesis. Staffordshire University, England (April 1998).
3. Ahmad I., Dalton R. and Clark D., "Unique Application of Microwave Energy to the Measurements on Ceramics". Material Research Society Symposium Proceedings, Vol. 269, pp. 611–16 (1992).
4. Clark D.E., Ahmad I. and Dalton R.C. "Microwave Ignition and Combustion Synthesis of Composites." *Mater. Sci. Eng.* A, **144**, 91–97 (1991).
5. Evans N.G. "The Dielectric Properties of Ceramics at 2.45 GHz and Their Influence on Microwave Firing"; Ph.D. thesis. Staffordshire University, England (April 1995).
6. Hamlyn M.G., Bowen A.L., Jones G. and Jackson S.M. "Microwave Firing Process for Earthenware Ceramics"; presented at High Frequency and Microwave Processing and Heating Conference, Arnhem (1989).
7. Kulbekov M.K. and Saibulatov S.Z. "Mechanism and Kinetics of Carbon Burnout During Calcinations of Resource-conserving Ash-based Ceramic Materials," *Izvestiya Akademi Nauk SSSR Neorganicheskie Materialy* **27** [5], 1074–78 (1991).
8. Mataxas A.C. and Meredith R.J. *Industrial Microwave Heating.* Peter Peregrinus, London (1983).
9. "The Microwave-Assisted Gas Firing of Ceramics," Future Practice Report 57, EA Technology, UK (1999).

Annealing and Self-Healing of Microwaved Ceramics

M.S. Morrow, D.E. Schechter, P.A. Eggleston, H.E. Huey and Q.S. Wang

Abstract

Several oxide and non-oxide ceramics were prepared and sintered by both conventional and microwave techniques to achieve finished dimensions. In order to achieve final net shape, grinding of the sintered ceramic pieces was necessary. Historically, an annealing treatment has been used for stress relief in the finished piece. Recent tests with 24 GHz microwaves have produced some very distinctive differences, when compared to traditional annealing methods and to 2.45 GHz microwaves. A variety of surface evaluations including scanning electron microscopy (SEM), fracture toughness and bend tests were performed. Preliminary results of the tests revealed significant improvement in mechanical properties. SEM observations reveal a significant reduction in the amount of surface and near surface micro-cracks. The authors have previously reported this effect, which is being investigated further in this report. This effect has not previously been observed using conventional annealing treatments for these materials.

Introduction

A "microwave effect" has been observed and reported [1] on the annealing of an oxide ceramic The result was a healing of the surface and near surface micro-cracks. This occurred on all surfaces exposed to the microwaves. A liquid phase sintering agent was used in the original hot pressing of the oxide ceramic. The details of the surface healing are not yet understood. Further discussion of the previous experiments is included herein. This is a status report of this topic.

Microwaves generally penetrate the ceramic material and heat internally, thus reducing the thermal stress of the pre-stressed samples. A 2.45 GHz, 6 kW multi-mode system available at Lockheed Martin Energy Systems Inc. Y-12 Plant, Oak Ridge (LMES), was utilized for the initial experiment. Oxide ceramic sample bars were enclosed in alumina fiberboard insulation. Attempts to heat the bars with 4-5 kW for a minimum of one hour were unsuccessful. In order to achieve uniform heating of the material, susceptors were not used.

A prototype 24 GHz, 5 kW Micramics, Inc. MGS24-5-101 microwave furnace based on a gyrotron [2] has been installed at the Y-12 Plant as part of a CRADA agreement. At 10 times higher frequency, than the 2.45 GHz unit, the system has demonstrated the ability to sinter pure alumina to full density without the use of hybrid heating [3]. At only 1 kW of average power, the 24 GHz microwave furnace was readily capable of heating the oxide ceramic to the desired annealing temperature of about 950°C. Upon annealing, SEM revealed a lack of micro-cracks and a morphological change down to a depth of about 150 microns. Flexural strength tests on a 4-pt bend, confirmed expectations that tensile strength was improved considerably by the surface healing. Tensile strength was improved by at least 50% and as high as 200% over conventionally treated samples.

The existence of a liquid phase sinter was believed to cause a major reduction of the penetration depth of the 24 GHz microwaves at the annealing temperatures. Then by a combination of capillary action, micro-arcs or ponderomotive force, the liquid phase sinter is believed to migrate over the surface and heal the micro-cracks. Thus, additional studies on an alternative ceramic without a liquid phase, such as, alumina (Al_2O_3, 99.5% purity) and a non-oxide ceramic, such as, boron carbide (B_4C) were suggested for further examination.

Heating Experiment

The B_4C was processed at the Oak Ridge, Y-12 plant, and sintered in the 2.45 GHz microwave furnace. Initial green density was approximately 50% theoretical density (TD) and final sintered density was typically at 96% TD. The 2.45 GHz system utilizes a 6 kW power supply connected to a multi-mode 100-liter volume chamber. Annealing of the B_4C was also performed in the 2.45 GHz system.

The prototype 24 GHz system was available as an alternative microwave system at the Y-12 Plant. The system consists of a 24 GHz, 5 kW gyrotron with a water cooled solenoid, power supplies, internal loop cooling water system, DAC/ADC units and computer controlled by a Micramics proprietary program using the National Instruments' LabVIEW for Windows NT system. The multi-mode chamber is a 24"-dia. x 26" length (200 liters) stainless steel chamber, see Fig. 1. A 2-color pyrometer was used for temperature monitoring. The pyrometer measures temperature from 800-2400°C. Thermocouples were used for temperatures below 800°C. The system was operated in the pulse mode at approximately 5 kW peak and 2 pulses per second repetition. Duty was adjusted nominally from 10 to 70 % to control the heating rate and temperature. The annealing temperatures were at 1400-1600°C for both the alumina and boron carbide samples.

The casketing technique [3, 4], normally consists of an inner crucible, boron nitride (BN), in this instance, and an outer crucible of alumina with alumina fiberboard housing around the outer crucible. The B_4C samples were placed in the inner BN crucible. Pumping down to 100 microns pressure and back filling with argon (Ar) purged the microwave system. A flow of Ar gas of approximately 0.05 to 0.1 scfm was maintained at about 1 psig.

Results and Discussion

The B_4C was annealed in both the 2.45 GHz and the 24 GHz microwave furnaces at approximately 1400-1600°C. The insulation setup was identical. SEM photographs of the surface and a machined surface (potted in epoxy and polished) were made. In the 2.45 GHz anneal, no discernable change was observed. However, at 24 GHz, there is a distinct change in the surface appearance, see Fig 2. On the fractured surface, the voids appear to be smaller and fewer down to about 300 microns, even deeper than the healing in the previously reported oxide ceramic. For the annealing process, the B_4C sample was set atop 3 other samples. Only the surfaces not covered by the neighboring B_4C showed healing, except for a region up to 0.5 cm within the covered surface.

The alumina samples were only heated in the 24 GHz furnace to about 1500°C. SEM did not reveal any changes to the surface morphology. This result may be attributed to the low microwave absorption of the material or the lack of a liquid phase sinter. Possibly higher temperatures or longer hold times may be necessary to gain the desired effect. Lower alumina content samples may also be considered.

Figure 1. The 24 GHz Micramics microwave unit used for this paper

At 1500°C, the penetration depth of 24 GHz is expected to be about 1-2 cm. At higher temperatures the penetration depth is expected to be reduced further.

It is not currently known whether the B_4C material has changed stoichiometrically, reacted with impurities, or reacted with a small amount of oxygen or water vapor. Measurements such as X-ray diffraction, and Fourier-transform infrared spectroscopy are being considered. One possibility is the formation of B_20_3 from O_2 or water present in the chamber. This is not considered a key factor, since the process is reproducible in the 24 GHz system and not in the 2.45 GHz system. If this were the case, then B_20_3 would also be formed during the sintering at 2.45 GHz.

The healing of the cracks may be due to a unique microwave effect at the surface and near the surface. Penetration depth of the electric field at the surface and near the surface certainly seems to play a role. The liquid phase material may play a direct role in this phenomenon in the oxide ceramic. Recent additional results on the previous oxide ceramic indicate that surface healing takes place as low as 500°C. Efforts will continue with the evaluation of a variety of parameter changes. Penetration depths of the 24 GHz microwaves at 500-950°C should be much larger than 100 μm in the oxide ceramic. Otherwise severe runaway should occur. No evidence of thermal runaway has been observed. The process at 24 GHz uses a pulse modulation with a repetition rate of 2 Hz. The peak power is varied from about 4 to 5 kW. Control of the average power is via the pulse duration, measured by the effective duty. The pulse duration is quite long for micro-arcs to occur, but other higher peak power experiments [5] should have observed this phenomena. It is possible in some cases that by reducing the peak power, but maintaining the same average power by increasing the duty may reveal threshold electric field intensity.

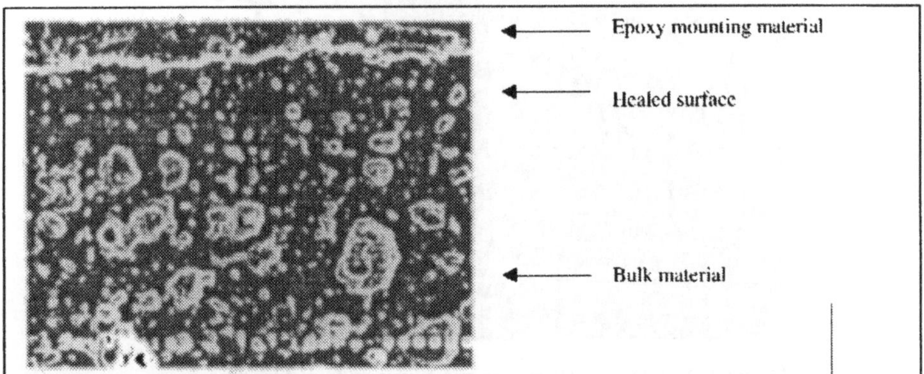

Figure 2. SEM of the boron carbide annealed at 24 GHz showing the surface healing effect

B_4C at the annealing temperatures is electrically conductive. The 24 GHz skin depth of B_4C at 1400-1600°C is expected to be less than the penetration depth in the oxide ceramic. However, the healing depth appears to be greater in the B_4C. Hold time for the B_4C was also less, only 17 minutes at 1600°C. Longer hold times may enhance the degree and depth of healing.

Conclusions

A comparison of conventional and 2.45 GHz microwave annealing with 24 GHz microwave annealing has produced unexpected benefits. An oxide ceramic with a liquid phase and a non-oxide ceramic without a liquid phase were annealed with 24 GHz microwave. Examination by SEM revealed a complete lack of surface micro-cracks. The morphology was altered on the surface down to a depth of about 100-300 μm. Penetration depth in B_4C of the morphological change appears to be greater than in the previously reported oxide ceramic with liquid phase sinter. Preliminary four-point bend tests on the oxide ceramic showed improved yield strengths with reduced variation, in agreement with qualitative expectations. The micro-crack healing has not been seen in any conventional heating tests at any temperature. Conventional heating methods as well as tests with 2.45 GHz are unable to reproduce this surface effect. Further investigation using the 24 GHz unit will continue.

ACKNOWLEDGMENT

We would like to acknowledge E.B. Ripley of Lockheed Martin Energy Systems Y- 12. Plant, Development Operations for his assistance.

REFERENCES

1. M.S. Morrow, D.E. Schechter, R. Simandl, H.E. Huey, Q.S. Wang, "A Comparison of Annealing Treatments of an Oxide Ceramic," *Ceram. Eng & Sci. Proceedings,* 20, [4], 207-212, (1999).
2. H.E. Huey and L.J. Craig, "24 GHz, a Forgotten Allocation," 1st World Congress on Microwave Processing, at Orlando Fla., Jan 1997, abstract book.
3. M.S. Morrow, D.E. Schechter, H.E. Huey & Q.S. Wang, "Multiple Compact Sintering at 24 GHz," Amer. Ceramics Society 100th Annual Meeting, Cinn. Ohio, May 1998, pg 395.
4. H.E. Huey, Q.S Wang, M.S. Morrow, D.E. Schechter, "Microwave Sintering at High Temperature," *Ceram. Eng & Sci. Proceedings,* in press.
5. Y. Setsuhara, M. Kamai, S. Kinoshit, N. Abe, S Miyake, T. Saji, "Advanced Ceramics Sintering Using High-Power Millimeter-Wave Radiation," In M.F. Iskander, J.O. Kiggins, J.C. Bolomey (eds), Microwave Processing of Materials V, Mat. Res. Soc Proc Vol. 430, 533-538.

Enhancement of Cutting Performance of Cemented Carbide Cutting Tools by Microwave Treatment

S. Aravindan, J. Ramkumar, S.K. Malhotra and R. Krishnamurthy

Abstract

Glass-fiber-reinforced plastics (GFRP) find wide applications, particularly in the aerospace and automotive industries. In order to fabricate structural components of GFRP to near-net shape, machining of GFRP composites, such as cutting and drilling are necessary. The heterogeneity/nonhomogeneous nature of GFRP makes machining a difficult task. The nonhomogeneous nature of these materials certainly poses problems, such as rapid tool wear and associated deterioration of the quality of the manufactured component. The present study attempts to enhance the cutting performance of tungsten carbide (WC) inserts by way of microwave postsintering treatment.

Introduction

Fiber-reinforced composites are finding structural applications due to their excellent mechanical and physical properties. Machining of GFRP is needed for the production of near-net-shape geometry. Production of damage-free surfaces in GFRP when machining is difficult due to contact of tool with relatively softer matrix and hard abrasive fibers of the composite material. Such variation results in rapid tool wear and reduced tool life. Improvement of cutting performance and quality of the machined surface are major requirements in cutting of GFRP [1, 2]. Microwaves are effectively used in the processing activities of metals, ceramics, semiconductors and composites by researchers [3-5]. Application of microwave radiation in the production of cemented carbide drills offers a great potential to the cutting industry by providing methods for shortening the process sequence [6-8].The present study emphasizes microwave-assisted postsintering treatment of cemented carbide (WC) drill inserts for possible enhanced metallurgical and mechanical properties of inserts, facilitating improved machining performance. Details are presented in the following sections.

Experimental Procedure

Cemented carbide drill inserts (K-20) were exposed to microwave-assisted post the sintering treatment, using a multimode microwave furnace (2.45 GHz), for microwave hybrid-heating technique. Three sets of samples (each sample consists of 5 cemented carbide inserts) were subjected to microwave exposure for 1800. The shank portion of drill inserts was effectively sealed from microwave exposure in order to avoid microwave reflection using alumina and silicon carbide fencing arrangement. Treated and untreated cutting tools were studied through optical microscopy and microhardness test to assess the mechanism behind the change in characteristics.

To evaluate the cutting performance of microwave-irradiated cutting inserts, drilling was carried out on GFRP composites. The trials were conducted and the best cutting speed and feed were fixed as 2000 rpm and 0.06 mm/rev. During drilling, thrust and torque were measured through a electrical resistive temperature-compensated strain-gauge-type tubular dynamometer. Acoustic emission (A E) signals emanated during drilling was acquired as a measure of the response of the work material. A.E. signals were acquired by an AET 5500 system using a broadband sensor (100 kHz–2 MHz). The signals from the sensor were magnified using a preamplifier and stored in a data logger. The acquired signals were analyzed by a suitable PC-based data acquisition system. Tool wear and hole geometric deviation using a toolmakers microscope were made as off-line measurements.

Results and Discussion

Microstructural Observations

Plain cemented carbide cutting tools consist of cemented carbide grains and cobalt (Co) as a binder. Microwave exposure of cemented carbide tools facilitated changes in metallurgical characteristics due to the absorption of microwaves by WC. Changes in microstructure of the WC inserts are observed owing to absorption of microwaves by WC grains. Liquification of Co, depletion of Co at the surface, upliftment of cemented carbide and formation of complex carbide phase at the surface are the major effects associated with microwave treatment of cemented carbide tools [9]. Typical optical micrographs of treated and untreated specimens are presented in Fig. 1. Compared to untreated, more cemented carbide grains are observed over the surface due to the depletion of surface Co resulting in formation of WC skeleton matrix. Pooling of WC is observed in the micrographs of the treated tool.

Microhardness survey

The influence of microwave treatment on mechanical characteristics of tools can be evaluated through hardness survey. Typical variation of hardness of both treated and untreated tools are presented in Fig 2. The possible reason for the enhancement in hardness of treated tools may be due to the WC grain upliftment and complex carbide formation.

Figure 1. Typical micrographs of (a) microwave-treated tool (b) untreated tool.

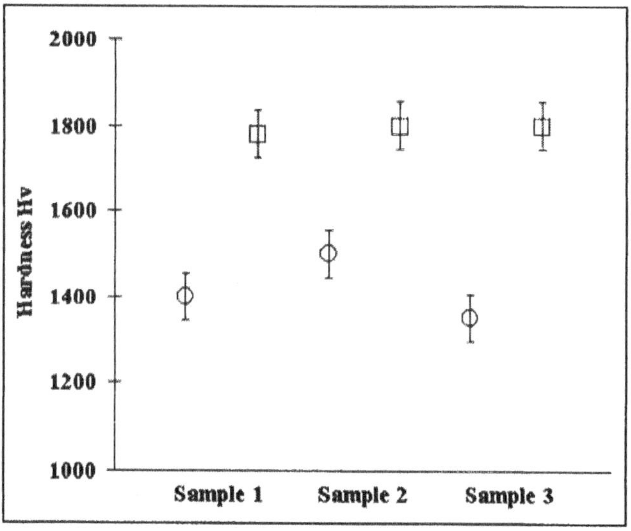

Figure 2. Typical variation of hardness of treated and untreated cutting tools (square form-treated, round form-untreated).

Performance of Treated Tools

Microwave-treated tools are subjected to metallurgical and mechanical/physical changes. Such changes enhance the cutting performance. The improved hardness over the surface and the softer bulk of cutting tools that results due to microwave treatment can be effectively used to machine heterogeneous composites. Treated tools are used to drill holes in GFRP. During drilling, measurement of thrust, torque and acquisition of acoustic emission signals were carried out. Measurement of tool wear and hole shrinkage were carried out as off-line quality assessment of produced holes. For comparison, normal untreated tools were also used to perform the drilling process.

Cutting Forces

Typical variation of cutting forces (thrust and torque) during drilling of GFRP using microwave irradiated and untreated tools with number of holes drilled is presented in Fig 3(a) and (b). Due to the tool wear, the cutting forces increase with the number of holes irrespective of treatment. There is considerable amount of reduction in cutting force components (thrust and torque) for a microwave treated tool, due to the improved properties of treated tools. Rate of increase in cutting force with number of holes for treated tools is comparatively at a lower order as illustrated in Fig. 3.

Tool Wear

Rapid tool wear is the major undesirable effect encountered during drilling of composites. Typical variation of tool wear of treated and untreated tools with the number of holes produced is illustrated in Fig. 4. Increase in tool wear is observed in both cases of treated and untreated tools, with the number of holes. The rate of increase in tool wear of untreated tools after around 250 holes is comparatively larger, as illustrated in Fig. 4, due to the accumulation of thermal-assisted material degradation. Tool wear of treated tools is on the order of lower magnitude and even after 250 holes the tool wear progresses steadily. During drilling of heterogeneous materials, cutting tools are often subjected to force variation because they encounter soft matrix and hard fibers. Such variation, in cutting forces results in reduced tool life due to increased tool wear. Combination of relatively softer bulk and harder surface of the treated inserts has given adequate strength to the cutting inserts to accommodate the fluctuation of cutting forces during machining of nonhomogeneous material.

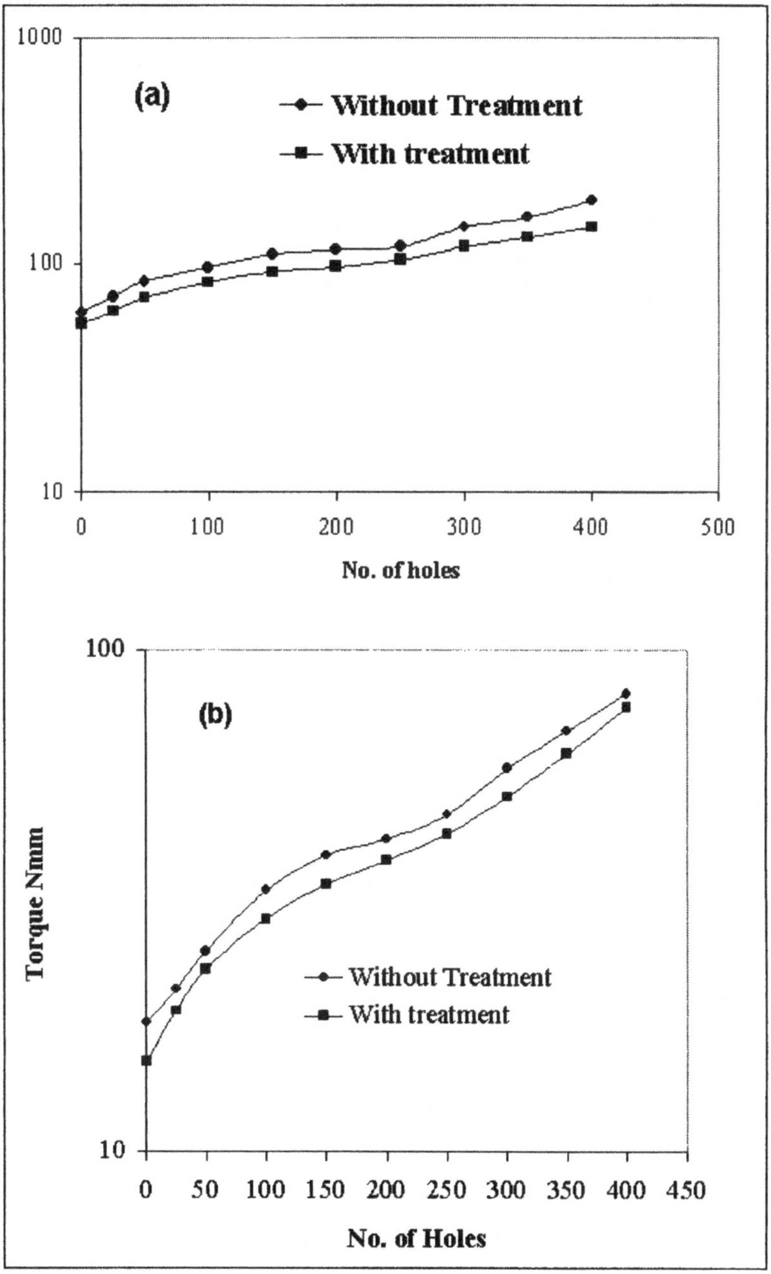

Figure 3. Variation of cutting forces (a) thrust and (b) torque) during drilling of GFRP.

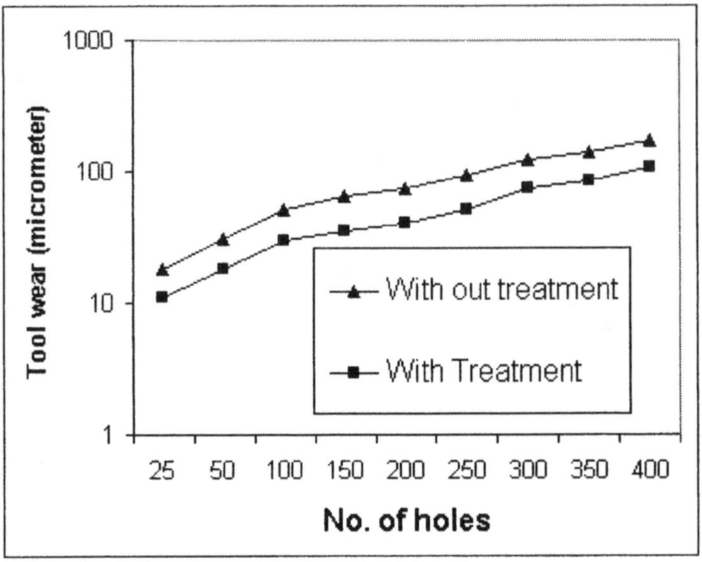

Figure 4. Variation of tool wear

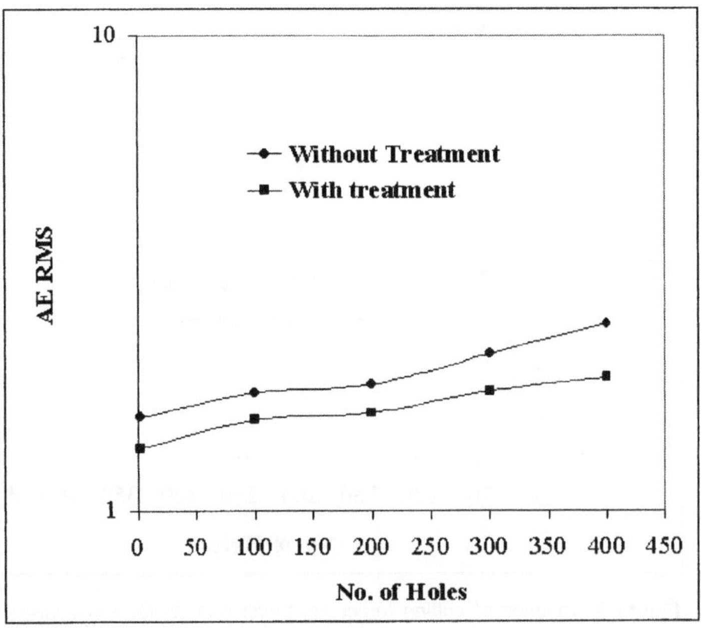

Figure 5. Variation of R.M.S. value of A.E. signal

Hole shrinkage

Quality of the surface produced is a good indicator of the machining process. The rapid tool wear resulted in deformation/deviation in hole size. Due to reduced tool wear of treated tools, the hole deviation/shrinkage of holes produced by treated tools is comparatively minimum. The maximum value for the hole tolerance allowed is 100 microns as per data reported. The untreated tool has reached the maximum value with 80 holes and the microwave-treated tool is found to have lower amount of hole shrinkage.

Monitoring of AE Signals

Acoustic emission is defined as the transient elastic wave generated by the rapid release of energy from a localized source or sources within a material when subjected to a state of stress. AE signals emanating from GFRP during drilling are acquired using AET 5500 system. Typical rms values of the acquired signals are presented in Fig 5. The increase in rms. values with the number of holes confirms the continuous mode of fracture of work material during drilling. Treated tools exhibit comparatively lower order rms. value, which confirms the quality of cutting tool and the manufactured component.

Conclusion

Microwave treatment of cemented carbide inserts has resulted in favorable material response. These effects resulted in softer bulk and harder surface of cutting inserts, which is highly suited for machining of heterogeneous material like polymeric composites. Microwave-treated tools exhibit enhanced cutting performance through reduced cutting forces, reduced tool wear and reduced hole shrinkage.

REFERENCES

1. Santhanakrishnan. R. Krishnamurthy and S.K. Malhotra, "Machinability Characteristics of Fiber Reinforced Plastic composites," *Journal of Mechanical Working technology,* **17**, 195-204 (1988).
2. W. Konig, Ch. Wulf, P.Grass and H. Willerschield, "Machining of Fiber reinforced plastics," *Manufacturing Technology CIRP Annals,* **34** [2] 537-548 (1985)
3. W.H. Sutton, "Microwave Processing of Ceramic Materials," *American Ceramic Society Bulletin,* **68** [2] 376-386 (1989).
4. D.E. Clark *et al.*, "Recent Developments in Microwave Processing of Ceramics," *MRS Bulletin,* Nov, 40-46 (1993).
5. K. Thompson, J.H. Booske, D.F. Downey, Y. Gianchandani and R. Cooper, "RF and Microwave Annealing for Ultra Shallow Junction Formation," *Rapid Thermal and Other Short Time Processing Technologies II Symposium, Proceedings of the Electrochemical Society,* Washington DC, March 24–28, 2001.

6. Willert-Porada *et al.*, "Microwave Sintering of Hard Metals," *International Journal of Refractory Metals & Hard Materials,* **16,** 409-416 (1998).

7. R.Roy, D.Agarwal and J.P. Cheng, "Extra-ordinary Successes with Microwave Fields for Synthesis and Sintering of Ceramics, Metals and Composites," *Proceedings of Symposium on Microwave Effects and Applications,* Kokushikan University, Tokyo, (2001).

8. J.P. Cheng, D.K. Agarwal, S. Komarneni, M. Mathis and R. Roy, "Microwave Processing of WC–Co Composites and Ferroic Titanates," *Material Research Innovation,* **1,** 44-52 (1997).

9. J.Ramkumar, S. Aravindan, S.K. Malhotra and R. Krishnamurthy, "Enhancing the Metallurgical Properties of WC Insert (k-20) Cutting Tool through Microwave Treatment," *Material Letters,* **53,** 200204 (2002).

Microwave Heating of Glass

U. Kolberg and H. Roemer

Abstract

A study concerning microwave heating of glass has been done. For that purpose dielectric properties of different glass compositions were measured. The investigations showed a strong correlation between electrical conductivity and microwave absorption of different glasses. From these results it could be concluded that glasses with high alkali concentration could be heated very effectively whereas glasses without alkali are difficult to heat with microwave energy. By heating a semiconductor containing glass with microwaves the chance of volume heating and also the problem of hot spot generating is shown.

Introduction

In the glass production process, heating is done with gas or by electric resistance heating elements. The heat is generated outside the glass and is transferred to the glass by heat conductivity, convection or radiation. Because of that there is always a lower temperature in the middle of the sample compared to the surface.

Using microwaves the heat is generated within the glass sample itself. Because of that it is possible to achieve very fast heating and to achieve a very effective process. It is even possible to have a lower temperature at the surface compared to the middle of the glass body.

Fundamentals

Special glasses have a wide variation in composition and because of this they are very different concerning the absorption of microwave energy. To characterize different glass compositions with respect to their microwave heatability, the dielectric constants have to be measured.[1,2,3] These dielectric constants depend on the microwave frequency and on the temperature.

$$\varepsilon_r' = \varepsilon_r'(\omega, T) = \text{real part of the dielectric constant}$$

$$\varepsilon''_{eff} = \varepsilon'' + \frac{\sigma}{\varepsilon_0 \omega} = \text{imaginary part of the dielectric constant}$$

If these data are known for a special glass, the penetration depth D_p and the absorbed power P could be calculated with the following formula:

$$D_p = \frac{1}{2\omega}\left(\frac{2}{\mu_r \mu_o \varepsilon_o \varepsilon_r^{'}}\right)^{1/2}\left(\left(1+\left(\varepsilon_{eff}^{''}/\varepsilon_r^{'}\right)^2\right)^{1/2}-1\right)^{-\frac{1}{2}} \quad (1)$$

$$P[W/m^3] = \sigma_{eff}|E|^2 = 2\pi f \varepsilon_o \varepsilon_r^{'} \tan\delta |E|^2 \quad (2)$$

$$\tan\delta = \frac{\varepsilon_{eff}^{''}}{\varepsilon_r^{'}} = \text{loss factor} \quad (3)$$

with:

$\mu_r = 1$

$\mu_o = 12{,}566 \times 10^{-7}$ Vs/Am

$\varepsilon o = 8{,}854 \times 10^{-12}$ As/Vm

$\omega = 2\pi f$

The penetration depth is an important value to estimate the efficiency of microwave heating for a glass sample. If the penetration depth is in the range of the sample thickness the whole volume could be heated. If D_p is large compared to the sample thickness only a small part of the microwave energy is transferred into heat within the glass sample, the effficiency of heating is low. If the penetration depth is low compared to the sample size the heat is generated within the surface region which is comparable to a conventional heating process.

Measurements for Different Glass Compositions

The dielectric constants have been measured for different glasses and frequencies as a function of temperature. Some measurements have been done at the University of Nottingham* and others at the MPN**.

All glasses showed an increase of the loss factor with increasing temperature (figure 1). In the range of the transition temperature T_g where the glass starts to soften, a sharp increase is observed for most glasses. An explanation for this sharp increase might be that the structure softens in this temperature range and the ions could be moved easily by the microwave field. Another interesting effect is a decrease in microwave absorption when crystallization starts during heating of a glassy glass ceramic material. An explanation for this might be the bonding of ions in the crystal lattice.

*Institute of Proflon Binner.

**Microwave Properties North, Ron Hutcheon, Canada.

A correlation is observed between the electrical conductivity (Figure 2) of different glasses and the loss factor tan δ (Figure 1). Glasses with a high electrical conductivity also show a high loss factor and therewith a high microwave absorption. It is known that the reason for the electrical conductivity is the alkali content and the bonding strength of the alkali ions within the glass structure.

The more an alkali ion is fixed in the structure the lower is the electrical conductivity. Because of this correlation it is assumed that the most important fact for microwave absorption within a glass is the alkali content and the bonding strength of these ions, too.

This assumption is confirmed by the investigations shown in Figure 3. There the loss factors of two different soda lime glasses are measured. The main difference between glass A and glass B is the alkali concentration listed in Table 1.

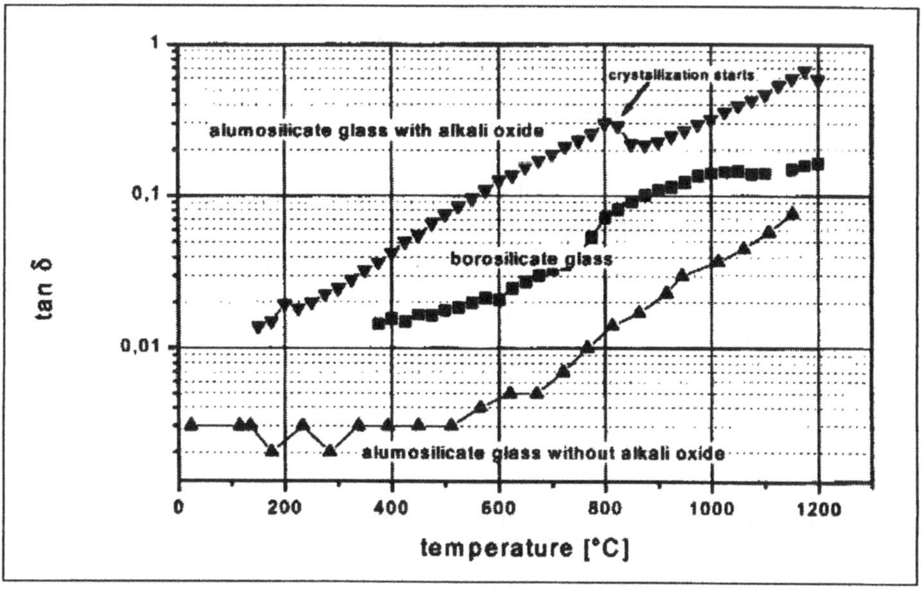

Figure 1. Dielectric loss factor as a function of temperature for different glasses measured at 2.2 GHz.

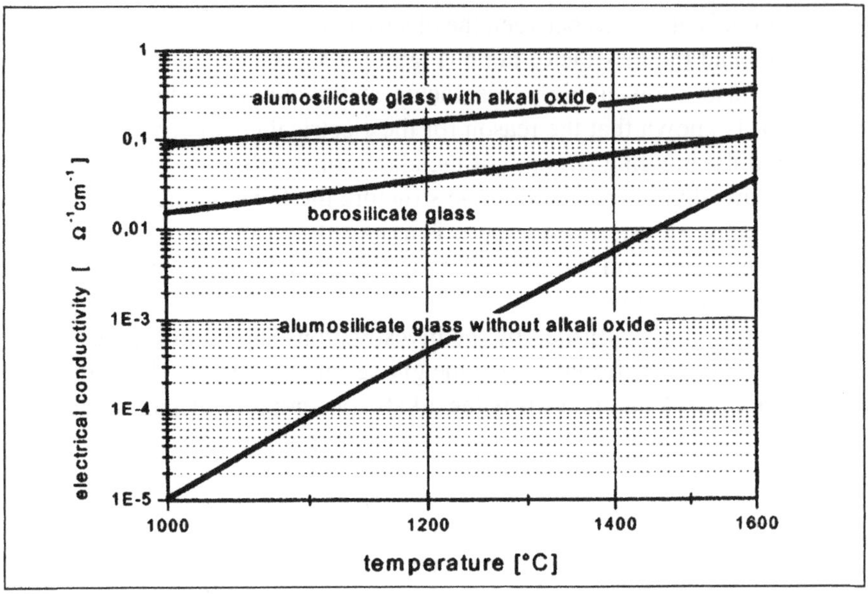

Figure 2. Electrical conductivity as a function of temperature (measured at 10 kHz).

TABLE 1: ALKALI CONCENTRATION OF THE INVESTIGATED SODA–LIME GLASSES

	Na_2O (wt%)	K_2O (wt%)
Glass A	12	3
Glass B	4	9

The A-type glass has a higher alkali concentration and therewith a high loss factor.

Experimental Results Concerning Heating Glass in a Microwave Field

There have been done some heating experiments with a semiconductor doped glass for sharp cut filters. This glass contains cadmium, sulfur and selenium. After melting and rapid cooling this glass is colorless or light yellow. During a heat treatment process, an ion diffusion could be started and growing of semiconductor crystals occurs, which leads to deep red coloration. Because of this change in color this type of glass is very useful to show the volume heating effect of the microwave field. Therefore some cubic glass samples are prepared and heated in a microwave field. After this heat treatment a deep-red colored region

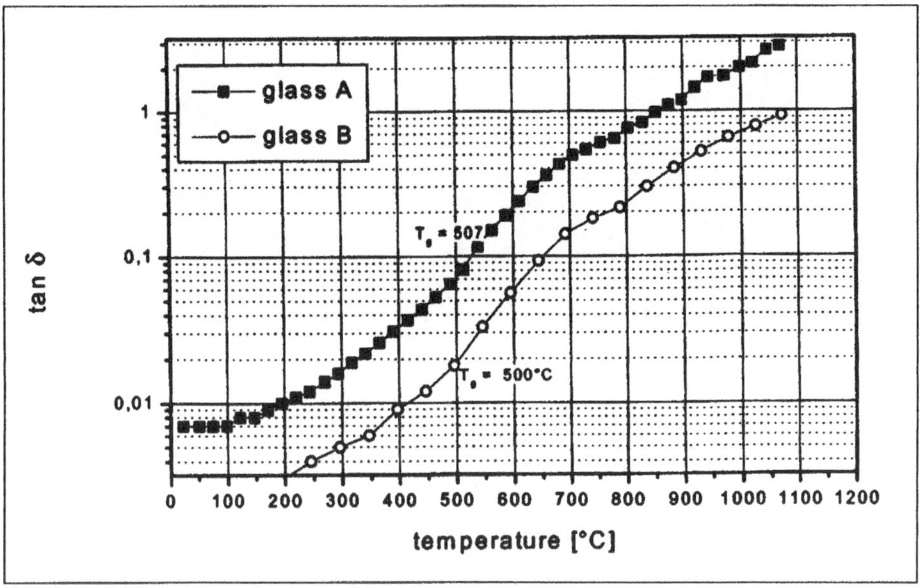

Figure 3. Dielectric loss factor for two different soda–lime glasses (measured at 2.46 GHz).

within the samples is observed whereas the surface of these samples remained in a much paler color (Figure 4).

All samples cracked during the heating process because of inhomogenities in temperature which lead to tension within the volume.

This example shows the chance and also the problem of microwave heating of glass. The great chance of this technology is the possibility to heat the volume of a glass sample which facilitates a rapid heating and an inverse temperature profile within the sample compared to conventional heating methods.

The problems of this heating method are the generation of temperature inhomogenities and especially the generation of hot spots which lead to cracking. The reason for hot spots based in the dependence of dielectric loss as a function of temperature. If there is a region within the glass which has a higher temperature then this region has a higher loss factor which leads to higher microwave absorption and therewith to higher temperatures within this region. This is a self boosting effect.

If homogenous heating of glass is necessary, a very homogenous microwave field must be generated to avoid hot spots. Industrial application of microwave heating in the glass industry will be very interesting after this problem is solved.

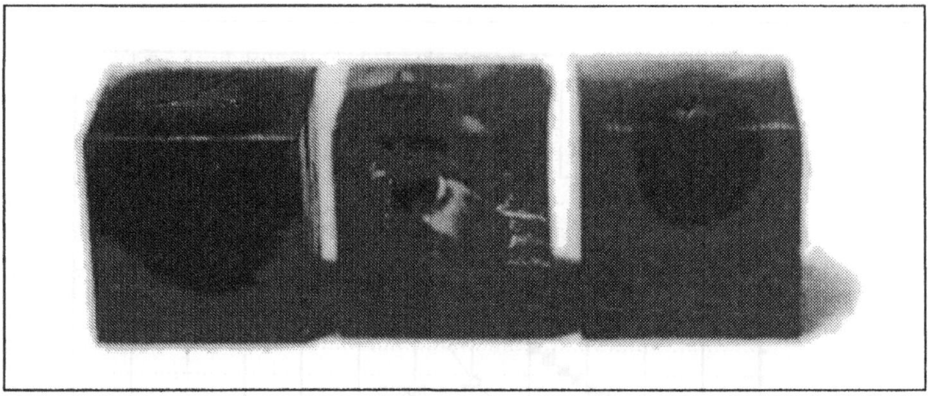

Figure 4. Semiconductor doped glasses heated in a microwave field, 3 cubic samples 25 x 25 x 25 mm^3.

TABLE 2. HEATING CONDITIONS FOR THE CUBIC SAMPLES SHOWN FIGURE 4

Preheating	Up to 500°C in a conventional furnace
Microwave heating:	From 500°C up to about 800°C
• Frequency	2.45 GHz
• Power	Max. 2.1 kW
• Time	From 1 minute (middle sample) up to 3 minutes (left sample)

Especially a combination of conventional heating with additional microwave heating will allow a rapid and homogenous heating if the hot spot problem is solved.

ACKNOWLEDGMENTS

The authors want to give thanks to Jon Binner and Ron Hutcheon who did the measurement of the dielectric constants and also to Monika Willard-Porada and Thorsten Gerdes who did the heating experiments.

REFERENCES

1. R.M.Hutcheon, M.S.DeJong, F.P.Adams, P.G.Lucuta, J.E.McGregor and L.Bahen / Mat. Res. Soc. Symp. Proc. Vol. 269/1992, p. 541-551.
2. A.C.Metaxas und J.G.P.Binner "Microwave Processing of Ceramics", Advance Cer. Proc. Techn., Park Ridge NY (1990), p. 285-367.
3. A.C.Metaxas "Industrial Microwave Heating" ISBN 0 906048893, 1993.

Surface Modification of Sodium Aluminosilicate Glasses Using Microwave Energy

Z. Fathi, I. Ahmad, J.H. Simmons, D.E. Clark and A.R. Lodding

Abstract

Microwave energy has been used in ion exchange reactions for four sodium-aluminosilicate compositions. The exchangeable cation in the glass (Na^+) was replaced at the surface by a larger atomic radii cation (K^+). Secondary ion mass spectroscopy (SIMS), electron microprobe analysis (EMP) and X-ray mapping were used to evaluate the effect of microwave heating on the extent and rate of ion exchange in comparison to conventional heating.

Introduction

Glasses have been known to man for thousands of years, and their applications have gone from rudimentary purposes to top-of-the-line technology. Today's world has broadened the use of oxide glasses in diverse and far-ranging fields such as telecommunication, electronics and aerospace. The intrinsic brittle behavior of oxide glasses and their catastrophic failure have been major drawbacks to their use in applications which require mechanical strength and durability.

In general, brittle materials fail to exceed more than a fraction of their theoretical strength. This characteristic behavior has triggered extensive research aimed at developing higher strength bodies. The desired strength can be achieved by various methods that can be split into two broad categories; bulk and surface techniques. The surface techniques aim at developing a compressive stress distribution on the glass surface. The resulting strength becomes the original strength of the glass plus the strength added by surface compression. These surface modification techniques can be summarized as the following: thermal strengthening, chemical strengthening, surface crystallization, thin coating, cladding, etching and related techniques [1].

Strengths greater than those obtained by thermal toughening can be reached by chemical strengthening. However, any significant strengthening is unlikely to be achieved within times compatible with the modern production rates [2], due to the slow reactions involved in the ion exchange process. For this reason, chemical strengthening techniques are applied where safety predominates over economics.

Within our microwave processing center at the University of Florida, we have been investigating techniques that could enhance ion exchange reactions. Processes by which deeper penetration of a selected ionic species into a glass matrix can be achieved when compared to the conventional techniques.

The present study was undertaken to evaluate the effect of microwave radiation on the ion exchange reactions of K^+ for Na^+ in four different compositions of sodium aluminosilicate glass. For the four chosen glass compositions, we are reporting the penetration depth and the relative concentration of the alkali ions achieved after processing. A comparison is made of the penetration depths accomplished by microwave processing and those achieved by conventional processing for each of the sodium aluminosilicate compositions.

The sodium aluminosilicate system was found to be the system of choice for surface modification using microwave energy. Indeed, this system exhibits major structural changes accompanied with significant dielectric properties differences when the ratio of alumina to alkali ion is taken as a variable.

Theoretical Basis

Ion Exchange Process

Ion exchange reactions in oxide glasses are limited substantially to alkali ions[3]. Ion exchange is a process by which at least two ionic species exchange with each other, as shown by

$$Na^+_{(g)} + K^+_{(s)} \rightarrow Na^+_{(s)} + K^+_{(g)} \tag{1}$$

where the subscripts represent glass and molten salt solution, respectively. The driving force for the reactions is a concentration gradient which is the natural phenomenon of mixing. A constructive approach to this phenomenon is to partially replace an ionic species pre-existing within a given glass matrix by another with a larger radius. The exchanged glass layer tries to expand. It cannot, due to restraint from the underlying bulk glass. The surface is thus put under compression.

The process mentioned above is diffusion controlled. The concentration of the selected species and its penetration depth are of primary importance and are related directly to the rate of exchange. The objective was to evaluate the effect of microwave heating on the extent and rate of ion exchange in comparison to conventional heating.

Microwave Processing

It is reported in the literature that an array of ceramic products have been heat treated in the presence of microwave radiation. Characterization of these materials revealed major microstructural differences when compared to those conventionally processed. It is our belief that the addition of activations other than those of a thermal nature, namely electrical and magnetic, account for the observed property differences.

Microwave heating is intrinsically different from conventional heating. It is a volumetric heating in which the sample itself generates heat due to the ability of the electric and magnetic fields to polarize molecules and the inability of the molecules to keep up with the rapid reversal of the fields [4]. The resulting dielectric losses account for the volumetric heating. Microwavc radiation falls within a range of frequencies at which the orientation polarization is the predominant polarization mechanism. There are losses from the other polarization mechanisms due to the extended tails of the loss curves which are different for each given system.

In glasses, the ionic dipole motions can be split into two categories: (1) deformation polarization in which permanent dipoles oscillate under an applied electric field and (2) defect associated dipoles which oscillate between several equilibrium positions. The temperature of glass plays an important role because, at high enough temperatures, the conductivity of the glass is such that a discharge of the induced dipoles occurs.

Experimental

The selected compositions in the sodium aluminosilicate system are shown in Table 1. The glasses that were studied had a constant concentration of alkali ions (15mol% of Na_2O). The molar ratio of alumina to sodium oxide, defined as Γ, was taken as a variable and was equal to 0.2, 0.5, 1.0 and 1.1.

TABLE 1. COMPOSITIONS OF THE GLASSES INVESTIGATED.

Components	Γ=0.2	Γ=0.5	Γ=1.0	Γ=1.1
Na_2O	15.0	15.0	15.0	15.0
Al_2O_3	3.0	7.5	15.0	16.5
SiO_2	82.0	77.5	70.0	68.5

The glasses were melted in alumina crucibles at 1500°C for 24 hours and annealed at 600°C for 5 hours. The alumina crucibles were cut to provide workable samples. Care was taken to disregard the glass areas that were in contact with the crucible walls. The faces that were to be ion exchanged were polished to a 1μm finish. The samples were coated with a potassium containing medium[1].

Electron microprobe analysis was used to evaluate the penetration depths involved in the ion exchange reactions. The change of X-ray intensity of the alkali species was found, by several investigators, to be related to time and to be proportional to the electron beam current, specimen temperature and bulk composition of the glass [5].

In the present study, we avoided the use of a stationary electron beam and conducted the analysis using a scanning beam sweeping a 100μm by 1.5μm area. The samples edges to be analyzed were polished to a 1μm finish.

The ion exchange reactions were performed in a multimode microwave oven[2] (2.45 GHz) using hybrid heating [6,7]. The samples were placed in an insulating material lined with a microwave susceptor which, in turn, heated the sample. The reactions temperatures were controlled with an error of ±6°C using an Inconel-shielded K-type thermocouple.

Results and Discussion

The results for the interdiffusion of $Na^+ \leftrightarrow K^+$ at 400°C for 30 min are shown in Figure 1. The general trend is an increasing potassium penetration in the different glass compositions as Γ increases from 0.2 and 0.5 to 1.0. As Γ goes above 1.0, interdiffusion appears to decrease.

If we define P^i_m and P^i_c as the respective penetration depths of microwaved and conventionally prepared samples, and $\Delta P^i_{m,c}$ as the difference between these penetrations for the (i) compositions. We found that P^i_m, P^i_c and $\Delta P^i_{m,c}$ all reach a maximum in the Γ=1.0 composinon.

Initially, as Al_2O_3 is introduced into a binary alkali-silicate glass, only AlO_6^- is formed [8] As Γ increases, there is more formation of AlO_6^- along with AlO_4^-. The ratio AlO_6^-/AlO_4^- decreases with further increase of Γ. The AlO_6^- concentration goes through a maximum, the point where the activation energy for alkali diffusion is at a maximum. With further increasing values of Γ, mainly AlO_4^- groups will be formed [9], and both the electrical and steric interactions of the alkali ion with the glass network decrease. This is reflected by a rise in the alkali ion diffusion coefficients. Increasing the amount of alumina in the alkali silicate glass reduces the number of oxygens which are only bound to one network former, either Si^{4+}or Al^{3+}. The number of nonbridging oxygen ions decreases to a minimum for Γ=1.0 in which all the alkali ions are positioned to the negative charge of the AlO_4^- groups [10]. The activation energy for diffusion reaches a minimum for Γ=1.0. For compositions having Γ>1.0, the alumina can no longer be present only as AlO_4 tetrahedra because of the absence of compensating positive charges. Thus, $AlO_{4,cl}$ triclusters groups will form. The increasing packing density makes it more difficult for the alkali ion to move out of or through the triclusters. This translates into a small diffusion coefficient [10]. This trend of interdiffusion is retorted by the conventional curves in Figure 1.

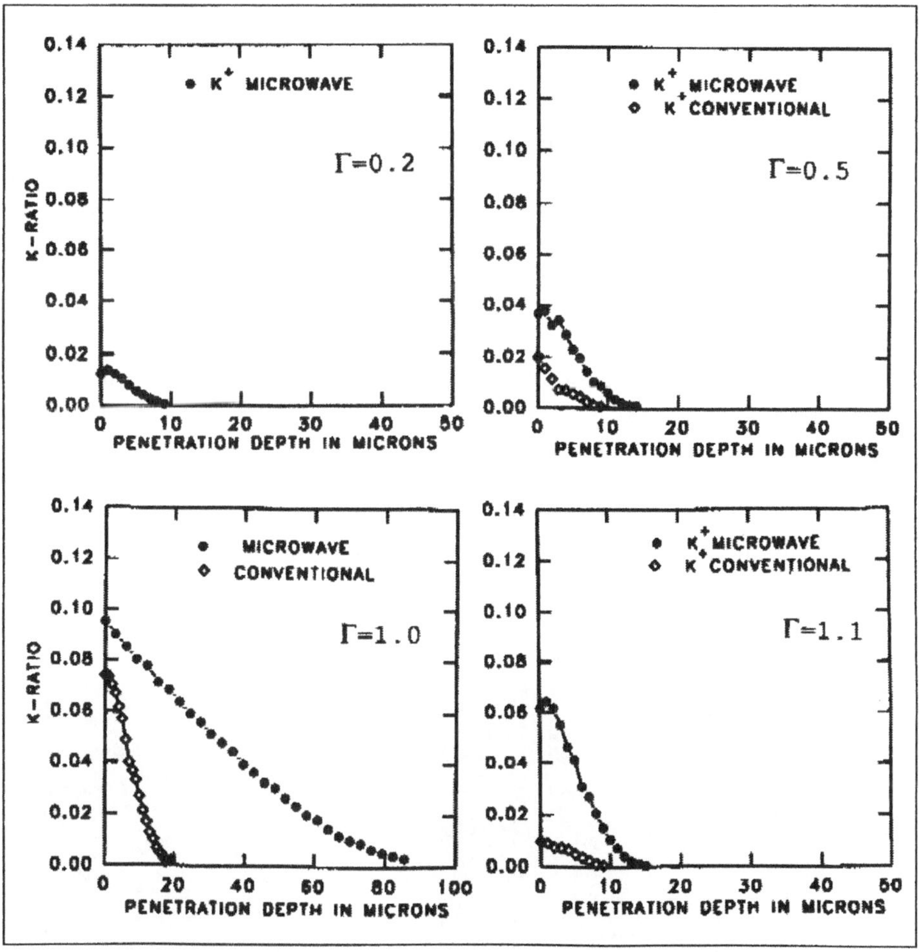

Figure 1. Penetration depths (in μm) of potassium into the glass compositions. K-ratios are relative to tugtupite for sodium and biolite for potassium.

The dielectric properties of sodium aluminosilicate glasses at microwave frequencies exhibit changes when the compositional factor, Γ, is taken as a variable [11]. It was found that the temperature-dependent losses were due to the broad tail of the migration loss mechanisms. The temperature-independent losses are due to the deformation losses and are not associated with the concentration of the non-bridging oxygens. Those losses were attributed to resonance with vibrational modes involving large atomic grouping.

The dielectric losses followed the structural model and were found to increase with increasing values of Γ up to Γ =1.0 at which the dielectric losses exhibit maximum values. When compared to conventionally heat treated samples, Pmi can be correlated to the loss tangent of each of the glass compositions [11].

Two characterization techniques were used to verify the findings of the electron microprobe analysis. Figures 2 and 3 correspond to X-ray mapping and SIMS depth profiling of the Γ=1.0 composition for microwaved and conventionally processed samples. As shown, there were higher surface concentrations and deeper penetration of the K^+ into the Γ=1.0 glass composition that were heat treated in the presence of microwave radiation.

Conclusions

It is premature to give definite reasons for the high interdiffusion in the case of microwave heating. We believe that the vibrational frequency of an ion within a potential well is increased due to the electric field of the microwave radiation. The rapid reversal of the electric field might be increasing the local temperature of the ion, thus making the ionic jump out of the well easier. More work needs to be done in order to understand fully how microwave radiation affects interdiffusion.

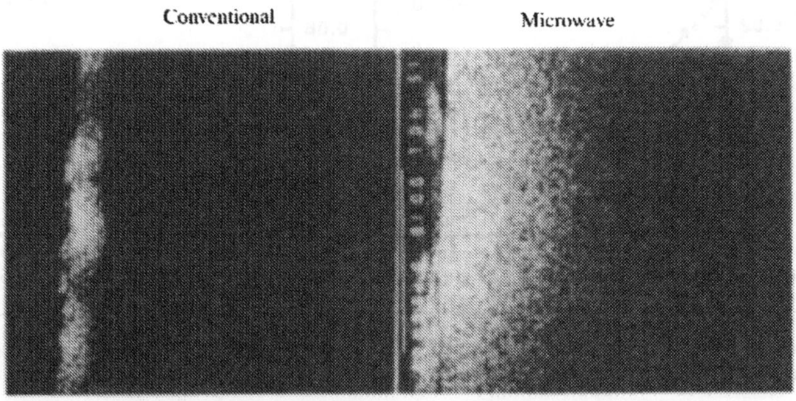

Figure 2. X-ray mapping of potassium in the Γ = 1.0 composition.

The ion exchange process using microwave energy provides a means by which deeper penetration and higher surface concentrations can be achieved, as compared to conventionally ion exchanged glass. The sodium aluminosilicate compositions having $\Gamma=1.0$ exhibited the deepest penetration of K^+, as compared to the compositions $\Gamma=0.2$, 0.5 and 1.1. A correlation can be drawn between the dielectric losses of the different compositions and their corresponding K^+ depths of penetration.

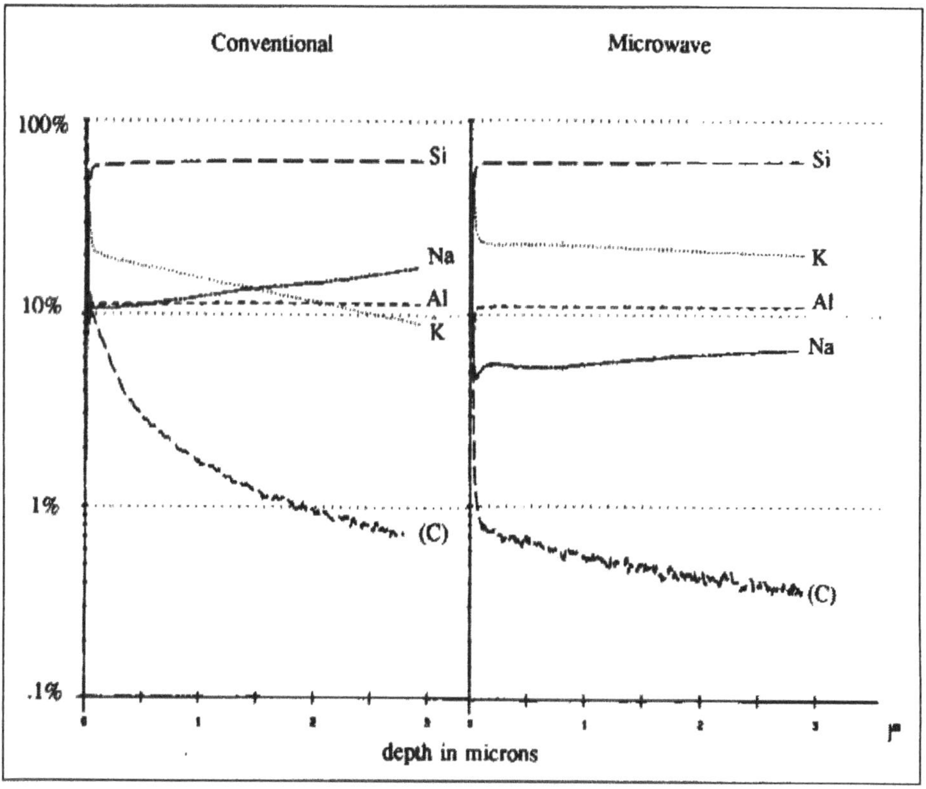

Figure 3. SIMS depth profiling on $\Gamma=1.0$ composition.

ACKNOWLEDGMENTS

The authors gratefully acknowledge support from the Space Research Institute, Florida High Technology and Industry Council (FHTIC) and the Defense Advanced Research Project Agency (DARPA).

The authors acknowledge W.A. Acree and the Major Analytical Instrumentation Center (MAIC) at the University of Florida for the electron microprobe analysis.

Special acknowledgement for H. Odelius and A. Lodding in the SIMS-laboratory at Chalmers University of Technology for their SIMS depth profiling.

REFERENCES

1. Donald, I.W., *J. of Materials Science,* **24** (1989) 4177-4208.
2. Doyle, P.J., *Glass Making Today,* Portcullis Press, Redhill.
3. Metaxas, A.C. and Meredith, R.J., *Industrial Microwave Heating,* Peter Peregrinius, Ltd. (1983).
4. Terai, R. and Hayami, R., *J. of NonCryst. Solids,* 18:217-264 (1975).
5. Clark, D.E., Hench, L.L. and Acree, W.A., *J. Am. Cer. Soc.,* 58:11-12, Nov-Dec., (1975).
6. De, A.S., Ahmad, I., Whitney, E.D. and Clark, D.E., *Effect of Green Microstructure on Microwave Processing of Alumina,* Presented at the 14th Annual Conference on Composites and Advanced Ceramics, January 14-17, 1990, Cocoa Beach, FL. Proceedings, 11[9-10] pp. 1743-1753 (1990).
7. Fathi, Z., Ahmad, I. and Clark, D.E. *Mat. Res. Soc. Symp. Proc.,* Vol. 180, 401-406, (1990).
8. Day, E. and Rindone, G.E., *J. Am. Cer. Soc.,* 45:579 (1962).
9. Day, E. and Rindone, G.E., *J. Am. Cer. Soc.,* 45:489-496 (1962).
10. Isard, J.O., *J. Soc. Glass Tech.,* 43, 113T-123T, (1959).
11. Topping, J.A., Isard, J.O., *Physics and Chemistry of Glasses,* Vol. 12, No. 6, Dec. (1971).

Microwave Crystallization of Glass

A. Boonyapiwat, D.C. Folz and D.E. Clark

Abstract

Microwave crystallization of $Li_2O{\cdot}2SiO_2$ was investigated. The crystal growth rate in microwave crystallization was greater than in conventional crystallization. The microwave heating rate of glass was compared to that of the fully crystallized glass-ceramics. Dielectric properties as a function of temperature of the glass-ceramic also were investigated. The glassy phase exhibited higher dielectric constant and losses than those of the fully crystallized glass-ceramics. For a nucleated glass sample, the dielectric properties changed with the phase transformation. Results from the analysis of microwave power distribution at the interface between the glass and crystal suggested that a multimode microwave applicator was suitable for uniform crystal growth.

Introduction

It has been observed that microwave processing can enhance the processing kinetics in several ceramic and glass systems [1,2]. These enhanced processing kinetics are thought to be the result of microwave-enhanced ionic transport. Since the phase transformation in glass-ceramic processing involves ionic transport across the interface between the glassy phase and the crystalline phase, the use of microwaves may offer advantages over conventional processing. Preliminary research has shown that the kinetics of glass crystallization may be accelerated in a microwave field [3,4].

Incorporation of a microwave processing unit into a glass-ceramic processing line would not be complicated. A possible heating schedule and the processing line with the incorporation of a microwave processing unit are shown in Figure 1. After the glass is melted at T_m, it is cooled between t_1 and t_2. During this time, the glass is formed into its final shape. The as-formed glass products are subjected to heat treatment in the temperature range of the phase transformation to glass-ceramics. In the conventional process, the conversion of glass to a glass-ceramic takes place in a conventional furnace or lehr. This unit can be replaced with a

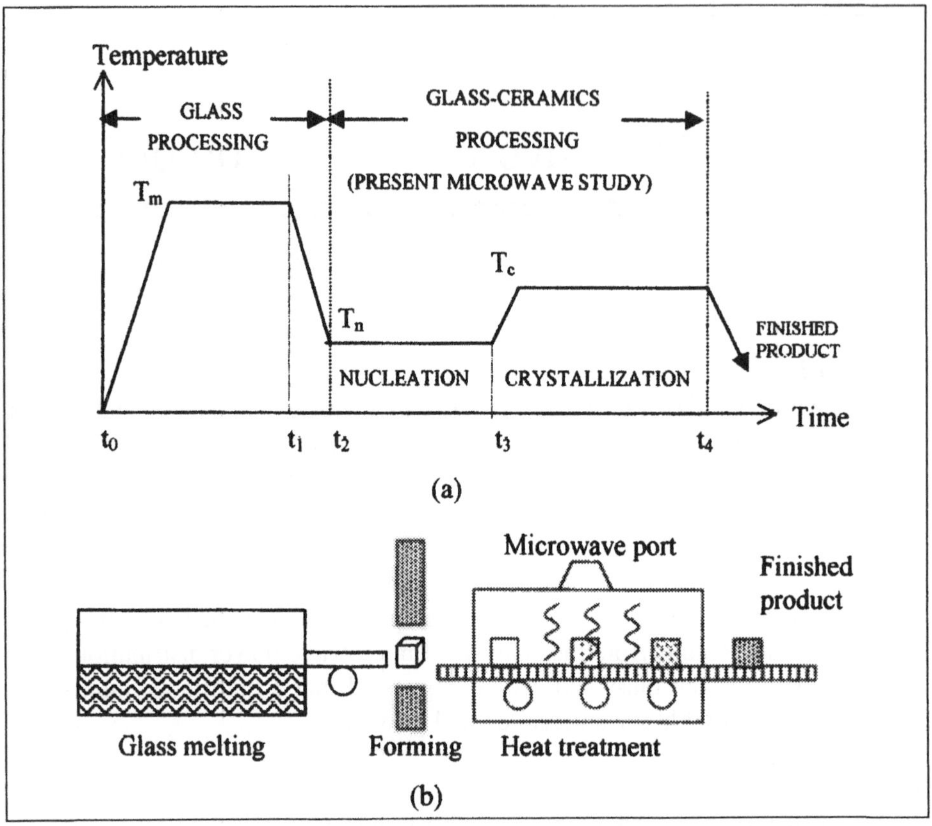

Figure 1. Glass-ceramic processing: (a) heating schedule and (b) microwave heat treatment unit incorporated into a glass-ceramic process line.

microwave heating unit, as shown in Figure 1 (b). No other sophisticated modifications to the glass-ceramic processing line are needed. There is no need for a microwave susceptor or the use of microwave hybrid heating (MHH) which generally is required to raise the temperature to a point where glasses and ceramics begin to absorb microwaves. Since the temperature of the as-formed products is still high, direct microwave heating is possible in glass-ceramic processing.

The goal of this study was to investigate the effects of microwave energy on glass-ceramic processing. The microwave effect on the transformation kinetics was determined through a comparison of crystal growth with microwave heating and conventional heating. The study of microwave interaction with different phases in glass-ceramics was based on the dielectric property analysis and a comparison of the heating rate during direct microwave irradiation.

The analysis of microwave power absorption near the interface between glass and crystal also was performed. Since a curved interface between materials of different dielectric properties can cause an uneven electric field distribution along the interface [5], one may suspect that the interface between crystal and glassy matrix can cause uneven electric field distributions around the crystal during microwave processing. This uneven electric field distribution can lead to uneven power absorption in the glass around the crystal. The information from microwave power distribution analysis near the interface between the glass and the crystal can be used to suggest a practical microwave processing method for glass-ceramics.

Experimental

Glass Preparation and Heat Treatment

The $Li_2O{\cdot}2SiO_2$ glass-ceramic system was examined in this study. A nucleating agent was not incorporated into the glass composition. The glass produced by melting the primary constituents, 5.0 μ Min-U-Sil SiO_2 and reagent grade Li_2CO_3, was obtained from Specialty Glass Company (Oldsmar, FL). After the glass composition was melted and homogenized in a fused silica crucible for 24 hours at 1350°C, the glass was wet-fritted in deionized water, then milled with alumina grinding media and sieved to –325 mesh.

The glass frit powder was remelted to produce the required specimen configuration. The frit was melted in a covered platinum crucible at 1400°C and was held at that temperature for 6 hours. Rods of 14 mm diameter and 60 mm length were cast by pouring the molten glass into a graphite mold. After casting, the rods were annealed in an electric furnace at 450°C for 1 hour before being furnace-cooled to room temperature. Glass specimens were produced by cutting the rods using a low-speed diamond saw.

Phase Transformation Kinetic Study

Two methods of heat treatment were employed in this study: microwave hybrid heating (MHH) and conventional heating. The reason for using MHH rather than direct microwave irradiation (even though this glass can be directly heated by microwaves) was that MHH allowed the precise temperature measurement using a thermocouple. (Based on previous experience, temperature measurements during direct microwave heating require a special temperature probe.) The details of the experimental set up for MHH and conventional heating were described elsewhere [4]. Quantitative stereology was used to evaluate the microstructure of the samples [4].

Heating Rate Study

A heating rate study was conducted to evaluate the possibility of using the direct microwave heating technique in glass-ceramic processing. This study was based on monitoring the increase in temperature of a sample during direct microwave heating (a microwave susceptor was not used). A variable frequency microwave oven, a Variwave® Model LT 1500 from Lambda Technologies, Inc. (Raleigh, NC) was chosen for this study because it was equipped with a special temperature probe that allowed monitoring the temperature during direct microwave heating. The Variwave® contains a travelling wave tube capable of sweeping microwaves over frequencies ranging from 6.5 to 18 GHz. The power of the oven is tunable from about 0 to 150 watts, depending on the selected frequency range. The cavity dimensions of the oven are 0.25 x 0.25 x 0.30 m^3. The temperature of the specimen was monitored by an optical fiber-based thermal monitoring system manufactured by Luxtron Fiber Optics (Santa Clara, CA). This system is based on the thermoluminesence property of the material at the tip of the optical probe. Since the maximum operating temperature of the optical probe is 300°C, the heating rate experiments were conducted over a range from room temperature to 250°C.

Dielectric Properties Measurement

Dielectric property measurements were performed at AECL Research (Chalk River, Ontario, Canada). The measurements were based on the cavity shift or cavity perturbation technique [6]. The measurement was done by first heating the sample conventionally to a specified temperature, and then rapidly moving the sample into a measurement cavity. The measurement was then performed before significant cooling occurred. After measurement, the sample was brought back to the furnace and was heat treated to another specified temperature. The dielectric property measurements were performed on samples while heating from room temperature to 600°C and during cooling-down to room temperature.

Calculation of Microwave Power Distribution

Quasistatic analysis is used in this study. This technique is based on the assumption that the applied electric field is electrostatic, even though the real applied electric field may not be electrostatic [5]. This technique can be used in the case where the area of interest is small compared to the wavelength of the applied electric field [5]. Because the dimension of the area of interest (a crystal and the nearby surrounding glassy phase in glass-ceramics which is in the micron range) is much smaller than the wavelength of the microwave in use (millimeter range), quasistatic analysis can be used.

Since $Li_2O{\cdot}2SiO_2$ crystals have a prolate spheroidal shape, the analysis is performed in prolate spheroidal coordinates. The details of derivation of equations and the analysis are described elsewhere [7].

Results and Discussion

The crystal radius as a function of time is shown in Figure 2. Crystal growth can be evaluated from the slope of the crystal radius versus time. While the results of crystal growth from conventional heating agreed with the results from Deubener et al [9], the crystal growth rate observed in microwave heating was greater than in conventional heating. Moreover, microwave heating results compared to conventional heating appeared to have shorter induction times. These results agreed with previous work [4]. The induction time was thought to be the result of very slow crystal growth rates (size-dependent growth rate) when the crystal sizes were in the range of critical nuclei [8,9]. The induction time for crystal growth is always observed in the single-stage heat treatment [8]. The results from this study agreed with the theory discussed above since the induction time only occurred for the as-cast and annealed samples (single-stage heat treatment). Samples that were heat treated at 470°C to develop stable nuclei did not exhibit an induction time. Both microwave and conventional heating showed similar results with regard to induction times.

The starting nuclei/crystal size or the volume fraction of crystals did not affect the crystal growth rate in microwave heating. However, the as-cast samples appeared to have a slightly faster growth rate than the samples having stable nuclei. For conventional heating, the crystal growth rate did not show an effect from pretreatment of the sample. Heat treatment (470°C) that was in the glass transition range (450–480°C) [10] may have resulted in a more dense glass structure. For the same composition, microwave absorption by the material having a dense structure tended to be lower than that of the material having a more open structure.

The microwave heating rate of glass as compared to the fully crystallized glass-ceramic is shown in Figure 3. The glass sample was heated from room temperature without using a microwave susceptor or insulation, while the fully crystallized sample did not heat at all. Higher microwave power resulted in higher heating rates in glass samples. Thus microwave processing of glass-ceramics was a self-limiting process since heating stopped when the glass was fully converted to a glass-ceramic.

The dielectric properties of the $Li_2O{\cdot}2SiO_2$ glass-ceramic system are shown in Figure 4. The glassy phase of $Li_2O{\cdot}2SiO_2$ had higher dielectric constant and dielectric loss values than the crystalline phase of the same composition. The values of the dielectric properties as a function of temperature for the glass and the

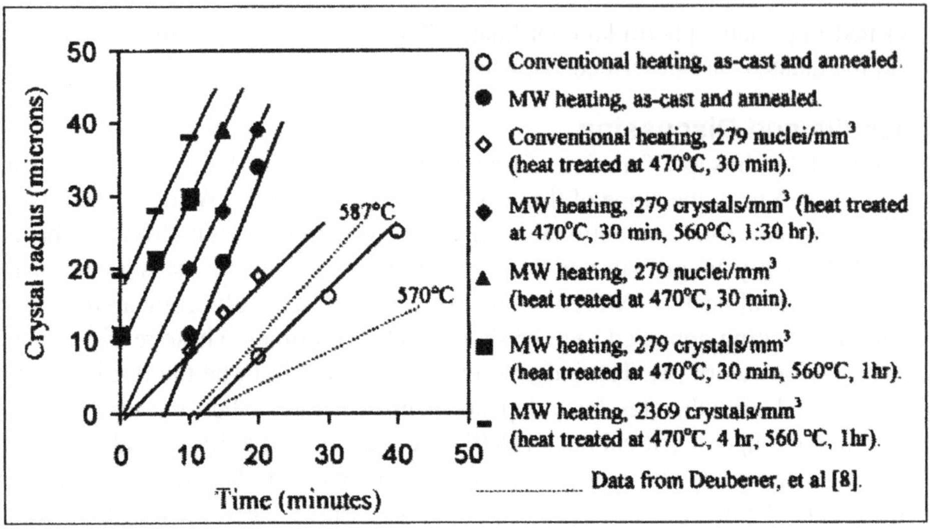

Figure 2. Crystal size versus processing time for conventional and microwave crystallization at 580°C.

fully crystallized glass-ceramics measured during heating were the same as the values measured during cooling. However, the prenucleated glass exhibited hysteresis behavior. The dielectric properties of this sample were the same values as those of the glass up to the temperature where crystallization occurred. As the sample was cooling down, the dielectric properties of this sample followed those of the crystalline phase. This behavior resulted from the growth of the nuclei into crystals. Microstructural analysis of samples performed after measuring the dielectric property confirmed this behavior.

The fact that the dielectric losses of the glassy phase had higher values than those of the crystalline phase can be used to explain the heating rate in Figure 3. The relative dielectric losses of the glassy phase at room temperature were four times higher than those of the crystalline phase (0.152 for glass and 0.038 for crystal). The microwave power absorption in a material (per unit volume) can be expressed as [11]

$$\frac{P}{V} = \omega\varepsilon_0\varepsilon''_{\text{eff}}E^2_{\text{rms}} \tag{1}$$

where: P = power absorb,
V = volume,
ε_0 = dielectric constant of free space
ε_{eff} = effective dielectric losses, and
E_{rms} = root mean square of the internal electric field.

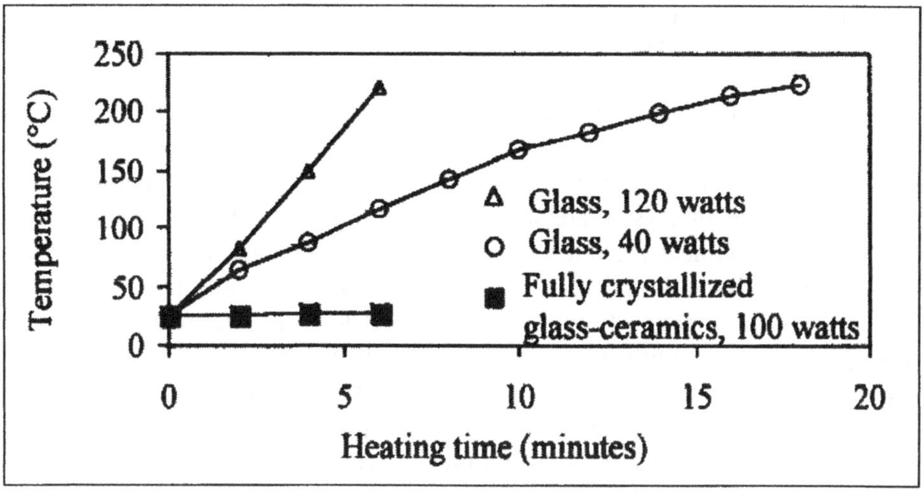

Figure 3. Temperature of the sample heated by direct microwave irradiation as a function of time (frequency = 9.5 GHz).

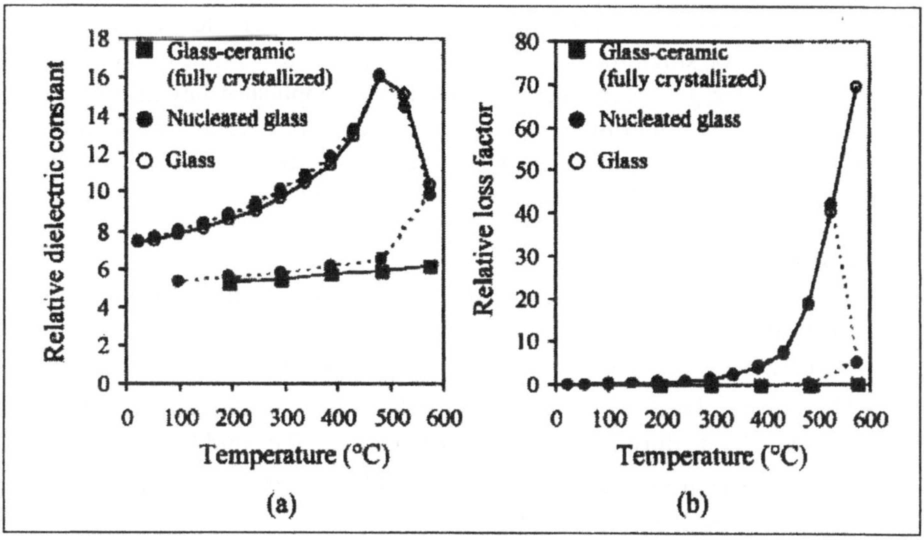

Figure 4. Dielectric properties of the $Li_2O{\cdot}2SiO_2$ glass-ceramic system at 2.467 GHz as a function of temperature: (a) relative dielectric constant and (b) relative loss factor.

As predicted by equation (1), the microwave absorption in the glassy phase at room temperature is about 16 times that of the crystalline phase.

A schematic illustrating the direction of the electric field and the positions where the microwave power absorption was evaluated is shown in Figure 5. The calculated distributions of microwave power absorption near the glass/crystal interface for two cases, applied electric field direction parallel and perpendicular to the major axis of crystal, are shown in Figure 6. The behavior of microwave power absorption near the interface in both cases were similar. The microwave power absorption at the position close to the glass/crystal interface varied with the angle of the direction from the applied electric field. The microwave power absorption became uniform as the distance from the glass/crystal interface increased. Generally, the lowest microwave power absorption was at the interface having angle 0° from the direction of the applied electric field, whereas the highest absorption was at 90° to the applied electric field. The microwave power absorption in glass was higher than that in the crystal (ranging from about 40 to 230 times), as seen in Figure 6. This resulted from the glassy phase having much higher dielectric losses than the crystalline phase.

It is suggest that the results shown in Figure 6 can be used to predict the behavior of crystal growth during microwave processing. Glass-ceramics processed in a microwave applicator having a single electric field direction (such as single-mode or single-path type applicators) tend to exhibit oriented crystal growth, while the glass-ceramics processed in an applicator having many electric field directions (multimode applicator) may tend to exhibit uniform crystal growth. All the experiments in this study were performed in multimode applicator types, so the oriented crystal growth was not observed.

Summary

- The $Li_2O{\cdot}2SiO_2$ glass can be heated directly using microwaves, while the crystalline phase of the same composition cannot be heated. Using higher applied microwave power resulted in higher heating rates in glass samples.
- The starting size and the number of nuclei or crystals did not effect the crystal growth rate during microwave heating under the conditions in this study.
- Crystal growth rates in a microwave field are higher than those produced using conventional heat treatments.
- The values of the dielectric constant and the dielectric losses of the glassy phase are higher than those of the crystalline phase.
- The values of the dielectric constant and the dielectric losses change during the phase transformation from glass to glass-ceramics.

• The microwave power distribution near the glass/crystal interface evaluated by the quasistatic method leads to the perception that microwave processing of glass-ceramics in an applicator having a single electric field direction may yield oriented crystal growth while processing in a multimode applicator yields uniform crystal growth.

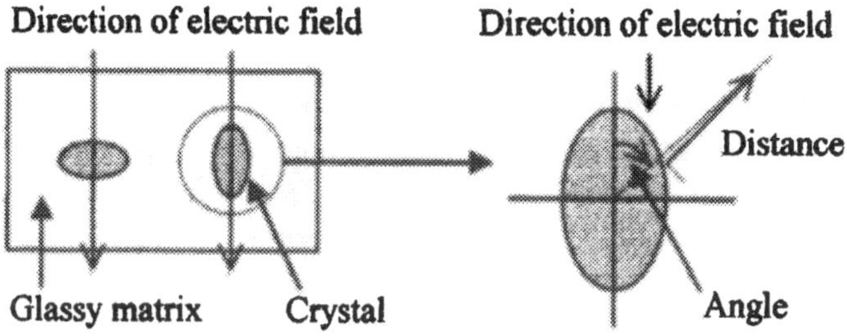

Figure 5. Schematic illustrates the direction of electric field and the positions where the microwave power absorption are evaluated. The case that the applied electric field parallel to the major axis of crystal is shown in detail.

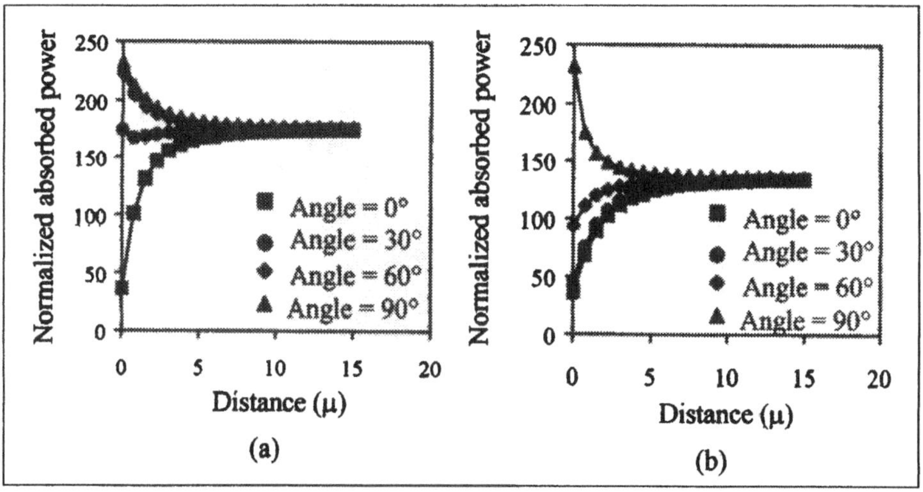

Figure 6. Normalized absorbed power (to the power absorbed by crystal) as a function of distance from the crystal surface, at several directions (crystal size: major axis length = 10 μ, minor axis length = 6.1μ, temperature = 550°C): (a) applied electric field parallel to major axis and (b) applied electric field perpendicular to major axis.

ACKNOWLEDGMENT

The authors wish to express their gratitude to R. M. Hutcheon for dielectric property measurements and EPRI for financial support (contract No. 08060-12A2).

REFERENCES

1. W.H. Sutton, *Am. Ceram. Soc. Bull.,* **68**, p. 376-386 (1989).
2. D.E. Clark and W.H. Sutton, *Ann. Rev. Mater. Sci.,* **26**, p. 299-331 (1996).
3. A.D. Cozzi, Z. Fathi, R.L. Schulz, and D.E. Clark, *Ceram. Eng. Sci. Proc.,* **14** (9), 856-862 (1993).
4. A. Boonyapiwat and D.E. Clark, p. 505-514, in *Microwaves: Theory and Application in Materials Processing,* edited by D.E. Clark, D.C. Folz, S. J. Oda, and R. Silberglitt, The American Ceramic Society, Westerville, Ohio (1995).
5. S. Ramo, J.R. Whinnery, and T. Van Duzer, *Field and Waves in Communication Electronics,* John Wiley & Sons, New York (1994).
6. F.P. Adams, M.S. deJong, and R. M. Hutcheon, *J. Microwave power Electromagn Energy,* **27**, 131-135 (1992).
7. A. Boonyapiwat, *PhD Dissertation,* University of Florida, to be published.
8. J. Deubener, R. Bruckner, and M. Sternitzke, *J. Non-Cryst. Solids,* **163**, 1-12 (1993).
9. K.F. Kelton and A.L. Greer, *J. Non-Cryst. Solids,* **79**, 295-309 (1986).
10. J.Y. Thompson, *PhD Dissertation,* University of Florida (1995).
11. A.C. Metaxas and R.J. Meredith, *Industrial Microwave Heating,* Peter Peregrinus, London (1988).

Microwave Joining of Engineering Ceramics

J.G.P. Binner, P.A. Davis, T.E. Cross and J.A. Fernie

Abstract

A TE_{102} single mode microwave applicator has been used to examine the potential for joining a range of engineering ceramics without the use of interlayers. Materials examined included silicon carbide, alumina and zirconia. Experimental variables included both temperature and pressure whilst joining time was kept to a minimum, typically ≤ 10 minutes. Successful joins were produced with reaction bonded SiC, medium and low purity Al_2O_3 and ZrO_2. Analysis of the results has led to the conclusion that it is possible to achieve successful microwave joins by both solid state and liquid phase mechanisms.

Introduction

The ability to form rapid, high quality joins between a range of ceramic materials using microwave energy has been demonstrated by a number of researchers[1-16] using two main approaches, viz direct and indirect joining. In the latter case, use is made of some form of interfacial layer to assist in achieving bonding. Ceramics which have been investigated to date include Al_2O_3[1-11], mullite[4], Si_3N_4[1,2,4,10] and SiC[11-16]. Rods, tubes and bars of varying size have all been joined.

Fukushima *et al.*[1,2] and Davis *et al.*[3] have investigated the microwave joining of different grades of alumina. In both cases, relatively low purity materials (< 96 %) could be joined readily whilst high purity ceramics (>99%) were insufficiently lossy to heat, making direct joining difficult. Characterisation of the joined samples showed little difference before and after joining; no grain growth or cracking was evident. Similar results were found by Palaith *et al.*[4] and in all cases evidence was found that the grain boundary phases had been melted during the process. Strengths of the joined materials were at least as strong as, and often stronger, than the base material[1,2,4]. The inability to achieve direct joining in high purity alumina led to experiments involving lower purity[1,2] interlayers. Using this approach, successful joins were achieved with strengths approaching

90% of the base material. Sealing glasses have also been tried and found to be successful[4-6] as have alumina gels[7-9]. The latter have the advantage that, at the joining temperature, the gel transforms into colloidal-alumina which subsequently sinters providing consistency across the join region. The use of alumina powder intermediaries[10] was less successful in that, although high strengths were achieved, results were more erratic due to shrinkage cracks caused during sintering within the intermediary. Attempts to simultaneously sinter and join green ceramics with a slip interlayer were more successful[11].

Successful microwave joins have also been produced in mullite[4] and silicon nitride[1,2,4,10]. Palaith *et al*[4] found that in the region of the join the hardness and toughness were both increased by 20-30%. This was attributed to a localised reduction in porosity caused by the melting of a glass phase in the mullite. Join strengths equivalent to those of the base material were found for a high dielectric loss, Fe-rich grade of Si_3N_4 whilst a lower loss Y/Al-containing grade only displayed strengths of ~20% of the base material[1,2]. The introduction of a thin Fe-rich Si_3N_4 sheet between the surfaces of the Y/Al-doped Si_3N_4 improved the quality of the resultant join and the strength increased up to ~70% of the base material1. Similarly, the use of powdered silicon nitride intermediaries[10]. between hot pressed Si_3N_4 samples resulted in a continuous bond with the parent material.

Reaction bonded SiC (RBSC) samples have been directly microwave joined[12,13] a Si-rich band being observed at the join line. The use of Al interlayers resulted in joins as strong as the base material itself[14]. Reaction bonded SiC has also been bonded to sintered SiC (SSC)[12,13]. The results suggested that although substantial cracking had occurred in the interfacial region, Si from the RBSC had wetted and bonded to the SSC. Although high strength microwave joins (400 MPa) have been produced between SSC samples by utilising conditions of 2050°C and 8 Mpa[15], indirect approaches have resulted in joins being produced under substantially less demanding conditions. Dense and homogeneous interfacial regions could be achieved when Si interlayers were used[16]. Attempts at using in-situ combustion synthesis of SiC/TiC interlayers were less successful due to oxidation of the metals prior to reaction and the formation of a porous reaction product[16].

The above results imply that some form of phase which softens or melts at the joining temperature is a requirement for microwave joining of engineering ceramics. In materials where this occurs naturally as a grain boundary phase the ceramics can be directly joined. In materials where no such phase exists, then some intermediary is required which will provide a similar function. Often this can be a different grade of the base ceramic itself, but other materials such as glasses, gels and silicon have been found to be successful depending on the composition of the ceramics to be joined. Two questions are left unanswered by such a simple approach. First, does the softening phase fulfil any other role besides what could be crudely described as a 'glue' and second, is there no chance of achieving successful joins when softening phases are absent, ie by solid state processes? This paper, based on work carried out at the University of Nottingham over the past 3-4 years, attempts to address these questions.

Experimental

A range of ceramics were chosen to investigate the effects of various material characteristics on microwave joining. The materials included 2 grades of SiC, 3 grades of Al_2O_3 and 1 grade of ZrO_2. Details of the materials used and the sample configurations are provided in Table 1 below. All butting faces were ground using a diamond wheel to a surface finish of ~0.5 μm and then cleaned in a solution of methanol, soap and warm water using an ultrasonic bath before drying.

TABLE 1. MATERIALS AND DIMENSIONS

Material	Density/gcm^{-3}	Grain size/μm	Dimensions*
Hot-pressed SiC (HPSC)	3.15	4	5 mm x 5 mm x 50 mm
Reaction-bonded SiC (RBSC)	3.10	10	Ø=15 mm, l=9 mm
85% alumina (AD85)	3.41	6	Ø=10 mm, l=50 mm
94% alumina (AD94)	3.70	12	Ø=10 mm, l=50 mm
99.8% alumina (AD99. 8)	3.96	3	Ø=10 mm, l=50 mm
8-wt%-yttria-doped partially stabilized zirconia (YPSZ)	5.98	0.4	Ø=7 mm, l=40 mm

*Ø is diameter; l is length.

A simplified diagram of the microwave system used to carry out the joining experiments is illustrated in Figure 1. Either a 300 W or 1 kW 2.45 GHz magnetron provided microwaves to a TE_{102} mode rectangular cavity. A matrix of experiments was performed for each material in which the joining temperature and pressure were varied. Heating rate, cooling rate and time at temperature were maintained as consistently as possible for any given material. To improve the uniformity of the temperature profile within the rods, they were encased in a foamed alumina insulation. A stainless steel support rig was constructed around the applicator to allow a compressive axial load to be applied to the ends of the rods during the joining cycle. The load was maintained at a constant level throughout the heating cycle by releasing the pressure as required by thermal expansion. The surface temperature of the sample was monitored using an

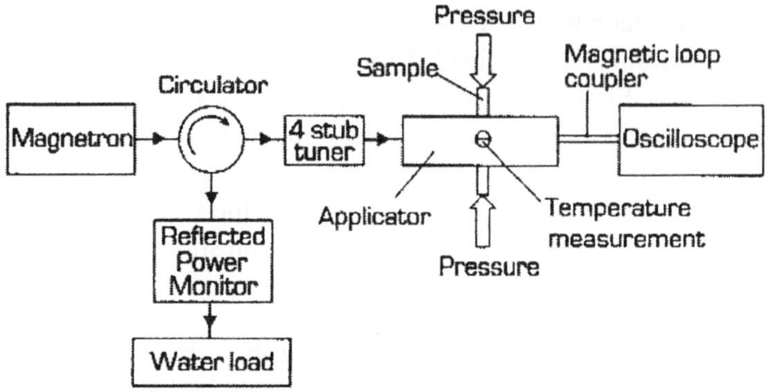

Figure 1. A simplified control diagram for the microwave heating system.

Accufiber lightpipe system connected via a microprocessor to a personal computer. The sapphire lightpipe was inserted through a choke in the narrow wall of the cavity and clamped into position 1 mm above the butting faces. The manufacturers quote an accuracy of ±5°C for the system as used.

After joining, the rods were sectioned and polished using normal preparation procedures. An SEM fitted with an EDX detector was used to characterise the join region. When very successful joins were produced, the only method of determining the location of the join was via a ~ 100 μm misalignment of the rods. A number of the samples were etched in hydrofluoric acid to highlight the grain boundaries. The flexural strength of some of the joints made between alumina samples were measured using four point modulus of rupture tests.

The dielectric properties of the ceramics were measured at 2.45 GHz as a function of temperature using a surface probe technique[17]. Approximate values were calculated for the glassy grain boundary phase present in the lower purity alumina ceramics using the refractive index mixture equation[18]

$$\varepsilon_m^{1/2} = \upsilon_1\varepsilon_1^{1/2} + \upsilon_2\varepsilon_2^{1/2}$$

where, ε_m is the permittivity of a mixture comprising two phases of values ε_1 and ε_2 and volume fraction υ_1 and υ_2 respectively. Values for both AD85 and AD94 were substituted for ε_m and for AD99.8 for ε_2 in separate calculations and average values for ε_1, the glassy phase, calculated over the temperature range 100-1200°C. The calculations thus assume that the AD99.8 is representative of the composition of the polycrystalline phase in the two lower purity ceramics.

Results and Discussions

Silicon Carbide

The processing parameters achieved are outlined in Table 2. Despite all attempts, including the introduction of an adjustable iris design, the HPSC could not be heated to a temperature much in excess of 1300°C. The RBSC ceramic could be joined under all conditions examined: however, the quality of join varied significantly. At low temperature and pressure, e.g., 1190°C, 0.5 MPa. a general lack of contact across the width of the sample was found. The butting surfaces were separated by a 10-20 μm gap which was only partially filled by free silicon (see Figure 2). As temperature and pressure were increased the width of the interfacial region reduced and became more completely filled with silicon. The gap was reduced to 5–10μm under conditions of 1235°C, 1 MPa, whilst by 1350°C, 1.5 MPa it had become 3–5 μm and was completely filled with silicon, see Figure 3. Increasing the processing conditions to 1376°C, 2.0 MPa produced a join which suggested that they were excessive. A region of enhanced porosity spanning the width of the sample was observed and free silicon was found on the surfaces of the rods indicating it had flowed out of the join region.

These observations indicate that the silicon grain boundary phase is softening and providing a 'gluing' effect. The optimum temperature is that which causes the silicon phase to soften sufficiently to fill the void between the butting faces, but not so extensively as to flow out of the join region. Within reason, the higher the pressure the narrower the interfacial region. This will lead to higher strengths in the resulting join.

No evidence for local rearrangement of the SiC grains across the join interface was found in any of the samples examined.

TABLE 2. MICROWAVE JOINING PARAMETERS FOR SILICON CARBIDE.

Material	HPSC-HPSC	RBSC-RBSC
Maximum or joining temperature (°C)	1312	1190–1376
Maximum power used (W)	600	300
Axial pressure (MPa)	0.5	0.5–2.0
Typical heating rate (°Cs^{-1})	0.6	2.0
Typical cooling rate (°Cs^{-1})	4.6	3.0
Holding time (min)		up to 2
Typical total processing time (min)	30	15

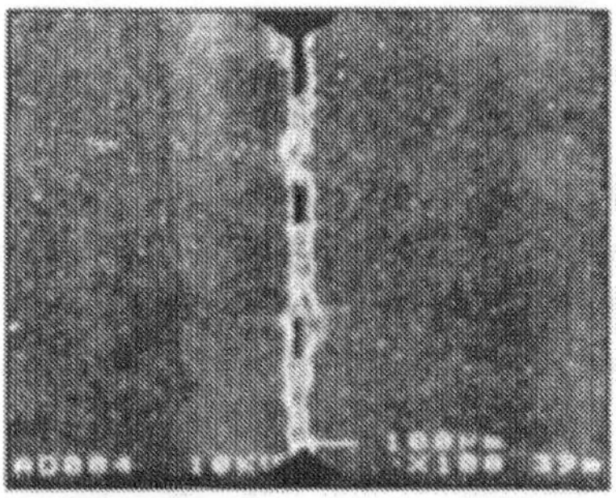

Figure 2. Scanning electron micrograph of RBSC joined at 1190°C, 0.5 MPa.

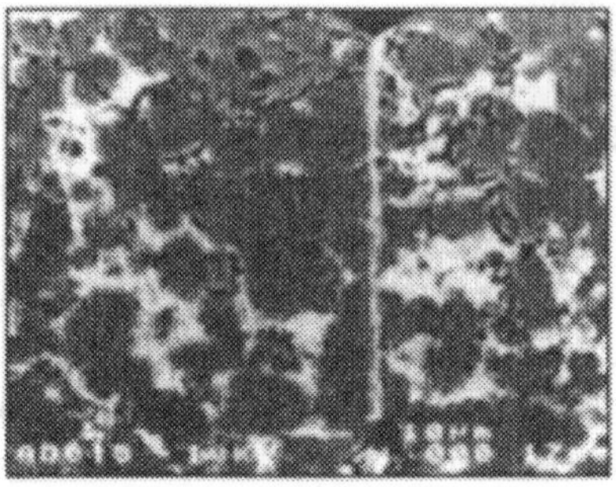

Figure 3. Scanning electron micrograph of RBSC joined at 1350°C. 1.5 MPa.

Alumina

The processing parameters achieved are outlined in Table 3. Attempts to join the AD99.8 to itself were unsuccessful due to an inability to heat the material to a sufficiently high temperature. This occurred because the material displayed too low a dielectric loss to absorb sufficient microwaves. Both the AD85 and AD94, however, could be joined successfully, as could the AD85 to the AD99.8. The conditions required for successful direct joins between 85% and 94% pure alumina ceramics are shown in Table 4. These observations indicate that the greater the quantity of glassy grain boundary phase present the less extreme the temperature/pressure conditions required to effect a successful join. Figure 4, which shows a back scattered electron micrograph of the join region for a specimen of AD85 successfully attached to one of AD99.8 (1600°C, 0.5 MPa), supports this theory. A clear line of glass, squeezed from the AD85, can be seen along the join line.

TABLE 3. MICROWAVE JOINING PARAMETERS FOR ALUMINA.

Material	AD85-AD85	AD94-AD94	AD99.8-AD99.8	AD85-AD99.8
Maximum or joining temperature (°C)	1400 – 1600	1500 – 1600	1400	1500 – 1600
Maximum power used (W)	250	250	250	250
Axial pressure (Mpa)	0.25 – 1	0.5 – 1	1	0.5 – 1
Typical heating rate (°Cs^{-1})	3.0 @ <1200°C 0.4 @ >1200°C	2.5 @ 1000°C 1.0 @ >1000°C	3.0	3.6 @ 1200°C 0.5 @ >1200°C
Typical cooling rate (°Cs^{-1})	3.0	3.7	4.0	3.5
Holding time (min)	10	10		10
Typical total processing time (min)	30	40	25	30

TABLE 4. DEGREE OF SUCCESS ACHIEVED IN JOINING ALUMINA RODS

	AD85 to AD85 alumina			AD94 to AD94 alumina	
Temp (°C) ↓ Press (MPa)→	0.25	0.50	1.00	0.50	1.00
1400	□	○	●		
1500	○	○	●	□	○
1600	●	●	●	○	●

□ = no/negligible join; ○ = joined but regions of poor contact; ● = successful join.

Figure 5 is a highly magnified view of an AD85 sample joined at 1500°C, 0.25 MPa and etched in hydrofluoric acid. This figure shows evidence of two features. First, at locations where the α-Al_2O_3 grains have come into contact from the two rods they appear to have sintered together by atomic diffusion across the interface. Second, the join shows evidence of grains crossing the line and hence suggest they are able to move in the reduced viscosity of the glassy phase. Extensive examination of joins has indicated that the glass tends to flow out of the join towards the edges of the sample under the applied load. This can result in an homogeneous structure in which no obvious glassy line exists at the boundary (see Figure 6).

Dielectric property results (Figure 7) indicate that the presence of the glassy grain boundary phase decreases the real part and increases the imaginary part of the permittivity. Also plotted in the figure are the values for the glassy grain boundary phase obtained using equation 1. It can be seen that above ~ 700°C it is the major contributor to the loss of the ceramics and its absence accounts for the inability to heat the AD99.8 from room temperature.

The flexural strength of the base and joined AD85 and AD94 materials are presented in Table 5. Examination of the fractured samples showed that all of the AD85 samples fractured away from the join, the cracks initiating from random locations on the tensile surface within the loading span. The strengths obtained suggest that fully homogenous joins have been produced. All of the AD94 samples, however, were observed to fail along the join line itself indicating that the temperature used was too low for this grade of alumina.

TABLE 5. FOUR-POINT FLEXURAL STRENGTH OF ALUMINA SAMPLES.

		Strength (MPa)
Material	Base material	joined @ 1600°C, 1 MPa, 10 min
AD85	264 ± 53	312 ± 31
AD94	282 ± 68	144 ± 61

Zirconia

As discussed earlier, the feature apparently required for obtaining good quality joins is a grain boundary phase which softens at the joining temperature effectively 'gluing' the butting surfaces together. However, Figure 5 appears to provide evidence of α-Al_2O_3 grains sintering together across the join interface. This suggests that, had the AD99.8 ceramic been sufficiently lossy to absorb microwaves, it is possible that it could have been joined by purely solid state diffusion processes. This tentative conclusion resulted in an investigation of

Figure 4. Scanning electron micrograph of AD85-AD99.8 joined at 1600°C, 0.5 MPa.

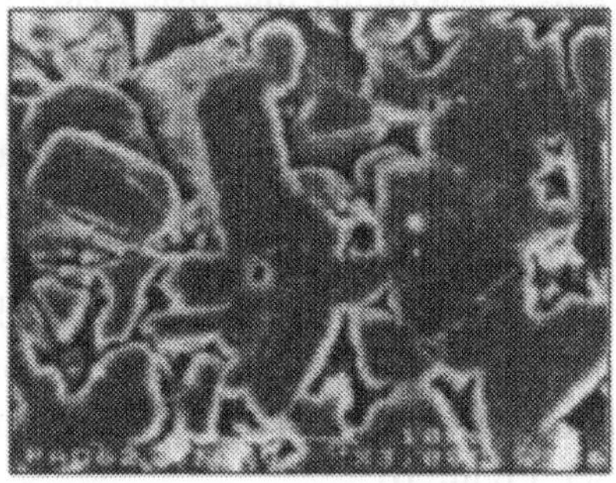

Figure 5. AD85-AD85 joined at 1500°C, 0.5 MPa and etched in hydrofluoric acid.

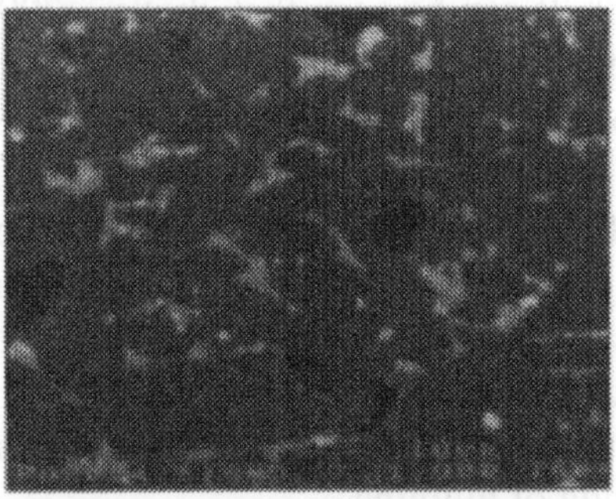

Figure 6. AD94-AD94 joined at 1600°C, 1.0 MPa, showing homogeneous join region.

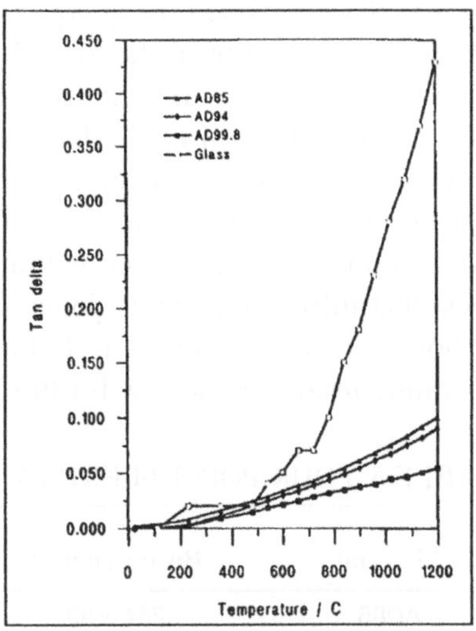

Figure 7. Dielectric properties of alumina ceramics as a function of temperature.

zirconia, a ceramic which has no grain boundary phases likely to soften at the joining temperature but which will heat readily in a microwave environment.

The processing parameters achieved are outlined in Table 6. Zirconia proved to be a difficult material to join due to the large thermal expansion associated with zirconia ceramics ($\alpha = 10.6 \times 100^{-6}\ ^{\circ}C^{-1}$). This made control of the applied pressure difficult. The thermal expansion also yielded severe problems with thermal shock, limiting heating rates. Nevertheless, it proved possible to produce joins which were undetectable by scanning electron microscopy, except via the sample misalignment mentioned earlier. Figure 8 shows such a join obtained at 1560°C, 2.7 MPa. Joins could be produced at temperatures as low as 1400°C but regions of isolated elongated porosity remained along the join line.

TABLE 6. MICROWAVE JOINING PARAMETERS FOR ZIRCONIA

Material	YPSZ-YPSZ
Maximum or joining temperature (°C)	1564
Maximum power used (W)	370
Axial pressure (MPa)	2.7
Typical heating rate (°Cs^{-1})	0.5@ < 900°C; 3.7@ > 1000°C
Typical cooling rate (°Cs^{-1})	3.5
Holding time (min)	10
Typical total processing time (min)	30

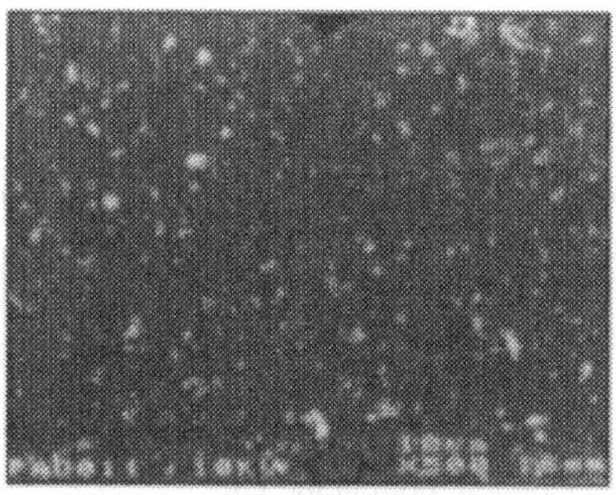

Figure 8. YPSZ-YPSZ joined at 1560°C, 2.7 MPa.

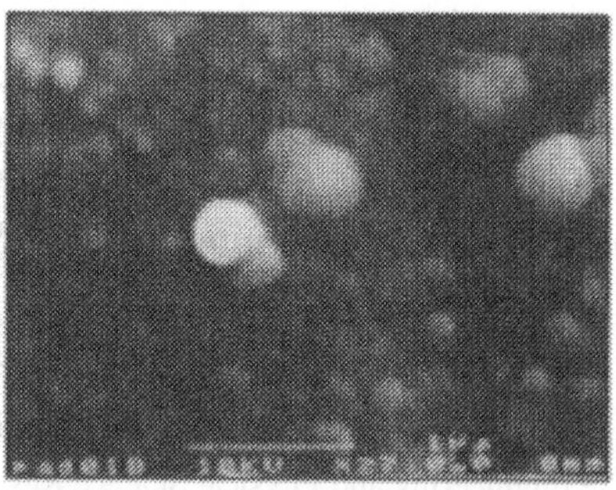

Figure 9. YPSZ-YPSZ join line etched using HF acid to reveal direct bonding of grains.

Figure 9 is a higher magnification back scattered electron micrograph obtained on a sample etched using hydrofluoric acid to highlight the grain boundaries. Since no glassy grain boundary phase existed in this ceramic, solid state diffusion must be responsible for the bonding obtained. Microhardness measurements made in the join region were comparable to those of the bulk untreated material.

Conclusions

A wide range of engineering ceramic materials can be joined using 2.45 GHz microwave energy with joining times less than about 10 minutes. Joining is most easily affected when (i) the ceramic is sufficiently absorbent of microwave energy and (ii) grain boundary phases are present in the base material which will soften at the joining temperature allowing the butting faces to be 'glued' together. For materials without one or other (or both) characteristic, appropriate interlayers can act as a suitable substitute.

For low purity alumina ceramics, the glassy grain boundary phases present appear to allow motion of the individual grains across the join interface resulting in an homogeneous structure in which no obvious glassy line exists at the boundary. For reaction bonded silicon carbide ceramics, the equivalent silicon grain boundary phase does not allow similar grain motion but rather forms a discrete and continuous layer at the interface under optimum joining conditions. The reasons for these two types of behaviour are not yet known. but are probably due to the different viscosities and wetting characteristics of the two phases. The size of the grains is not thought to be relevant since both the AD94 and RBSC ceramics had similar sized grains.

Work with zirconia, a ceramic which absorbs microwaves readily but which does not possess glassy grain boundary phases, has confirmed that it is possible to join ceramics via purely solid state diffusion processes. This may open up a number of new and exciting joining possibilities previously thought unsuitable.

ACKNOWLEDGMENTS

The authors would like to acknowledge the financial contribution made to this project by TWI and the Engineering and Physical Science Research Council. both in the UK.

REFERENCES

1. H. Fukushima, T. Yamanaka & M. Matsui, in *Microwave Processing of Materials,* MRS Proc 124, ed. WH Sutton, MH Brooks and IJ Chabinsky, MRS, Pittsburgh, pp. 267-272 (1988).
2. H. Fukushima, T. Yamanaka & M. Matsui, J. Mater. Res. **5** [2] 397-405 (1990).

3. P.A. Davis, J.G.P. Binner, T.E. Cross & J.A. Fernie, in *Microwave Processing of Materials IV,* MRS Proc 347, ed. WH Sutton, MH Brooks & IJ Chabinsky, MRS, Pittsburgh, pp. 655-660 (1994).

4. D. Palaith, R. Silberglitt, C.C.M. Wu, R. Kleiner & E.L. Libelo, in *Microwave Processing of Materials,* MRS Proc 124, ea., MRS, Pittsburgh, pp. 255-266 (1988).

5. T.T. Meek & R.D. Blake, US Patent No. 4,529,857 (1985).

6. T.T. Meek & R.D. Blake, *J. Mat. Sci. Letts* **5** 270-274 (1986).

7. S. Al-Assafi, 1. Ahmad, Z. Fathi & D.E. Clark, in *Microwaves: Theory & Application in Materials Processing,* Ceramic Trans 21, ed DE Clark, FD Gac & WH Sutton, ACerS, Westerville, pp. 515-521 (1991).

8. S. Al-Assafi & D.E. Clark, in *Microwave Processing of Materials III,* MRS Proc 269, ed. RL Beatty, WH Sutton & MF Iskander, MRS, Pittsburgh, pp. 335-340 (1992).

9. S. Al-Assafi & D.E. Clark, in *16th Annual Conference on Composites & Advanced Ceramic Materials,* Ceramic Engineering & Science Proceedings **2** 1073-1080 (1992).

10. P.S. Apte, R.M. Kimer & M.C.L. Patterson, Microwave heating for ceramic joining, Report prepared for the Ontario Ministry of Energy, Project #600049, Alcan International Ltd, Kingston Research & Development Centre, Kingston, Ontario, Canada (1988).

11. X. D. Xu, V. V. Varadan & V. K. Varadan, in *Microwaves: Theory & Application in Materials Processing,* Ceramic Trans 21, ed DE Clark, FD Gac & WH Sutton, ACerS, Westerville, pp. 497-505 (1991)

12. I. Ahmad & W.M. Black, in *16th Annual Conference on Composites & Advanced Ceramic Materials,* Ceramic Engineering & Science Proceedings 1 520-527 (1992).

13. I. Ahmad, R. Silberglitt, W.M. Black, H.S. Sa'adaldin, J.D. Katz, in *Microwave Processing of Malerials III,* MRS Proc 269, ed. RL Beatty, WH Sutton & MF Iskander, MRS, Pittsburgh, pp. 271-276 (1992).

14. T. Y. Yiin, V.V. Varadan, V.K. Varadan & J.C. Conway, in *Microwaves. Theor,v & Application in Materials Processing,* Ceramic Trans 21, ed DE Clark, FD Gac & WH Sutton, ACerS, Westerville, pp. 507-514 (1991).

15. T. Sato, T. Sugiura & K Shimakage, *J. Ceram. Soc.* Japan **101** 422-427 (1993).

16. R. Silberglitt, D . Palaith, H. S. Sa'adaldin, W.M. Black, J. D. Katz & R .D . Blake, idem, pp. 487-495.

17. M. Arai, J.G.P. Binner, G.E. Carr & T.E. Cross, in *Microwaves. 7heory & Application in Materials Processing II,* Ceramic Trans 36, ed DE Clark, WR Tinga & JR Laia Jr., ACerS, Westerville, pp. 483-492 (1993).

18. S.O. Nelson & T-S. You, J. Phys. *D: Appl. Phys.,* **23** 346 (1990).

3. P.A. Dalis, J.G.P. Binner, [illegible] MRS [illegible] (1991).

[illegible]

17. M. [illegible] Microwave [illegible] Materials, [illegible] (1994).

18. S.O. Nelson, T.S. [illegible] (1960).

Auto Ignition Synthesis and Microwave Sintering of Nanocrystalline Ceramics

S.B. Bhaduri, W.R. Tinga, J.G. Huang, E.H. Zhou and S. Bhaduri

Abstract

An "Auto Ignition" process was used to synthesize nanocrystalline ceramics with the specific example of CaO doped ZrO_2 powders. Suitable metal nitrates were used as oxidizers and urea was used as fuel. Results showed that the solid solution (t-ZrO_2) was produced with CaO up to 10-mol% content, which was stable at room temperature. Microwave sintering has been demonstrated to be one of the effective techniques for densifying nanocrystalline oxide powders without the occurrence of significant grain growth. In order to determine the microwave coupling effectiveness of the as-synthesized powders, the high temperature dielectric properties of some compositions were determined. The grain size of as-synthesized powders ranged from 5 to 20 nm as detected by both X-ray and TEM. Subsequently the as-synthesized powders were cold compacted using CIP at 300 MPa and sintered in a microwave furnace. The as-sintered compacts were further characterized by XRD and SEM. Results showed that nanocrystallinity can be maintained after densification.

Introduction

Nanocrystalline materials are single-phase or multiphase polycrystals, with a crystal size less than 100 nm (typically 1-50 nm) [1,2]. This small crystal and interface size changes the mechanical and physical properties of nanocrystalline materials significantly compared with those of their coarse-grained counterparts. Specifically, it has been found that ceramics can be made ductile by restraining grain sizes in the above range [3,4,5]. Such enhanced ductility can allow these ceramics to be formed in near-net shapes by means of deformation processing methods. But the challenge is to be able to densify nanocrystalline ceramics to highly dense components without any grain growth at high sintering temperatures. In order to meet this challenge, it is essential to synthesize powders with microstructural tailoring to obtain an inherent resistance to grain growth.

In this paper, ZrO_2 is chosen as the candidate material. The main theme of this paper is to synthesize and densify n-ZrO_2 while retaining fine grain size. In the literature, synthesis of n-ZrO_2 has been reported by several groups of researchers. Hahn [6] and Skandan et al. [7,8] synthesized n-ZrO_2 by the gas condensation process. The as-prepared ZrO, showed monoclinic phase with either tetragonal or orthorhombic form as a second phase. After sintering [6] or sinter forging [7,8], near theoretical density can be achieved with nanocrystalline grain size but monoclinic phase. This is not desirable since monoclinic phase is unstable during thermal cycling.

To stabilize the desirable tetragonal phase, a stabilizer such as yttria addition to zirconia is effective. Several references reported sol-gel processing of yttria stabilized zirconia nanoceramics. Compacts of these powders, when sintered between 1050°C and 1100°C for several hours yield 100% dense nanoceramics.

A combustion synthesis route has been utilized in preparing magnesia as well as yttria stabilized nanocrystalline zirconia powders [12-15]. A fast firing process suppressed the surface diffusion and promoted the volume diffusion. This led to overall shrinkage rather than grain and pore growth. A final density of higher than 99% of the theoretical density with final average grain size smaller than200 nm was achieved at 1300–1500°C with dwell times of 1-5 min [13-15].

Previous studies mostly focused on synthesizing pure zirconia or yttria doped zirconia. Not much attention was given to choosing suitable dopants and optimized compositions of dopant addition. Hwang and Chen [16] argued that the greater the difference in valence and size, the greater is the segregation of the dopant in ZrO_2 and the smaller is the extent of grain growth. They proved their point by critically examining the microstructures of ZrO_2 doped with various oxides. Going by their hypothesis, it can be concluded that CaO is the best possible dopant (better than yttria) in minimizing grain growth.

In the first part of the paper, a variation of the combustion synthesis process called the auto ignition process [17,18] was used in synthesizing nanocrystalline zirconia with CaO. In this process single or several metal nitrates (as oxidizers) and appropriate amounts of urea or glycine (as fuels) are mixed with de-ionized water to form a homogeneous aqueous precursor. These precursors are then ignited within a very short time leading to non-equilibrium phases and unique microstructures.

In the second part of the work, microwave sintering was used as an effective means to densify nanoceramics without much grain growth. The advantages associated with utilizing microwave energy as compared to conventional heating include: (1) reductions in manufacturing costs due to energy savings and shorter processing times, (2) improved product uniformity and yield, (3) improved or unique microstructure and properties, (4) synthesis of newer materials and (5) generation of heat without electrodes leading to less contamination. While all of the above criteria are important, the most important one in the present case is the development of nanoscale microstructure while achieving full density.

To the best of our knowledge there are very few papers on microwave sintering of nanoceramics. Freim et al. [20] sintered α-Al_2O_3 (nanocrystalline) by microwave sintering at temperatures of 1200°C-1300°C with a holding time of approximately 5 minutes. The sample density was 82% of the theoretical value with an average grain size of approximately 100 nm Mckittrick et al. [21] sintered ultrafine Al_2O_3-ZrO_2 powders at 1600°C and the microstructure was composed of nanocrystalline grains of ZrO_2 intragranularly present in the micrometer sized Al_2O_3 grains. Bykov et al. [22] achieved 99% density for TiO_2 by microwave sintering at 975°C. They also reported the grain growth behavior of Al_2O_3-ZrO_2 nanocomposites in the range of 700°C to 1500°C. Fundamental to the understanding of microwave sintering is a knowledge of how and to what extent these materials absorb microwaves as a function of temperature. As a material's temperature increases, its dielectric constant and loss factor typically increase causing the microwaves to couple more effectively resulting in even faster heating, which may lead to a catastrophic situation called thermal runaway. This may result in grain growth and gross melting. Knowledge of a material's dielectric properties also allows estimates of starting shrinkage temperature, power requirements and heating rate to be made. As a result of a lack of expertise or equipment, high temperature dielectric measurements are usually not conducted by material scientists.

A primary objective of the work reported here is to synthesize nanocrystalline powders using ZrO_2 as an example. A second objective is to study the high temperature dielectric behavior of these materials. Finally, it is intended to examine the feasibility of densifying nano ZrO_2 doped with 6 and 10 mol% CaO. Two facts are to be noted here. First, ZrO_2-CaO couples readily with microwaves. Second, the addition of dopants such as CaO tends to reduce the grain growth in ZrO_2, since the dopants segregate at the grain boundaries. Because of these reasons, CaO was used as dopant rather than Y_2O_3.

Experimental Procedure

Zirconium nitrate (Wah Chang, Albany, OR) and calcium nitrate (Fisher Scientific, Pittsburgh, PA) were used as oxidizers. For the synthesis five compositions 2, 4, 6, 8, and 10 mol% CaO with appropriate ZrO_2 were used. Urea (Sigma Chemical Co., St. Louis, MO) was used as fuel. In order to maximize the exothermicity of the reaction, the ratio of the oxidizer and the fuel was determined using the method of Jain et al. [23]. In a typical experiment, the constituents were mixed and dissolved in about 3 ml of distilled water, and placed in a fumace (Fisher Scientific, PA) maintained at 600°C. Some water boiled off leaving behind a sol which then ignited producing white foamy powders which were ball-milled for fine dispersion before densification.

The phases and grain sizes of both the as-synthesized powders and as-sintered compacts were characterized by X-ray diffractometer (Siemens D5000) at 40 kV and 30 mA settings using a CuKα radiation. Crystal size was calculated using the Warren-Averbach method after subtracting the instrumental broadening. Fourier Transform Raman spectra were measured using a Perkin-Elmer (P-E) System 2000 spectrometer equipped with a diode-pumped Nd: YAG laser (λ = 1064 nm, maximum output = 1.6 W), a room temperature InGaAs detector, and an optimized Raman beam splitter. Samples were analyzed in a standard powder accessory. Sample spectra were collected at 8 cm^{-1} resolution over the wavenumber range Δv = 200-3600 cm^{-1} by accumulating approximately 500 scans using a laser power of 20-50 mW (depending on the sample).

For high temperature dielectric measurements of as-synthesized powders, a unique single frequency dielectrometer developed by Tinga [24] was employed. This method uses only a small 5 mm diameter sample about 3 cm long which is

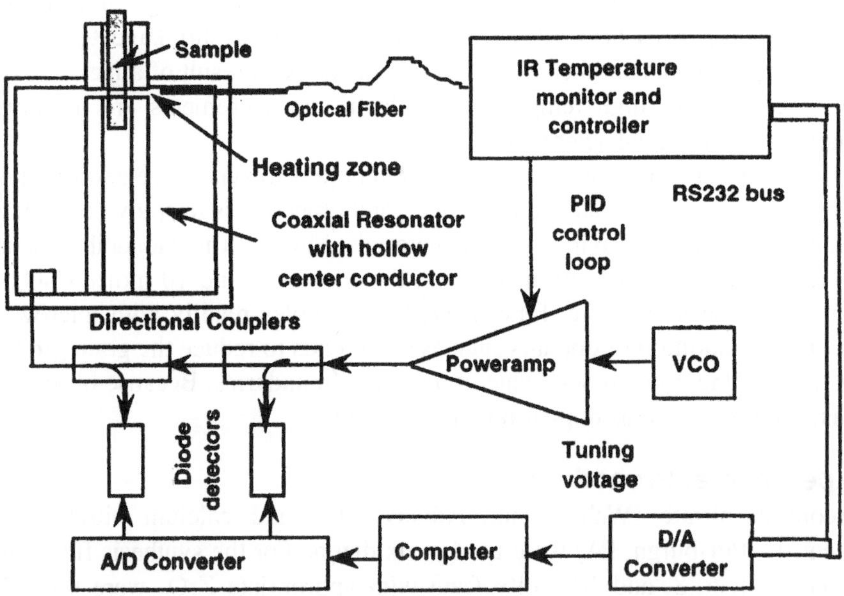

Figure 1. Single-frequency dielectrometer system configuration.

placed into a 7 mm OD quartz tube. A 2.5 mm cross-sectional slice of a sample is heated by microwave absorption while the changing microwave resonator characteristics are simultaneously measured and used to calculate the sample's dielectric properties. Xi and Tinga detemmined the method's measurement uncertainties to be +/- 5% and +/-15% for the dielectric constant and loss factor respectively [25]. Fig. 1 shows the dielectrometer's system configuration.

The powders were uniaxially pressed at 200 MPa and then CIPed at 300 MPa level pressure. Sintering of CaO-ZrO_2 was carried out at (1000-1150) °C for (5-30) mins at ramp rates of 100-300°C/min. in a 0 to 3 kW 2.45 GHz furnace (Microwave Materials Technology, M.M.T, Knoxville, TN) which is coupled to a large "overmoded" cavity ensuring reasonably uniform temperature distribution in the samples. The microwave source is coupled to the cavity by means of a wave-guide via an isolator and a circulator, as is common in microwave high power systems. Temperature was measured using a sheathed K-type thermocouple.

Results and Discussions

The as-synthesized powder was characterized using X-ray diffraction (XRD) technique and the XRD pattems of all compositions are shown in Fig. 2. None of the pattems showed any indication of the presence of CaO phase and only three broad ZrO_2 peaks are observed. These patterns show a cubic structure. However based on the JCPDS files it is possible that the two split tetragonal peaks overlap each other to form a single peak when the peaks are very broad. In the present cases the observed peaks are very broad, thus the crystal structure of the as-synthesized powder is ambiguous.

The Raman spectra of the as-synthesized powders as well as the wave numbers of each Raman mode are shown in Fig. 3. Every spectrum with wavenumber values between 200 to 800 cm^{-1} clearly exhibited five Raman modes in agreement with the spectra of tetragonal zirconia without much background. Reduction of symmetry from cubic to tetragonal should give rise to six modes. The mode at 149 cm^{-1} was not within the range of current measurements.

Thus we can conclude that all of the as-synthesized powders have a tetragonal structure. The broadening of the modes increases with the CaO content while no significant change of the wavenumber was observed, a result of cation substitution and the formation of anion vacancies introducing disorder into the ZrO_2. Similar observations were reported in ref [26-28] for YTZP and Ce-TZPs.

The absence of CaO peaks in both XRD pattems and Raman spectra, and the increasing broadening of Raman modes with CaO content indicate the formation of solid solution of CaO in ZrO_2 and that all of the CaO dissolves into the ZrO_2. Based on the most recent phase diagram of the ZrO_2-CaO system, the maximum solid solubility of CaO in ZrO_2 is about 8 mol% in the tetragonal phase which is only stable at temperatures above 1204°K. Therefore, the present processing suggests that the solubility of CaO in tetragonal ZrO_2 has been extended and this zirconia is stable at room temperature. This result supports the results of Yashima et al. [28] and, based on their definition, the ZrO_2 containing 10 mol% CaO can be denoted as t'-phase while the ZrO_2 containing 2 to 8 mol% CaO can be denoted as t-phase.

Work conducted by Venkatachari et al. and our research group on synthesis of nanocrystalline Y_2O_3 and CeO_2 stabilized zirconia, reported a decrease in zirconia crystallite size with increasing dopant content. However in the current research the crystal size did not show much change with CaO content probably because the crystal size obtained was too small to show a significant trend while the dopant content increased. It is interesting to note that the crystal size reported in the two prior investigations are larger than the crystallite size of our current research. For example the

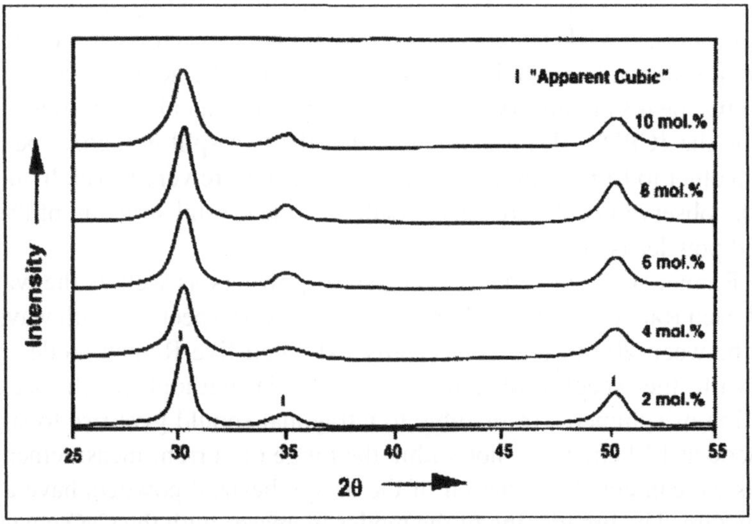

Figure 2. X-ray diffraction (XRD) patterns of all compositions of the as-synthesized powder.

crystal sizes of ZrO_2- 4 mol% Y_2O_3 and CeO_2 are both 10 nm while the size of ZrO_2- 4-mol% CaO is 3.4 nm. Supporting the results from Hwang and Chen [16] who found that, in stabilized zirconia, solutes of a lower valence and a larger radius correlate with smaller grain sizes. The valence and radius of Ca, Y, and Ce cations are 2^+ with 0.112 nm, 3^+ with 0.102 nm and 4^+ with 0.097 nm respectively.

Figure 4 gives the dielectric temperature response at 925 MHz for two compositions of CaO in ZrO_2 powder. Dielectric properties for ceramic materials typically do not show a strong frequency dependence, permitting the use of this lower frequency data to be used in interpreting the 2450 MHz sintering data. Near 650°C, the loss factor for both compositions starts to increase rapidly indicating that at this temperature significant energy absorption occurs. This fact correlates with the shrinkage starting temperature of 650°C for nanocrystalline ZrO_2. Dielectric data also show that repeated heating of the sample increases the dielectric constant and loss factor indicating densification of the sample with time.

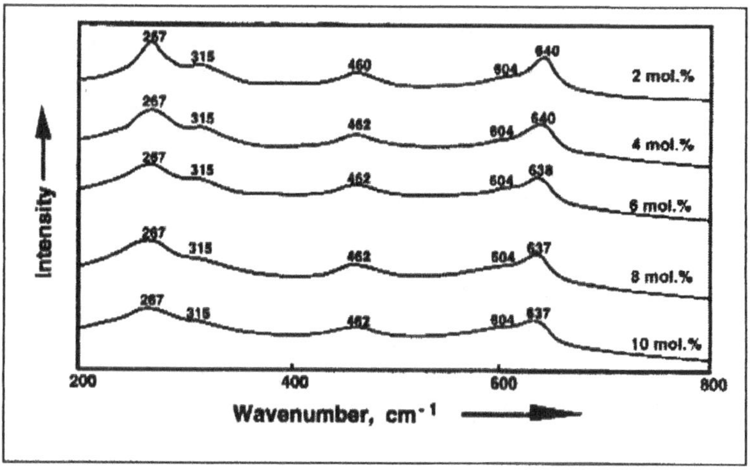

Figure 3. Raman spectra of the as-synthesized powders as well as the wave numbers of each Raman mode.

Figures 5 and 6 present the XRD data for the microwave sintered samples of two compositions chosen for these preliminary studies. According to the phase diagram 6 mol% CaO represents the composition where the ZrO_2 should exists as tetragonal phase, whereas the 10 mol% CaO composition lies in the t+c phase field. In the as synthesized powders, they were both identified as tetragonal phase. However, after these powders were sintered in a microwave furnace, the 10-mol% CaO containing powders seem to have changed entirely from a non-symmetrical (tetragonal phase) to a symmetrical (i.e. cubic) phase. This may be

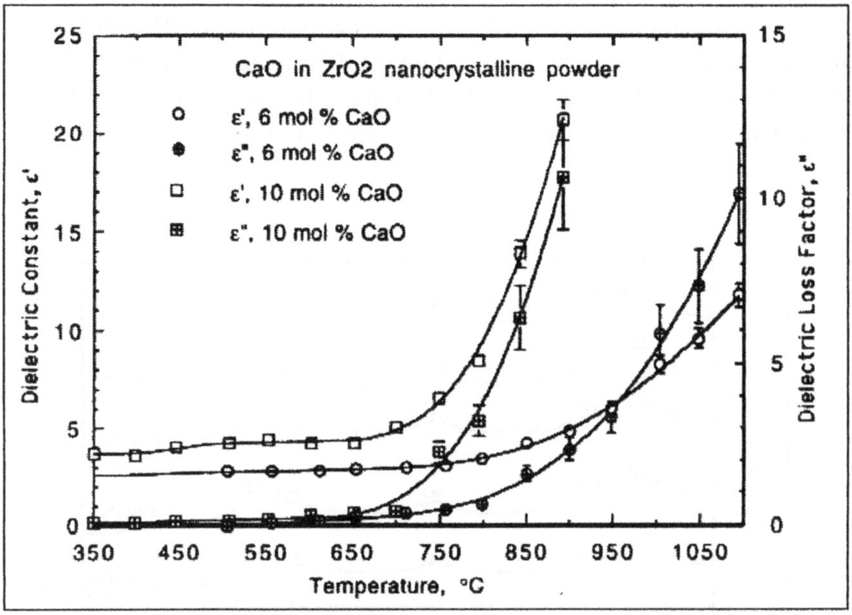

Figure 4. Dielectric constant and loss factor versus temperature at 925 MHz for two compositions of CaO in ZrO_2 powder. Density=1.11 g/cm^3.

the result of the broad nature of the peaks (fine crystallite size retained) or it could be that with more diffusion, the lattice changed from non-symmetrical to symmetrical. On the other hand, the 6 mol% CaO containing material has split peaks around 2θ = 35 and 60 °. These are clear signatures of tetragonal structure. In order to prove the matter further, a slow scanned XRD procedure was carried out on sintered compacts of these compositions. Fig. 4 confirms that the 10-mol% CaO compact is cubic while the 6-mol% CaO compact is tetragonal. The densities achieved in both cases were about 80 % of the theoretical. Hence higher temperature microwave sintering needs to be performed.

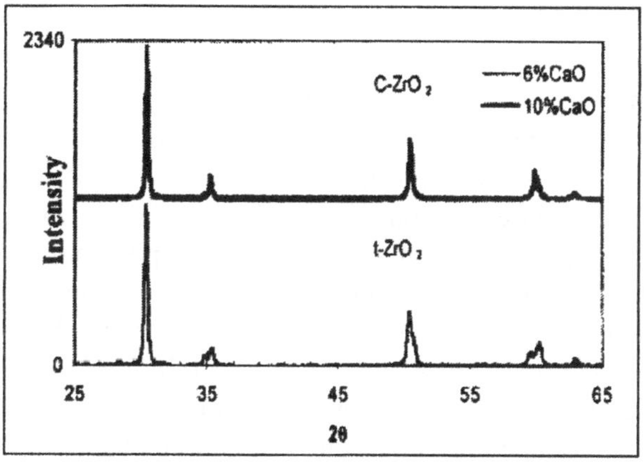

Figure 5. XRD data for the microwave-sintered samples of two compositions.

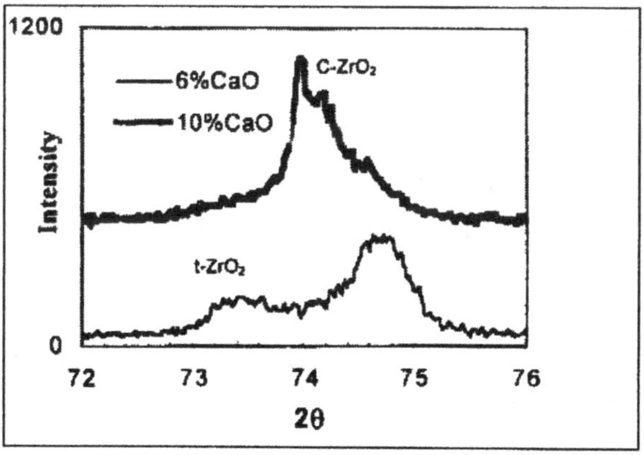

Figure 6. XRD data for the microwave-sintered samples of two compositions.

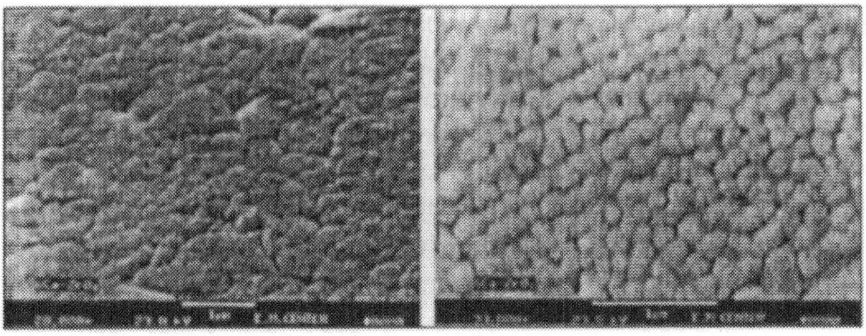

Figure 7. (a) and (b). SEM micrographs as-sintered ZrO_2 (6 mol%).

Conclusions

In conclusion, CaO doped nanocrystailine ZrO_2 powders were synthesized by an ignition process. The as synthesized powders were identified to be tetragonal phase. This suggests an extension of solubility of CaO in ZrO_2 The crystallite sizes were in the range between 3 to 5 nm. Two compositions (6 mol% CaO and 10-mol% CaO) were microwave sintered at 1150°C. While the former retained its tetragonal identity, the latter composition became cubic. However, the density achieved was only up to 80 % of the theoretical. In future, the samples will be sintered at a higher temperature. High temperature dielectric data predicted the shrinkage starting temperature and indicated continuous microwave absorption over the sintering range.

ACKNOWLEDGMENT

This work is supported by NSF grant number DMII-9800009.

REFERENCES

1. H. Gleiter, *Nanostructured Mater.,* **1** (1992) 1.
2. R.S. Averback, H.J. Hofler, and R. Tao, *Mater. Sci. and Eng.,* **A166** (1993) 169.
3. J. Karch, R. Birringer, and H. Gleiter, *Nature,* **330** (1987) 556.
4. J. Karch and R. Birringer, *Ceram. Inter.,* **16** (1990) 291.
5. R.H.J. Hannink and M.V. Swain, *Ann. Rev. Mater. Sci.,* **24** (1994) 359.
6. H. Hahn, *Nanostructured Mater.,* **2** (1993) 251.
7. G. Skandan, *Nanostructured Mater.,* **5** (1995) 111.
8. G. Skandan, H. Hahn, M. Roddy, and W.R. Cannon, *J. Am. Ceram. Soc.,* **77** (1994) 1706.

9. S.D. Ramamurthi, Z. Xu, and D.A. Payne, *J. Am. Ceram. Soc.,* **73** (1990) 2760.

10. T.L. Wen, V. Hebert, S. Vilminot, and J.C. Bernier, *J. Mater. Sci.,* **26** (1991) 3787.

11. M.J. Mayo, D.C. Hauge, and D.-J. *Chen, Mater. Sci. & Eng.,* **A166** (1993) 145.

12. N.A. Dhas, and K.C. Patil, *J. Mater Sci. Lett.,* **12** (1993) 1844.

13. K.R. Venkatachari, D. Huang, S.P. Ostrander, W.A. Schulze, and G.C. Stangle, *J. Mater. Res.,* 10 (1995) 748.

14. D. Huang, K.R. Venkatachari, S.P. Ostrander, W.A. Schulze, and G.C. Stangle, *J. Mater. Res.,* 10 (1995) 756.

15. K.R. Venkatachari, D. Huang, S.P. Ostrander, W.A. Schulze, and G.C. Stangle, *J. Mater. Res.,* 10 (1995) 762.

16. S.-L. Hwang and I.W. Chen, *J. Am. Ceram. Soc.,* **73** (1990) 3269.

17. S. Bhaduri and S.B. Bhaduri, *Nanostructured Mater.,* **8** (1997) 755.

18. S. Bhaduri, S.B. Bhaduri, and E Zhou, I.R. Lewis and P.R. Griffiths, *Intl. J. SHS,* **7** (1998) 317.

19. S. Bhaduri, S.B. Bhaduri, and E. Zhou, *J. Mater. Res.,* **13** (1998) 156.

20. J. Freim, J. Mckttrick, J. Katz, and K. Sickafus, *Nanostructured Mater.,* (1994) 371.

21. J. Mckittrick, B. Tunaboylu, and J. Katz, *J. Mater. Sci.,* **29** (1994) 2119.

22. Y. Bykov, S. Gusev, A. Eremeev, V. Holoptsev, N. Malygin, S. Pivarunas, Sorokin, and A. Shurov, *Nanostructured Mater.,* **6** (1995) 855.

23. S.R. Jain, K.C. Adiga, and V.R. PaiVernekar, *Combustion & Flame,* **40** (1981) 7.

24. W.R. Tinga, in MRS proceedings 269, edited by W.B. Snyder, W.H. Sutton, M.F. Iskander and D.L. Johnson, MRS, Pittsburgh (1992) 505.

25. W. Xi, and W.R. Tinga, *J. of Microwave Power & Electromagnetic Energy,* **28** (1993) 93.

26. C.M. Phillippi and K.S. Mazdiyasni, *J. Am. Ceram. Soc.,* **54** (1971) 254.

27. V.G. Keramidas and W.B. White, *J. Am. Ceram. Soc.,* **57** (1974) 22.

28. M. Yashima, M. Kakihana, K. Ishii, Y. Ikuma, and M. Yoshimura, *J. Mater. Res.,* **11** (1996) 1410

The Tools

Designing Industrial Microwave Heating Systems for Safe Operation: Batch Ovens

R.F. Schiffmann

Abstract

While a great deal of effort in the design of industrial microwave heating equipment is concentrated upon heating of the load and the achievement of the desired material properties, often not enough is focused upon its safe operation. This paper will focus upon batch type ovens and examine various aspects of equipment safety and suggest ways in which they may be achieved.

At the last AMPERE, the author presented a paper "Fires in Microwave and RF Heating Systems: Causes and Prevention". There are many other safety concerns that are of serious potential harm to the personnel, equipment and facility. Among the items to be discussed:

- .The construction of access doors and windows.
- Designing openings for the introduction of hot air and other gases.
- Interlocking for access and cleaning.
- Personnel safety requirements in large batch ovens.
- Arc detection requirements.
- Dummy load requirements.
- Reducing exposure to microwave leakage.
- Compatibility of materials of construction and processing.

Introduction

Industrial microwave processing systems have been in use since the mid 1950's and their applications have spanned a wide range of industries including food, rubber, pharmaceutical, ceramics, chemical, and many more. When designing these systems, microwave engineers focus upon:

- Efficient heating of the load,
- Maximizing energy coupling to the load,
- Achieving the desired product quality,

- Maximizing throughput,
- Costs of the system,
- Application of auxiliary heat sources,
- more.

While significant attention is paid to safety issues regarding worker exposure to microwave radiation, often there is not an equally serious concern for other worker safety issues.

Considerable attention has been paid to issues regarding the safe use of microwave heating for many years. Two early publications (1,2) concerned safety aspects for domestic microwave ovens. As for industrial microwave systems, the safety literature is very limited. A paper delivered at the Second World Congress discussed "Fires in Microwave and RF Heating Systems: Causes and Prevention" (3) that was the result of two serious fires occurring in one microwave and one RF industrial heating system. Subsequently, serious design flaws were uncovered by this author, regarding the safe operation of a large microwave batch oven. This has led to the consideration of the many nonradiation safety issues that should be considered when designing industrial microwave systems, and this paper focuses upon industrial microwave batch ovens. These ovens are characterized as multimode resonant cavities with one or more access doors. They are designed to have the load placed in the cavity; the door closed (either manually or automatically); the microwave power energized for a specific amount of time, and then shut off; the door opened and the load removed from the oven. The next load may be placed in the cavity immediately thereafter, or after some time. Such ovens often employ auxiliary heating including hot air, infrared or steam, or combinations thereof. They may also use atmospheres other than air and even vacuum. Most batch oven cavities are only a few times larger than the load and so, may be approximately one or two meters long on a side. However, there are also very large batch ovens that may be several meters on each side, large enough for a workman to enter the cavity and even for a fork lift truck to be driven into the cavity to deposit or remove the load. Such very large ovens present unusual safety problems, as described below.

Objective

The overall objective of this publication is to provide information for the avoidance and elimination of injury to workers; damage to the equipment and facility; and, liability for the manufacturer. It will not deal with issues such as improving microwave distribution or coupling efficiency.

General Equipment Requirements

Cavities and Waveguides (4)

The multimode resonant cavities generally used for batch ovens are constructed of steel–either stainless or carbon, or aluminum. The choice is largely based upon processing conditions and construction costs. There is an exception in situations such as the U.S. Food and Drug Administration's (FDA) requirement for only stainless steel.

Aluminum has excellent electrical and microwave characteristics, but does not tolerate temperature well. Even at temperatures as low as 120°C, aluminum will distort over time, but this may be years rather than hours. Therefore, durability is a consideration where elevated temperatures are involved for constructing cavities of aluminum.

Stainless steel has excellent temperature and mechanical characteristics but is somewhat microwave lossy. It distorts readily with temperature, so welding can be a challenge. However, it remains the material of choice for a wide range of high temperature applications.

Carbon steel suffers from lower electrical conductivity, and easily forms unattractive oxide coatings. It can, however, be coated with a variety of materials that enhances its use as a microwave construction material.

All seams and joints must be welded, preferably continuously rather than spot welding. This is to ensure that joints do not separate which can cause serious microwave leakage problems. It is suggested that these be internal welds for better cavity efficiency.

Interface points for such things as waveguides and chokes must be flat, flush and electrically conductive to prevent microwave leakage. They may be bolted in place, with bolts being placed on less than half-wavelength centers.

Waveguide sections should be mechanically strong and supported, if necessary, to avoid undue stress on joints and flanges, as these may break causing microwave leakage.

Circulators

Also called iso-circulators, or junction circulators, these microwave ferrite devices are placed between the power source and the cavity and provide total protection to the magnetron from even 100% reflected power, which otherwise would damage or destroy the tube. It is important, however, that the circulator and power source be matched in terms of capacity. For example, a 75-kilowatt generator requires a circulator of a 75-kilowatt capacity. If, for example, a 30-kilowatt circulator were used, very high reflected power could overload and damage the circulator and ultimately the power source. (4)

(Note that the nomenclature for circulators is somewhat misleading. Circulators often need a load for magnetron protection; together they become an isolator.

Quite often they are specified as to total power handling capacity, so a "75-kilowatt circulator", under this nomenclature, could only tolerate 25 kilowatts of reflected power with a 50 kilowatts of forward power.)

Access Doors

These are constructed of the same material as the rest of the cavity. Their function is to allow an operator to place into and remove from the cavity, the load to be processed. There are two major classifications of doors: manual and automatic.

Manual Doors

The doors must be structured to prevent leakage of microwave energy. Such structures as ferrite-filled quarter wave chokes are useful. (5) It is important that these chokes operate successfully after long time use. They must be kept clean, and a periodic cleaning schedule, commensurate with the processing schedules, should be employed.

When auxiliary heat is used in the cavity, the door must be of sturdy enough construction to prevent warping, which can increase microwave leakage and worker exposure. A strong, positive latch system, such as used in industrial refrigerators, is useful to maintain good door to cavity/choke contact. Many of the doors employed in batch ovens are quite heavy, so a sturdy hinging system must be employed to ensure tight door closure even after thousands of hours of use.

An absolutely essential feature is to have at least two independently wired interlocks that will intercept the microwave power when the door is opened. They also prevent the microwave power from being turned on unless the door is firmly closed and latched. This is usually shown on the generator as a warning or "Ready" light. By being independently wired, should one interlock switch fail, the other will continue to operate. A preferable system may be to follow the design required for domestic microwave ovens in the United States (6): which requires that each oven be provided with two interlocks and an interlock monitor, commonly a third "monitor" interlock. The first two, primary and secondary interlocks, are usually normally open switches, while the monitor interlock is a normally closed switch. In this configuration should either of the first two fail, the other will short and blow the fuse, to prevent microwave power generation.

Automated Doors

Beside most of the same requirements as manually operated doors, these doors have another important requirement: that they stop closing when they contact any part of the operator's body or any other object placed in their path. They must react, in a manner similar to elevator doors, to a minimal amount of force, opening fully until they are re-activated. The stop mechanism should have at least two independent interlocks.

This problem is further exaggerated in very large walk-in type ovens, where the doors are very large and extremely heavy–perhaps weighing several tons, and could kill or severely injure a worker in its path when closing, unless a positive stop mechanism is employed. It is also useful to employ an audible warning and/or flashing lights as the door closes.

Another potential problem encountered with very large batch ovens in that the choke seal, contacted by the door when it closes is likely to be located in the floor. There is a danger of damage to this choke structure if forklifts, rolling carts, or similar are employed in order to bring loads in and out of the oven. Ramps over the choke should be employed in these situations. Also there is a danger of damage to the choke due to external forces that may damage the floor. These floor chokes must be cleaned frequently to remove any dirt or foreign material which may interfere with its functioning. Finally, it is essential that microwave leakage be monitored frequently, over the entire length of the choke with the door fully closed and the oven energized with a minimal load.

Windows

Windows allow an operator to view the product in the oven as it is being microwaved. In order to prevent microwave leakage through the window, they usually employ a thin perforated metal sheet with small holes, although there have been other structures employing metal tubes of one inch or slightly larger cross section, about 6 inches in length. All of these structures must keep the microwave leakage level at acceptable levels, e.g., 1 milliwatt per cm^2 at a distance of 5 cm from the windows surface (this is only a suggestion, not a standard).

In almost every case, the metal screen is covered, usually inside and outside with glass, quartz or high temperature clear plastic. This serves multiple functions:

- Prevents insertion of foreign objects through the screen.
- Prevents loss of hot air through the screen. When the ambient temperature in the cavity is high, this could be a significant personnel hazard.
- Prevents the loss of moisture or other vapors, some of which may be dangerous, through the screen.

When there is high temperature inside the oven during processing, it is necessary to use tempered glass or quartz to deal with the mechanical stresses due to heat-induced warping, expansion and shrinking, as well as the heat itself.

Introducing Gases, Instruments and Materials

It is often necessary to introduce materials, gases, or objects into the cavity, for example, temperature probes. When doing this it is important to prevent microwave leakage, which can cause worker exposure to microwave radiation. The simplest and most common way to do this is to use a circular or rectangular waveguide beyond cutoff, typically, approximately one inch or narrower in diameter and four to six inches long. (It may be wider for ovens operating at 915 MHz). While this is effective there are certain safety precautions:

- Metallic devices such as thermocouples or screwdrivers should not be inserted in the tube as these create a coaxial structure that may cause serious microwave leakage problems. To prevent this the tube can be capped when not being properly used. Or, instead of being a straight tube, it may have one or more right angle bends making it impossible to insert these foreign objects.
- Gases, steam, vapors, etc. may escape through the tube. In order to prevent this, the end should be capped when not being properly used for temperature measurement, for example. This cap, which may also include a sealing bladder, may be permanently installed and have an opening of a size to fit the device being properly inserted into the tube.

Safe Access into the Cavity

It is often necessary for workers to have access to the interior of the cavity: to place or remove product to be processed, or for cleaning or repair. This will be done through the open access door(s). These doors must be properly interlocked, or described above, in order to prevent microwave exposure. All appropriate OSHA (in the U.S.A.) lockout/tag out procedures must be followed. (4)

Any cleaning that is done should not use abrasive cleaning agents to avoid scratching the interior surface of the cavity. Should flammable solvents be employed, the cavity must be fully vented before microwave power can be restored. Any standing water in the oven must be removed and the cavity left completely dry; a water drain in the floor of the oven can be used for this purpose.

Safety Issues with Large (Walk-In) Ovens

Large, walk-in ovens are employed for the processing of large or numerous items. Workers frequently enter these ovens in order to place or remove the load, which may be on a rolling cart. Sometimes the worker may bring in or remove the load by means of a forklift. There are other occasions such as cleaning, maintenance or repair in which one or more workers may need to be inside the oven cavity. All of these situations present unique problems to avoid workers being accidentally exposed to the microwave power.

As a first line of defense, there should be warning lights to indicate to other plant personnel that someone is in the oven. There should also be a positive shut-off to prevent the doors closing and the generators being energized. The safety interlocks on the doors will do this, but an independent backup is best. Emergency "Stop" buttons should be placed both inside and outside the oven.

In the unlikely, but by no means impossible event, that the oven door closes and the power is turned on while the worker is inside, it is imperative that the worker be able to shut off the power and get out of the oven before suffering injury. An easily accessible shut off mechanism is required. It is likely to be very dark in the oven so safety switches are not useful. A very useful device is a rope or several ropes along the entire length of the interior cavity walls, at easily accessible heights. By pulling on any one rope the worker is able to shut off the power, open the door, sound an audible alarm and energize flashing danger lights. It would also be sensible to have an automatic time delay of 10 second, for example, between the time the door is fully closed and the time at which the microwave power can come on.

Two other serious problems encountered with this type of oven is a result of the very large heavy doors and the use of the bottom choke seal in the floor. Both these issues have been dealt with earlier in this text.

Many of these ovens employ auxiliary heating systems which require special safety precautions:

- All auxiliary heating must be shut down and the internal temperature reduced to a safe ambient temperature before the door may be opened to allow a worker to enter. In the United States, OSHA limits the maximum surface temperature of parts to 60°C so they may be manually handled. Only in serious emergencies, such as a fire, should it be possible to open a hot oven, and that should require a separate set of controls.
- In some cases, atmospheres other than air, or vacuum, may be used inside the oven during processing. In such cases, the oven cavity must be fully purged to restore the ordinary air atmosphere before the door is opened and a worker may enter.

Designing for High-Voltage Safety

The voltages required to power modern day microwave power sources are high voltages. These voltages can be lethal. Sufficient provision should be designed into each piece of high voltage gear to assure that accidental exposure to high voltage is prevented. Typically a double redundant safety interlock system is employed in a microwave power source that will inhibit personnel exposure to the high voltage conditions. All servicing of high voltage microwave power sources should be left to properly trained high voltage technicians or the manufacturer's authorized service group. (4)

All high voltage components should be sufficiently isolated so that it is impossible for accidental contact or exposure. Enclosures housing the high voltage devices must be interlocked so that if anyone opens the enclosure, the power will immediately interrupted. In many cases, the generator housing is locked with a key in order to prevent any unauthorized personnel from opening the enclosure.

Another important consideration is to prevent water from entering or condensing inside the high voltage enclosure. Many plants, especially for food processing, use large quantities of hot water for cleaning. Unless the high voltage cabinet is made especially watertight, this can create a serious hazard. The most common and effective technique is to locate the generators remotely from the oven, in a separate room. This is typically done with microwave bacon cookers, where the generators are located on a floor above the cavity and interconnecting waveguide brings the microwave power into the oven.

As for condensation, industrial magnetrons are water-cooled and this can cause water to condense within the cabinet. To prevent this, it is important that the cooling water and its quality be maintained within the manufacturer's specifications. Cooling water that is too cold is likely to condense inside the generator cabinet; water that is too warm will not provide sufficient cooling.

Still another potential problem is faulty placement of components. In one case I investigated, an under-capacity circulator was placed above the high voltage power supply. The circulator was water-cooled and had it failed due to large reflected power, it was possible that the cooling components would rupture spilling large volumes of water on to the generator. It is imperative that such construction be avoided. Note that some manufacturers include the circulator and water load inside the generator cabinet. The head of engineering for one firm stated, "I would prefer to engineer my way around a leak scenario than have the circulator outside the generator."

Compatibility of Materials of Construction and Processing

There are a number of issues that must be considered in the design of a microwave processing oven and one of them relates to the compatibility of the materials being processed and the materials of construction. For example, in the United States, it is mandatory that all food processing cavities and metal structures within the cavity be manufactured of stainless steel. Another requirement is that any glass that is present, in the door windows, for example, be tempered and shatter proof to prevent glass pieces falling in the food.

In ovens where a corrosive atmosphere is present, the cavity must be constructed of a non-corroding metal, such as stainless steel (7), and it may even be necessary to coat the interior walls with a corrosion resistant coating. Volatiles may condense on the inner walls of the cavity and other internal structures. These may represent a fire, corrosion or other hazard. It is imperative that the interior of the oven be examined daily and any condensate be removed. This includes the waveguide windows that may otherwise be blocked and reduced power into the oven or create intense heating at that surface. (7)

Polymeric materials are frequently used as mechanical supports for the load. (8) Teflon® or similar materials are clearly the choice. They are microwave transparent, tolerate temperatures over 200°C and are reasonably priced. All polymers exhibit runaway heating when they soften due to temperature, therefore, it is advisable to avoid lower temperature plastics. Even polypropylene, which may be useful up to 120°C can be a problem after a long time since its glass transition temperature decreases overtime. This is described in greater detail in reference 3.

Most importantly, never use Nylon® in a microwave cavity. Both its permittivity and loss tangent increase dramatically with temperature, leading to runaway heating and, since it is flammable, fire! (9)

Dealing with High-Temperature Systems

Some ovens are used to process materials at high temperatures. This can cause the exterior wall of the oven to be above the safe temperature limit (60°C in the United States) for contact by the workers. In these cases it is usually necessary to double-wall the oven, with insulation between the walls, to prevent worker injury.

Note that insulation may be used at lower temperatures, 50°C for example, to prevent heat loss and warming the processing area.

Arc Detection

The presence of arcs within an oven can cause damage to the cavity or internal structure, and may even lead to a fire. In order to prevent serious problems it is advisable to place arc and fire detectors within the oven, capable of scanning the entire work area. Detectors of the ultra-violet type are recommended, as they are able to see through smoke.

When an arc is detected, the microwave power should immediately be shut off and the source of the problem investigated and removed. Both visible and audible alarms should be installed to alert workers of the problem.

Controlling Worker Exposure to Microwave Leakage

Several actions should be taken on a daily basis:

• All access doors, surfaces and components must be designed to keep leakage at or below acceptable standards,

- Choke seals and contact point around doors should be kept clean,
- Any damage to the door or choke should be repaired at once or the defective part replaced
- Leakage testing should be done on a frequent mandatory basis

Miscellaneous Issues

Flammable and explosive solvents should be avoided, even when cleaning. If used, they must be thoroughly removed before powering up the system. The cavity work area must be kept clean and free of tramp metals and other sources of arcing.

Conclusion

It is not possible to define all the issues that may create hazards when operating an industrial microwave processing system. However, the issues described above are all to be taken seriously and the safety suggestions followed. In my forty plus years of working with industrial microwave processing I have encountered nearly all described. Several led to catastrophic results including serious worker injury, the total destruction of the processing system and the processing plant. Hopefully, following these suggestions will avoid such calamities in the future.

REFERENCES

1. R.F. Schiffmann, Food Product Design, Weeks Publishing, Northbrooke, Illinois (1994).
2. R.F. Schiffmann, Leaders Product Liability, Law and Strategy, New York, NY (1994).
3. R.F. Schiffmann, Proceedings of the Second World Congress on RF and Microwave Processing, The American Ceramic Society, Westerville, Ohio (2000).
4. M. Yonnone, private communication (2002)
5. R. Meredith, Microwave heating applicators 1: multimode ovens 5.8.2 Door chokes, pp 184, Engineering Handbook of Industrial Microwave Heating, The Institution of Electrical Engineers, U.K. (1998).
6. The Code of Federal Regulations (CFR) Part 1030.30, "Performance standards for microwave and radio frequency emitting products", Center for Devices and Radiological Health.
7. R.V. Decareau and R.A. Peterson. Applicators and processing systems 6.6.7 Materials of construction, pp 186, Microwave Processing and Engineering, Ellis Horwood Ltd., Chichester England (1986).
8. R. Meredith, and Microwave heating applicators 1: multimode ovens 5.10.1 Conveyor types and materials, pp 190, Engineering Handbook of Industrial Microwave Heating, The Institution of Electrical Engineers, U.K. (1998).
9. A.R. von Hippel, Dielectric constant and loss vs. temperature, pp 412, Dielectric Materials and Applications, The M.I.T. Press, Cambridge, Massachusetts, (1954)

A Technical Review of Equipment and Methods for Waveguide Power Measurement in Microwave Heating Applications

J.F. Gerling

Abstract

Various methods for waveguide power measurement were developed in the 1930's that have changed little over the years and are still in common use today. However, more recent improvements and advances in measurement signal processing technologies offer significant enhancements to the practical utilization of traditional power sensing methods. Before choosing a power sensing method for a particular application, the importance of coupling factor and directivity as performance parameters should be well understood. Measurement accuracy depends on a) the accuracy of calibration of the stated coupling factor, and b) the error introduced by the effects of reverse power as a result of finite directivity.

Three waveguide power sensing methods stand out as the most commonly used in laboratory and industrial heating systems, each having characteristics preferred for different applications.

Aperture Array Coupler—The two basic types of aperture array couplers are a) Broadwall couplers in which two parallel waveguide sections are coupled to each other along the broad wall by an array of apertures spaced 1/4-guide wavelength apart, and b) Cross-guide couplers in which two perpendicular waveguide sections are coupled to each other by a single aperture.

Resistive Loop Coupler—Waveguide power is coupled magnetically and electrically to a sensor element consisting of a resistor-terminated loop and a probe.

Multi-probe Reflectometer—Three or four non-directional antenna probes detect voltages in the waveguide and enable the determination of standing wave reflection coefficient and relative phase angle as well as forward and reverse power.

The detected power signal must be converted to any of several forms as may be preferred for utilization in the application. The most popular signal conversion methods are diode detectors and thermocouple devices. Recent advances in detector circuitry offer significant improvements by combining a variety of useful features in cost-effective packages. Methods of calibration, an important factor in the accuracy of any measurement, are briefly reviewed.

Introduction

Measurements of microwave power became necessary simultaneously with the advent and use of high power microwave energy as early as the 1930's. Several methods for power measurements were devised then and in the decades to follow, most of which are quite obsolete by today's standards. Of these methods, only a few are suitable for high power measurements (i.e. a few hundred Watts and above). For applications in microwave heating, the selection is further narrowed to methods suitable for operation within the ISM (Industrial, Scientific and Medical) frequency bands centered at 915 MHz (896 MHz in the UK) and 2450 MHz.

The methods of choice for most microwave heating applications are mainly those utilizing waveguide components for power coupling, along with various types of sensors for power signal processing. More recently, the integration of microprocessors and sophisticated electronic circuitry have greatly simplified the power measurement task and reduced costs while maintaining or improving accuracy and reliability. Fundamental differences in performance, reliability, size and cost necessitate careful consideration when selecting the measurement method best suited for a particular heating application.

High Power Coupling Methods

Water loads were devised very early and quickly became the most widely used method for high power calibration. The primary advantage is that power is absorbed directly into the water itself, thus acting as both a load and a means for reasonably accurate power measurement by calorimetry. However, water load calorimeters are mostly unsuitable for industrial heating applications due to their bulkiness and the complexity of ancillary equipment.

An interesting yet mostly impractical method is that of force-operated power indicators. Such instruments operate by interaction of the electromagnetic field with a moving object, such as a rotating vane. Several designs and methods have been devised, although most are highly sensitive to external effects of mechanical vibration making them unsuitable for general use.[3]

Various types of waveguide power couplers were devised as a means to divide the power between the main waveguide line and a branch line, ideally with no losses or reflections. A sensing device is then used to convert the microwave field in the branch line to a dc signal having a known mathematical relationship to microwave power. Waveguide couplers generally are either directional, meaning preferential to power flowing in one direction, or non-directional.

Another common waveguide power measurement device is the multi-port reflectometer. Detailed discussions of multi-port reflectometers can be found in early literature, but they did not gain popularity as waveguide power measurement devices until the 1970's with the advent of inexpensive desk-top computers.[3] Designs having three or four non-directional probes to sense the voltage at different points in the waveguide, along with appropriate signal processing, deliver information relating to the standing wave reflection coefficient and relative phase angle in addition to forward and reverse power.

Waveguide Couplers

Non-directional couplers typically employ a simple probe that protrudes into the waveguide to sense the e-field voltage at a single point in the waveguide. The voltage detected is the vector sum of voltages due to both forward and reverse power, thus the non-directionality of such devices. One can then see the inherent sensitivity to the relative phase of the standing wave resulting from the presence of both forward and reverse power. Consequently, non-directional couplers are best used in the absence of reverse power or where the relative phase of an anticipated standing wave is constant.

Early designs of directional couplers included relatively simple arrangements between adjacent waveguide sections connected by one or more apertures as well as resistive-loop type couplers. Three directional coupling types as shown in Figure 1 have proven to be quite popular for microwave heating applications.

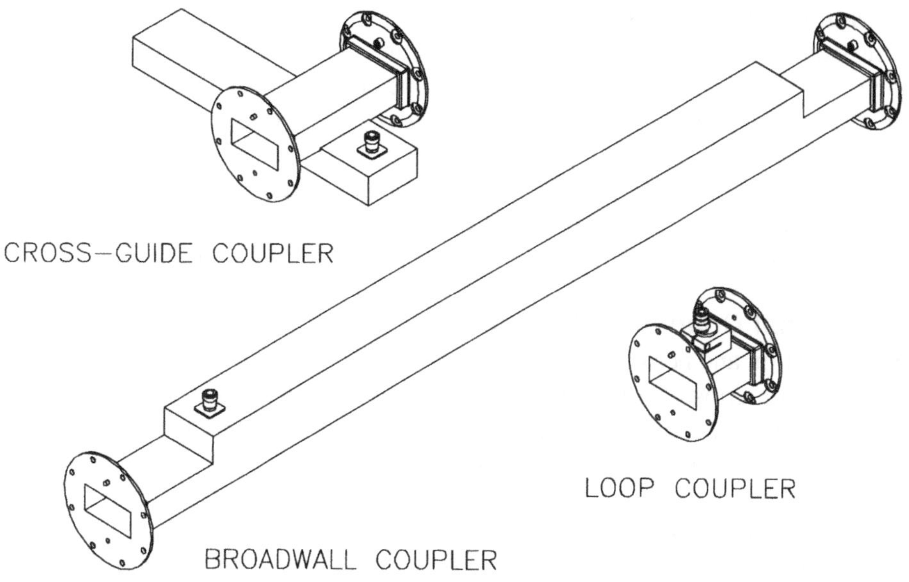

Figure 1. Common waveguide directional coupler types.

Cross-Guide Couplers

Cross-guide couplers consist of two perpendicular waveguide sections which are coupled to each other by a single aperture (two apertures for bi-directional couplers). The aperture is located such that a unidirectional circularly polarized H-field wave in the main waveguide couples a similar circularly polarized H-field in the branch waveguide which then excites unidirectional wave propagation.[4] Directivity is obtained by the fact that H-field polarization is opposite in direction between forward and reverse propagating waves, preventing a reverse wave from coupling through an aperture located for the forward wave and vise versa. Crossguide couplers offer superior flatness of coupling factor over a broad band, although directivity is generally less than other coupler types.

Broadwall Coupler

In broadwall couplers, two parallel waveguide sections are coupled to each other along the broad wall by an array of apertures spaced1 1/4-guide wavelength apart. Energy propagating in one direction in the main waveguide section couples through each aperture and propagates initially in both directions in the branch waveguide. However, as a result of the aperture spacing, reverse propagating energy in the branch waveguide cancels while forward energy is additive.[2,3,4] Broadwall couplers offer superior directivity of 40 dB or more, but their large size is a significant disadvantage in most industrial applications.

Resistive Loop Coupler

A resistive loop type coupler is where the branch line is a sensor element consisting of a small strip of metal terminated at one end by a resistor and the other end by the output connector. Again taking advantage of the change in polarity between forward and reverse waves, the loop is oriented in such a manner as to couple only to the transverse magnetic field polarized in the preferred direction.[2,4] Loop couplers typically have the smallest size of all waveguide coupler types and offer reasonably good coupling factor and directivity performance.

Directional Coupler Performance

The operation and performance of a waveguide directional coupler are generally expressed in terms of (a) accuracy and frequency sensitivity of its coupling factor, and (b) directivity. It is important to understand the potential error in coupling factor that can be introduced due to finite directivity.

Performance Parameters

Coupling factor is defined as the ratio between the measured (forward) wave in the main waveguide and the coupled output wave.[2,4] When connected as shown in Figure 2(a) to a source and perfect loads, coupling factor is expressed in decibels (dB) as

$$CF = 10 \log(\frac{P_F}{P_{3F}}) = 20 \log(\frac{V_F}{V_{3F}}) = 20 \log(\frac{1}{S_{31}}) \quad (1)$$

where P, V, and S represent power, voltage and s-parameter measurements, respectively. This equation represents a positive value for coupling factor, although it is often given as a negative value by inverting the logarithm function.

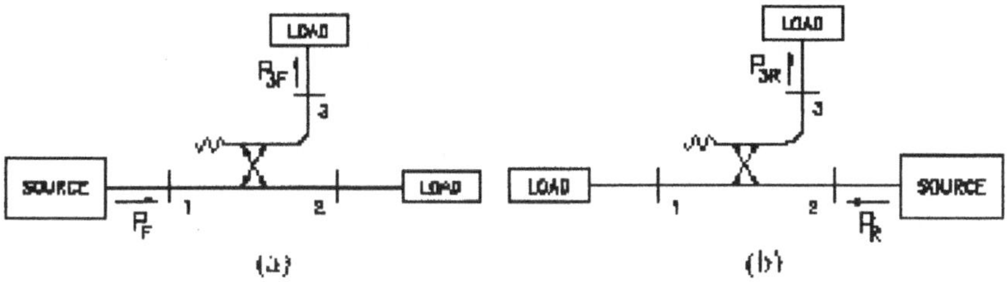

Figure 2. Directional coupler connections for measuring (a) coupling factor and (b) directivity.

Directivity is then defined as the ratio between the forward coupled and reverse coupled waves.[2,4] When connected as shown in Figure 2(b) and assuming P_R is equal to P_F, directivity is expressed in decibels as

$$D = 10 \log(\frac{P_{3F}}{P_{3R}}) = 20 \log(\frac{V_{3F}}{V_{3R}}) = 20 \log(\frac{S_{31}}{S_{32}}), \text{ or } D = CF_R - CF_F \quad (2)$$

Error due to Directivity

As mentioned earlier, the finite directivity of a directional coupler gives rise to the potential for power measurement error due to interference from a reverse propagating wave. The indicated reading from the output of the branch line is derived from the vector sum of the forward coupled and reverse coupled wave. The vector diagram in Figure 3 shows that the reverse wave can be at any phase relative to the forward wave. Thus, the reading is sensitive to both relative phase of the reflection and coupler position along the waveguide.

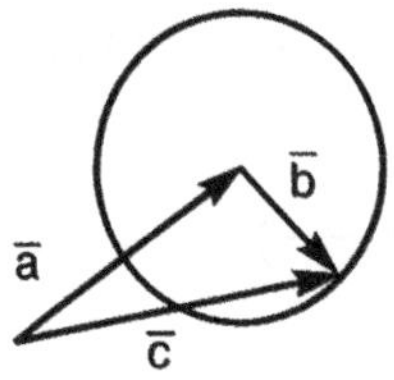

$\bar{a}$ = voltage at coupler probe output due to true forward power

$\bar{b}$ = interfering voltage from reverse power due to finite directivity

$\bar{c}$ = total actual voltage at coupler probe output

$\bar{c} = \bar{a} + \bar{b}$

Figure 3. Vector diagram of directional coupler output voltage.

From Figure 3 it is easy to see that the limits of the true reading can be found by adding and subtracting the absolute value of the interfering voltage from the actual output voltage. Thus,

$|\bar{c}|$ is calculated from the indicated power reading

$|\bar{b}|$ is calculated from known reverse power and coupler directivity

Limits of $|\bar{a}| = |\bar{c}| \pm |\bar{b}|$

Assuming waveguide impedance is constant, the actual voltage at the coupler probe output is calculated from the indicated forward power as

$$V_{Ind} = VF \cdot 10^{(\frac{-CF}{20})} = \sqrt{P_F} \cdot 10^{(\frac{-CF}{20})} \tag{3}$$

If reverse power is present and known (and assumed correct), the interfering voltage is calculated as

$$V_{Ind} = V_F \cdot 10^{(\frac{-(CF+D)}{20})} = \sqrt{P_F} \cdot 10^{(\frac{-(CF+D)}{20})} \tag{4}$$

The limits of actual forward power can then be found by adding and subtracting the interfering voltage from the indicated voltage,

$$P_{F\,max} = V_{F\,max}^2 = (V_{Ind} + V_{Int})^2 \cdot 10^{(\frac{CF}{10})} \tag{5}$$

$$P_{F\,max} = V_{F\,max}^2 = (V_{Ind} - V_{Int})^2 \cdot 10^{(\frac{CF}{10})} \tag{6}$$

A quick analysis of the exponents reveals that the term for coupling factor (CF) can be eliminated, implying that the error in power measurement due to finite directivity is entirely independent of coupling factor.

Figures 4 thru 6 illustrate the effects of the interference and magnitude of the resulting measurement error for typical cases encountered in microwave heating processes. Figure 4 represents a typical case where the load is reasonably well matched. The range of error is minimal when using a coupler with 40 dB directivity and within acceptable limits with a 25 dB directivity coupler.

Figure 5 represents a case where the load is poorly matched, as may be the case during initial startup and tuner adjustment. Again, the error limits are minimal for a 40 dB directivity coupler, but the limits for a 25 dB directivity coupler are significantly greater. The system operator should not be surprised to see large changes in the forward power reading during tuner adjustment.

Power reflected from the process load is often measured on a continuous basis. If the load is well-matched and reflected power is minimal, the error caused by interference from the forward wave can be many times the indicated reading as illustrated in Figure 6. A common mistake especially in research applications is to infer a process effect based solely on a change in the reflected power reading without considering the potential magnitude of the interference.

Table 1 provides a comparison of performance characteristics between the three common coupler types. Note that the frequency response of the cross-guide coupler is much better than that of the loop coupler, but directivity is not as good. As such, the cross-guide coupler is preferred for well matched, broad band applications such as telecommunications whereas the loop coupler is preferred for heating applications. The broadwall coupler offers superior performance in both coupling factor and directivity compared to the other coupler types, but size and cost limit its usefulness and desirability.

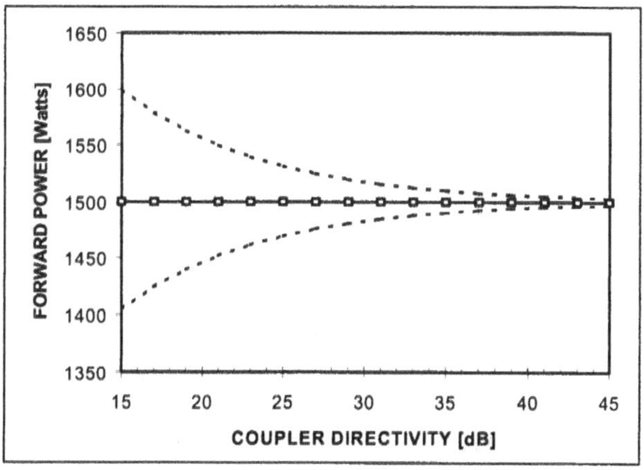

Figure 4. Minimum and maximum true forward power limits for a typical case with a matched load (1500 Watts indicated forward power, 50 Watts true reverse power)

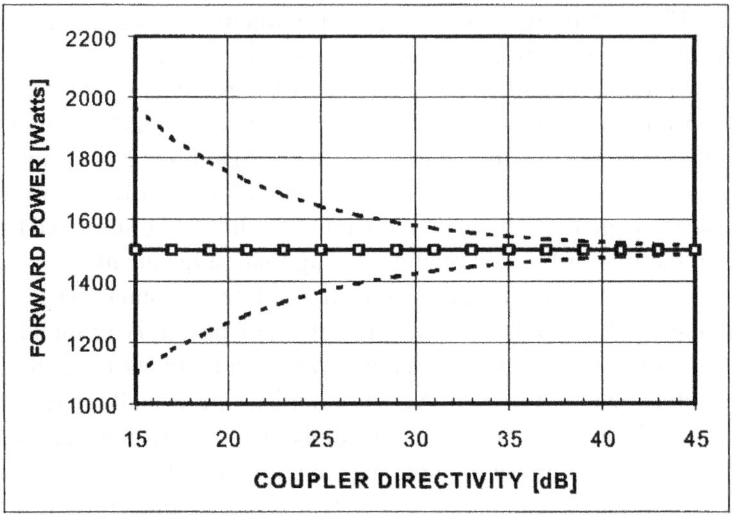

Figure 5. Minimum and maximum true forward power limits for a typical case with a poorly matched load (1500 Watts indicated forward power, 1000 Watts true reverse power)

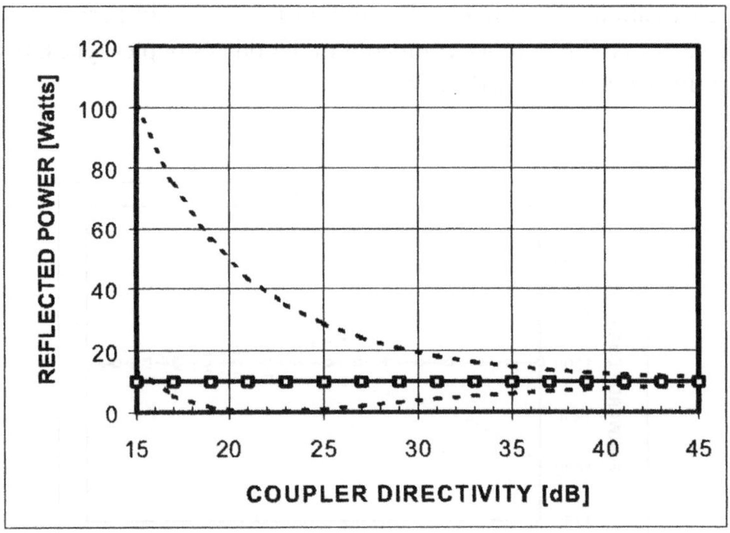

Figure 6. Minimum and maximum true reflected power limits for a typical case with a well matched load (10 Watts indicated reflected power, 1500 Watts true forward power)

TABLE 1-PERFORMANCE COMPARISON FOR
WR284 WAVEGUIDE DIRECTIONAL COUPLERS

Coupler Type	Mfr	Operating Band (GHz)	Coupling Factor Frequency Response Mfr Specification +/- dB	% of Band	Within ISM Band +/- dB	% Error	CF Mean Accuracy (+/- dB)	Minimum directivity (dB)
Loop	A	2.60-3.95	0.75	25%	0.22	5.3%	0.5	25
	B	2.60-3.95	1.0	20%	0.37	8.9%	N/A	25
	C	2.4-2.5	0.2	100%	0.20	4.7%	0.2	28
Cross-guide	A	2.60-3.95	1.0	100%	0.07	1.7%	1.0	20
	B	2.60-3.95	1.5	100%	0.11	2.6%	0.5	20
Broadwall	A	2.60-3.95	0.5	100%	0.04	0.9%	1.0	40
	B	2.60-3.95	0.5	100%	0.04	0.9%	0.5	40

Sensors

The microwave output signal from the branch line of a directional coupler is of little use for power measurement purposes until it has been converted to a form which can be easily displayed and/or recorded. Several types have been devised, each having characteristics preferred for specific applications.

Bolometric types, including thermistors, operate on the basis of the dc resistance temperature coefficient of a power absorbing element. Bolometers are favored for use as laboratory measurement standards due to their inherent accuracy and stability.[3] Thermistors have been commonly used commercially in conjunction with electronic power meters, but more recently they have been replaced by diode and thermocouple devices which offer greater sensitivity, dynamic range and power handling capability.[1]

Detector diodes operate by rectifying the microwave signal directly to provide a pulsed dc voltage signal. At the low end of the operating range, the output signal follows a "square-law" transfer function which is conveniently proportional to waveguide power. Diodes are quite popular in commercial and industrial applications for their simplicity and low cost. However, as semiconductor devices they have an inherently high temperature coefficient which limits their accuracy under varying ambient conditions.[1,3]

Thermocouple sensors are popular for high performance measurement requirements due to their higher sensitivity, ruggedness and accuracy. The principle of operation is based on the voltage generated due to the differential temperature between hot and cold junctions of dissimilar metals. The effect is inherently "square-law" in operation and truly averaging, making thermocouple sensors ideal for all types of signal waveforms. The only drawback to these sensors is high cost which limits their practicality in typical industrial heating applications.

Refectometer

Multi-port reflectometers have gained popularity in industrial heating applications for their ability to provide information about the complex impedance of a load in addition to forward and reverse power. Commercially available types utilize three or four non-directional probes to detect the voltage at different positions along the waveguide. Various models with integrated electronics for signal detection and conditioning deliver the individual probe output signals, while others utilize microprocessors for signal analysis. Several mathematical methods are available for complex impedance analysis using three and four probe models with equal or random probe spacing.[5,6] The most sophisticated products are used in conjunction with automatic impedance tuners and predictive sparameter algorithms for very high speed performance.

Calibration

A critical factor in the usefulness of any directional coupler is its calibration accuracy. Reasonably accurate calibration at low power is possible using a standard network analyzer having suitably high dynamic range. However, high power calibration requires the use of a high power measurement standard. The

National Institute of Standards and Technology (NIST) maintains standards up to I kW at up to 1 GHz, but in the I GHz to 98 GHz frequency range the available standards are limited to 10 mW.

Thus, alternate methods must be employed for high power calibration at 2.45 GHz using NIST-traceable standards. A simple method is by substituting a previously calibrated coupler with the coupler under test. A more cumbersome yet highly accurate method is the "cascaded-coupler" technique.[7] Commercially available calorimeters utilize the resistance substitution method which compares the electrical heat loss to microwave power absorption.

A calibration routine should be established depending on the type of coupler used and requirements for measurement accuracy in a particular application. Aperture array couplers have fixed geometries and thus require no periodic calibration. Resistive loop couplers are adjustable and therefore should be routinely calibrated at least once every 1-2 years. Devices employing electronic circuitry to derive the output signal are more subject to drift and should be calibrated a minimum of once every year.

Summary

Several methods for high power measurements in waveguide are available, but only a few are recommended for industrial heating applications. Directional and non-directional couplers offer a range of performance characteristics for simple waveguide power measurements, while multi-port reflectometers add the capability to deliver information on the complex impedance of a load. Directivity and interference from reverse power should be well understood when interpreting the output reading from a directional coupler. Calibration must be routinely maintained to ensure measurement accuracy.

REFERENCES

1. "Fundamentals of RF and Microwave Power Measurements," HewlettPackard Application Note 64-lA, 1997
2. C.G. Montgomery, "Technique of Microwave Measurement," MIT Radiation Laboratory Series, Vol. 11, McGraw-Hill, New York, 1947
3. A. Fanton, "Radio Frequency & Microwave Power Measurement," Peter Peregrinus Ltd., London, 1990
4. R. Meredith, "Engineers' Handbook of Industrial Microwave Heating," Institution of Electrical Engineers, London, 1998
5. P.I. Somlo and J.D. Hunter, "Microwave Impedance Measurement," Peter Peregrinus Ltd., London, 1985
6. R. Caldecott, "The Generalized Multiprobe Reflectometer and Its Application to Automated Transmission Line Measurements," IEEE Transactions on Antennas and Propagation, Vol. AP-21, No. 4, July 1973
7. J.W. Allen, "Switched-Coupler Measurements for High-Power RF Calibrations," NIST

Sintering of Ceramics Using Low-Frequency RF Power

J.B.O. Caughman, D.J. Hoffman, F.W. Baity,
M.A. Akerman, S.C. Forrester and M.D. Kass

Abstract

Sintering with low frequency rf power (~50 MHz) is a new technique with unique capabilities that has been used to sinter a variety of ceramic materials, including zirconia-toughened alumina, alumina, silicon carbide, and boron carbide. Processing with low frequencies offers many advantages compared to processing with conventional microwave frequencies (915 MHz and 2.45 GHz). Because of the longer wavelength, the rf electric field penetrates materials more than microwaves. This effect allows the processing of a wider variety of materials and allows for an increase in the physical size of the material being processed. In addition, the material is heated in a single mode cavity with a uniform electric field, which reduces the occurrence of hot-spot generation and thermal runaway effects. This technique has been used to sinter large crack-free alumina samples (3″ square) to >97% density. The sintering and/or annealing of a number of carbide materials has been demonstrated as well, including silicon carbide, boron carbide, tungsten carbide, and titanium carbide.

Introduction

A great deal of work has been done in the past ten years with the use of microwaves in the sintering of ceramics [1,2]. An advantage of microwave sintering over conventional sintering is related to the volumetric interaction of the electromagnetic fields with the ceramic. This volumetric coupling to the ceramic leads to a higher heating efficiency and faster processing times than those achievable in conventional furnaces. However, because of its small penetration depth, direct microwave sintering (not hybrid heating) is mostly limited to materials with relatively small volumes and/or low dielectric loss tangents.

The heating of the ceramic is obtained by the coupling of the electric field with the dielectric losses of the material. The role of the electric field and the applied frequency in sintering can be seen by looking at the power deposition in a material. The amount of power per unit volume that is coupled into the ceramic by electric fields is given by [3]

$$P = \frac{1}{2}\omega\varepsilon(\tan\delta)E^2$$

This equation shows that the losses are proportional to the frequency (ω), the loss tangent ($\tan\delta$), and the square of the electric field (E). In the typical microwave range, the frequency is high (1010 Hz), the electric field is low (<100 V/cm), and the loss tangent is generally high (10^{-2}).

The distance that the electric field can penetrate a ceramic is given by [4]

$$x = \frac{\sqrt{2(1+\sqrt{1+\tan^2\delta})}}{\omega\tan\delta\sqrt{\mu_0\varepsilon'}}$$

where x is the *e*-folding penetration depth, μo is the material permeability (assumed free space), and ε' is the permittivity. An example of the penetration depth as a function of frequency is shown in Fig. 1 for boron carbide (B_4C) at 1100°C and alumina at 23°C. Only low frequencies can penetrate high-loss materials (such as B_4C). Sintering of relatively large samples of high-loss materials with a uniform electric field can be accomplished when samples have a characteristic length on the order of 0.2λ. These considerations led to an exploration of the application of rf energy to the sintering process.

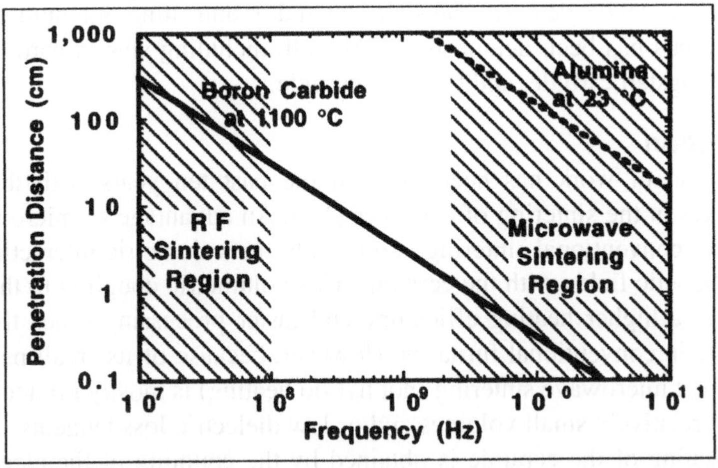

Figure 1. Depth of electric field penetration as a function of frequency for a low-loss and a high-loss material.

RF Sintering

The advantage of rf sintering over microwave sintering is related to the higher electric fields and greater sample penetration depth obtained with sintering at rf frequencies. For rf sintering, the frequency is lower (108 Hz), the loss tangent is generally lower (10^{-3}), and the electric field is higher (1000 V/cm). Thus, compared to most microwave techniques, rf absorption is reduced by three to four orders of magnitude in the product of frequency and loss tangent, but electric field effects are up to four orders of magnitude higher because the field is 10-100 times greater. Therefore, rf sintering has roughly the same total power deposition as microwave sintering but has a much higher electric field and can be used to determine whether the critical component of electromagnetic energy sintering is the strength of the electric field or the value of the excitation frequency. In contrast to microwave sintering, the wavelengths of the rf waves are considerably greater than the dimensions of the sintering cavity. This generally results in a single permissible mode of operation with a uniform electric field, requiring tuning and impedance matching circuitry not commonly found in microwave fumaces.

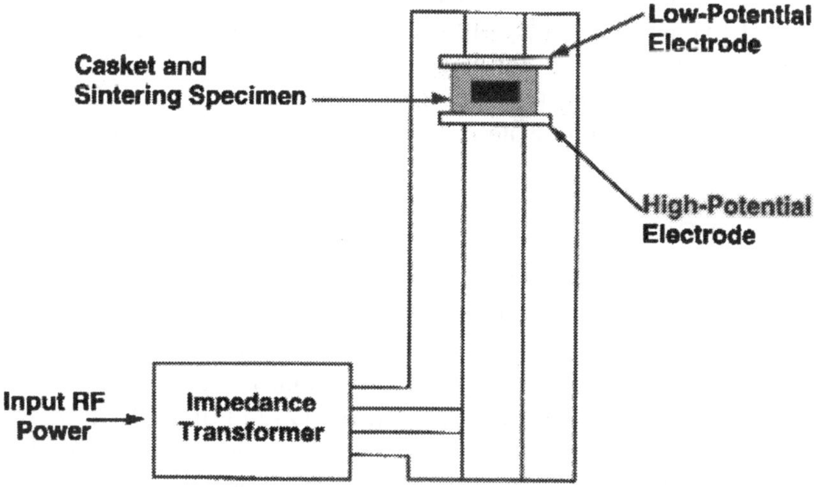

Figure 2. Schematic of the rf sintering fumace. The casket assembly is placed in the high-field region between the electrodes.

A schematic of the rf sintering furnace is shown in Fig. 2. The cavity is composed of a quarter-wave coaxial waveguide resonator, with the center conductor grounded at one end and an electrode placed at the high voltage end. An additional electrode (near ground potential) completes the circuit. The rf power is coupled through a variable impedance transformer near the base of the cavity, which keeps the cavity tuned and eliminates reflected power during the run. The resonant structure is designed to produce a high electric field at the specimen to be sintered. The sintering specimen is typically packed in insulation inside a casket. The casket assembly is then placed in the high field region between the electrodes and is electrically equivalent to a lossy dielectric between the two plates of a capacitor. Since the cavity is coaxial, an equivalent circuit was analyzed by using a lossy transmission line model that included dielectric losses in the ceramic and resistive losses in the cavity. For a 5 cm thick lossy dielectric slab placed between the electrodes, calculations have shown that electric fields of 1-5 kV/cm can exist for a loss tangent variation of 10^{-2} – 10^{-4} and an input power of 2 kW. An electrostatic analysis of the geometry similar to that shown in Fig. 3 was performed in which a test specimen (dielectric constant, $\varepsilon = 8.9$) is surrounded by loose dielectric fiber ($\varepsilon = 2.3$). The analysis showed that the field is uniform to within a few percent and is orders of magnitude higher than achievable microwave fields.

To date, two rf sintering chambers have been made. A smaller chamber that operates between 50-55 MHz is used for small samples. The electrodes have a diameter of 13 cm and a spacing of 5-9 cm. A larger chamber that operates at 35-40 MHz is generally used for larger samples or where larger amounts of insulation are needed. The electrodes have a diameter of 36 cm and a spacing of 11-15 cm. Several different materials have been processed and are summarized in the next section.

Processing Results

Zirconia-Toughened Alumina

The density as a function of processing temperature has been determined for the rf sintering of zirconia-toughened alumina (ZTA). This material was chosen so that comparisons could be made to previously published results [5]. The samples were composed of 40 wt % zirconia (Tosoh TZ-2Y) and 60 wt % alumina (Sumitomo AKP 50) and prepared as previously described. The mass of the samples were typically 18-20 g. The rf processing frequency was 52 MHz, and the processing atmosphere was flowing nitrogen at slightly over 1 atm of pressure.

The samples were arranged in a casket as shown in Fig. 3. They were placed inside a 7.62 cm diameter, 4.76 cm thick boron nitride casket and surrounded by a 50 wt % mixture of alumina and zirconia fiber (Zircar). The casket was placed between the electrodes of the smaller rf sintering furnace. Either graphite or alumina fiberboard plates were placed on both sides of the casket. The temperature of the sample was measured optically by looking at the sample through a sight tube. An Inframetrics 600 Infrared camera was used for measuring temperatures below 1100°C and an Ircon 2-color pyrometer was used for temperatures > 1100°C. The densities were measured using the Archimedes technique in 200-proof ethanol.

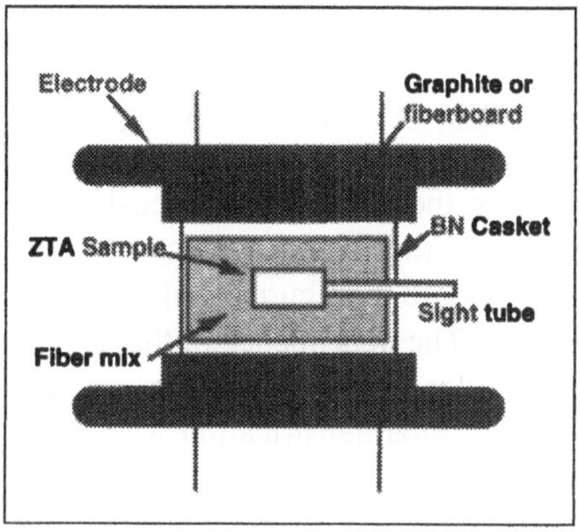

Figure 3. Arrangement of the casket between the electrodes.

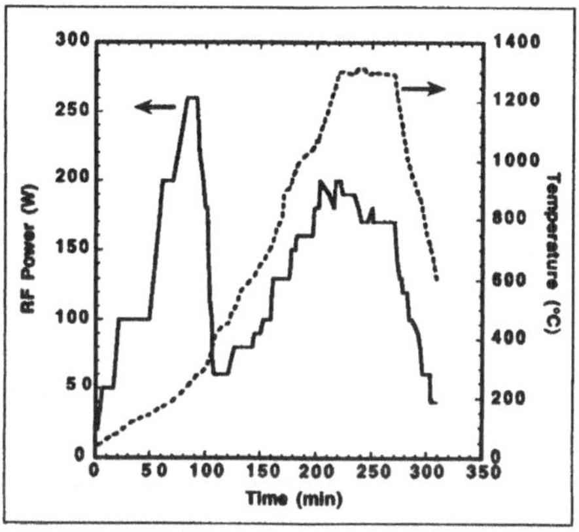

Figure 4. RF power and sample temp during a ZTA sintering run.

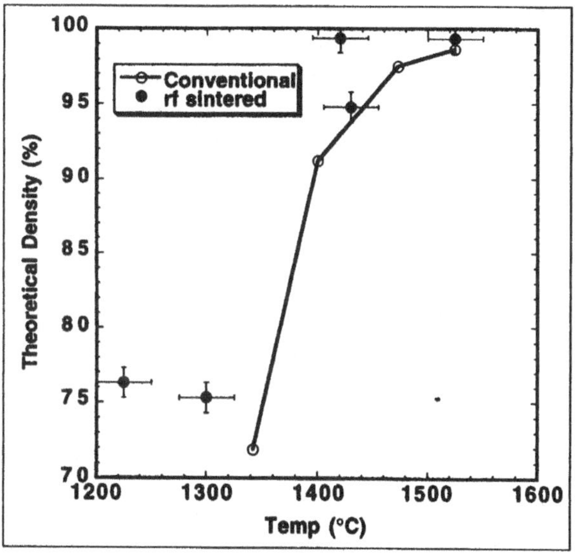

Figure 5. Density as a function of processing temperature for rf sintered ZTA.

Heating rates and power levels were dependent on the temperature of the sample/casket assembly. A plot of rf power and temperature as a function of time for a typical run is shown in Fig. 4. Initial heating rates of the sample were roughly 2.5°C/min until dielectric relaxation occurred, which was between 250-350°C. After this point, the power required to maintain the temperature dropped, and the heating efficiency increased significantly. Heating rates increased to roughly 10°C/min up to the desired temperature. The samples were held at the target temperature for 1 hour.

The density as a function of temperature is shown in Fig. 5. For comparison, the sintering curve for conventional processing [5] is also shown in the figure. The uncertainties in the temperature come from the difficulty in focusing the pyrometer through a narrow sight tube. In addition, the center of the sample will be hotter than the edge, and the temperatures shown in the figure should be interpreted as a lower limit for the actual temperature. The dependence of the density on temperature for the rf sintered samples is similar to the temperature dependence of the conventionally sintered material. The final densification was >99% for the higher temperature samples. Because of the uncertainty of the temperature, however, it is difficult to conclude that there is a significant reduction in the sintering temperature for these samples.

Alumina

Sintering of alumina has also been achieved using low frequency power at 52 MHz and 38 MHz. The samples were 99.5% pure alumina (Coors 995) and were baked to 800°C for 8 hours in air to remove the binder before being sintered. The size and mass of the green bodies varied and ranged from 40 g (4.5 cm x 4.5 cm x 0.9 cm) to 380 g (9.0 cm x 9.0 cm x 1.8 cm). The initial density was about 55%. As with the ZTA, the alumina was processed in a flowing nitrogen atmosphere at slightly over 1 atm.

TABLE I. CARBIDE AND OTHER MATERIALS PROCESSED TO DATE.

Material	Process	Temp (°C)	% Density	Comments
Silicon carbide (Cercom)	Sinter	>2000	84	12.5 g sample (Limited by current insulation scheme)
Boron carbide (2.5 wt % C [7])	Sinter	>1700	97	Repeatability limited by current insulation scheme
Zirconia	Sinter	1410	98.4	9.2 g sample
Titanium carbide	Anneal	1500		31 g sample
Tungsten carbide (6 wt % Co)	Sinter	1500	97	14 g sample

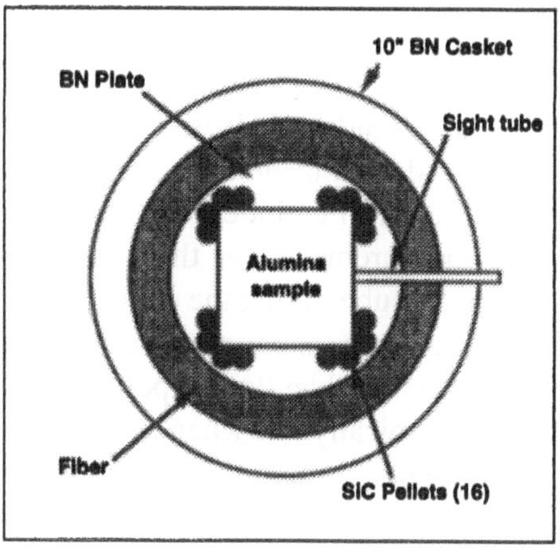

Figure 6. Arrangement of the sample and casket for alumina runs.

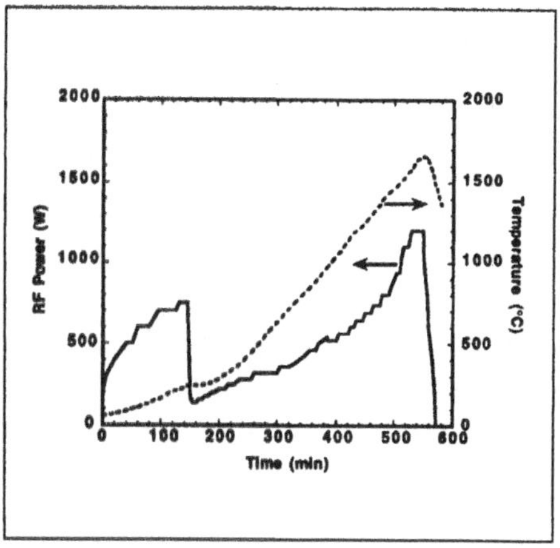

Figure 7. RF power and sample temp during a 330 g alumina run.

The samples were arranged in a casket as shown in Fig. 6. The tiles were placed on a thin BN plate (0.32 cm) inside a 25.4 cm diam BN casket (11.4 cm high). For the larger samples, a thin layer of alumina powder was used as an interface with the BN plate to reduce surface friction. Silicon carbide pellets (1.3 cm diam, 1.3 cm thick, 5.3 g) were placed at the corners of the sample. The pellets help during the initial heating of the tile and were placed at the corners to reduce the thermal gradient in the sample. A mixture of alumina and zirconia fiber (50/50) was used as the insulation, and the temperature was measured through a sight tube as described above.

The power and temperature during a typical run are shown in Fig. 7 for an experiment with a 330 g sample. Initial heating was slow because of the low loss tangent of the alumina. The presence of the silicon carbide helps in the initial heating phase, but a different arrangement may be needed to optimize the low temperature performance. After relaxation occurs, the heating efficiency increases and the sample readily absorbs power and heats. The heating rate was kept to less than 5°C/min to reduce the chance of cracking in the larger samples. Samples were typically heated to at least 1600°C and held there for 30 minutes before cooling. For a 330 g sample, a power of 1200 W was needed to heat the sample to 1650°C. The samples typically densified to 96-98%. Larger samples (800 g, 4.75 cm x 4.75 cm, x 2.4 cm greenbodies) are currently being attempted.

Carbides

One of the advantages of rf sintering is the ability of the electric field to penetrate high-loss materials, such as silicon carbide and boron carbide. Both of these materials have been sintered using rf power. Other carbides, such as tungsten carbide and titanium carbide have been either sintered or annealed. The results are summarized in Table I. Both silicon carbide and boron carbide have been sintered to > 83% dense. Accurate temperature measurements for these samples have not been made because of clogging of the sight tube during the experiments. However, the densities that have been obtained are consistent with heating the material to >2000°C. An insulation package that can handle temperatures in excess of 2000°C is also an issue. Alumina fibers typically fail when the temperature exceeds around 1700°C. Recent experiments using a combination of zirconia fiber and a mixture of BN powder with glassy carbon [6] have shown promising results.

Summary

The use of low frequency (< 55 MHz) rf power to sinter ceramics has been demonstrated. Sintering with lower frequencies than microwaves has the advantage of being able to penetrate large or high-loss materials with higher electric fields. Because the rf wavelengths are longer than the dimensions of the furnace, the mode of operation can be controlled to ensure nearly uniform electric fields in the region of interest. Processing to temperature up to 1700°C is now routine, and progress is being made to extend the temperature range to > 2000°C. The sintering of ZTA and large (> 330 g) alumina samples (without cracking) has been accomplished, along with the sintering and/or annealing of high-loss carbide materials, such as boron carbide and silicon carbide. This low frequency operating regime offers a unique opportunity to determine the effects of sintering due to electric field interactions with materials.

ACKNOWLEDGMENT

Research sponsored by the U.S. Army Tank Command under Interagency Agreement 1969-C101-A1 under Martin Marietta Energy Systems, Inc., contract DE-AC05-840R21400 with the U.S. Department of Energy.

REFERENCES

1. *Microwaves: Theory and Application in Materials Processing II,* edited by D. E. Clark, W. R. Tinga, and J. R. Laia, Jr., Ceramic Transactions **36** (1993).
2. *Microwave Processing of Materials IV,* edited by M. F. Iskander, R. J. Lauf, and W. H. Sutton, Materials Research Society Symposium Proceedings, Vol. 347 (1994).

3. W. H. Sutton, "Microwave Processing of Ceramic Materials," *Ceram. Bull.* **68** (2) 376-386 (1989).

4. A. R. von Hippel, "Descriptions of Dielectrics by Various Sets of Parameters," in *Dielectric Materials and Applications,* edited by A. R. von Hippel, Technology Press of MIT/John Wiley & Sons, New York, p.12 (1954).

5. H. D. Kimrey, J. O. Kiggans, M. A. Janney, and R. L. Beatty, "Microwave Sintering of Zirconia-Toughened Alumina Composites," *Microwave Processing of Materials II,* Materials Research Society Symposium Proceedings, Vol. 189, p. 243 (1991).

6. C. E. Holcombe, N.L. Dykes, and M.S. Morrow, "Thermal insulation for high-temperature microwave sintering operation and method thereof," U.S. Patent Application in Progress (1993).

7. C.E. Holcombe and N.L. Dykes, "Ultra High-Temperature Microwave Sintering," *Ceramics Transactions* **21**, p. 375 (1991).

Materials Processing Via Variable Frequency Microwave Irradiation

R.S. Garard, Z. Fathi and J.B. Wei

Abstract

Variable frequency microwave (VFM) irradiation is a new technique applied to advanced materials to overcome the reported problems with uniform scale-up and repeatable control of microwave energy inside a material. A broadband, variable frequency microwave source is used to provide tuning to an optimum incident frequency and then sweeping around that centeral frequency, as required, for uniform distribution of the microwave energy. This approach has been employed in evaluating microwave processing benefits for several advanced matenal systems. Results of these trials plus the comparison of variable to fixed frequency processing is reviewed.

Introduction

Alternative processes are being sought to improve on the cost effectiveness and the processing cycles of various advanced materials. Microwave irradiation has been investigated as an alternative processing means by which short process cycles are often obtained. The body of work carried out up to date demonstrates several advantages brought about by the presence of a microwave electric field. High frequency heating is an efficient means by which short process cycles, fast heating rates and enhanced reaction kinetics are obtainable. Microwave heating is more effcient than conventional thermal processing because microwave (electromagnetic) energy is directly coupled into the material at the molecular level, bypassing the air medium in the cavity chamber.

Research teams worldwide are investigating the use of microwave energy in the area of materials processing—all of which use either single-mode or multimode, single-frequency microwave processing techniques. The demonstrated "global advantages" of microwave heating over conventional heating include direct and volumetric heating, high selectivity, fast heating rates, and reaction kinetics enhancement. This has been demonstrated in numerous systems on a laboratory scale. Unfortunately for advanced materials, these potentials have not yet been attained on an industrial scale.

Using variable frequency irradiation, the processing of high-performance materials has been targeted such as polymers, ceramics and monolithic composites. Variable frequency irradiation provides solutions to the inherent scale up limitations of fixed-frequency microwave irradiation while simultaneously producing high energy efficient coupling to the material to be processed.

The Variable Frequency Microwave Concept

The concept behind variable frequency processing involves the control and distribution of microwave energy throughout the material being processed. There are four controllable processing parameters that distinguish variable frequency microwave (VFM) from conventional, single frequency microwave based technologies. These controllable processing parameters include; central frequency, frequency bandwidth, frequency sweep-rate and power. Variable frequency processing enables precise frequency tuning to optimize the energy coupling efficiency during initial heating. During heating, continuous and/or selective sweep through several cavity modes within a period of a few milliseconds results in a time-averaged uniformity of heating throughout the load. Frequency sweeping improves heating uniformity through the superposition of many modes within the microwave cavity. This uniformity facilitates process scale up either to large components or to large batches of small components. The bandwidth required to achieve microwave energy uniformity inside the cavity depends on the operating frequency, the material chemistry (dielectric properties) and the size of the cavity. At high central frequencies, only a narrow bandwidth may be required to provide uniform energy distribution.

During processing, the central frequency irradiated inside the microwave cavity can be tuned to increase the coupling efficiency with the material to be processed. This frequency tuning capability is especially important for materials with frequency sensitive dielectric behavior. Thus, the heating rate can be increased without changing the power. The combination of bandwidth and sweep rate around the selected central frequency, utilized during processing, provides the necessary distribution of microwave energy to carry out uniform heating throughout the workload. The sweep rate is electronically controlled and can be varied from minutes to milliseconds. The microwave incident power can be pulsed or continuously varied to provide additional control over the heating profile of the workload.

The dielectric loss measurements, of a given material, over the range of temperatures and frequencies of interest provides the necessary feedback for frequency selection during heating in order to optimize the processing of different materials. In general, the orientation polarization loss peak (maximum losses) shifts to higher frequencies as the temperature of the material is elevated. Thus, there is a definite advantage in having the ability to initially tune and then vary the frequency of the irradiated microwave signal as the material being processed is heated.

The shift in frequency of the loss peak maxima associated with materials is temperature dependent. While many polymers and ceramics do not exhibit a strong dependence on the incident frequency, there are definite advantages in frequency tuning capability when processing strong microwave absorbers as is the case of drying applications and/or in a composite form such as processing of graphite fiber reinforced polymeric materials.

Materials Processing and Results

An array of materials have been processed using variable frequency irradiation with emphasis on epoxies and polymer matrix composites. Drying, binder burnout as well as ceramic and ceramic composite, high temperature sintering are also being investigated. Of special interest in the ceramic field is the co-firing capabilities offered by controlling variable frequency irradiation. Microwave assisted CVD and other plasma based processes are also of interest for applying variable frequency microwaves. The ability to control plasma location and/or generate multiple plasma balls by launching multiple incident frequencies has generated interest in various industrial applications.

Experimental. The current microwave source used for variable frequency irradiation is a traveling wave tube (TWT). Variable frequency microwave systems ranging from 100 to 1,500 watts and over the frequencies of 0.9 to 18 Ghz, typically in one to two octave bandwidths are available. Model LT-0200 VFMF system has been utilized to process polymers and polymer composites in the 2.5 - 7.5 GHz frequency range. This furnace can be operated at a fxed frequency, or over the entire bandwidth, or any portion thereof may be swept at interval rates of 100 msec to 2 minutes. The furnace cavity is nominally a 12″x12″x10″ stainless steel enclosure. A thin sheet of polyimide film was placed over the microwave launcher to prevent any volatiles from condensing in the waveguide, and potentially contaminating the power source and/or sensors.

Epoxies. A stoichiometric mixture of an epoxy resin ERL-2258 (Union Carbide) with M-phenylenediamine hardener (Dupont) (25 g MPDA per 100g resin) was cured using variable frequency irradiation. The MPDA was dissolved into the resin by mixing at 65 °C. The properties of the processed materials are reported elsewhere [1]. The recommended conventional curing for these epoxy samples is illustrated in Figure 1. The heating in this particular system is hard to control and slow heating rates are used because of the reaction exotherm, avoiding materials degradation and temperature excursions. In general, microwave energy can be pulsed as to eliminate exothermic reactions without slowing down the heating rate, and has been shown to substantially shorten the process cycle [2]. The variable frequency approach was found to provide a means by which uniform and controllable heating can be achieved. The curing cycles utilized during variable frequency processing were 8 times faster than the conventional cycles. To illustrate uniformity, eight samples were configured as shown in Figure 2. The temperature was monitored in one representative sample. The

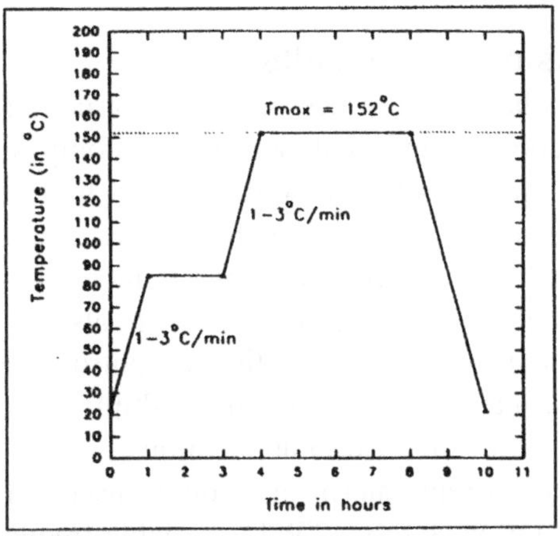

Figure 1. The recommended conventional curing cycle for the Union Carbide Epoxy Resin — ERL 2258.

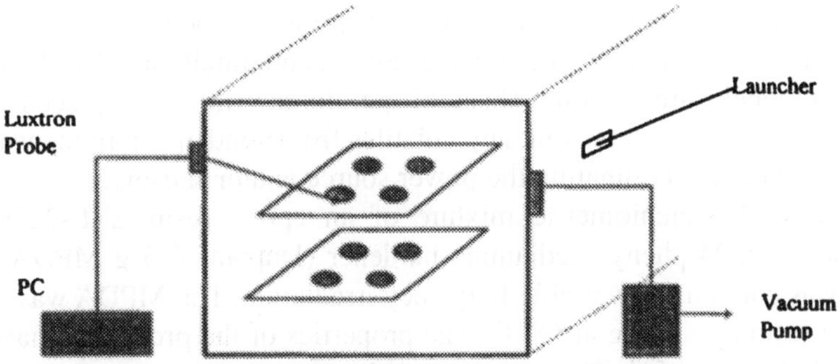

Figure 2. The variable frequency microwave setup. Eight samples have been configured as illustrated; the temperature monitoring was performed in one representative sample.

temperature versus time profile is illustrated in Figure 3. The curing was uniform for all eight samples and within each and every sample.

As a scale up illustration, an epoxy sample weighting 530 grams has been controllably cured using variable frequency irradiation. Figure 4 illustrates the microwave cured 10.5" x 6.5" x 0.6", 530 g epoxy sample.

Polymer Matrix Composites. Glass and graphite fiber reinforced polymer matrix composites were successfully processed using variable frequency microwave energy.

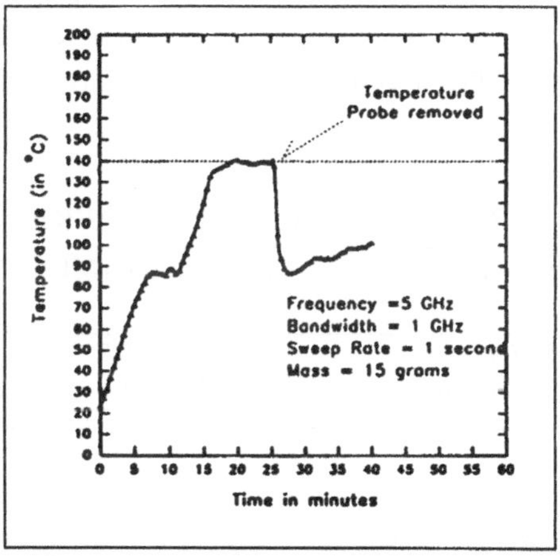

Figure 3. The thermal profile during processing at 5.0±0.5 GHz variable frequency.

Figure 4. Illustration of a 530 grams ERL 256 epoxy processed using variable frequency.

The chemistry of an isocyanate/epoxy composite was evaluated. Specific physical and chemical charactenstics (glass transition temperature, flex modulus, flex strength and dielectric behavior) of the conventionally cured PMCs have been reported [3,4]. The conventional post-curing cycle requires 8 hours at temperatures of up to 240°C. Post-curing cycles of 50 minutes have been demonstrated with variable frequency microwave energy. Variable frequency heating of a series of plate configurations were performed, including 1,2,3,4, and 8 plates stacking. The mechanical properties of the samples processed using variable frequency microwave energy are comparable to those of the samples processed using conventional heating (refer to Table 1).

Scale-up benefits of variable frequency was demonstrated by processing 8″ x 8″ PMC plates. The variable frequency heating was directly contrasted with fixed frequency heating. The fixed frequency irradiation at 4.5 GHz (no mode

TABLE 1. MECHANICAL BEHAVIOR

	Flex modulus(GPa)	Flex strength (MPa)	Impact Strength (KJ/m^2)
Microwave	5.77 ± 0.15	197 ± 15.5	5.7
Conventional	5.75 ± 0.14	197 ± 11.5	5.7

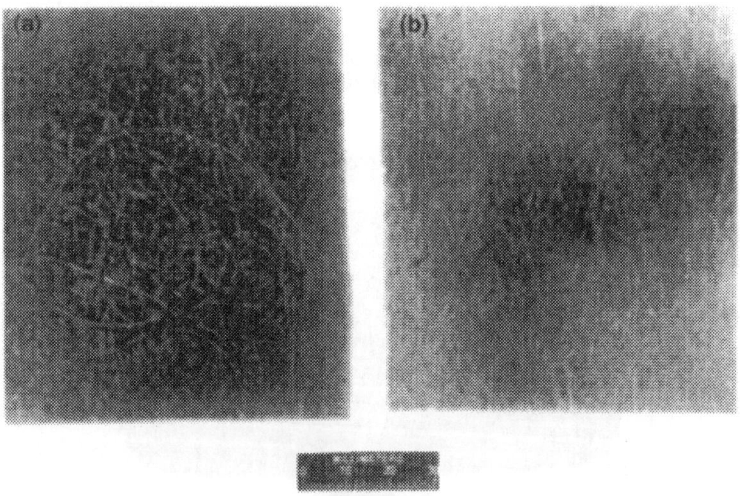

Figure 5. Fixed frequency versus variable frequency illustration of 4″ x 4″ glass reinforced isocyanate/epoxy composites: (a) variable frequency processing and (b) fixed frequency processing,

stirring) of a 4″ x 4″ sample leads to non-uniform curing, as is illustrated in Figure 5a. The non-uniform curing is the results of a non-uniform distribution of the electric field within the microwave cavity. The sample processed using variable frequency (4.5±0.5 GHz) is fully uniform as shown in 5b.

Unidirectional and crossply graphite-reinforced PMCs were also investigated. Carbon Fiber Composite (CFC) of up to 8″ x 8″ and up to 32 ply have been successfully processed with variable frequency. Figure 6 illustrates the heating profile of 4″ x 4″ unidirectional and crossply samples. As illustrated, the heating rates generated within unidirectional samples are generally higher than those achieved within crossply samples. To investigate the uniformity of cure after heating with variable frequency microwave energy, glass transition measurements were performed throughout the samples dimensions. The glass transition values summarized in Figure 6 reveal uniform curing throughout the sample.

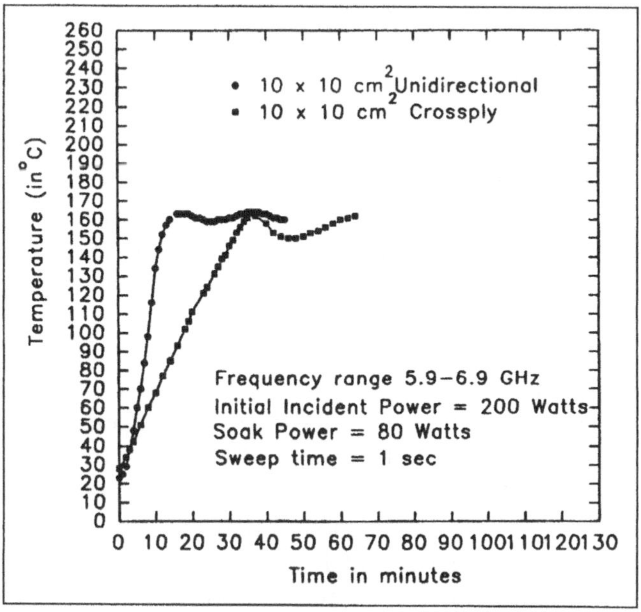

Figure 6. The heating profiles of 4" x 4" unidirectional and crossply graphite reinforced PMC along with the glass transition measurements as a function of location within the variable frequency microwave processed samples. Uniform curing is achieved.

Plasma Processing. Plasma generation and plasma position control have been investigated using variable frequency microwave energy. By changing the excitation frequency of a plasma in a multi-mode chamber, the locations of the maxima and minima of the power density within the chamber are also changed. This allows localization of the plasma discharge in the precise processing area desired in order to increase the process' efficiency. In a similar fashion, sweeping of the processing frequency during the process cycle can theoretically "scan" the plasma across an arbitrarily-shaped target surface. A series of experiments was conducted to evaluate the flexibility of variable frequency microwave energy as a plasma process tool. The same basic LT 0200 variable frequency source was used with a bell jar adapter and vacuum system. The gas used for this study was a solution of 50% ethanol and 50% water vapor. Pressure control was maintained between 15 and 25 Torr. The results of these tests are presented elsewhere [5].

Clay Drying. Two series of clays with different water contents (57.18 % and 46.42%) were dried under different experimental conditions over the 2.5 - 7.5 GHz range. The benefits of varying the frequency of the incident signal were clearly illustrated. The weight loss (dehydration) and heating rates were derived for several samples and were found to be directly related to the incident frequency. The microwave absorption characteristics of water depend on both temperature and frequency [6]. The experimental results suggest that the moisture removal is more efficient at higher frequencies and powers. The various clay samples were weighed and placed inside microwave transparent plastic jars, the samples were then inserted inside the microwave furnace for drying. A Luxtron temperature probe was used to monitor temperature during irradiation. The temperature and time were then recorded for each of the samples, the thermal profiles generated were then plotted as a function of time. The samples were weighed at the end of the drying process, and the water loss was then calculated for each of the samples. Figure 7a illustrates the temperature profile generated within clay samples under different microwave incident frequencies. The heating rates as well as the weight loss increase with increasing frequency. Figure 7b illustrates the weight loss as a function of frequency for clay samples processed under otherwise similar experimental conditions.

Pending & Future Work. Preliminary work in the areas of binder-burnout reveal a good potential for effective microwave/materials coupling via variable frequency microwave energy. The details of this work will be revealed in future publications. The heating and sintering of Alumina, SiC and Si3N4 are being investigated under controlled and uncontrolled atmospheres. The high temperature processing of B_4C (1" diam x 1" long), porous tantalum capacitors, and several samples of tungsten alloy (WNiFe) have been achieved.

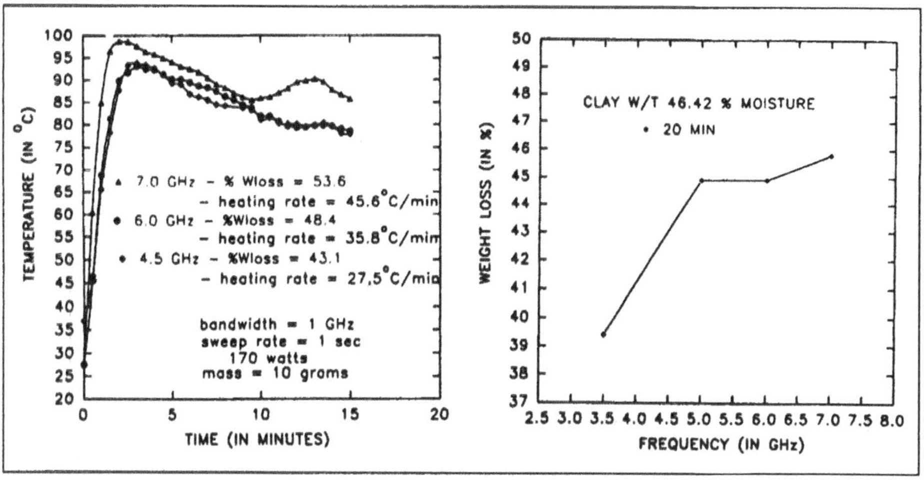

Figure 7. (a) Illustration of the thermal profiles generated during drying of clays as a function of incident frequencies and (b) the accompanying weight loss.

Conclusions

The results discussed in this paper illustrate the use of variable frequency microwave energy as a controllable and flexible processing tool. Variation of the processing frequency has shown to generate uniform heating and curing profiles within various materials. These results indicate that it is possible to rapidly and uniformly process a variety of polymer matrix composites using variable frequency microwave energy. The benefits of frequency sweep to achieve uniform heating has been demonstrated. The benefits of frequency tuning has been demonstrated to be limited to specific materials, but of particular advantage for addressing materials of complex composition. The benefits of frequency agile processing for clay drying was clearly demonstMted. Vanation of the processing frequency was shown to generate intense, controllable, localized water-ethanol vapor plasmas at pressures of 5 to 10 Torr in virtually every location within a quartz bell jar. By increasing the swept-frequency bandwidth about a center processing frequency, the ability of variable frequency microwave energy to generate uniform, large-volume plasmas was shown. Future work will expand into processing high temperature advanced materials.

REFERENCES:

1. A.D. Surrett, R Lauf, F. Paulauskas and A.C. Johnson,"Polymer Curing Using Variable Frequency Microwave Processing", *Microwave Processing of Materials IV,* eds. M.F. Iskander, R.J. Lauf, and W.H. Sutton, MRS **347**, 691(1994).

2 Jianghua Wei, Ph.D. dissertation, "Microwave Processing of Epoxy Resins and Graphite fiber/Epoxy Composites in a Cylindrical Tunable Resonant Cavity", Department of Chemical Engineering, Michigan State University, (1992).

3. DeMeuse, M.T., and Johnson, A. C., "Variable Frequency Microwave Processing of PMCs", *Microwave Processing of Materials IV*, eds. M.F. Iskander, R.J. Lauf, and W.H. Sutton, MRS **347**, 723 (1994).

4. Zak Fathi, Mark. T. DeMeuse, John Clemens and Craig Saltiel, "Characterization & Numencal Modeling of Variable Frequency Microwave Processed Materials, Spring 95 ACerS Conference, (1995).

5. A.C. Johnson, R.A. Rudder, W.A. Lewis, and R.C. Hendry, "Use of Variable Frequency Microwave Energy as a Flexible Plasma Tool", *Microwave Processing of Materials IV,* eds. M.F. Iskander, R.J. Lauf, and W.H. Sutton, MRS **347**, 617, (1994).

6. W.H. Sutton,"Key Issues in Microwave Processing Technology", *Microwaves: Theory and Application in Materials Processing II,* eds. D.E. Clark, J.R. Laia, and W.R. Tinga, Ceramics Transactions, **36**, 3, (1993).

MM-Wave Processing of Ceramics

G. Link, W. Bauer, A. Weddigen,
H.-J. Ritzhaupt-Kleissl and M. Thumm

Abstract

Sintering tests with structural ceramics as well as with functional ceramics, were performed using mm-wave radiation in a Gyrotron Installation. The materials under investigation have been Si_3N_4, piezoelectric ceramics of the PZT-type and TiO_2-ZrO_2-MgO composites. The results with respect to processing parameters, final density, mechanical properties and functional parameters are compared to those generated by conventional sintering in a resistance-heated furnace. It could be demonstrated that mm-wave processing leads to enhanced sintering and improved properties of the sintered bodies.

Introduction

One of the most crucial steps in the processing of modern high performance ceramics is the sintering process. Due to the low penetration depth of the IR-radiation in a standard resistance heated sintering furnace, temperature gradients in the ceramic parts are induced. This may result in inhomogeneities in the sintered body or even in its destruction, if not an optimized time-temperature program with small heating rates is applied. Establishing such an optimized sintering run is rather time consuming. An interesting possibility to overcome these mentioned difficulties is the application of microwaves, due to the possibility of a volumetric heating. This results in the possibility to run very fast heat up rates and so to markedly shorten the processing time. An optimized thermal isolation with an appropriate powder bed and a box of alumina fiber boards allows small temperature gradients in the ceramic body during the sintering process. So mechanical stresses and material inhomogeneities can be reduced. Furthermore, the densification process of ceramic bodies seems to be enhanced by sintering in a microwave field, what has been demonstrated by several authors [1,2]. This means a possible reduction of the maximum sintering temperatures, and soaking times compared to the processing in a standard sintering furnace.

Gyrotron Installation

At the Forschungszentrum Karlsrohe a compact Technological Gyrotron System has been established since 1993. This system, designed in the Institute of Applied Physics in Nizhny Novgorod, Russia, includes a 30 GHz gyrotron with a maximum continuous wave output power of 10 kW. The electronic efficiency of the gyrotron working in the TE_{02} cavity mode at the second harmonic of the electron cyclotron frequency is about 25%. The maximum design values of the energizing parameters are: beam voltage 25 kV and beam current 2 A.

The millimeter-wave radiation generated by the gyrotron passes through the boronnitride tube output window and a quasi-optical transmission line to the microwave furnace. The transmission line with an efficiency of about 90% consists of an in-waveguide mode converter and two mirrors of special geometry focusing the circular symmetric Gaussian beam into the microwave furnace [3]. By using a quasioptical transmission line it is possible to diminish the influence of the radiation reflected back from the furnace upon the operation of the gyrotron to an admissible level. The microwave furnace has a cylindrical symmetry with 50 cm in diameter and 60 cm in height and is loaded through a hemispherically shaped lid. A rotating mode stirrer and a diffuse applicator surface in the area where the beam is reflected first is used in order to improve the field homogenity [4]. The temperature, measured with a type S thermocouple at the sample surface, is computer controlled along a given time-temperature program with an accuracy not greater than ± 5°C.

Experimental Results

In order to reduce temperature gradients in the ceramic bodies due to radiation losses from the sample surfaces, the samples were placed in an optimized thermal isolation box. This reduces mechanical stresses and material inhomogeneities. Such a box consists of alumina fiber boards for all ceramic materials under investigation. Inside this box the sintering experiments with TZM ceramics were performed with four samples of different composition simultaneously. The samples were placed between two alumina ceramic plates as shown in Figure 1a. For Si_3N_4, single samples were sintered in an appropriate powder bed of Si_3N_4 and BN (Figure 1b). The thermocouple is placed in an alumina ceramic sheath in order to avoid chemical reactions with the powder bed. Due to higher mm-wave absorption in the PZT-powder, PZT-samples were sintered in a powder bed smaller than the bed of Si_3N_4 powder shown in Figure 1b.

Piezoceramic

One of the most interesting classes of functional ceramics are piezoceramics based on the lead-zirconate-titanate (PZT) system. As lead has a rather high vapor pressure and tends to evaporate during normal sintering processes, shortening of the processing time and a possible reduction of the sintering temperature may help to avoid these difficulties. Studies with two types of PZT-powders were performed in order to optimize the sintering conditions of PZT-ceramics with respect of density and electromechanical properties. One of the powders has been produced by a wet chemical process hereinafter named as PZT1 and the second is a mixed oxide powder hereinafter named as PZT2. Cylindrical shaped samples with 6.4 mm diameters and approximately 6 mm in height were axially pressed to 64% of the theoretical density (%TD)

Figure 2 shows the densities of both PZT-types, sintered with mm-waves compared to the densities reached in a standard resistance-heated sintering furnace. As can be seen from this figure, sintering of PZT1 starts at about 100K lower temperatures in the mm-wave device compared to conventional sintering. Also the final density of about 97%TD is reached already at 1000°C, which means at 150K lower temperatures compared to sintering in a resistance heated furnace. With PZT2. no conventional sintering was performed. Although for this material a holding time of 2-3 hours at a temperature of 1230°C is recommended in a conventional furnace. With mm-waves the final density is reached after 10 min. holding time at a sintering temperature of 1200°C.

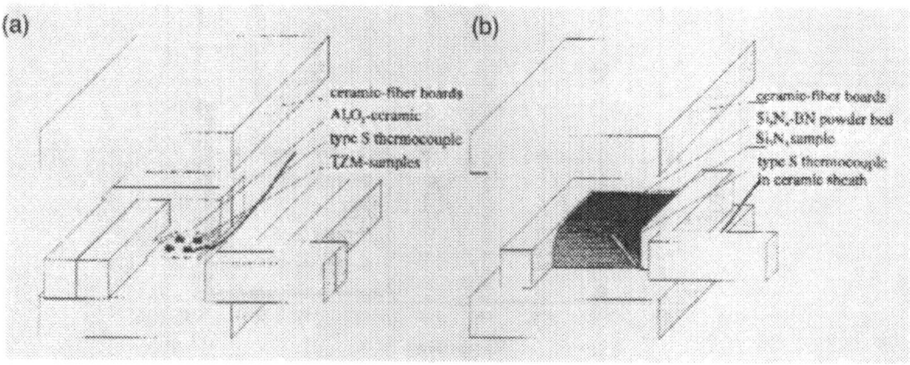

Figure 1. (a) Thermal isolation for mm-wave sintering of Si_3N_4 ceramics.

Figure 1. (b) Thermal isolation for mm-wave sintering of Si_3N_4 ceramics.

As the piezoelectric properties, especially the piezomodulus d_{33} and the electromechanical coupling factors, show a strong dependence on the density, also these factors can be improved by mm-wave sintering. First results of the measurements of the piezoelectric and electrical properties of PZT1 samples are given in Table 1. The relative permittivity ε' and the dielectric loss tan δ were measured at 1 kHz. The coupling factor, k_υ, and the frequency constant, N, were determined using the resonance method, the piezomodulus (d_{33}-mean value and standard deviation) was measured directly by using a Berlincourt-piezometer. The piezoelectric voltage constants g_{33} were calculated from d_{33} and ε'. The properties of ceramics are functions of the preparation process, and are subject to systematic and statistical fluctuation within a given batch. Nevertheless, measured coefficients of ceramics of defined composition and density converge to values which vary not more than 5% for elastic, 10% for piezoelectric, and 20% for dielectric constants [5].

Therefore, comparing the piezoelectric values given in Table 1 it seems that the piezoproperties of the gyrotron sintered samples are more or less the same as those gained by conventional sintering. Since those properties are mainly, but not only dependent on the material density but also on the grain size [6], still an improvement can be reached by optimizing the gyrotron sintering process with

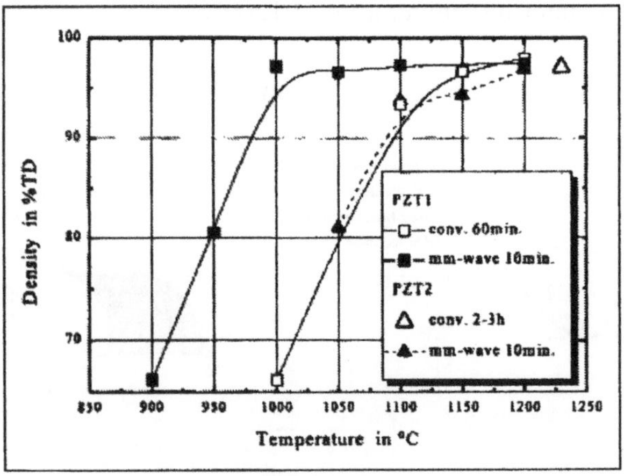

Figure 2. Comparison of the sintering conditions and final densities of PZT-ceramics treated in a Gyrotron Installation and in a conventional sintering furnace.

respect to an optimized microstructure. Due to a 200°C lower sintering temperature and a six times shorter holding time compared to conventional sintering technique a noticeable reduced grain growth could be expected with mm-waves. However, grain sizes below 5μm result in a decrease of the piezomodulus d_{33}.

Silicon Nitride

In the field of high performance structural ceramics, silicon nitride (Si_3N_4) is one of the most promising materials. But, because of its covalent bonding, Si_3N_4 is rather difficult to sinter to high densities at ambient conditions. Normally, sintering temperatures are so high, that nitrogen overpressure must be applied in order to avoid the decomposition of the material. So, processing of Si_3N_4 is generally rather complicated and expensive. To evaluate the possibilities for sintering Si_3N_4 in a mm-wave field, first tests in the Gyrotron System have been performed. The material investigated was silicon nitride (Ube SN E10) with only small amounts of sintering additives (5 w% Y_2O_3 and 1.5 w% Al_2O_3). Small bars were prepared by slip casting. After drying, they were sintered using the following conditions: heating rate = 30 C/min; maximum temperature and soak time of 1650 C and 20 min., respectively; nitrogen atmosphere at 1bar. These first tests resulted in densities of 3.13 g/cm^3 or 96% of the theoretical density.

TABLE 1. MATERIAL DATA OF CONVENTIONAL AND MICROWAVE SINTERED PZT1.

Sintering conditions	Conventional 60 min, 1200°C	mm-wave 10 min, 1000°C
Electromechanical properties		
Coupling factors, k_t	0.34	0.34
Piezoelectric charge constant, d_{33} (pC/N	350±45)	350±45
Piezoelectric voltage constant, g_{33} (10^{-3} Vm/N)	19.1	14.7
Frequency constants, N_p (Hz m)	2320	2330
N_t (Hz m)	2220	2200
Electrical properties		
Relative permittivity, ε'	2070	2300
Dielectric loss tanδ, (10^{-3})	32	38

Figure 3a. shows a micrograph of one of the mm-wave sintered samples. The most important fact, which can be seen from this picture, aside from the high density, is the grain size of the material, which is in the submicron range.

For comparison, a micrograph of a conventionally sintered sample of the same composition is given in Figure 3b. Small grain sizes in a ceramic material are a presupposition for high fracture toughness and so it is not surprising that this type of submicron silicon nitride shows values of the fracture toughness coefficient K_{IC} of about 7.5 $MPam^{0.5}$ and Vickers hardness values of about 1500, derived by the Vickers indentation method using an indentation load of 100N. Standard deviations are about 3% for the Vickers hardness and 5.5% for the fracture toughness coefficient.

Other Ceramics

Another interesting ceramic material, which can be used as a functional ceramic for special applications is a ceramic with the composition TiO_2-ZrO,-MgO (TZM). The sintering kinetics of different types of TZM, as listed in Table 2, have been investigated. The different types correspond to compositions characterized by three numbers, e.g., TZM 80/20/20 where the first two numbers characterize a mixture of 80 mol% TiO_2 and 20 mol% ZiO_2 and the third number provides the amount of MgO in mol% added to the mixture of TiO_2 and ZrO_2. The samples' size was similar to the PZT samples. Sintering tests in a conventional sintering furnace revealed that high densities of all compositions can only be reached by sintering at temperatures as high as 1500°C and sintering times as long as 24 hours followed by rapid quenching [7]. When using mm-wave sintering full density was reached at 1400°C after 20 min. holding time (Figure 4 and Table 2).

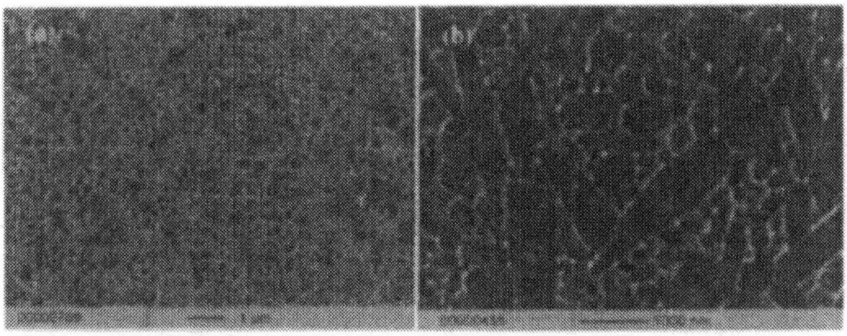

Figure 3. (a) Micrograph of a silicon nitride sample sintered at 1650°C by mm-waves with 10 minute hold time and (b) conventionally sintered for 1 hour.

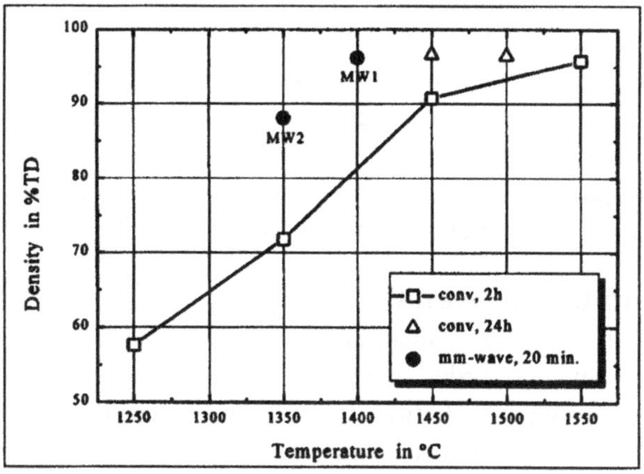

Figure 4. Densification behavior of different mixtures of TZM-ceramics conventionally sintered compared to mm-wave sintered.

TABLE 2. DENSITY, VICKERS HARDNESS AND FRACTURE TOUGHNESS OF TIO_2-ZRO_2-MGO CERAMICS SINTERED WITH MM-WAVES COMPARED TO CONVENTIONAL SINTERING

Type	Sintering conditions	Theor. density (g/cm^3)	Density (%TD)	Vickers hardness	K_{1C} [Mpam$^{0.5}$]
TZM 80/20/20	Conv.	4.24	97	515 HV0.5	n.m.
	MW1		100	872 HV5	2.2
	MW2		97	839 HV5	2.4
TZM 80/20/10	Conv.	4.45	96	866 HV0.5	n.m.
	MW1		95	806 HV5	2.4
	MW2		83	711 HV5	2.8
TZM 80/20/5	Conv.	4.54	97	903 HV0.5	n.m.
	MW1		95	794 HV5	2.8
	MW2		84	458 HV5	3.1
TZM 70/30/10	Conv.	4.61	96	944 HV0.5	n.m.
	MW 1		96	814 HV5	2.4
	MW2		87	628 HV5	2.8

* Abbreviations: conv. = conventional sintering for 24h at 1500°C with rapid quenching; MW1 = mm-wave sintering for 20 min. at 1400°C; MW2 = mm-wave sintering for 20 min. at 1350°C; and n.m. = not measured, sample was too brittle.

The most remarkable differences are the microstructure and the mechanical properties of the samples treated in these different ways. Whereas, the conventionally sintered samples show very coarse grains and are so brittle, that only microhardness and no fracture toughness coefficients could be measured, the microwave treated samples are fine-grained and show K_{IC}-values of about 2-3 $Mpam^{0.5}$. A comparison of the microstructure is shown in Figure 5 and the comparison of the mechanical data is represented in Table 2.

Summary and Conclusions

Sintering tests in a gyrotron installation at the Forschungszentrum Karlsruhe have been performed in order to study the densification behavior of several ceramics under mm-wave radiation. It could be shown that—with respect to the investigated types of ceramics—mm-wave sintering gives some advantages compared to conventional heat treatment. This means that the necessary processing times to reach the desired densities are markedly shorter and the maximum temperatures are lower. These changes of the sintering conditions, as could be shown with the silicon nitride and the TZM-ceramics result in a more fine-grained microstructure of the ceramics, which is responsible for improved mechanical properties.

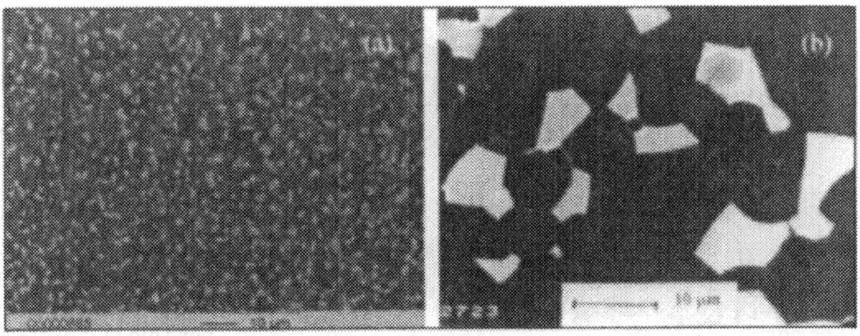

Figure 5. Micrographs of (a) mm-wave sintered TZM-ceramic (1400°C, 20min.) and (b) conventionally sintered sample (1500°C, 24h).

REFERENCES

1. M.A. Janney, H.D. Kimrey; Diffusion Controlled Processes in Microwave Fired Oxide Ceramics; MRS Symp. Proc., Vol. 189, Microwave Processing of Materials II, ed. by W.B. Snyder, W.H. Sutton, Pittsburgh, PA, pp. 215-227 (1991).
2. G. Link, V. Ivanov, S. Paranin, V. Khrustov, R. Böhme, G. Müller, G Schumacher, M. Thumm, A. Weisenburger; A Comparison of mm-Wave Sintering and Fast Conventional Sintering of Nanocrystalline Al_2O_3; MRS Symp. Proc., Vol. 430, Microwave Processing of Materials V, ed. by M.F. Iskander J.O. Kiggans, J.-Ch. Bolomey, Pittsburgh, PA, pp. 157-162 (1996).
3. A. Möbius, C.del Rio, M. Sorolla, M. Thumm; A Hollow Cone Gaussian Beam Quasi-Optical Transmission Line for a Gyrotron Technological System; 21st Int. Conf. on IR and mm-Waves, Berlin, paper AT 11 (1996).
4. L. Feher, G. Link, M. Thumm; The MiRa/THESIS3D-Code Package for Resonator Design and Modeling of Millimeter-Wave Material Processing; MRS Symp. Proc., Vol. 430, Microwave Processing of Mat. V, ed. by M.F. Iskander J.O. Kiggans, J.-Ch. Bolomey, Pittsburgh, PA, pp. 363-368 (1996).
5. IRE Standards on Piezoelectric Crystals: Measurements of Piezoelectric Ceramics, 1961; Proc. IRE, Vol.61, pp. 1161-1169, July (1961).
6. B. Jaffe, W.R. Cook, H. Jaffe; Piezoelectric Ceramics, Non-Metallic Solids, Vol. 3, editors J.P Roberts, P. Popper, Academic Press London (1971).
7. P. Hanselmann, unpublished results (1995).

Temperature Measurement in a 2.45 GHz Microwave Furnace

D.J. Grellinger and M.A. Janney

Abstract

A major concern in conducting microwave furnace experiments is to ascertain that temperature is properly measured. Because the sample is heated directly by the microwave energy (and not by indirect methods such as radiation or convection), temperature measurement must be made directly within the sample, and not simply in the vicinity of the sample as is often done in conventional furnaces. This investigation made simultaneous measurements of temperature in ceramic samples (alumina and zirconia) using a thermocouple, an optical fiber probe, and a 2-color infrared pyrometer. Agreement among the three types of measurement (± 20°C) was well within the range typically achieved for conventional furnaces.

Introduction

The importance of accurate temperature measurement inside a microwave furnace is manifested in the reliance on temperature to describe system behavior. Whether it be the sintering of a green powder compact, or a description of microstructural evolution, temperature has become a defining measurement in the field of ceramic processing. Furthermore, temperature measurements have also become important in describing the comparison between microwave and conventional furnaces—and more importantly the advantages of one over the other.[1,2] Therefore, it is necessary to ensure that temperature is measured correctly in a microwave furnace.

Background

The area of microwave processing of ceramics has demonstrated a potential for rapid, efficient processing of ceramic materials through microwave heating. The difference in processing between conventional and microwave furnaces is directly related to the fundamental difference in their heating mechanisms. In microwave furnaces, electromagnetic energy is absorbed by the ceramic part producing heat throughout the volume of the sample. In contrast, conventional furnaces electrically heat resistive heating elements that heat the surface of the part producing a thermal gradient which, in time, heats the entire sample. Therefore, in conventional furnaces it is sufficient to make temperature measurements in the vicinity of the ceramic sample. Temperature measurements in a microwave furnace need to be made on or within the sample.

Typical methods of measuring temperature in both a conventional and microwave furnace include the use of thermocouples, radiation pyrometers, and optical fiber probes. Thermocouples used in this study were made of reference grade wire with an accompanying Certificate of Compliance. Likewise, the optical fiber probe and radiation pyrometer were calibrated by the manufacturer. In high temperature (>800°C) ceramic processing, all three instruments have suitable measurement ranges in both furnaces. In any of these cases, the accurate measurement of temperature inside a microwave furnace is a non-trivial problem.

Materials and Methods

In this study, conditions were optimized for the temperature measurement of ceramic parts in a 2.45 GHz microwave furnace. Measurements were also made in a conventional furnace under similar conditions to facilitate a direct comparison of measured temperature in the two different environments. In addition, two different materials (dense alumina and dense zirconia) were studied to examine the effect of material properties and heating behavior on temperature measurement.

Conventional Furnace Approach

The temperature measurement comparison in an electrically-heated conventional furnace served as a reference to establish a reasonable temperature measurement variation. For this comparison we used a NEYTECH 85P furnace* to heat an 85g sample of dense alumina. This sample was cylindrical in shape with a diameter of 3 cm and a height of 5 cm. The alumina part was also machined to allow for the insertion of two type-S thermocouples (one sheathed in platinum and the other in inconel) and an Accufiber® lightpipe[+] directly into the sample. The lightpipe was covered with a stainless steel tube to prevent light emitted from the heating elements from affecting the measurement. To further reduce the "shine-through" from the heating elements, the entire sample was covered with a 0.5 cm thick sintered SiC optical shield. A diagram of the conventional furnace system appears in Figure 1. The conventional furnace was programmed to follow a temperature profile that provided temperature measurements from all three devices at several temperatures between 600-1000°C. A steady-state temperature was maintained at each set point for at least thirty minutes. The tips of all three temperature measurement devices were no more than 2 cm apart.

* Neytech, Yucaipa, CA.

† LUXTRON, Accufiber Division, Beaverton, OR.

Microwave Furnace Approach

For processing in the microwave furnace, ceramic samples were prepared from die-pressed alumina (Baikowski CR30)‡ and zirconia (TOSOH TZ-8Y)§ powder. For both materials, using a stack of three parts allowed the machining necessary to fully embed the thermocouple and lightpipe within the middle part of the sample stack. A similar hole was machined in the top part of the sample stack to produce a blackbody cavity necessary for accurate pyrometer measurements. The sample stack was cylindrical in shape with a diameter of 5 cm and a height of 6 cm. The ceramic samples of both materials were sintered in a conventional furnace to full density prior to temperature measurement experiments.

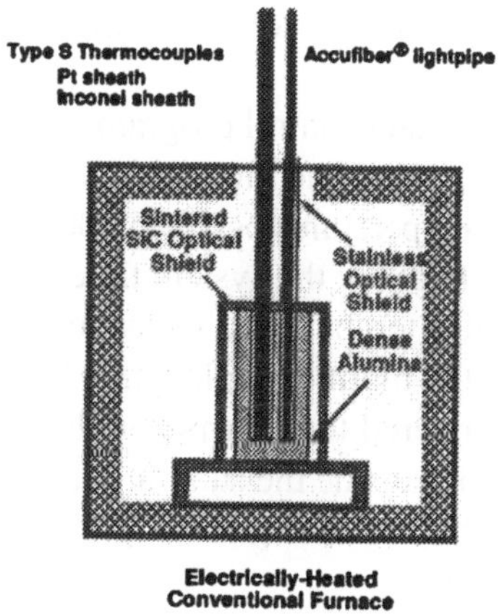

Figure 1. Conventional furnace temperarature measurement system.

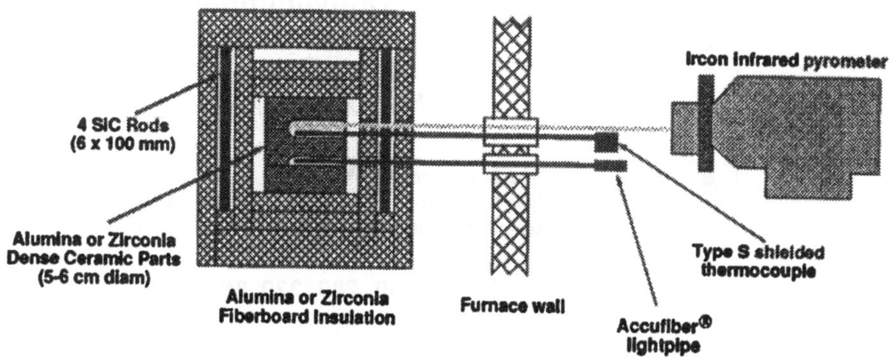

Figure 2. Microwave furnace temperature measurement system.

‡ GTE Products Corporation, Exeter, NH.
§ TOSOH Corporation, Japan.

Each sample stack was enclosed in an insulation system composed of refractory fiberboard to prevent heat loss from the sample. The fiberboard of the insulation system was of the same material as the sample stack to ensure similar heating characteristics and to eliminate unwanted temperature gradients. In addition, in both the alumina and zirconia systems a "picket fence" of 0.6 cm diameter SiC rods was used to provide indirect heating of the sample stack at lower temperatures (<500°C).[3]

The microwave furnace used a 6 kW, continuously variable, 2.45 GHz microwave generator. The microwave cavity consisted of a water-cooled aluminum vessel 70 cm in diameter by 120 cm long. The cavity had ports for both the thermocouple and lightpipe as well as an optical window for an IRcon® 2-color infrared pyrometer.¶ A diagram of the microwave furnace system appears in Figure 2.

A computer data acquisition and control program was used to control and record the furnace temperature and power. The furnace was programmed to follow a temperature profile that held at temperatures between 800-1600°C. A steady-state hold at 100°C intervals was used to allow the system time to equilibrate.

To prevent arcing and false temperature measurements the thermocouple had a platinum sheath to shield the thermocouple wires from the microwave field. In addition, an alumina tube covered the thermocouple, as well as the lightpipe, to reduce material discontinuities inside the sample stack. Additionally, for the thermocouple, the alumina tube also helped to prevent arcing and subsequent temperature misrepresentations.

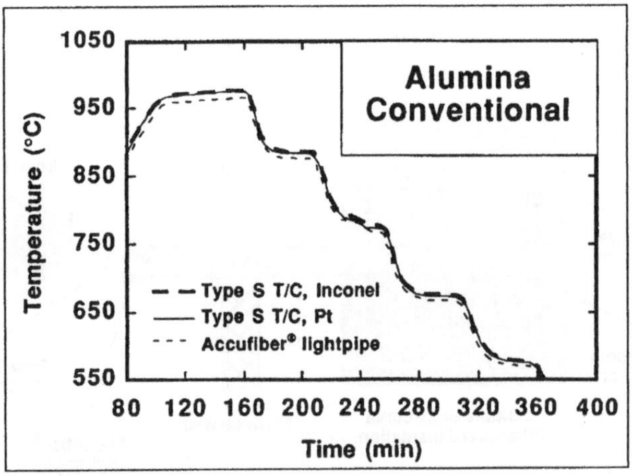

Figure 3. Temperature measurement of an alumina sample stack in a conventional furnace.

¶ IRcon Inc., Niles, IL.

TABLE 1. COMPARISON OF STEADY-STATE TEMPERATURES (30 MIN HOLD) MEASURED IN AN ALUMINA BLOCK HEATED IN A CONVENTIONAL FURNACE.

Type S T/C Pt sheath (°C)	Type S T/C Inconel sheath (°C)	Accufiber® Lightpipe (°C))
978	979	967
882	882	873
784	783	777
683	683	678
587	586	582

Results and Discussion

Comparisons between the temperature measurement devices in both the conventional and the microwave furnaces were reasonably close. In the dense alumina sample that was heated in the conventional furnace, both of the thermocouples and the lightpipe gave readings within 12°C of one another (see Figure 3, Table 1). Over the entire temperature range, the two thermocouples were in very good agreement—within 2°C. The lightpipe, inserted into a different hole in the ceramic sample, deviated slightly from that of the thermocouple as the temperature of the furnace increased. Precautions taken to prevent "shinethrough" from the heating elements were effective in eliminating disturbances in the measurement of the lightpipe. The temperature range of the conventional furnace limited measurements to 1000°C.

In the microwave furnace, temperature measurements were made with a comparable degree of variation. Temperature measurement variation among the three devices in the microwave furnace was generally within ±20°C (see Figure 4&5, Table 2). The pyrometer showed the largest variation from the other two devices probably due to the emissivity properties of alumina and zirconia. More importantly, the thermocouple and lightpipe measurements were in accordance with the 12°C maximum temperature differential observed in the conventional furnace.

In the microwave furnace, systematic differences were noted between the alumina and zirconia samples. As in the conventional furnace, temperature measurement differences in the alumina sample increased with increasing temperature. The opposite behavior was observed in the zirconia stack where the difference in temperature measurements decreased with increasing temperature.

The difference in behavior between the alumina and zirconia samples can be seen by considering the readings from both the lightpipe and the pyrometer with respect to the thermocouple in Table 2. For the alumina stack, the lightpipe reading increased from 3°C to 9°C above the thermocouple's reading as the furnace temperature increased from 1000°C to 1400°C. The pyrometer reading increased from 4°C to 16°C below the thermocouple's reading over the same temperature range. The opposite behavior occurred for the zirconia samples. In the zirconia samples, temperature deviation from the thermocouple reading decreased with increasing furnace temperature with the pyrometer reading above and the light-

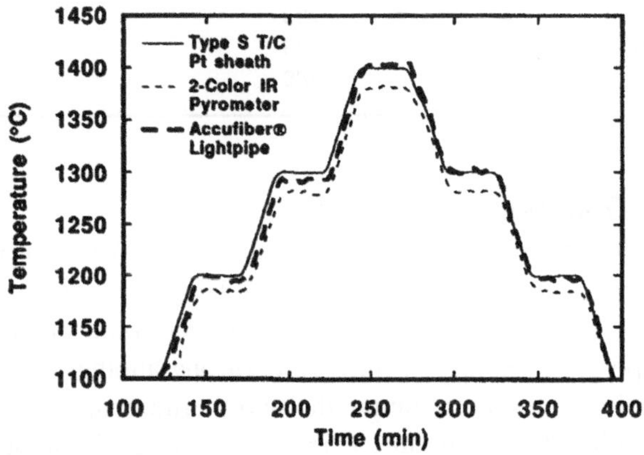

Figure 4. Temperature measurement in an alumina sample stack in a microwave furnace.

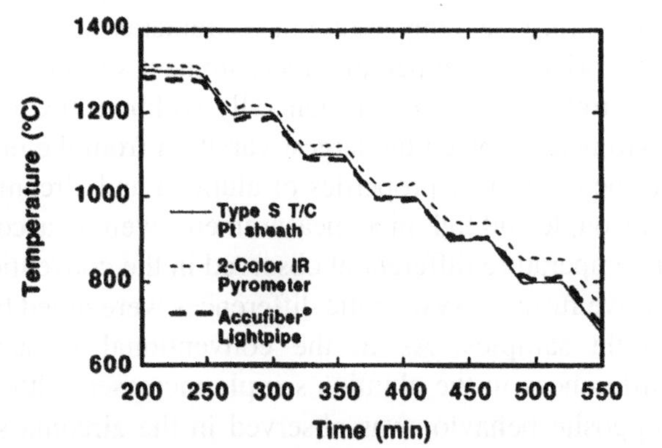

Figure 5. Temperature measurement in a zirconia sample stack in a microwave furnace.

pipe reading below the thermocouple reading. Generally, the thermocouple reading was between that of the lightpipe and the pyrometer indicating that the thermocouple is not grossly inaccurate.

In all cases, the heating behavior of the alumina and zirconia systems was consistent in all microwave heating experiments. Furthermore, the relationship between temperature difference and furnace temperature was independent of the staircase direction of the programmed temperature profile. However, measure-

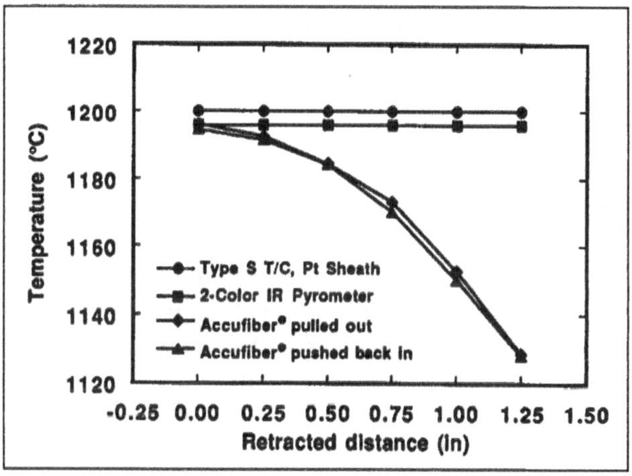

Figure 6. Temperature measurement comparison versus movement of the lightpipe to the sample surface and back.

TABLE 2. COMPARISON OF STEADY-STATE TEMPERATURE MEASUREMENTS IN A MICROWAVE FURNACE.

	Zirconia		Alumina	
Type S thermocouple (°C)	Accufiber® Lightpipe (°C)	Infrared Pyrometer (°C)	Accufiber® Lightpipe (°C)	Infrared Pyrometer (°C)
1400			1409	1384
1300	1280	1312	1303	1285
1200	1187	1219	1200	1187
1100	1093	1126	1102	1091
1000	990	1038	1003	996
900	913	940		

ment agreement generally improved during the steady state plateau and the ramp-down as compared with the ramp-up to temperature.

As a final observation, the lightpipe was moved from its buried position at 0.635 cm (1/4 inch) intervals toward the surface of the sample stack—and back in again. As the tip of the lightpipe approached the surface of the sample stack the measured temperature decreased by approximately 70°C, with the readings of the two other devices remaining constant. This behavior was repeated when the lightpipe was moved back into the sample (see Figure 6). This demonstrates the importance of measuring temperature directly within the ceramic part in a microwave furnace. These observations are also in accordance with calculations and measurements made at other laboratories.[4,5]

Summary and Conclusions

It has been demonstrated in this study that with careful planning good agreement between temperature measurement devices is possible inside a microwave furnace. This includes the use of thermocouples, radiation pyrometers, and optical fiber probes. It should be emphasized that careful attention must be paid to the proper shielding of temperature probes such as thermocouples and lightpipes to prevent inaccurate measurements caused by arcing and "shinethrough", respectively. In addition, proper positioning of the temperature measurement devices inside of the heated sample is necessary to avoid significantly low measurements.

ACKNOWLEDGMENT

Research sponsored by the U.S. Department of Energy, Fossil Energy Advanced Research and Technology Development (AR&TD) Materials Program, under contract DE-AC05-840R21400 with Martin Marietta Energy Systems, Inc.

REFERENCES

1. T.N. Tiegs, J.O. Kiggans and H.D. Kimery, "Microwave Processing of Silicon Nitride," in *Microwave Processing of Materials II,* W.B. Snyder, W.H. Sutton, D.L. Johnson, and M.F. Iskander, eds., MRS Symp. Proc., **189**, 267-72 (1991).
2. M.A. Janney, H.D. Kimery, M.A. Schmidt, and J.O. Kiggans, "Alumina Grain Growth in a Microwave Field," *J. Am. Ceram Soc.,* **74** [7] 1675-81 (1991).
3. M.A. Janney, H.D. Kimery, and J.O. Kiggans, "Microwave Processing of Ceramics: Guidelines Used at the Oak Ridge National Laboratory," in *Microwave Processing of Materials, III,* R.L. Beatty, W.H. Sutton, and M.F. Iskander, eds., MRS Symp. Proc., **269**, 173-85 (1992).

4. J.R. Brandon, J. Samuels, and W.R. Hodgkins, "Microwave Sintering of Oxide Ceramics," in *Microwave Processing of Materials, III,* R.L. Beatty, W.H. Sutton, and M.F. Iskander, eds., MRS Symp. Proc., **269**, 173-85 (1992).

5. Arindam Dé, Iftikhar Ahmad, E. Dow Whitney, and David E. Clark, "Microwave (Hybrid) Heating of Alumina at 2.45 GHz: I. Microstructural Uniformity and Homogeneity," in *Microwave Processing of Materials, III,* R.L. Beatty, W.H. Sutton, and M.F. Iskander, eds., MRS Symp. Proc., **269**, 319-28 (1992).

Temperature Measurement in Microwave-Heated Silicon Wafers

K. Thompson, J.H. Booske, R.F. Cooper, Y.B. Gianchandani, and S. Ge

Abstract

Results of experiments comparing the efficacy of thermocouple, pyrometer, lightpipe, and thermally-activated chemical paint sensors are described. Pyrometer sensors were found to be superior. Thermally-activated chemical paint sensors provided independent verification of the pyrometer data, and provided the first data on temperature spatial variation. Comparison of steady state temperatures with absorbed microwave power indicates good agreement with a simplified model assuming radiative energy dissipation.

Introduction

High temperature processing dominates many areas of semiconductor manufacturing. Just a few of these are dopant drives and activations, oxide growths, dielectric reflow, metal sintering, MEMS bonding, and CVD. The temperatures of some of these processes typically reach 1100°C on timescales ranging from minutes to hours. Being generally accepted that high temperatures adversely affect the more sensitive structures on the wafer, it is preferable to reduce the total time spent at high temperatures whenever possible.

This is especially true for the case of dopant activation. Advances in rapid annealing have shown that at temperatures of 1000°C only a few seconds are necessary to achieve dopant activation [1-3]. The main goal of rapid thermal annealing has been to reach these temperatures as quickly as possible (ramp time >100°C/sec), soak for a few seconds, and then immediately cool the wafer. Sometimes, in a process termed "Spike Annealing", the soak step is bypassed altogether. [4,5]

One possible way to achieve these ramp rates is through microwave heating. Besides high ramp rates, microwave heating offers several advantages over conventional furnaces. The processing is very efficient, which means lower energy costs, and unlike furnaces the actual process chamber is maintained at room temperature.

Thermal uniformity can be manipulated by tailoring the field profile and thermal shocks are reduced because the heating is volumetric. Finally, it is possible to selectively heat conducting films on the wafer [6].

Investigations of microwave annealing of Si have commenced at the University of Wisconsin. The initial studies have utilized a single-moded, tuned-resonant cavity at 2.45 GHz. Figure 1 includes a picture of the apparatus, along with camera images of 75 mm diameter wafers heated in TM111 and TM011 modes.

For microwave heating to be characterized, reliable temperature diagnostics must be established. An ideal probe must be fast (with respect to the temperature fluctuations), reliable, accurate both locally and globally, non-reactive with the sample, and non-perturbing to the environment or the process. This paper discusses the applicability of both direct and indirect temperature measurement techniques. Examples of direct methods include thermocouples, optical pyrometers and lightpipe sensors, and thermoelectric sensors. Examples of indirect methods include chemical reaction indicators, impurity diffusion, and oxidation growth.

Experimental

While thermocouples represent a mature, well-characterized temperature measurement technique, they were found to be unsuitable for the microwave environment used in this study for several reasons. First, the response time was confirmed to be too slow for investigations of rapid thermal annealing where temperature transients of 100°C/s or greater

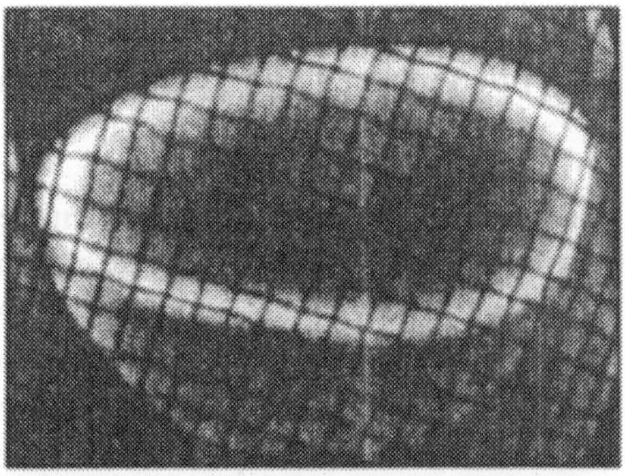

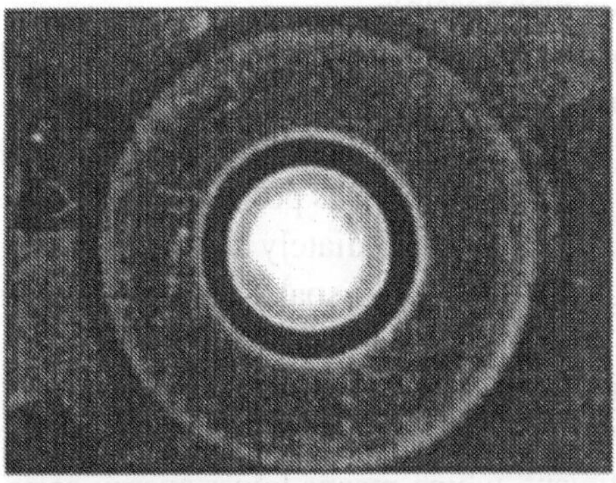

Figure 1. (a) Wafer heated by TM111 mode (viewed from above). (b) Wafer heated by TM011 mode (viewed from the side). (c) microwave heating apparatus.

are not uncommon. Figure 2 illustrates the typical transient response of a k-type thermocouple measured on a calibrated hot-plate. The voltage was recorded directly into a fast oscilloscope to eliminate any losses of response time associated with bandwidth-limited processing circuitry. Converting the voltage to temperature using standard conversion tables confirms that the thermal response time of a point-contacting thermocouple is too slow (< 30°C/sec) to keep up with some of the fastest desired heating rates (>100°C/s). This limits its use in transient analysis. Its ability to measure under steady-state conditions was also tested during microwave operation. This failed as well. The probe tip created an intense electromagnetic hot-spot wherever it touched the wafer. Even at low powers (~100 Watts), this spot heated to several hundreds of degrees higher than the rest of the wafer. At high powers (>700 Watts) a stable, localized atmospheric plasma ignited on the wafer surface where it was contacted by the probe and eventually melted the silicon. During cool-down, this hot spot cooled rapidly causing a thermal shock and cracking the wafer. Finally, the microwave fields interacted directly with the probe. Positioning the thermocouple such that it did not contact the wafer but measured only the relatively cool ambient just behind the wafer still produced unrealistically high temperatures (e.g., > 1000°C at 100 Watts when the wafer temperature was clearly less than 500°C based on the absence of visible light emission).

Optical measurements with a pyrometer or light-pipe would appear to provide a suitable alternative to thermocouples They are immune to microwave interference, provide a faster response time, and are not directly in contact with the wafer.

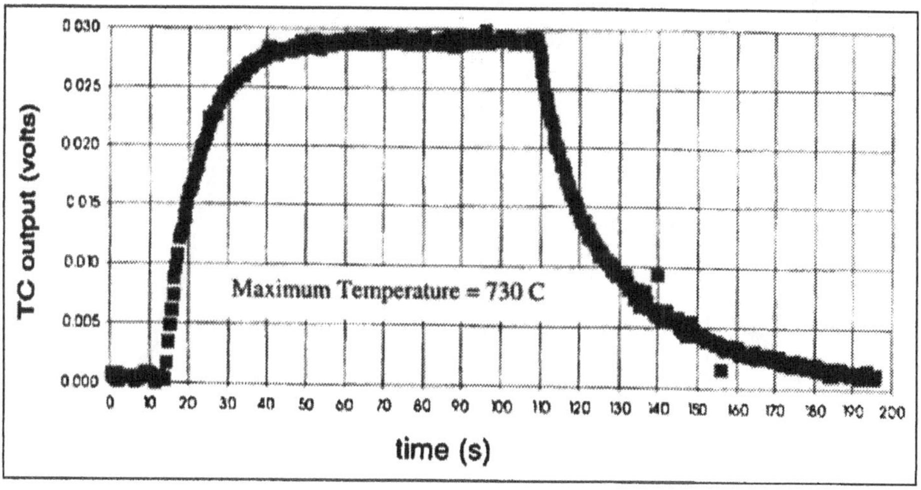

Figure 2. Illustrative response time of k-type, 1/16-inch-radius, 8-inch length thermocouple.

Although the light-pipe used in our experiments needs to be within a few millimeters of the sample, direct contact is not required. The pyrometer takes this one step further. With its focusing lens, it can be placed safely outside the chamber.

Potential problems with optical techniques revolve around the collection mechanism of the probe. Optical collectors have a certain solid angle over which they collect light. In the absence of a remote focusing lens, as the collector is pulled further from the source, the surface area corresponding to this angle widens. If the probe is placed too far from the sample, the probe will average the radiation intensity from the source with the "cooler" objects which fall inside this collection area. This results in an artificially lower reading, reducing the sensor accuracy. External sources (other than the hot sample) of infrared radiation sensed by the optical sensor, e.g., from reflection or transmission will increase the pyrometer readings. However, since the cavity walls for this application are cold, such concerns should be negligible.

The emissivity, reflectance, and transmittance of the silicon wafers are also an issue. Typical radiation-based thermal sensors operate in the 0.8-10 μm range, where thermal radiation is substantial, and numerous practical window and light-guide materials have significant transmittance. Between 0.9 and 1.0 μm, the emissivity of silicon has a fairly temperature-independent value of ~ 0.65 [7], which means that it will not radiate as a perfect black body. This must be accounted for in the analysis of the actual temperature. The fact that Si has a high transmittance for infrared radiation can be a problem if objects behind the sample are being heated as well. Radiation from these objects may pass through the

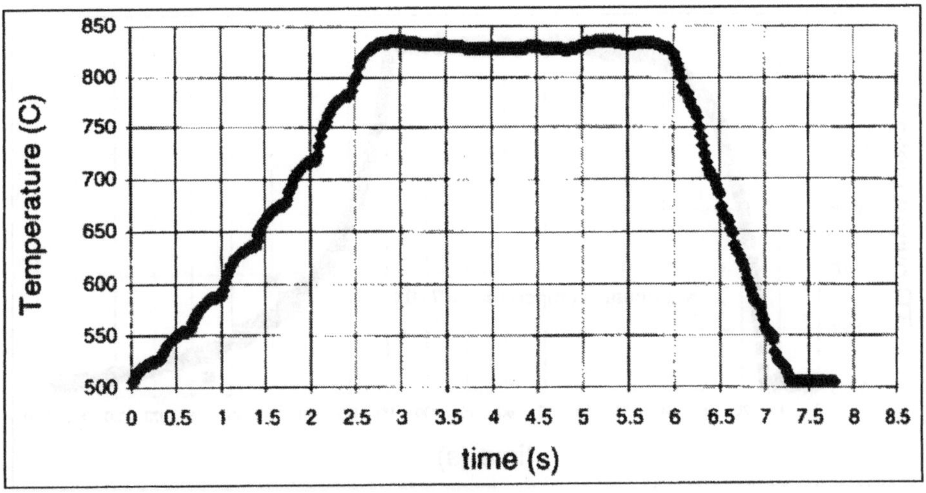

Figure 3. Illustrative temperature response of pyrometer.

sample and be collected. Finally, the high reflectivity of polished silicon and certain films deposited on the wafer will reflect radiation from other sources into the collection sensor of the light-pipe/pyrometer. These effects may act to increase or decrease the amount of light collected and consequently cause an offset in the measured temperature relative to the actual temperature of the Si wafer.

The response time of an available pyrometer and light-pipe were measured on a hot-plate by opening and closing a shutter. Illustrative results for the pyrometer are shown in Fig 3. The temperature readings from the pyrometer and light-pipe sensor for this test configuration were indistinguishable from each other, and thus only the pyrometer results are shown. A transient thermal response greater than 100°C/sec was obtained. The main time delay comes from the analysis circuitry [7], but the model 100A from Luxtron can measure fluctuations in excess of 500°C/sec. The gain-bandwidth product of the photodetector-and-current-amplifier circuitry used to convert infrared photons into a measurable electronic voltage will ultimately determine the transient response of pyrometric-type sensors.

The accuracy of the pyrometer when measuring silicon was calibrated against a thermocouple inside a typical 75 mm diameter diffusion furnace The silicon sample was aligned vertically so that looking down the tube, the pyrometer was focused normal to its surface. This should minimize interference from radiation reflections. In addition, to minimize the effects of infrared radiation transmission through the sample, nothing was behind the silicon sample for at least 30 cm. Finally, the analysis was configured for an assumed sample emissivity of 0.65. The thermocouple was placed directly behind the sample but was removed during the pyrometer measurement to minimize interference. The results are shown in Table 1 below.

The pyrometer results were very close to the thermocouple results, but it is apparent that not all sources of error can be eliminated. Residual sources of error can be expected to include slight deviations of the emissivity from the assumed

TABLE 1. CALIBRATION RESULTS IN A DIFFUSION FURNACE. ALL TEMPERATURE VALUES ARE GIVEN IN DEGREES C.

Run	Thermocouple	Pyrometer	Deviation
1	633	642	9
2	637	648	11
3	667	665	-2
4	714	699	-15
5	663	652	-11
6	614	616	2

value of 0.65, failure to place the focal point of the pyrometer focusing lens directly on the wafer surface, and slight errors in the calibration of the thermocouple. This data, however, shows that the pyrometer can provide an accurate substitution to the thermocouple as long as transmission or reflection of extemal sources are minimized and any deviations in the emissivity of the material are accounted for.

Next, the response of the light-pipe and pyrometer were tested during microwave heating of a 75 mm diameter silicon sample. The microwave cavity walls were water cooled to prevent any external radiation sources. Because the silicon was the only object in the cavity that radiated, any interference from transmission or reflection effects was minimized. The pyrometer was placed outside the chamber and focused normal to the wafer surface, onto a spot at the wafer center through an opening in the top of the cavity. The light-pipe was positioned to view the bottom of the wafer, but due to physical constraints was 75 mm from the wafer (outside its optimal range). This was expected to result in reduced collection efficiency and, in turn, a reduced temperature measurement. As a further verification of these results, thermally-activated, chemical-sensor paints were applied in several thin, parallel "stripes" across the wafer. The commerciallyavailable paints have a high degree of thermal sensitivity and react when a specific temperature is reached. Six different paints were applied, with reaction temperatures of 500, 600, 700, 800, 900, and 1000°C, respectively. The results are shown in Table 2 below.

TABLE 2. RESULTS OF TEMPERATURE DIAGNOSTICS DURING MICROWAVE HEATING

Power (Watts)	Lightpipe	Pyrometer	Thermocouple	Paint/Color change
120	520	570	1515*	500
220	590	660	MAX*	600
475	715	820	MAX*	700, 800
900	840	1002	MAX*	900, 1000

Temperature values are specified in degrees C. The asterisk indicates that the thermocouple was strongly influenced (and rendered unusable) by the microwave field.

The thermocouple data were taken from a separate run conducted under identical conditions, but were added to this table to illustrate its inability to measure temperature in the presence of intense, microwave-frequency, magnetic fields. The pyrometer readings were verified by the chemical reactions of the thermally-sensitive paints. The light-pipe results were significantly lower than the pyrometer. This was expected because the light pipe was placed beyond its optimal distance from the sample. Finally, by placing the thermochemical sensor paints in several different locations on the wafer, a rough idea of the heating uniformity could be obtained. For example, when heated by a TM111 mode, there were two spots near the wafer edges that were markedly cooler than the hot central "hourglass" region characteristic of the TM111 field distribution (see Fig 1). Using the thermochemical paint sensors, it was established that these cooler edge regions were approximately 800°C when the wafer center was approximately 1000°C.

The net microwave power (incident minus reflected) required to reach a steady state temperature was measured using both the pyrometer and the light-pipe. At temperatures above 600°C, the dominant thermal dissipation mechanism is expected to be radiation. Assuming that most of the microwave power delivered to the cavity applicator is absorbed in the Si wafer, a simple power balance model would therefore predict that the net absorbed microwave power should scale with the fourth power of the steady state wafer temperature (in °K). Net microwave power versus temperature (°K) is plotted in Fig. 4 using temperatures as reported by both the pyrometer and the light-pipe sensor. Also included in each plot is the prediction of a model (normalized to the maximum measured power), that relates net microwave power to measured temperature, raised to the fourth power. The theoretical model values were scaled to obtain agreement at the maximum measured microwave power point. In both cases (pyrometer and light-pipe) good agreement is obtained, confirming the applicability of the model assumptions. Better agreement is obtained with the pyrometer-measured temperatures, consistent with the expectation that the light-pipe values were lower than the true temperatures due to the light-pipe being placed outside its optimal distance.

Both optical techniques are found to be acceptable diagnostics of microwave heating as long as care is taken to reduce outside interference and the probes are used within their specified capabilities. Neither probe type perturbed the cavity or reacted with the sample. They were found to provide accurate (when placed properly) information and a relatively quick (~100°C/s) response to fluctuations in the temperature.

Work in Progress

Another diagnostic technique involving the inter-diffusion of impurities is currently in progress. It is anticipated that it will verify current results and provide an extended capability. A baseline of Boron diffusion profiles after heating at several steady state temperatures is currently being established in a calibrated, conventional furnace. The boron is implanted in the surface of the wafer and then driven by the furnace. Once this baseline is completed, an identical boron diffusion (impurity drive only) will be performed by microwave heating. A comparison of the profiles will infer the temperature obtained by the microwave heating. This approach, like the thermal paint is an excellent method for calibration verification, and has been used with great success in prior experiments involving microwave-sintered ceramics [8,9].

A variation of this technique, also in progress, is dopant activation. First, a significant dose (~$10^{15}/cm^2$) of a specified dopant is implanted into the wafer surface. The dopant is then "spike annealed" by microwave heating. A measurement of the sheet resistance and comparison to previously published results [5,6] is expected to give a rough idea of the activation temperature. While not as rigorous as the full-length diffusion profile measurements, it is cheaper and offers a much faster turn-around time.

Summary and Conclusions

Several temperature measurement techniques, both direct and indirect, were compared. The thermocouple was found to be unsuitable for microwave heating because of its slow response time, its coupling with the microwave fields, and its perturbation of the system. Both the optical pyrometer and light-pipe were found to be excellent substitutes for the thermocouple. When extemal radiative sources were eliminated from the collection area, and when the emissivity of silicon was accounted for in the analysis software, the temperature readings were well calibrated to the thermocouple results in a conventional furnace. When used in a microwave cavity, the ability of the pyrometer to accurately measure the wafer temperature was verified through thermally-sensitive paint. The light-pipe did not produce accurate results in the system used for these studies because physical constraints required it to be placed beyond its optimal length from the wafer.

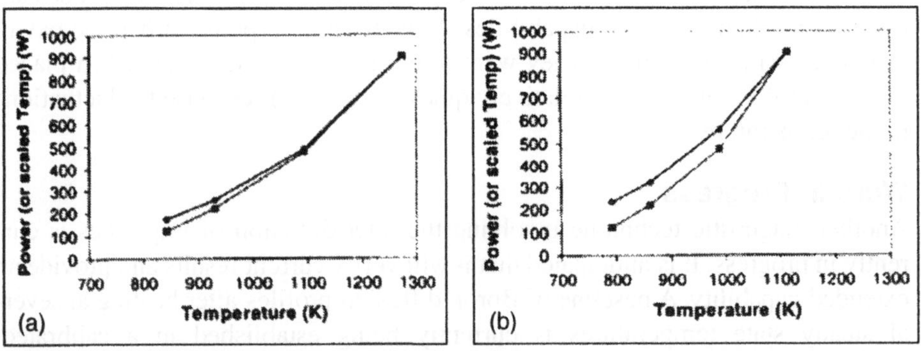

Figure 4. Microwave power (P) versus steady-state temperature (T, in degrees K) measured by pyrometer (left) and light-pipe (right) compared to an expected model scaling of P ~ T4. In both figures, the lower curve represents the power and temperature data taken from the sensor, and the upper curve is the model prediction, scaled to agree with the maximum microwave power point.

Future diagnostics that are currently in development include a comparison of impurity diffusion profiles.

As the demands of the semiconductor industry continue to grow, novel methods of heating silicon both faster and more selectively will need to be developed. Before these processes can be developed the proper diagnostic tools necessary to monitor temperature must be in place.

ACKNOWLEDGMENTS

This work was partially supported by a grant from the U.S. National Science Foundation and by the Wisconsin Alumni Research Foundation through a grant administered by the University of Wisconsin-Madison Graduate School. Special acknowledgements to Gerling Applied Engineering, Inc, for generous technical support.

REFERENCES

1. Alan Bratschun, *Journal Electronic Materials,* Vol. **28,** No. 12, pl328-1332. (1999)
2. A.T. Fiory, K.K. Bourdelle, *Journal Electronic Materials,* Vol. **28** No. 12, p.1345-1352. (1999)
3. L. Soliman, M. Benzohra, M. Masmoudi, K.Ketata, F. Boussaid, A. Martinez, M. Ketata, *Journal Electronic Materials,* Vol **28,** No. 12, p.1353-1357. (1999)
4. Aditya Agrawal, Hans-J. Gossmann, Anthony T. Fiory, *Journal Electronic Materials,* Vol. **28** No. 12, p.1333-1339. (1999)
5. Daniel F. Downey, Scott W. Falk, Adam F. Bertuch, Steven D. Marcus, *Journal Electronic Materials,* Vol. **28** No. 12, p. 1340-1344. (1999)
6. N. K. Budraa, H.W. Jackson, M. Barmatz, W.T. Pike, and J.D. Mai, Proc. 7th Int. AMPERE Conf. Microwave and High Frequency Heating (Valencia, Spain), pp.463-466 (1999).
7. Luxtron/Accufiber Application Note EAS 100-01. (1988)
8. J.H. Booske, R.F. Cooper, S.A. Freeman, K.R. Binger, *MRS Symp. Proc.* Vol. **527**, p.525-5323 (1998).
9. J.H. Booske, R.F. Cooper, K. Ruffer, S.A. Freeman, Proc. 7th Int. AMPERE Conf. Microwave and High Frequency Heating (Valencia, Spain), pp. 407-410 (1999).

Scanning Dielectric Analysis—Another Analytical Tool for Ceramics Studies

R.M. Hutcheon, J. Mouris, C. Pickles, and P. Hayward

Abstract

The automation of our multifrequency, broad temperature range (25–1400°C), complex dielectric constant measurement system (based on the cavity perturbation technique) has increased the relative accuracy and detail of the data to the extent that it now serves as a tool for analytical studies of ceramics, minerals, glasses, polymers, organic fluids and composites of these materials. Rapid scanning over a selected group of frequencies at each of a sequence of selected temperatures - scanning dielectric analysis (SDA)—yields the complex dielectric constant of a single sample over a programmed temperature versus time curve in any selected cover gas.

SDA is a thermal analysis technique that is related to differential thermal analysis (DTA), thermogravimetric analysis (TGA), differential scanning calorimetery (DSC) and dynamic mechanical analysis (DMA), but is based on measurements of dielectric property changes during thermal ramping, rather than weight change, heat evolution or mechanical damping. For a stable, thermally reversible material, SDA measures the two numbers of dielectric response (the real, ε', and absorptive, ε'', parts) as a function of temperature. For a thermally irreversible process (phase transformation or thermal reaction), not only are the initial and final state properties measured, but the dielectric response due to transient species may also be seen.

The real part of the rf/microwave dielectric constant, ε', is especially sensitive to small changes in the distance and distribution of neighboring ions or atoms and to the hindering of the reorientation of a molecule or ion having a electric dipole. The absorptive part, ε'', is strongly dependent on the electron/hole or ionic conduction characteristics (bandgap energy, band population, impurity concentrations, collision rates), or the number and relaxation rate of any dipole moments.

SDA measurements on a few specific materials will be compared with standard DTA and TGA results, and an attempt will be made to demonstrate the utility of both the common and complementary aspects of the techniques.

An Introduction to Scanning Dielectric Analysis

Automation of the previously reported,[1] high-temperature, microwave-frequency dielectric properties measurement system has increased the reproducibility and reliability of the data, and in practical operation allows much smaller temperature steps to be used, yielding more detailed information. A computer-driven linear translator rapidly moves the sample (~ 3.5 mm diameter, ~ 12 mm long, held in a long silica tube) vertically between a furnace and a multimode microwave cavity mounted directly under the furnace. The lapsed time between the sample leaving the furnace and the completed cavity measurement is ~ 1.5, making for minimal and calculable sample cooling. The sample is immediately returned to the furnace. The furnace is ramped in a step/hold mode, with ~ 5 min soak time at each programmed temperature.

The system measures both the real, ε', and the absorptive part, ε'' of the relative complex dielectric constant, $\varepsilon = \varepsilon' - j\varepsilon''$, at a selection of frequencies, as the temperature is sequentially stepped over a user-specified range, all in a selected cover gas. This refinement has made possible the use of scanning dielectric analysis (SDA) as an analytical tool both in the study of the properties of thermally stable systems and also in tracking the progress of a reaction from beginning to completion—much the same way as one would use differential thermal analysis (DTA) or thermogravimetric analysis (TGA) to determine the chemical phase composition of a sample or to determine the response of a reacting material, respectively.

During a solid-state phase transformation or a chemical reaction, the change in atomic or ionic positions and spacings generally changes the deformability of outer-shell electron clouds, and usually produces a net change in the electronic and ionic polarizability and, thus, in ε'. The change in spacing and arrangement usually influences the electron and/or hole bandgaps and their population or changes the mobility of a conduction ion, producing observable changes in the microwave absorption (*i.e.*, in ε''). As well, any change in the environment of an electric dipole that influences its ability to reorient itself under the influence of the microwave electric field will contribute to changes in both ε' and ε''. Thus, there are many effects that can cause a change in the dielectric constant and give a quantifiable signature of a particular transition or reaction. Identifying the specific mechanism for any specific material often requires a detailed knowledge of the material or reaction properties, but a literature of experience with dielectric properties is slowly accumulating. In any case, noting and quantifying a change is often all that is required for analytic purposes.

In this paper, a simple formalism is outlined with which to start the process of interpreting the SDA spectra. Then several SDA spectra are shown for thermally reversible polycrystalline powder samples that are traditionally used for temperature calibration of DTA systems. Also are shown spectra of well-known oxide and ferroelectric phase transformations showing rate-hindering temperature hysteresis.

Two examples are given of monitoring the progress of a decomposition reaction: (1) binder burnout in alumina compacts; (2) an irreversible phase transformation, with evidence of intermediate species that are not present in the final state.

Frequency and Temperature Dependence of Dielectric Constant and Dielectric Loss Mechanisms

The relative complex dielectric constant, ε, of a single crystallite of a material is represented by

$$\varepsilon = \varepsilon' - j\varepsilon'' \qquad (1)$$

where ε′ is the usual real "dielectric constant" and depends on the material polarizability and ε″ the "absorptive" part and is a measure of the rate at which thermal energy is deposited in the material by an internal ac electric field. At frequencies between 10 MHz and 100 GHz, the polarization contributions primarily are (a) frequency-independent electronic and ion–ion polarizabilities (electron cloud deformation, change in mean spacing of oppositely-charged ions), and (b) frequency-dependent permanent dipole rotation (e.g., water). The primary energy loss mechanics are electron (and hole) conductivity, ion conductivity, polaron relaxation (rare, mainly in ferroelectrics) and permanent dipole relaxation collisions.

For general data analysis and interpretation of the interaction mechanism, the following approximate expressions can be used:

$$\varepsilon' = 1 + \Delta\varepsilon'_{\text{electronic}} + \Delta\varepsilon'_{\text{ionic}} + \frac{\Delta'_D}{(1 + \omega^2\tau_D^{\,2})} \qquad (2)$$

and

$$\varepsilon'' = \frac{\sigma^{dc}_{el}}{\omega\varepsilon_o} + \frac{\sigma^{ac}_{ion}}{\omega\varepsilon_o} + \Delta\varepsilon''_{\text{ionic}} + \frac{\omega\tau_D\Delta'_D}{(1 + \omega^2\tau_D^{\,2})} \qquad (3)$$

where the final term in each of Eqs. (2) and (3) has the classical Debye relaxation form, and Δ'_D is the low-frequency contribution to the dielectric constant by relaxing "dipoles", τ_D the relaxation time of ac electric-field-driven dipole structures, $\sigma^{dc}_{electronic}$ the frequency-independent, "free" electron conductivity (S/m), σ^{ac}_{ion} the frequency-dependent, (usually $\propto \omega^{-n}$, $n \le 1$) ion conductivity (S/m), $\Delta\varepsilon''_{ionic}$ is the lattice vibration collision or polaron collision loss contribution, ε_o the permittivity of free space (8.85×10^{-12} F/m), ω is the frequency multiplied by 2 *p* (ie., $\omega = 2{\cdot}\pi{\cdot}f$).

Although there are exceptions (water, for example), the terms are introduced here in what is usually decreasing order of importance at high frequencies (>10 MHz). The term $\Delta\varepsilon'_{electronic}$ has no explicit temperature dependence, while the electronic conductivity usually has a standard, thermally activated, dependence, with a specific activation energy for each phase. The ionic susceptibility, $\Delta\varepsilon'_{ionic}$, normally has only a weak temperature dependence, related to the thermal expansion coefficient[(2)]. However, in the region of a ferroelectric or antiferroelectric transition, the relaxation frequency of the soft lattice vibration modes can drop down into the microwave range, and the values of ε' and ε'' are influenced by the presence of strong, very-temperature-dependent, "soft" ion lattice oscillation modes, which can have either resonant or relaxational character. The signature of this situation is a value of ε'' that increases with increasing frequency to frequencies well above 10^9 Hz.

In contrast, the signature of pure "free electron" conduction is a value of ε'' that is inversely proportional to frequency, with the following temperature dependence:

$$\varepsilon''(T) \approx \frac{\sigma^{dc}_{electronic}}{\omega\varepsilon_o}(T) = \frac{\sigma_{ref}}{\omega\varepsilon_o}\,\frac{T^{3/2}}{T_0^{\;3/2}}\exp\left(\frac{-E_g}{2\,k_B\,T}\right) \tag{4}$$

where s_{ref} is a constant reference value conductivity, k_B the Stefan–Boltzman constant, T the absolute sample temperature, T_0 a reference temperature, both in Kelvin E_g the semiconductor conduction band gap energy (which is twice the activation energy in the Arrhenius activation formalism, $E_{Arr} = E_g/2$).

Water, in sample pores or in loosely bound states, or molecules with charged functional groups are the usual sources of permanent dipole terms, although the interfacial polarization mechanism can have the same functional dependence. The temperature dependence of Δ'_D and τ_D depends on the source of the dipole moment. For liquids, the number and strength of permanent dipoles (Δ'_D) is not strongly temperature dependent, but the relaxation time (the inverse of the collision rate) usually decreases rapidly with increasing temperature, and often shows a thermally activated dependence.

At first glance, the dielectric constant seems to have so many contributing terms that extracting information would be impossible! However, in practice, the loss factor (ε'',) usually has only one prime contributing mechanism over a modest temperature range, and the frequency dependence often indicates the mechanism. As for the real part of the dielectric constant, ε', a frequency-independent, temperature-independent, constant low value at relatively low temperatures (150–250°C) can usually be attributed to the electronic and ionic polarizability components, and these can be extrapolated to high temperatures and subtracted to leave the components from other mechanisms. For example, moisture in a sample usually contributes a significant component between room temperature and 150°C, but most lightly bound water (Van der Waals surface binding) evaporates by 150°C, and the value of ε' drops to the frequency/temperature-independent electronic/ionic component. This initial "bump" may be used to quantify the amount of moisture in a sampl–after calibrations have been done. This separation technique is not exact, for the individual polarizabilities do not contribute to the total dielectric constant in a linear addition process. However, the approximation is often good enough to gain a reasonably quantitative measure of an effect.

Polycrystalline powder samples have "effective" dielectric constant values that are lower than the single-crystal values, as expected of a gas–solid mixture. If one wishes to extract a single-crystal (i.e., a full crystalline density) value from the polycrystalline measurements, then the shape and dielectric anisotropy of the crystallites must be determined, and then the appropriate mixing formula chosen. However, the transition temperatures and the character of individual phases may be determined without this step, since the single-crystal and "effective" dielectric constant values are monotonically related when the second component is passive (as with air).

Comparison of SDA and DTA for Standard DTA Calibration Materials

A DTA system provides a continuous measurement of the temperature difference between a sample and a known, thermally stable, reference material when both are placed near each other in an oven whose temperature is ramping at a constant rate. The measured temperature difference is proportional to the latent heat (enthalpy change) associated with a phase transition in the sample or with the specific energy released by a chemical reaction, and it is larger for higher ramp rates. Temperature calibration of DTA systems is often done by measuring the response of materials that have fast, reversible, first-order phase transitions. The first four materials listed in Table I are DTA calibrators, and their DTA spectra are shown in Fig. 1.

The DTA "difference" signal occurs only when the heat is released (or absorbed) during the actual transition. The signal polarity is also governed by the direction of the ramp: with increasing temperature, an endotherm gives a negative signal while an exotherm gives a positive one. These polarities are inverted if the temperature is ramping down (see $PbTiO_3$, Fig. 1). If a transition is slow (e.g., some reconstructive transitions are "sluggish"), the DTA does not give very good results because use of a very slow ramp to accurately define the transition temperature allows the reaction (transition) heat to dissipate as fast as it is being generated—thus giving a very weak or zero DTA peak that is indistinguishable from background noise.

In contrast, SDA measures the properties of a state of the material, before and then after the transition, as demonstrated by the frequency-independent step change in dielectric constant, ε' , at the reconstructive transition at 928°C in strontium carbonate, $SrCO_3$ (Fig. 2). The values of ε' before and after the transition reflect the change in electronic and ionic polarizability between the two states; the usual increase in average lattice constant in going to the higher-temperature state produces an increased deformability and larger ε'. The loss factor values, ε'', have an inverse frequency dependence typical of either a conductivity or interfacial polarization mechanism, but only by fitting to an exponential thermal activation dependence can the small change in conductivity at the transition be seen. Note that a sluggish transition can be located accurately by using very small temperature steps and long soak-holds over the region where the transition is known to occur.

TABLE I. LIST OF PROPERTIES OF SELECTED SOLID-STATE PHASE TRANSITIONS.

Compound	Transition onset temperature (°C)	Transition type and description	Crystal configuration low temp. → high temp.
$SrCO_3$	928	Reversible, diffusionless	Orthorhombic → hexagonal
K_2CrO_4	665	First order (hysteresis), anisotropic conductivity	Orthorhombic → hexagonal?
K_2SO_4	582	First order, ordering of SO_4 group, semiconducting in high phase	Orthorhombic → hexagonal
SiO_2	571	Network puckering, displacement/rotation of SiO_4 tetrahedra	α(trigonal) → β(hexagonal)
ZrO_2	~ 1150	Displacive martensitic, 9% volume contraction	Monoclinic → tetragonal
$PbTiO_3$	$T_c = 490$	Ferro→paraelectric, soft modes underdamped	Tetragonal → cubic
$KNbO_3$	$T_c = 434$	Ferro→paraelectric soft modes overdamped	Tetragonal → cubic

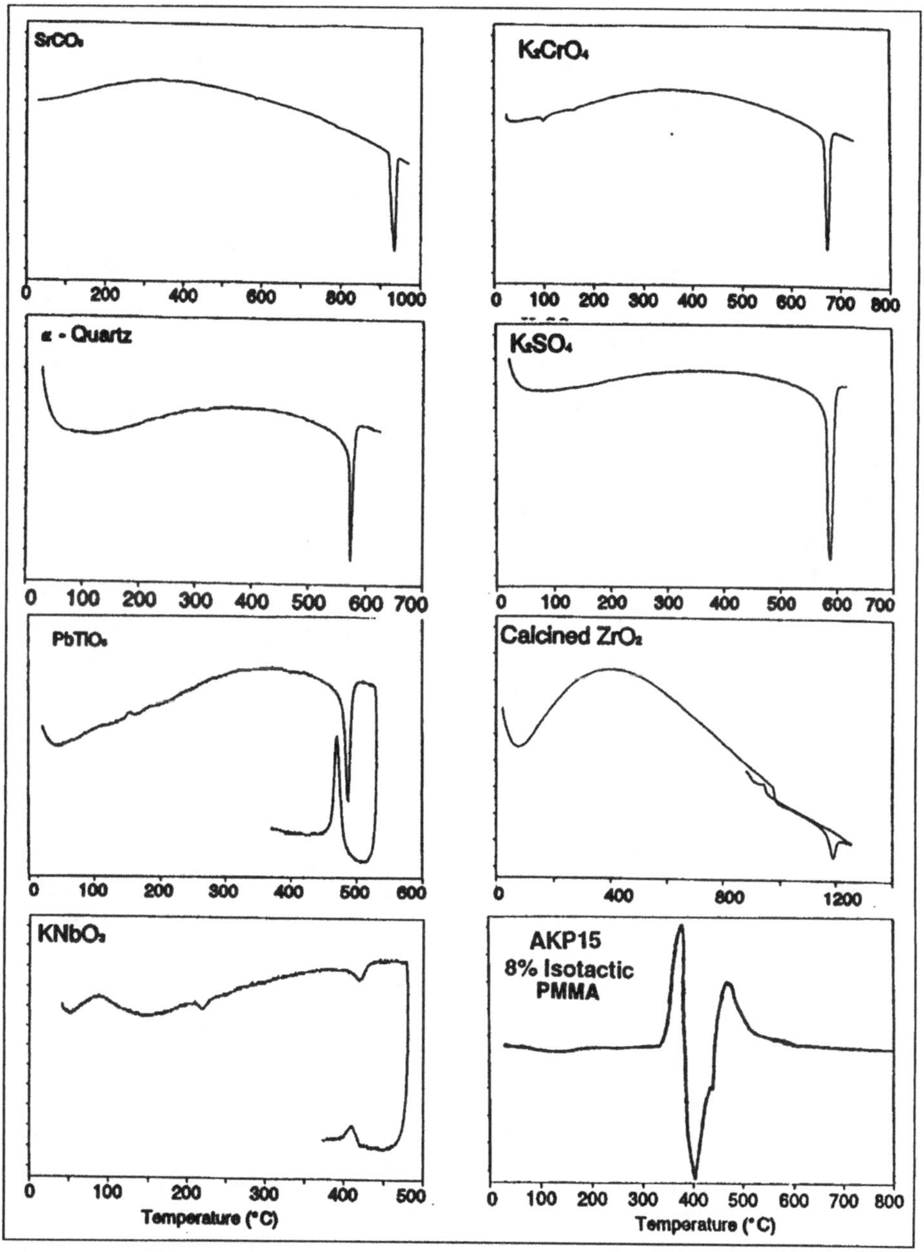

Figure 1. DTA plots of differential temperature (amplified) versus temperature. All are of polycrystalline powder specimens and are the second thermal measurement cycle, except the AKP15 alumina-isotactic PMMA sample, which was a densely pressed powder compact with binder burnout during the heating cycle.

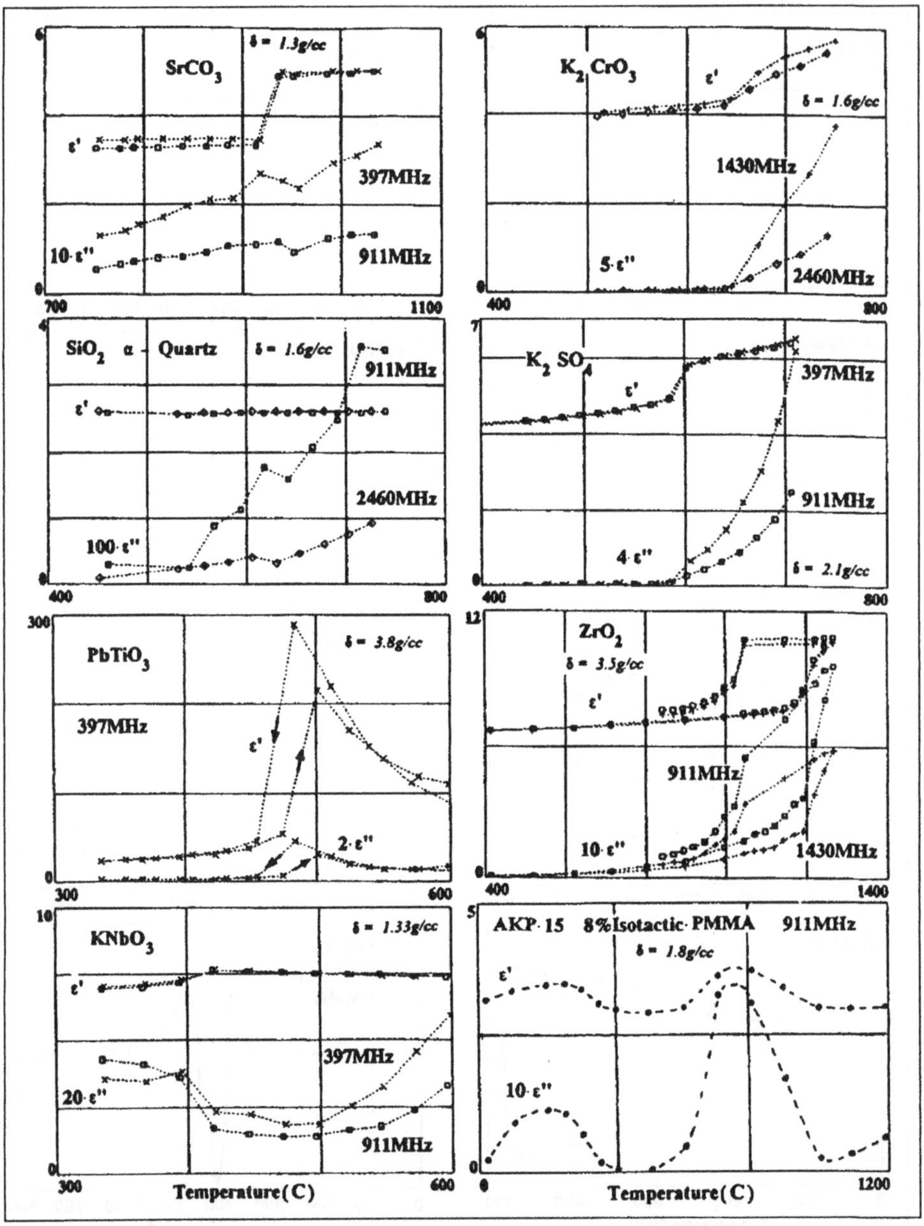

Figure 2. Scanning dielectric analysis (SDA) spectra of the permittivity (ε′,ε″) of poly-crystalline powder samples. The values of ε″ are multiplied by an arbitrary factor so they can be plotted on the same graph as the ε′ values. Most spectra were taken on the second thermal measurement cycle. The AKP15 alumina plus binder sample was a pressed powder compact, ramped at ~20/min.

The SDA spectra for potassium chromate (K_2CrO_4) (Fig. 2) provide a clear demonstration of how the rearrangement of atoms can change the dielectric loss factor. The change in the slope of ε' at 650°C (onset temperature) indicates the existence of a transition, with an on-going development at higher temperatures (possibly some sintering). The dramatic increase in ε'' (which has the inverse frequency dependence characteristic of "free" electron conductivity) suggests a factor of 9 increase in dc electron conductivity (0.0020 → 0.018 S/m) at the transition.

The SDA spectra for potassium sulfate (K_2SO_4) (Fig. 2) is a classical case of a change in semiconducting properties at single well-defined phase transition. The slowly increasing values of ε' are a result of the slowly increasing deformability produced by thermal expansion. The step change at the transition is caused by atom position changes. The values of ε''/w against inverse temperature (Fig. 3) show a clear thermally activated, "free" electron conductivity with a measured band gap value E_g = 2.3 eV.

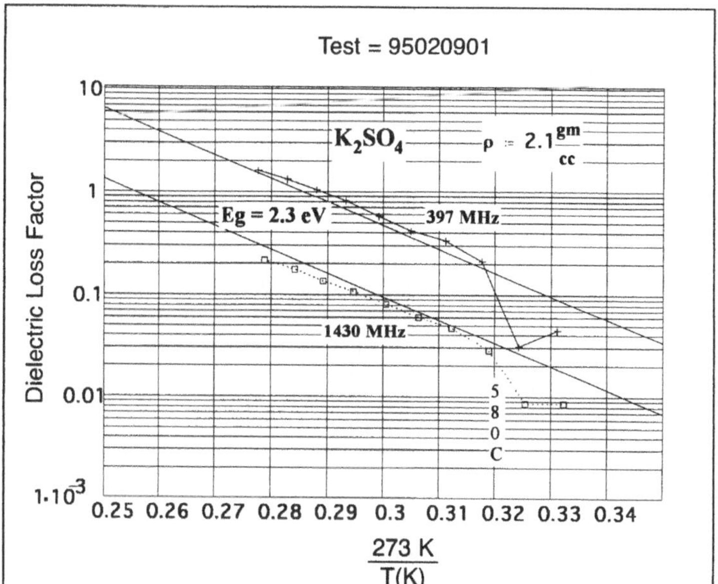

Figure 3. Measured values of dielectric loss factor, ε'', for K_2SO_4 powder. The solid lines are Eq. 4 with T_0 = 273 K, and σ_{ref} = 4000 S/m . The "free" conductivity value at 636°C is 0.01 S/m. The transition at ~ 580°C is clearly visible.

To make the SDA spectra appear more like the DTA spectra, one could differentiate the ε' curve with respect to temperature and produce a differential spectrum with a peak at each step change in dielectric constant. However, this process actually discards the information on the relative dielectric constants of the two phases, and is usually inapplicable to analysis of the ε'' values.

The final example, the $\alpha \rightarrow \beta$ transition in quartz (SiO_2) at 571°C (onset) (Fig. 2), demonstrates the problem of using the SDA technique in isolation; the transition produces essentially no change in ε' ($\leq 5\%$) and, within experimental error, has no readily apparent threshold in ε''. The reason for this lies in the nature of the transition. It has been termed a network puckering transition, one where the silicon-oxygen tetrahedra only rotate slightly, without change in coordination or oxygen bonding, and with only a 1% increase in volume. Clearly, the polarizability of the atoms is unaffected by this transition.

SDA and DTA on Ferroelectric Materials

SDA and DTA spectra were run on 99.5% purity lead titanate ($PbTiO_3$) and high-purity (99.999%) potassium niobate ($KNbO_3$) powders obtained from Johnson–Matthey (Figs. 1 and 2). Both materials have ferroelectric to paraelectric (tetragonal to cubic) transitions, and both are known to have "soft" modes in the ferroelectric state. The dielectric responses of the two systems are clearly quite different.

The lead titanate shows large hysteresis (~25°C) and has a very large paraelectric response just above the Curie point, decreasing with temperature with an approximate $(T-T_c)^{-1}$ response. The potassium niobate, however, shows only a small increase in dielectric constant over the transition (compatible with only small changes in electronic polarizability from phase to phase) although it is known to have a large, low-frequency paraelectric response just above the Curie point. This suggests strong damping of the soft lattice modes. The dielectric-loss response, although different in magnitude for the two materials, shows the same inversion in frequency dependence over the transition. In both cases, the ε'' frequency dependence in the paraelectric cubic phase suggests a standard thermally activated electronic conductivity.

These examples demonstrate that ferroelectric transitions are readily seen in SDA spectra and that some information concerning the fundamental properties of any specific transition can be obtained.

Spectra for the Tetragonal–Monoclinic Transition in Zirconia

Zirconia has three phases: a high-temperature cubic phase, which, on cooling, transforms at 2370°C to tetragonal phase, which, in turn, transforms to a monoclinic phase as it cools, via a well-studied transition at ~1170°C. The latter transition is a diffusionless shear process (martensitic-like) with thermal hysteresis between heating and cooling cycles, and an accompanying 9% increase in the atomic volume as the system cools and transforms to the lower-temperature monoclinic phase. This unusual situation of abrupt expansion on cooling made the manufacture of large sintered zirconia pieces impossible until the discovery that the addition of small amounts of some particular oxides (e.g., yttria) will stop the tetragonal phase from transforming to the monoclinic phase even at room temperature.

The DTA spectrum for a standard-purity, calcined, zirconia powder (Fig. 1) shows the tetragonal-to-monoclinic transition, with ~160°C hysteresis between heating and cooling. The SDA spectrum (Fig. 2) shows the expected hysteresis in ε', but the increase of ε' (and thus of the polarizability) in the more dense tetragonal phase is not the usual trend. A sample of very-high-purity (99.995%) zirconia had the same increased value of ε' in the tetragonal phase, suggesting the change in ε' is a fundamental property of the repositioning of the basic atoms. However, the ε'' value of the high-purity material was a factor of 20 smaller value of ε'' (although with the same typical "free" electron conductivity dependence), suggesting the ε'' values were dependent on impurity levels.

As an example of the response of a well-known stabilized zirconia, a sample of 7%-yttria-stabilized zirconia (ie., 7% impurities) showed the same inverse-frequency dependence of ε'', but with a much larger loss (the value of ε'' was ~100 at 400 MHz, yielding a predicted (dc) "free" electron conductivity of ~ 0.4 S/m at 800°C).

Binder Burnout from Alumina Compacts

A previous study[6] used SDA to assist in understanding the process of polymethylmethacrylate (PMMA) binder burnout during the low-temperature stages of sintering of pressed alumina pellets. The DTA data (Fig. 1) show an initial exotherm, followed by an endotherm and at least one succeeding exotherm. The endotherm disappears and the exotherms coalesce into one peak if the temperature ramp rate is very low, suggesting that the endotherm is associated with oxygen starvation in the pellet core and resulting pyrolysis of the binder. The second exotherm is believed to result from carbon burnout in the pellet core. The SDA spectrum (Fig. 2) suggests the generation (between 150 and 250°C) of high-dielectric loss components in the sample (relaxation of polar PMMA pyrolysis fragments?), whose disappearance is complete at 400°C. At ~ 600°C another, thermally activated, loss mechanism appears (carbon-based electron conductivity?), which slowly disappears as the carbon burns away at high temperatures.

The differences between DTA and SDA results are very evident here. The DTA signal is proportional to the rate of evolution of energy in the sample, while the SDA response reflects the state of the system at the time of the measurement. Thus, the low-energy thermal dissociation of the PMMA, seen in SDA at 200°C, is hardly noted in DTA, while the large exothermic spike at 370°C in DTA corresponds with the rapidly decreasing SDA value, reflecting the decreasing amount of reactant.

A Thermally Activated De-watering Reaction—The Goethite (FeOOH) to Hematite (Fe_2O_3) Transformation

The thermally induced transformation of goethite to hematite between 200 and 400°C is commercially important as a source of various forms of hematite. Although it has been extensively studied, the solid-state reaction mechanism still is not completely understood, for various intermediate states exist, and the final hematite can have varying structures and amounts of hydroxyl radical remaining. The basic reaction is

$$2\alpha\ FeOOH \rightarrow \alpha Fe_2O_3 + H_2O(g) \text{ (endothermic, enthalpy of} \sim 58 \text{ kJ/mol)},$$

but, the actual solid-state reaction involves diffusion of thermally activated ions to the surfaces of each crystallite, where they recombine to form water in gaseous state. The remaining defected crystal structure readjusts itself to the changing local ratio of atomic species, forming first protohematite and then hydrohematite, as the amount of hydroxyl radical decreases.

The SDA spectra for the reaction, in air, at a ramp rate of ~ 3.2°C/min, with a 45 min hold at 380°C, are in Fig. 4. There appears to be no significant reaction until ~300°C, at which point both ε' and ε'' increase rapidly as the dissociation reaction proceeds. The fact that both increase means the effect is not due to thermally activated "free" electron conductivity (this phenomena has very little influence on ε'), but rather to a permanent dipole relaxation or ion motion related mechanism. After the hold at 380°C, the primary dewatering reaction is largely complete, and most of the transient microwave absorption mechanism has disappeared. The remaining protohematite sample has a thermally activated microwave absorption mechanism at least up to 700°C.

The comparison of the SDA and the time-derivative of the TGA data in Fig. 5 (known as DTGA) certainly suggests that the underlying mechanisms must be related. The most likely mechanism of dewatering is the diffusion of protons and/or hydroxyl radicals through the lattice until they reach a crystallite surface, where they combine into water and are released immediately as steam (an actual water molecule is likely too large to diffuse through the lattice). Steam, at

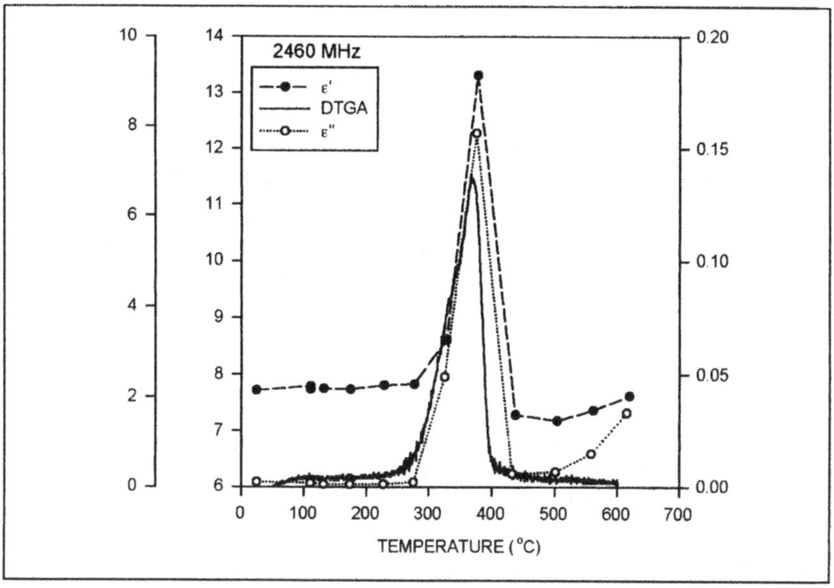

Figure 5: Comparison of the Derivative Thermogravimetric Analysis (DTGA) results with the real (ε′) and imaginary permittivities (ε″). The heating rates were ~ 7.5°C/min. for TGA, ~ 10°C/min. for SDA.

atmospheric pressure, has too low a density to produce a significant contribution to the dielectric response, so the amount of mobile transient species must be the source of the dielectric response—as is substantiated by the similarity with the derivative of the TGA curve (ie., proportional to the dissociation rate). If the Debye relation for a single relaxing permanent dipole species (the last term in each of Eqs. (2) and (3) is applied to the part of dielectric response that is related to the reaction, one can extract an approximate value for the relaxation time of the species as a function of temperature. This yields a relaxation time of $\tau \approx 10^{-9}$ s at the reaction peak.

General Conclusions

The basic premise that SDA can be used as an analytic tool, much as one uses DTA, DSC or TGA, seems well substantiated. In general, a step change in the real part of the dielectric constant indicates a phase transition, while the change in the absorptive part provides additional information on changes in the relaxing electric dipoles and/or ac conductivity mechanisms. The presence of defects or impurities is generally seen and can be quantified in SDA data if comparison data with pure or known materials are available. Time-dependent reactions over a large temperature range, in selected atmospheres and at moderate pressures, can be followed and quantified as they occur, generally in a nonreacting environment.

As with any new technique, one must learn how to interpret the data, and, in this respect, SDA is in early stages, for, although a great amount of literature exists on dielectric property measurements, very few articles are available to help with practical interpretation of real data.

We have started to apply SDA as a tool to assist in understanding specific ceramic synthesis reactions[7] and polymer–composite processing[8]. Over the next few years, we hope to develop a general methodology for integrating SDA data as part of a general analytic study of a material.

REFERENCES

1. R.M. Hutcheon, M.S. de Jong, F.P. Adams, P.G. Lucuta, J.E. McGregor and L. Bahen, "RF and Microwave Dielectric measurements to 1400∞C and Dielectric Loss Mechanisms", MRS Symposium Proceedings, **269**, pp. 541–551. *Microwave Processing of Materials III,* Edited by R.L. Beatty, W.H. Sutton and M.F. Iskander (1992).
2. W. Xi and W.R. Tinga, "Microwave Heating and Characterization of Machinable Ceramics", ibid., pp. 569–577.
3. S. Bringhurst, O. Andrade and M.F. Iskander, "High-Temperature Dielectric Properties Measurements of Ceramics", ibid., pp. 561–568.
4. M. Arai, J.G.P. Binner, G.E. Carr and T.E. Cross, "High Temperature Dielectric Property Measurements of Engineering Ceramics", Ceramic Transactions, **36**, pp. 483–492. *Microwaves: Theory and Application in Materials Processing II,* Edited by D.E. Clark, W.R. Tinga and J.R. Laia (1993).
5. W.W. Ho, "High-Temperature Dielectric Properties of Polycrystalline Ceramics", MRS Symposium Proceedings, **124**, pp. 131–148. *Microwave Processing of Materials,* Edited by W.H. Sutton and M.H. Brooks, I.J. Chabinsky, (1988).
6. E.H. Moore, DE. Clark and R.M. Hutcheon, "Removal of Polymethyl Methacrylate Binder from Alumina using Conventional and/or Microwave Heating", Ceramic Transactions, **36**, pp. 325–332 in Ref. 4.
7. M.D. Mathis, D.K. Agrawal, R. Roy, R.H. Plovnick and R.M. Hutcheon, "Microwave Synthesis of Aluminum Titanate in Air and Nitrogen", Ceramic Transactions, **59**, pp. 557–564. *Microwaves: Theory and Application in Materials Processing III,* Edited by D.E. Clark, D. Folz, S. Oda and R. Silberglitt. (1995).
8. X. Fang, R. M. Hutcheon and D. A. Scola, "A Study of the Kinetics of the Microwave Cure of a Phenylethynyl-Terminated Imide Model Compound and Imide Oligomer (PETI-5)," *Journal of Polymer Science Part A: Polymer Chemistry*, **38** , Issue [14], 2526–2535 (2000).

Profiting from Microwaves

Microwave and RF Energy Utilization—Expert and Audience Perspectives

W.R. Tinga

Introduction

To maximize the interchange of ideas, over one hundred scientists, engineers, industrial users, equipment manufacturers and a panel of selected experts discussed and debated their views on microwave (MW) and RF energy utilization by specifically addressing the list of questions posed below. To be sure, not all of these questions have simple answers nor was there enough time to have an exhaustive discussion on all of them.

1. When Does Using Microwave And RF Technology Make Sense (Cents)?
2. What Are The User's Needs?
3. When To Consider A Combination Of MW, RF And Conventional Energy?
4. Scale-Up From Lab To Production - What Are The Rules?
5. How To Generate An Expert-Knowledge Database For This Technology?
6. Microwave Effects - Fact Or Fiction?
7. What Are The Secrets Of Coupling MW Or RF To A Process?
8. What Are The Limits On Energy Leakage Design?

Nonetheless, the wide variety of interests presented by the attendees gave fair breadth to this forum and offered new insights into the benefits and problems inherent in utilizing this technology. What follows is a summary of the views and opinions expressed by a panel of experts and the audience.

When Does Using Microwave and RF Make Sense (Cents)?

Primarily, the use of highly ordered energy sources such as MW or RF makes sense when other process techniques can't do the job. As a result of the higher cost of these sophisticated energy sources, mostly products with high dollar value per unit weight should be considered. Highly ordered energy forms such as RF or MW will always be more expensive per Joule then the raw fuel energy. A MW-RF solution may be reasonable

- When volumetric heating, selectivity and greater energy uniformity are desired.
- When it makes business sense, i.e., when the return on investment is positive.
- When the MW-RF approach offers new advantages or benefits.
- When there is a "Green" or environmental advantage.

For example, in the chemical industry, cents may not be the problem but the environmental issue is and, therefore, a greener process is an advantage. In another example, a customer used mercury seals to make contact with ceramic fibers to be electrically heated. Using microwave energy eliminated the electrical contact requirement, and therefore the mercury, saving the processor a lot of environmental headaches. Consequently, energy efficiency or process cost may not be as important as the avoidance of hazardous materials in the process. When the cost of polluting with a given process is less than the cost of the "green" process, it won't be adopted by industry, even if it can greatly reduce environmental hazards.

Noting this environmental concern context, various participants were disappointed at the lack of representation from the chemical, pharmaceutical, wood, paper and other large industries. One reply was that there must be a visible economic benefit to adopting any given technology.

Industry will resist change unless it is forced on them by economic neccessity or by government legislation. In the future it is likely that environmental factors will play an ever more dominant role in dictating what type of technology will be used and thereby counteract an industry's inherent inertia to change. Others commented that high-tech waste generation treatment can lead to recovery of saleable byproducts.

Lack of interest from certain large industries in this technology is often caused by ignorance of the management about emerging technology areas which could benefit their companies. One thing that will motivate management to look for other answers is when they have trouble, for example, getting an FDA approval for a pharmaceutical, chemical or food product.

This leaves us with the question of "How does one make decision makers aware of alternate possibilities?" Maybe one way to get a few key decision makers to come to conferences like this one is to give them a role to play! One large chemical company did have a representative present who stated that his company (Dupont) did value this technology but emphasized the careful definition of the application. By succeeding in one application, the experience gained is used to build other applications to the benefit of the company. Other reasons for adoption.

- when building new or updating old processing plants.
- when a more hygienic approach is needed.
- when fossil fuel costs increase significantly as caused by pollution control costs or resource depletion.

What Are the User's Needs?

Defining the term "user" is all-important in answering this question. There are different types of users which can be targeted. One user may be a process machine manufacturer wanting to incorporate microwave technology in one of the process machines that he sells. For example, a manufacturer of pharmaceutical drying equipment or one who manufactures machines for making tablets. The motivation is that, if he succeeds, he can sell a new piece of equipment. Another user may be a processor of food, pharmaceutical or other products who is mainly interested in a return on investment, the only driving force for any industry to exist at all. Still another user could be the consumer who can be convinced that a certain new product, for example, a microwave clothes dryer, is better than the conventional product. Clearly, it is important to identify the level of user you wish to convince and to communicate with accordingly. How not to communicate with the consumer, for example, is humorously stated at the end of this paper.

> *Return on investment is the only driving force for any industry to exist.*

As to what the user needs, still another viewpoint emerged, namely that of the electrical utility companies, dubbed the "silent majority." Small and medium sized manufacturers need someone to help them solve their manufacturing problems. Too often one focuses on problems that, to the manufacturer, are not seen to be problems. The challenge is to go out and find out what those problems are rather than asking the question, "Why are these manufacturers not here?"

Frankly, from their perspective, solutions to their everyday problems aren't going to be solved by coming to a meeting such as this. It is we who need to go to them to identify and understand what their manufacturing problems are and then to apply our know how of the technology, its benefits and its limitations to see if it can be used to address those problems. Finally, when we think we have found a solution, we must present it to them in a manner that they can understand. Don't say, we can do it faster or we can do it selectively or even, we can do it more efficiently. Reduce the answer, if at all possible, to financial terms and return on investment or payback period. Then they will listen. Still, one needs a champion within that company to help sell the new approach or technology. "Hit them hard right at the beginning with something that hits home to them."

Food and Drug Administration (FDA) requirements can be encapsulated in that one word: Control.

Greater process control is a foremost concern in the pharmaceutical world. We need to get a better handle on what chemical transformations take place under microwave processing conditions. Pharmaceutical industry representatives stated that ignorance of the MW-RF technology is a main cause of its lack of use in their industry. Education of energy users is clearly very important in addressing this problem. Time to market is another important aspect of effectively solving a processing problem - apparently only three months from lab to full-scale in generic pharmaceutical companies "who are leaner, meaner and very fast on their feet." They also have a need to protect proprietary information.

When to Consider a Combination of MW, RF and Conventional Energy?

A simple answer is that you combine technologies when the whole is greater than the sum of the parts. In other words, when in doing so it causes synergy such as better control, lower cost, new product. One example given was that of fluid heating. Results are very flowrate dependent. If the flow is relatively stable, good temperature regulation is possible with conventional, e.g., IR heating. However, the use of microwave energy allows excellent temperature regulation even with unstable flows as a result of the lower thermal inertia of the electromagnetic energy. But the cost of microwaves alone for high flow rate applications is an order of magnitude greater than the cost of IR. When the bulk heating is done with the lower cost energy source, the temperature control is achieved by trimming the energy input with the microwave source. At an incremental cost of about 10%, an order of magnitude improvement in temperature control is achieved.

Rules of thumb: Bulk heat with conventional energy and trim with RF or MW. When the intrinsic cost or value of the product is of the order of \$1/lb, then MW or RF energy is probably not competitive.

Scale-Up from Lab to Production—What Are the Rules?

Scale-up thinking should start at the beginning before the R&D is even started. Which applicator type will be used? Will it be a conveyor mode, single or multimode system? Which frequency will be used? In one case, a lab prototype applicator was modelled at 2450 MHz but with the intent of scaling the production unit to 915 MHz. By verifying the model in the lab at 2450 MHz and then using the model to predict the 915 MHz production performance, it was possible to confidently predict success for the production version.

McMillan Bloedel's Parallam process was prototyped in the lab but it went through a second prototype stage to duplicate the process to make full-scale real product but in commercially insignificant quantities. This stage was used to prove out the applicator technology for about two years. Such a long a development cycle may not work in the fast-moving pharmaceutical industry but in the 20-30 year lifetime forest product industry this was acceptable. Also note that this was an entirely new product with no conventional competition and a full-scale plant would cost hundreds of millions of dollars. This long proving out process was necessary to minimize risk. Other advice included rules of thumb.

- Don't scale power up more than 10 times or else use an intermediate stage.
- Don't do research on a singlemode applicator and then try to scale it up to a multimode applicator.
- One must understand dielectric property changes when changing frequencies since these data affect penetration depth and energy uniformity.
- Scaling up is a system matter. It involves a lot more than just the microwave aspect of the design. Successful application of microwaves to a process is often a result of good overall systems engineering.
- Think production while researching in the lab. Something as simple as water-cooling is not always available in a production environment.
- Decide which things you want to learn at each step of the investigation.
- Scale-up is generally linear, with respect to power use, when the load is a good electromagnetic (EM) absorber. For low-loss materials this rule may not hold.
- Whether the process is known or new will make a difference in the approach taken.
- Economy of scale with respect to volume depends on the particular industry. In pharmaceuticals, the active ingredient is generally less than several kilograms and typically has a very high dollar value per unit weight. Final

equipment cost will be much higher than the per watt cost of a domestic microwave oven because highly automated manufacturing lines are economic for the high volume consumer market. Industrial equipment markets are much different.

- Use outsourcing of certain aspects of the process design to leverage expertise of smaller specialized companies saving both time and money.
- Anticipate a change of a clean to dirty environment when going from lab to pilot plant. Environmental issues are generally not simulated in a lab model. For example, power outages, mechanical vibrations, dust buildup, high temperatures, and equipment cleaning procedures.
- Scaling actual microwave equipment design is different and is often not a linear process. See a later section on "What Are The Secrets Of Coupling MW Or RF To A Process?"

How to Generate an Expert-Knowledge Database for This Technology?

Where is the knowledge now? From the user's point of view, the expert knowledge is not concentrated very well. Resources available are scattered throughout the scientific literature. These include things such as dielectric property data, simulation or modelling codes (computer programs), component design information, microwave engineering design books, the Journal of Microwave Power and Electromagnetic Energy (JMPEE), IEEE Transactions, Transactions of the American Ceramics Society and the Materials Research Society, a large number of older, out of print, books, and so on. Internet technology should be put to more effective use in enabling access to these varied information sources. However, some motivation and funding is needed to make this transition to ready information access a reality. Copyright problems will need to be addressed when making previously published informtion available via this new medium. Apparently expired, but still useful, patent information is already available via Internet access even now [Micropatents]. Recently IBM announced that it would make its huge patent files available on the Internet also.

> *For starters, the International Microwave Power Institute could put JMPEE, etc., on Compact Discs (CDs) since many of their useful publications are out of print. Academic researchers could possibly facilitate this process by making more reference information available on their homepages.*

Serious system design using an expert system is not realistic at this time, but to use such a system to transfer useful data is indeed very possible. Expert systems need databases. There is a clear need to collect dielectric data in a more useable database form. Historically, the MIT Insulation Lab reports contain a lot of useful data. Many of these reports are still available from the NIST. (USA). Caution was advised with regard to use of much of the newer dielectric data, for example, the high temperature ceramic data, where repeatability of the data is often questionable because it is difficult to reproduce experimental conditions.

> *Dielectric measurement technology is, and probably will remain, an expert's domain.*

Another point made was that it should be possible to readily add one's own data to that of an existing database to insure that the database grows and holds current relevant data. Nonetheless, there will always remain a need to measure the dielectric properties of a particular, often unique, process material, indicating a need for more ready access to tools to measure the dielectric data inhouse. In recent years, there appears to be an increase in the number of labs where these measurements can be made.

Microwave Effects—Fact or Fiction?

There was little disagreement that microwave effects are fact. but what is the real question? Is there an electromagnetic effect? Yes! In fact, some would say, many! Is there a non-thermal effect? Yes! What exactly is a microwave effect? Is it non-thermal? A steak in a microwave oven exhibits a microwave thermal effect, it gets hot! However, a UV photo absorption process in a polar organic molecule is a non-thermal effect because during this process there is no equilibrium distribution of energy states. Moreover, just because an energy-material interaction is non-thermal doesn't necessarily mean that it is a microwave effect. In the vapor filtration process, use of microwave energy causes a reverse temperature profile in the material. This is not a nonthermal effect but it is a microwave effect and possibly a very valuable one at that. So the question becomes a definition issue.

> *One proposed definition is that a microwave effect is: "A response of the system that does not follow the (known) chemical or physical rules."*

Of course, a customer's view is often, "Who cares?" He is interested in a return on investment. On the other hand, for the microwave designer a microwave effect is important if it can help to design better systems. This doesn't mean that the scientific perspective on these questions is not important but the potential microwave energy user has a process problem to solve and doesn't care about the semantics of interpreting a question such as this. What he does care about is the application of an energy transfer process.

At the atomic level, non-equilibrium conditions cause temperature to change and any energy absorption at this level is a non-equilibrium mechanism. To some, what is more important to know is whether a given non-equilibrium energy distibution will persist long enough for some diffusion effects to take place or to affect the outcome of a chemical reaction or physical process. In other words, is the time scale on which this non-equilibrium temperature condition exists sufficient for other physical phenomena to occur before that energy can dissipate into the rest of the atomic or molecular structure? The consideration of what happens to the interaction of the microwave energy with the vapor phase above the material being processed would be important as it will affect the system pressure, and hence the outcome, especially in chemical reactions.

It may be worthwhile to observe in this context that an electric field is a force field (Coulomb's law) and that this force is explicitly independent of frequency.

It was mentioned that waste material bacteria are killed by the use of an EM energy process. But, replied another speaker, "Microwaves heat the water to create steam and it is the hot steam that killed the bugs." From another point of view, though, we should encourage people to look for particular effects achievable with microwaves or RF which can be useful in processing.

Workers in this area ought to use great caution and care in establishing a proper control and repeatability.

What Are the Secrets of Coupling MW or RF to a Process?

Fit the right field patterns (or modes) to the product's geometry, size and the process needs. This most definitely requires fairly sophisticated engineering knowledge. Electromagnetic fields can be formed, directed and channelled much like airflow can be. Electric fields are force fields just like "wind fields" are. Immense flexibility is inherent in the wave nature of the fields but the design options are not always easy to implement or decide on. Modern electromagnetic modelling tools will become more and more sophisticated and indispensible.

Nonetheless, most designers would agree that most of the time, "Simpler is better."

Interaction of EM fields with process materials is shape, density, composition and size dependent. If the wavelength used is comparable to the size of the object it needs to interact with, then wave phenomena will largely govern the behavior of the energy deposition. For example, at 915 MHz, the wavelength is about 30 cm while at 2450 MHz it is about 12 cm. When the wavelength used is much smaller than the object being treated, the interaction can often be modelled by a plane wave (as opposed to standing, resonant waves) travelling into and through the object. In the latter case, the object's geometry and that of the surrounding metal boundaries have only a secondary effect. This last observation also holds when the material is a strong EM absorber. That is why power scaling for good electromagnetic absorbing materials can be done linearly.

Still another case exists when the object being treated is much smaller than the wavelength used, quite typical in RF applicators where the wavelength is of the order of a meter. In that case, the geometry of a small object will affect the energy absorption only marginally but the surroundings of the object, presence of metal walls, electrodes, etc., will cause standing waves along with the inescapable non-uniform energy patterns.

Coupling of EM energy to a material is always dependent on its dielectric or magnetic properties and the associated electromagnetic energy dissipation mechanisms. Magnetic fields set up currents in proportion to the conductivity of the material containing the fields. Therefore "resistive losses" will be the cause of heating by such currents.

Another lesser used method of coupling is the use of slow-wave electromagnetic structures. For example, using a meandering or serpentine waveguide one can effectively slow down the energy velocity through the structure while traversing a large surface area of a material. This wave energy deceleration has the effect of increasing the power deposition per unit volume.

Thus, if RF is thought not to be suitable because the power absorbed at the RF frequency is less than that at a microwave frequency, often a factor of 100 higher, than the use of a slow-wave applicator, with a deceleration factor of 100 at the RF frequency, can attain the same power absorption per unit volume as at the higher microwave frequency.

What are the limits on energy leakage design?

From an engineering point of view, the number of design options for creating electromagnetic (EM) energy traps, or chokes, are many and varied. One of the fundamental design techniques is to use the quarter-wave principle. This principle simply states that an open circuit condition, such as at the end of an open coaxial line, can be effectively transformed into a non-radiating short circuit condition by placing a section of transmission line between the open circuit and the point at which energy enters this transmission line. This is how noncontacting door-seals for a microwave oven door are designed.

> *This bit of design magic, makes the open gap at the non-contacting door to oven interface appear to the electromagnetic waves as a short or an open circuit which causes the waves to be returned to the oven interior.*

Openings in the EM equipment which are larger than a wavelength are difficult to choke in this manner but other absorbing and reflecting traps can often be designed in such cases. Zero EM energy emission is only possible for completely "closed and welded" applicators and EM generators. Costs of limiting the EM energy leakage go up with the complexity of the energy trap design and with the desired percentage amount of energy trapping desired. Perforated metal screens, e.g., metallic window screens, are very effective in blocking EM wave propagation as long as the perforation diameter is very much less than a wavelength. Rotating belt drive shafts can be extended through a microwave tunnel wall by incorporating properly designed non-contacting sleeves at the protruding ends of the rotating shaft.

Conclusions

Some may have wondered "Where are the young people doing work in this field?" We were all encouraged to help train and reward students in this field, e.g., by scholarships or industrial sponsorships, so that our current knowledge base will not only be preserved but expanded and used in the future.

What has been reported above is a summary of what was a useful and interesting discussion. As to why MW and RF processing utilization is at present only a small industry is best summed up by paraphrasing B. Krieger's closing remarks. We are working at a technology which is at the bottom of the barrel in terms of industrial processing. A dismal picture indeed. Yet, almost every home has a microwave oven. Some say, that is because it is inexpensive whereas industrial microwave ovens are much more expensive. True, but in comparing a domestic natural gas range with an industrial natural gas installation, such price difference remains. So what is the problem?

The problem is marketing.

"Let me illustrate. Let us say that a housewife wants to cook a hamburger. So she goes to the appliance store clerk and says to him, "I want to cook a hamburger. " The appliance guy asks, "What is the moisture content? Wet or dry? Do you know the dielectric loss factor because I have a network analyser. And if you don't know that let me start out with a small meatball and then scale it up to your hamburger. "

We need to focus on making useful things for our customers and therefore create markets by using technology. So maybe what we need in the future is a World Congress on Marketing.

ACKNOWLEDGMENTS

Special thanks goes to the audience and all participants who made frank and insightful contributions to the discussion. Also a sincere thank you to the panelists who were: John Booske, University of Wisconsin, USA; Mark Churchland, McMillan Bloedel, Canada; John F. Gerling, Gerling Applied Engineering, Inc., USA; Peter Jones of EA Technologies/Peal Jay Consultants, UK; Harold Kimrey, Microwave Materials Technology, USA; David Lewis, IBM T.J. Watson Research Center, USA; Ricky Metaxas, University of Cambridge, UK; Gary Walzer, EPRI-R&D Applications, USA. This discussion was chaired by Wayne R. Tinga, University of Alberta, Canada.

Dielectric Heating: EPRI's Perspective on the Market and the Technology

C. Gellings

Abstract

Dielectric heating enabled by microwaves and radio frequencies has been applied to a variety of industrial processes. In each application, the result is enhanced product quality and productivity. In addition, overall environmental improvement and reduced energy needs usually result. However, despite these benefits, dielectric heating has been slow to achieve substantial market penetration. Now, by reducing the cost and networking with business and industry, dielectric systems can finally bring wide-spread benefits.

Dielectric Heating

There are enormous opportunities to cut costs and improve product quality in the heating and drying operations of non-conductive materials with radio frequency (RF) and microwave heating. Both RF and microwave heating transfer energy directly into the materials, giving uniform heating through the entire thickness. This means that, by using RF or microwave heating, manufacturers of plastics, rubber, ceramics, textiles, and wood products can avoid the great expense and slow production rates frequently incurred in heating and drying operations.

Electric Process Heating

Dielectric heating embodies the application of only one of a range of possible uses of the electromagnetic spectrum to form materials, for medical diagnostics, and to heat and cure materials. Collectively, all of these applications involve electric process heating.

Few of mankind's endeavors impact the world in the way energy production and consumption do. Electricity is often the most visible energy form, since it has the most complex production, delivery, and utilization system. Because of the energy conversion that occurs in producing electricity, many policy makers have argued that a unit of electricity saved is equivalent to preserving three units of primary energy. While this can be the case, there is a broad assumption that electricity use is inefficient and adversely impacts the environment. In extreme cases, some policy makers have incorrectly begun to encourage the use of fossil fuels instead of electricity.

The real focus of the effort in understanding energy efficiency should be on total resource requirements. Electricity is a uniquely valuable form of energy, offering unmatched precision and control in application, and, hence, in efficiency. This is particularly true in dielectric heating. In addition, electricity offers unrivaled benefits in comfort and in its environmental advantages. Because of these unique attributes, in almost every case, new electric appliances and devices require less total resources than comparable natural gas of oil-fired systems.

Attributes of Electricity

The advantages of electricity as an energy form offer society a second dimension to efficiency. As an energy form, electricity has greater "form value" than any other energy. Form value affords technical innovation with enormous potential for economic efficiency. The attributes of form value include three dimensions: Technical, Economic, and Resource Use [1].

Technical

Technical attributes include: Electrical Phenomena, Input Energy Density, Volumetric Energy Deposition, Controllability, and Synergistic Combinations [2].

Electrical Phenomena. Residential appliances, building energy systems, and industrial processes frequently involve the interaction of energy and matter to modify materials, pump refrigerants or fluids, or to transform them from one form to another. Three types of electrical phenomena can be involved in these transformations: electromotive. electrothermal and electrolytic. These phenomena are all unique to electricity as an energy form and contribute to its form value.

Input Energy Density. In combustion processes using chemical fuels (e.g, oil and gas), the maximum achievable temperature is thermodynamically limited to the "adiabatic flame temperature," a practical limit of about 3000 F for fossil fuels burned in air. When heating material electrically, there is no inherent thermodynamic limit on the temperature. Typical temperatures of 10,000 F and higher are achieved routinely in arc-produced plasmas.

Volumetric Energy Deposition. Electrothermal phenomena are volumetric, e.g., generating heat within the material itself, a significant advantage in many processes. In using fossil fuels or in using electric infrared convection ovens, to heat material, heat is usually imposed slowly at the surface by radiation and convection.

Controllability. Electricity is often referred to as an "orderly" form of energy, in contrast to thermal energy, which is random in nature. This orderliness means that electrical processes can be controlled much more precisely than thermal processes.

Synergistic Combinations. In some processes, electrolytic, electrothermal, and electromotive effects combine in an advantageous way.

Economics

Economic attributes include. Fixed Cost, Flexibility of Raw Material Base. and Product Quality and Yield.

Fixed Cost. High energy density and precise control typically result in increase rate of production, which, in most cases, reduces fixed costs per unit of product. Electrical process equipment is typically economical in smaller sizes than combustion equipment, since fuel handling and environmental control are not required. Electrical production technology can give rise to an "economy of reduced scale," in that smaller equipment requires less space, and facilities can be decentralized and situated in proximity to diverse raw material sources and product markets.

Flexibility of Raw Material Base. High energy intensity and precise control permit a greater degree of flexibility with regard to raw material resources than is possible with most conventional processes.

Product Quality and Yield. In overall production economics, product quality and yield are critically important. Electrical processes typically provide improvements in both areas.

Resource Use

Resource use attributes include: Flexibility of Fuel Supply; Domestic Resources, Balance of Payments and National Security; Environmental; and Energy Consumption.

Flexibility of Fuel Supply. Combustion-based processes are highly dependent on the availability of specific fuel sources, since combustion equipment in general is not very adaptable to changes in fuel type. The shift to electrically based processes places the choice of fuel on the electric utility, where fuel diversity can be optimized.

Domestic Resources, Balance of Payments, and National Security. As a result of the overall efficiency of electrically based systems, total resource requirements (and, therefore, the need for imported fossil fuels) decline with the increased use of electricity. Decreased dependence on fossil fuels reduces dependence on politically unstable foreign suppliers, and, thereby, enhances national security.

Environmental. Electric processes and systems are indisputably the most environmentally benign at the point of end use. Recently, it has also become evident that the application of electrotechnologies can mitigate what would otherwise be adverse environmental impacts in transportation, industrial, municipal, and medical waste applications.

Energy Consumption. At first thought, it would seem that primary energy consumption with electrical processes should always be higher than that with conventionally fueled systems, especially in light of the conversion ratio of fuel to electricity at the power plant. The opposite is true in most cases as a result of exploiting the form value advantages of electricity.

Total Resource Efficiency

To analyze the total resource efficiency of an end use, it is appropriate to consider the total energy "system" from the raw resource through conversion, delivery, and utilization (see Figure 1).

To illustrate the comparative efficiency of electrically based end uses with fossil fueled end uses, the total system must be examined. To illustrate this point, the system illustrated in the following figure was modeled. It illustrates one unit of energy being derived from an oil or gas well. In an electrical system, it is converted to electricity at an assumed heat rate of 11,000 BTU/kWh. This results in 0.31 units of electricity from the original 1.0 unit of fossil fuels. Applying a steady-state loss of 8% in transmission results in 0.29 units of energy being delivered to the electric end use. This low efficiency is not typical of new electric production facilities. New installations of combined cycle combustion turbines typically are over 50% efficient. Typically, gas transmission and distribution systems lose 10% of their energy in pumping and leakage. This results in 0.90 units of energy being delivered to the gas end use from the original 1.0 available at the well head.

An example follows to describe the total energy system efficiency of dielectric textile heating as compared to use of a natural gas system, and thus reveal the breadth of the benefits of these uses of electricity.

Textile drying is typically done with antiquated equipment, most typically employing the use of steam. This technology is not nearly as efficient as dielectric heating and uses large supplies of fossil fuel as an energy source [3].

Frequently, a "hot house" is used to dry textile goods and garments. Similar to a giant sauna, these hot houses can consist of a 3,000-square-foot room with steam heaters that control the indoor temperatures 24 hours per day, seven days per week. Garments are rolled into the house on racks of trays, taking an average of 36 hours to fully dry. When any of its fans, belts or blowers require repair, or when no garments are dry enough for packing, production stops. Other methods typically used are just as inefficient. They include use of steam cans, tray dryers, or tumblers.

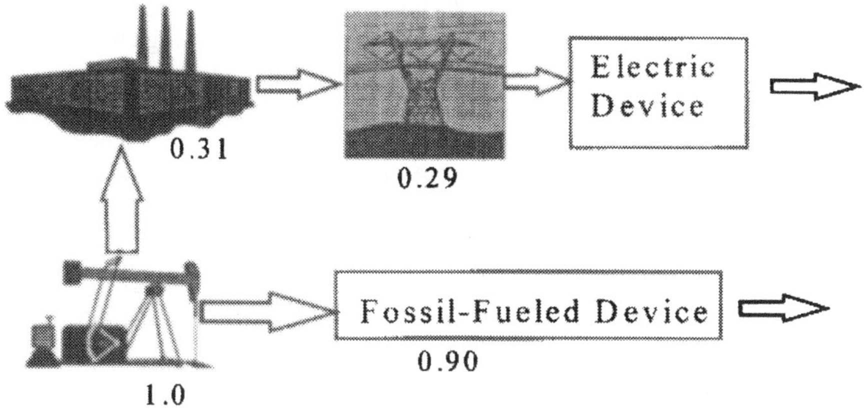

Figure 1. Total system efficiency.

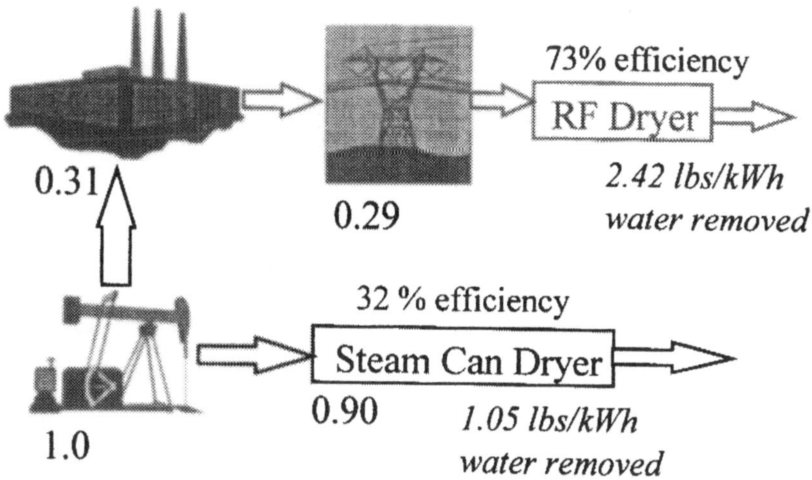

Figure 2. Schematic showing how radio frequency drying may be more efficient than fossil-fuel drying.

Radio frequency drying offers high volume, high speed and high quality drying. Because RF waves concentrate in the wettest, densest portion of the garment, RF drying has a leveling effect, practically eliminating any problems of uneven shrinkage and over drying. Radio frequency drying times depend on the type of fabric, the percentage of moisture in the fabric going into the dryer, and the degree of dryness desired. Radio frequency dryers can dry natural and synthetic fabrics within 1% of the desired moisture content.

The Figure 2 compares steam can drying to RF drying for a typical textile drying application. The RF systems are very efficient in transferring energy directly into the fibers. For each kilowatt hour of energy input, 2.42 pounds of wastes are evaporated. Thermally heating textiles to evaporate water requires a lot of energy in this case, a net of 1kWh for every 1.05 pounds of water. When the total systems are compared, the electric based process has more than twice the overall efficiency as the natural gas based process. In addition, applying RF drying in the textile industry yields superior quality products and enhances productivity overall.

The net effect of wider use of electricity for dielectric heating and industry electrotechnology applications could be an increase in electricity use by as much as 525 billion kWhs in the U.S. in the year 2010. Due to electricity's excellent "form value," this increase in electricity use is estimated to lead to a net reduction in total energy use of 4.7 quadrillion BTUs (quads) and a reduction in carbon dioxide emissions of over 290 million tons. Improvement in existing applications when electricity is already used could reduce consumption by 330 billion kWhs in the year 2010, with reductions in energy use of 3.5 quads and CO_2 of 225 million tons.

If comparisons are made of the impacts of increased electrification achieved through new electrotechnology applications against the impacts of electricity sav-

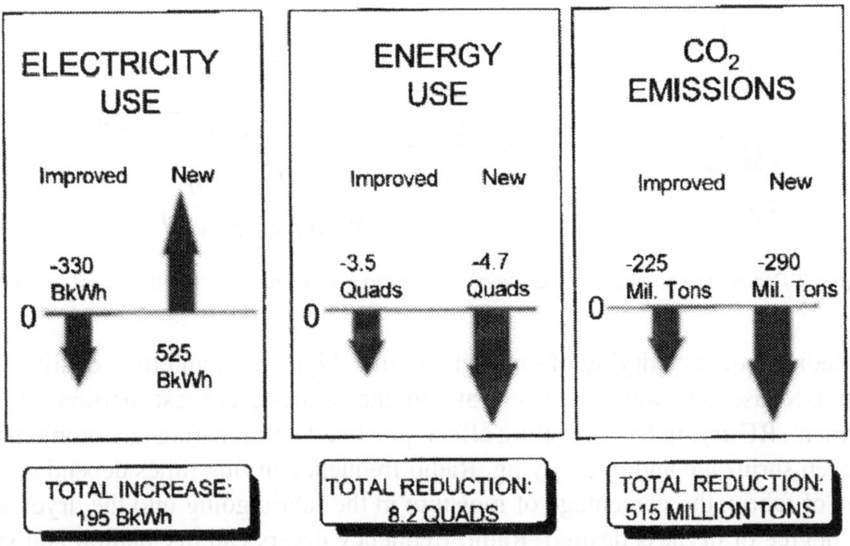

Figure 3. Benefits to be gained by selecting dielectric heating and other advances by 2010.

ings achieved by replacing old equipment, the results are that both measures will lead to reduced CO_2 emissions, although new uses will increase electricity use, and upgrading existing electric technologies will reduce electricity use. Thus, the combination of the new and improved uses of electricity could reduce total energy use by over 8 quads and CO_2 emissions by 515 million tons in the year 2010.

Productivity

From the customer's perspective, the most important aspect of dielectric heating is its increase in productivity (see Figure 3). Dielectric heating is faster, offers better product quality, and, when properly specified, can be more cost effective. This is best illustrated by examples [4].

RF Drying of Textiles

Referring to the energy saving example in the previous section, there are other advantages of the use of RF in drying textiles:

Faster Drying. Drying time reduced from 24 to 72 hours to only 45 minutes to 3 hours.

Faster Production. The average time from raw goods to finished product can be cut from 10 days to less than 8 days.

Increased Labor Productivity. Hot houses can keep 2 to 3 workers busy; the RF dryer may only require 1 worker. Many types of garments skip being placed on trays and racks, and can go directly in bins from the extractor to the RF dryer to packaging.

Less Work in Progress. Because of faster drying and continuous output from the RF dryer, 50% less work is in production.

Less Down Time. Hot houses need frequent and time consuming repairs. Cooling for 2 to 3 hours is necessary before maintenance personnel can work in it. After repairs, regaining operating temperature can take up to 8 hours. When the RF dryer requires occasional repair, work can begin immediately, and production can begin after a brief 5 minute warm-up period.

Lower Maintenance. Maintenance time spent to repair rolling racks, belts, fans and motors can be reduced by approximately 75%, or even eliminated.

Space Savings. Over 30% of floor space can typically be saved.

RF Vacuum Lumber Drying

Through the use of electrotechnologies, new methods are merging in the wood processing and wood manufacturing industries that improve and hasten production, save energy, reduce breakage and loss, and address environmental issues. Several of these specialized techniques are based on new concepts; other represent modifications of current wood processing and wood manufacturing procedures. These advances are of vital importance to the wood producing industry, which consumes over $600 million in electric energy to produce products valued at over $21 billion [5].

For plywood veneer, RF drying is a highly viable electrotechnology. The conventional redrying process results in a 15-20% veneer breakage rate and often results in over drying. The RF method is attractive because it provides moisture redistribution from wet spots to over-dried areas. In this way, moisture leveling occurs and the moisture profile can be controlled, minimizing veneer breakage. Plywood operations with high dry losses, as well as plants with limited drying capacity are good candidates for RF drying.

Radio frequency drying is also a faster method for drying green wood. But, because operating costs are usually higher, it is most suited for niche applications such as high value hardwoods. It is also appropriate in cases where drying speed is critical to production schedules or where very thick lumber is involved.

Applications

Many applications take advantage of the benefits offered by RF and microwave heating. Microwave and RF heating uses are firmly established in the textile, food processing and wood furniture making industries. Other typical applications are highlighted here:

Baking Foundry Cores

Application. RF heating bakes foundry cores used to make internal cavities in metal castings.

Process. The cores are made with resin binders, water, and sand. Heating is necessary to cure the resins and boil off water.

Competing Process. Fuel fired, hot air circulating ovens at 500°F heat unevenly and take over an hour to cure the cores. The long cycle time causes the cores to slump and warp. Non-uniform heating and curing cause stress cracking.

Benefit. The RF heating bakes the cores in 5 minutes at 230°F with little slumping and warping. Stress cracking is reduced and surface appearance is enhanced.

Preheating Plastics

Application. RF energy preheats epoxy preforms prior to molding [6].

Process. Rubber must be heated under pressure with additives to increase strength and resiliency and reduce stickiness and odor. The entire thickness must be heated to the vulcanization temperature.

Competing Process. Steam autoclaves or hot salt baths would preheat rubber as batch processes. Salt contamination is a concern.

Benefit. Microwave energy preheats the rubber to a uniform temperature throughout its thickness in a continuous process. Vulcanization produces more uniform properties, and there is no risk of salt contamination.

Drying Ceramics:

Application. Microwave energy heats and dries molds and castings for bathroom fixtures (soap dishes and paper holders).

Process. Ceramic slip is cast into molds, which absorb the excess water. The ceramic green bodies are dried, glazed, and fired to form final products.

Competing Process. The green bodies air dry for 24 hours before gas firing. The molds air dry for 12 hours before reuse. These long drying times prevent high production rates.

Benefit. High production rates, 1600 pieces per hour, are achieved with microwave drying of the slip in the mold. The setting of the piece in the mold is accelerated, and the heated mold dries in minutes for reuse. The entire processing time is 30 minutes.

Why isn't Dielectric Heating More Widespread?

This naturally leads to the question, that with all of these advantages, why isn't dielectric heating in more widespread use? Its market penetration is inhibited by the following four factors:

1. Higher capital cost—up to $4,000/kW. This cost is generally offset by energy savings, productivity enhancement and/or environmental compliance features.
2. Minimal market penetration—few installations. Each installation has to be specifically engineered to its application.
3. Few packaged technologies available. Technologies are almost always custom built and not as efficient or economic as they could be if some economics of scale were introduced.
4. Not well understood by the industrial community. The customer is very unwilling to try something new—certainly not something that has not been widely adopted.

In order to overcome these impediments, two key areas need to be addressed: improvements in technology, and enhancements of the market infrastructure. Regarding technology, to make dielectric heating more desirable, there needs to be an increase in the efficiency of RF and microwave generators, enhancements in controllability, and a reduction in cost. Regarding the market infrastructure, three actions are necessary: first, the industry needs to demonstrate high valued technologies that meet customer needs and to publicize these demonstrations; second, developers, scientists and engineers need to engage with effective partners that have access to successful distribution channels; and, third, the industry needs to leverage infrastructure to catalyze the market.

Technology Enhancements

The enhancements necessary are best illustrated with an example from the plastics industry [6].

Radio frequency energy can be used effectively to fabricate materials that are electrical and thermal insulators. Its use increases production speeds, allows greater control, improves product quality, and reduces costs. Yet, many new opportunities exist for RF heating in plastics. A few examples are:

- *Acrylics:* Skylights, tail light lenses.
- *Epoxies:* Primer coatings, electrical laminates.
- *Melamines:* Circuit breakers, counter tops.
- *Polyesters:* Microwave cookware, boat hulls.
- *Poly Vinyl Chloride (PVC):* Siding, packaging.

Radio frequency heating is ideal for both thermoset plastics, those that cure to a permanent hardened state upon heating, and thermoplastics, those that soften or melt each time they are heated. Because RF energy heats the plastic directly, it offers decided advantages over conventional heating methods, such as infrared or convection ovens with high or low velocity hot air.

The Market

Successful technologies require a solid base of research, adequate demonstration, and proper delivery channels. Developers of new dielectric technologies should never take their technologies out of the laboratory unless they are allied with a partner that has adequate distribution and marketing channels.

An example of how partners are essential to successfully introduced new technologies involves a microwave process focused on the steel industry's Hot Strip Mill (HSM) Sludges. These sludges typically contain sand, dirt, oil and metallic particles. Most HSM sludges are disposed in a landfill at a total industry cost of over \$50 million per year. Microwave systems separate sludge into reusable oil and metallic particles. It is estimated that the cost to industry to operate these microwave processes would be \$20 million per year. In the meantime, resale of recovered oil and metallic particles could further improve economics by \$15 million per year. Thus, the microwave process is more economical with zero waste disposal, and eliminates future liability issues associated with land filling [7].

In order to make this project a success, EPRI has enlisted as partners the following: Bethlehem Steel, U.S. Steel, and Inland Steel. With these industries at its side, the successful development and demonstration of this technology is substantially more likely to succeed.

Networking to Deliver

Now, more than ever, electric utilities and their industrial customers have closely tied mutual interests. In business to thrive economically, industrial customers are interested in improving productivity and reducing costs. Customer well being is also to a utility's advantage, since customer prosperity usually leads to more jobs, increased spending locally, and generally greater economic good for the community and area. For every primary job created, there are as many as three support jobs created. The resulting community is a healthy one that can help retain and attract new business. Cost effective compliance with environmental quality standards is another goal shared by utilities and their customers.

To promote prosperity, a utility can work with its industrial customers to identify electrotechnologies that improve productivity, enhance quality, or address environmental issues at customer facilities. These electric powered technologies often save overall energy while enhancing industrial process capabilities. But, most utilities cannot afford the depth of staff to keep up on state-of-the-art technical developments in a wide variety of industries. They need ways to make best use of their existing staff by learning from the experiences of others. To address this need, EPRI has established a network of regional centers and offices, each with a group of industries as its specialty.

These centers pull together experts in various industries and locate them where the utilities and customers need them. So, for example, EPRI established its Center for Materials Fabrication (CMF) in Columbus, Ohio, in the heart of the industrial Midwest; and, nearby, in Pittsburgh, EPRI formed the Center for Materials Production (CMP) not far from many steel making and other materials production companies.

Electric Power Research Institute operates 30 offices and centers that can tap into a wide range of expertise to help utilities help their customers and themselves; and, 11 of these focus on the industrial sector [8].

Conclusions

The use of radio frequency heating, curing, and drying technology in industrial applications has become well established. Drying of foodstuffs to enhance quality and texture, curing of rubber, drying and setting of adhesives, drying of wood and wood products, creation of synthetic coatings on substrates, chemical synthesis, and clean-up of toxic spills are among the many emerging and proven applications for radio frequency and microwave technology. There exists today a significant opportunity to increase the industrial market penetration of dielectric technology throughout the world. Two obstacles must first be overcome: the cost of the deployment of the technology; and education of industrial end-users concerning the benefits of dielectric heating, curing, and drying over conventional technologies.

REFERENCES

1. Gellings, "Saving Energy with Electricity," Electric Power Research Institute, Summer Seminar, August, 12 (1991).
2. "Electricity and Industrial Productivity," P. S. Schmidt, University of Texas, EPRI Report EM3640 (1984).
3. "RF Drying of Textiles," TechApplication, Volume 2, Number 2, Electric Power Research Institute (1990).
4. "Dielectric Heating: RF and Microwave," Tech Commentary, Volume 4 Number 1, Electric Power Research Institute, Reprinted (1993).
5. "Electrotechnology Advances in Wood Processing," TechCommentary, Volume 6, Number 1, Electric Power Research Institute (1990).
6. "Radio-Frequency Heating of Plastics," TechCommentary, Volume 4, Number 2, Electric Power Research Institute (1987).
7. "Microwave Treatment of Infectious Wastes," TechApplication, Number 1, Electric Power Research Institute (1991).
8. Gellings, "EPRI's Center and Office Network: The Center of Attention for Utilities and Their Industrial Customers," Public Power, American Public Power Association, Washington, D.C. January/February (1996).

Commercialization—Steps to Successful Applications and Scaleup

B. Krieger

Abstract

Successful applications of Microwave Heating Technology result from much more than successful engineering, modeling and pilot demonstration. There must be beneficial fulfillment of a customer's need!

Understanding that need requires thorough knowledge of the industry being served and all of the process and conventional heating alternatives. We hamper ourselves with excellence in physics and chemistry and only superficial experience in ceramic casting and manufacture. The microwave supplier must change his focus if he is to realize commercialization.

Introduction

The keynote speaker, Robert F. Schiffmann, brought home the reality that the industrial microwave heating industry, which was initiated with such great promise more than thirty years ago, has in reality commercialized relatively few substitutive applications. Tempering of meats, high power production of precooked bacon and the vulcanization of rubber, are most prominent. It should be pointed out however, that the concept of microwave heating has been so intriguing over these years that at the same time it has stimulated many thousands of ideas, equal numbers of laboratory and pilot systems, and even more inventions.

Why have so many innovations resulted in so few commercialized successes? Will the application of microwaves to the processing of ceramics follow the same route? The fact that we are attending this session on steps to successful applications seems to indicate our own concerns.

Discussion

The session chair asked me to address this issue based on my personal experience and from my own perspective. I must tell you frankly that I see the challenges not deriving from Maxwell, but rather from Peter Drucker. Peter Drucker is one of our nation's most prominent management consultants, and one of the creators of "The Marketing Concept." He chaired the Management Science Department at the Graduate School of Business at New York University years ago when I was a student. Drucker explains that the purpose of a business is "to create a customer." This means that commercialization is dependent upon our industry providing something for which a customer will exchange his purchasing power to acquire or not to acquire, considering all of the alternatives which are available to him. Marketing and the creation of customers is not merely selling. It is far deeper and more complicated. It requires very thorough knowledge of the industry to be served, their problems, processes, management culture, competitors and objectives. Commercialization results not nearly from solving technical challenges, but more importantly, from satisfying the wants of customers. In this room we have more than adequate physics, electronics, chemistry and engineering. What we lack is marketing science.

I find that what we are doing with microwave sciences is pushing technology, rather than responding to the pull of the needs of customers. The steps to successful scaleup are therefore largely marketing related.

We desperately need the "science" of creating customers. We continue to believe that if we can get something to work technically, customers will come to us to buy it and therefore, we will have commercialization. Consequently, we develop microwave technology in response to the challenge of heating ceramics, a material with dielectric properties that intrinsically defy heating with microwaves at room temperature. We, therefore, create pre-heaters, reflectors, develop microwave receptive additives, go to higher power levels, special resonant cavities, gyrotrons and even variable frequency high powered traveling wave tubes. Through these techniques we can get ceramics to heat very rapidly with microwaves. Is that enough to get customers to buy it? Obviously not!

Why isn't more effort put into understanding what would really satisfy our customers? If it is ceramics that is our objective, why aren't we in the ceramics plants to determine real needs? Why do we stuff our own staff with microwave scientists and consultants, when we should really be adding ceramics processing technologists?

Perhaps the problem is even larger than that. Maybe we do not even know who the customer is, and we have the wish that if we process faster and better, the customers will find us. We think that if EPRI or DOE, or similar funding agencies, provide money for research that they are the customer. Yet today we talk about commercialization. These agencies are not commercial!

If we continue pushing through technology rather than pulling through customers, I believe that there is little chance to commercialize. The answer is not in the microwave laboratory, it is someplace else; a place very remote from us.

Let us assume for this discussion that the manufacturer of ceramics materials, who is interested in making higher technology parts, is the real customer. (We may very well be wrong—it could be the aerospace manufacturer or the automotive plant who are looking for higher performance engines and components, or even the computer manufacturer who is really looking for smaller size, greater efficiencies and better data processing concepts). The identification of the real customer is a major part of the science of marketing. Let us stick for the moment with the model of the ceramics manufacturer and let us examine what we are offering him through the microwave sciences.

- *Speed:* Microwave heating is energy transfer, not just heat transfer. We generate "cold" microwave energy with microwave generators and put them into a microwave oven, which we call an applicator and the ceramic material takes that energy and converts it into heat within itself. Microwave ovens, therefore, are rated in kilowatts and not in temperature setting. The microwave energy excites the material to be heated through the process of dipole rotation or ionic conduction. Since it is a process of energy conversion, rather than heat transfer through conduction or convection, very high speeds can be obtained by the application of incremental amounts of microwave power. From the users point of view, however, high speed is usually equated with savings in floor space. If a machine can produce product faster, it can do it in shorter space if you look at it on an individual part basis. So, if we can produce sixty parts in an hour with microwaves, and the conventional process required an hour per part, we believe that we have made significant innovation. However, if the conventional production line is longer, or if the customer puts sixty parts in a batch oven for an hour, then production rates are really the same. The real trade off is in terms of real estate. If the customer's plant has absolutely no room to expand, the value of this microwave attribute is therefore high. Conversely, if there is plant space for multiple conventional processes, the value becomes much lower.
- *Penetration:* We know that the principles of microwave heating are through molecular activity, therefore the microwave energy penetrates deeply into the material. The poorer the microwave receptivity, the deeper the penetration. This attribute of microwave heating is perhaps the most important feature, since it can result in higher quality products for the ceramic manufacturer. We speak of better grain structure and other properties. These attributes can be distinct advantages, but are they practical? Are they economical? Is this technology an investment that has the flexibility needed in a manufacturing plant, not only for the products being manufactured today, but also for those which will be manufactured tomorrow? Will the customer

be willing to add microwave receptive materials to his product for the sake of his microwave process? I doubt that! Probably more than 90% of the manufacturer's cost relates to the cost of his materials. I suspect that less than 10% of the total cost is in processing, and, of course, less than that in heating. He will not add cost per pound, even though he may get higher efficiencies, unless there is very, very dramatic advantage. Can we quantify these advantages? Are they meaningful?

- *Electronic Control.* Microwave heating, of course, is an electronic process As such, it ties into automation and process control. Feedback systems, instrumentation and data logging are all means for achieving high quality repeatable precise production results. Additionally, this form of electronic heating is quiet, not heat intensive and extremely clean and efficient. Microwaves have more of these attributes than alternative means of processing. However, alternatives are by no means without the capability for advanced process control. Will the microwave advantages be worthwhile to the customer if he has to preheat by conventional means, or be involved in changing frequencies along with all of the attendant RFI implications? Additionally, is he willing to pay the high cost for microwave capital equipment and solve the perplexing problem of uniformity of heating, which results from differences in position, shape, materials, etc. within the microwave cavity?

Coming back to Drucker and the marketing concept, if we know that what a customer buys is satisfaction, we must realize that we are competing with all sorts of different alternatives. Let us look at some of the more easily defined competitive alternatives.

- *Other Microwave Companies.* The least of our problems is competition from within our own industry. Our problem is not competing with each other; it is creating customers. If anything, that is a mutual objective, because the customers are not now using microwave technology. Therefore, the companies which you may normally think of as competitors, should really be allies at this stage of our marketing science. Indeed, I see that as we sit together in this room as a microwave industry at a microwave symposium.
- *Conventional Heating Alternatives.* Consider the competition from direct fire gas heat, convection hot air, conduction, steam, infrared, UV, electron beam or even leaving something out in the sun to dry. Conventional technology, of course, is the enormous competitor. How much do we know about any of these alternatives? Have we considered where their attributes

are best used, either alone or in conjunction with microwave? Are we merely trying to justify microwave, or are we trying to justify the wants of the ceramic manufacturer? Do we realize that only if we do the latter, can we commercialize?

- *Non-Heating Alternatives*. If we know that what the customer buys is satisfaction, then we must realize that we must compete with all sorts of different alternatives in the ceramics plant which may take a higher priority for funds in the customer's mind. This can include things which serve much different purposes, such as pressers, parking lots, office furniture, etc. We all know the disappointment that comes when we work hard to achieve a technical concept and find out that in the end there are no funds available from the customer. The funds were obviously spent by the customer on something else that gave him greater satisfaction. Just because we can do it with microwaves. does not mean that he will buy it. The dimension of this aspect of alternative purchase competition is enormous and should impress upon us the importance of being closely tuned in to customer's needs.

Conclusions

We are here at a microwave symposium today which we prepared for and anticipated for two years. Are we sitting with our customers? I do not think so! We are, however, at a meeting of the American Ceramic Society. Many thousands of people from the ceramics industry who may be the customers that we are looking for, are at this event and are, perhaps, next door, but I doubt that they are in this room. This room is focused on our technology and not on our customers, and yet our customers are the only people that can make commercialization possible. They are next door—do we understand them? Are we talking to them and do we realize that satisfying them is the purpose and essence of our science?

Commercializing Microwave Systems: Paths to Success or Failure

R.F. Schiffmann

Abstract

Although the application of microwave power to industrial processing began 40 years ago, the actual adoption by processors has been disappointingly small. Approximately 90% of all installed microwave processing systems in the US are devoted to three applications—meat tempering, bacon cooking and rubber vulcanization. Among the reasons are: unrealistic expectations for the microwave process; lack of resources and capital to develop commercial processes, high cost of microwave processing equipment, and competitive conventional processes. In order to improve the opportunities for success the following questions must be addressed: (1) what is the real problem; (2) what are the true economics of the situation; (3) is the effect of the microwave system truly unique; would direct or indirect labor charges be reduced; what is the best combination of the energy sources for the new application?

Historical Perspective

It is now 50 years since Percy Spencer of Raytheon first conceived of a heating device which would use electromagnetic energy to cook foodstuffs (1), and shortly thereafter described the first microwavable food product—popcorn (2), and then a means for conveyorizing materials through a microwave oven (3). It awaited the development of higher power magnetrons and Cryodry's development in 1962 of practical choke systems for large tunnel openings for the use of microwave power for industrial heating to become useful. Predictions for the adoption of microwaves by industry were bullish while most technical people did not forecast a similar rosy future for domestic microwave ovens. After all, industrial heating processes could be revolutionized by this new technology whereas there would be little interest in a cooking appliance which would not brown, crisp or toast foods and would cost hundreds of dollars. How wrong we all were. Table 1 compares the domestic microwave ovens and industrial heating markets in the US in terms of numbers of units, their total selling price and the number of installed megawatts of output power.

TABLE 1. A COMPARISON OF DOMESTIC MICROWAVE OVENS AND INDUSTRIAL MICROWAVE HEATING SYSTEMS IN THE UNITED STATES

	Domestic ovens	Industrial systems
Number of units installed	150 million	<1000
Total sales value	$22.5 Billion*	$500 Million†
Installed megawatts	75,000‡	100

The following assumptions were made: *$150 per oven †$5,000/kilowatt installed ‡500 watts per oven.

Obviously, these numbers are only estimates since most of the industrial data is proprietary; however, they are useful as a measure of success or lack thereof. They indicate that most of the experts were wrong. They never realized how important convenience is to a consumer especially when tied to the great social changes of working parents, and more diversity among families which lead to changes in people's eating habits. Even the food industry did not anticipate the impact of the microwave oven since it did not really begin providing microwavable foods until there was already a 40% saturation level of ovens in our homes.

As for industrial microwaves, the 1960's and 1970's saw a number of major companies join the bandwagon that they thought would take them to success: Litton Industries, Bechtel, Varian, Raytheon, Armour-Cryodry. All invested a great deal of money in personnel, laboratory and pilot plant facilities and conducted hundreds of tests for companies in many industries. Many of the tests were quite successful but few commercial systems resulted from them. Even those which showed great promise, such as potato chip drying, chicken precooking, plywood veneer redry and donut processing did not last (see Table 2). That is not to say that they did not work, but for very different reasons each onc disappeared to be replaced by other technologies. Some of the reasons are spelled out below.

By the 1980's, of the companies mentioned, only Raytheon remained in the business with a small but successful industrial application group. All the other players were small companies with annual sales well under $10 million, and that is the situation which prevails today.

TABLE 2. MAJOR EARLY APPLICATIONS OF MICROWAVE POWER IN INDUSTRY NUMBER OF INSTALLATIONS

Potato chip drying	~100
Donut proofing	24
Donut frying	6
Chicken precooking	3

Not only are the number of installations limited but the number of current successful applications is very small as seen in Table 3.

TABLE 3. CURRENT INDUSTRIAL INSTALLATIONS OF MICROWAVE HEATING NUMBER OF INSTALLATIONS

Applications	# Installations	Average Power
Meat tempering	400	75 kW
Bacon cooking	25	200 kW
Rubber vulcanization	<200 (U.S.) 600 (worldwide)	18 kW
Pasta drying	20	75 kW
Pasteurization/sterilization	>10 (Europe)	25kW
Pharmaceutical drying	>25 (worldwide)	10kW
Plasma generation		3-4 MW total use
Miscellaneous including:		
Wood curing		
Foundry core drying		
Potato chip cooking		

Unless otherwise noted, the numbers are for the United States. All numbers are approximate since all data is considered proprietary. Plasma generation refers to systems used for etching, ashing, and chemical vapor deposition for the semiconductor industry, as well as diamond deposition. While not heating applications, they represent a significant area for the use of microwave technology.

The list is by no means complete—many more single custom systems exist. However, what is apparent is that at least 90% of the numbers of installations and installed power is the result of three major applications: meat tempering, bacon cooking and rubber vulcanization. This is not to say that many other applications are not technical and commercial successes, rather, they have not been adopted to any large extent in their respective industries and many that were installed ended up being shut down, and never perpetuated themselves.

The Elements of Success

Several authors have outlined the criteria for successful adoption of a microwave processing system. One of the earliest and most quoted is Jeppson (4) and this is summarized in Table 4.

On the other hand, Krieger (5) has suggested a different set of criteria as seen in Table 5.

Whereas Jeppson's list can be applied to many different technologies, Krieger's is focused particularly upon the benefits provided by microwave heating. However, even if these criteria are met, there may still be other barriers to the adoption or continued success of specific microwave systems. These may become the overriding factors in some cases.

TABLE 4. CRITERIA FOR SUCCESSFUL ADOPTION OF A NEW MICROWAVE APPLICATION.*

Product cannot be produced any other way
Cost is reduced
Quality is improved
Yield is higher
Raw material can be cheaper
The resulting product is more convenient to the consumer
Production throughput is increased †

*(M.R. Jeppson, 1968)*Added by this author

TABLE 5. CRITERIA FOR SUCCESSFUL ADOPTION OF NEW MICROWAVE APPLICATIONS.*

1. Energy transfer, not heat transfer
   a. High speeds
   b. Short line lengths (less floor space)
   c. Less scrap
   d. Energy conservation because only the material is heated
2. Deeply penetrating heat from molecular agitation
   a. Better quality from more uniform heating
   b. Greater yields
3. Selective heating based on dielectric properties
   a. Unique solutions to problems
   b. Ability to heat within a package
4. An electronic mechanism
   a. Automation
   b. Conntrol and data logging
   c. Environmentally clean and quiet
   d. Quick set up and fast changeover to new products

*(B. Krieger, 1993)

Timeliness. In order to be adopted, the technology must come at the right time. Microwave donut processing offered major advantages to bakeries but its introduction in 1969 was so early that once other factors occurred it rapidly disappeared. The common appearance of microwave oven in homes has created a familiarity which can be beneficial in this regard, although it may also raise unrealistic expectations for the processor.

Reluctance to be First. True innovation is frightening—it may lead to failure. Very few users are willing to be the first to adopt a new technology unless there is security in its success which can only result from a great deal of work, and it will still be a gamble. In 1972 Freedman (6) described a survey done for Raytheon relating to the acceptability of microwave heating equipment. The primary answer to the question of the conditions under which the company would purchase microwave equipment was: when the application is proven.

Need for Support Technology which may not be Available. The absence of ferrite circulators in early microwave systems permitted the destruction of numerous magnetrons and klystrons due to sudden mismatches during processing. This was a major reason for the failure of chicken processing. Rapid in-package pasteurization of foods is rapid only in how rapidly the food is brought to the pasteurization temperature—the hold and cooling times are the same—perhaps someone can invent a much faster cooling procedure so the overall time is shorter and the process and results will be more attractive.

Ability to Provide a Suitable Microwavable Product. Often, with processed products such as foods, it is necessary to modify the formulas in order to make them microwave compatible. As an example, the sharp decline in conventional dough proofing time of 45 minutes to only 4 minutes in the microwave proofer demanded the development of entirely new science of dough formulation, a task which took over one year. By comparison, the microwave hardware was simple and straight forward. Such research and development is expensive and time consuming and requires corporate dedication which must be founded on a solid cost benefit analysis.

Can Another Technology do the Job as Well as or Better than Microwaves? In the case of potato chip drying conventional frying systems were improved and were largely responsible for the disappearance of microwave systems. Sometimes innovation spurs further innovation in a competitive technology.

Understanding the Real Benefits. Speed by itself is rarely the single reason to use microwaves. What is needed is a careful analysis of those things which will improve profits—increased throughput, reduction in labor costs, greater yield, etc. Even these must be studied in depth. For example, yield improvements may sound good but is it commercially useful? Improving the yield of chicken when it is sold by the piece may prove not to be beneficial.

Comparison of the Cost of the Microwave System to the Conventional System. In determining the former it is important to evaluate both capital and operating costs and these should be in terms useful to the customer—cost per hour or per unit of product, for example. The capital cost analysis should account for the time over which the system is to be amortized and the cost of money over that period Operating cost analysis should include: utility, tube replacement, maintenance, cooling water and floor space plus any other unusual system demands. On the other hand, it is sometimes difficult to assess the cost of a presently operating system as a comparison.

Effect of the Microwave Process on the Rest of the Production System. If a microwave process is capable of tripling the output of a product, for example, then all other parts of the process, up and down stream, will have to operate three times faster This may lead to an unacceptable increase in capital expenditures.

Psychological Comfort of the Operators. This is perhaps no longer as important a factor today as when domestic microwave ovens were rare. However, an operator who is afraid of microwaves can do harm to himself or herself or the system.

Responsibility for Developing the System. This is perhaps the single factor which has led to the adoption of certain microwave technology and not others. The manufacturers of microwave hardware are expert at that and generally not in the users technology. Users tend to know little about microwaves (although this is changing in some industries). So, who does the R&D which may take one or more years and cost a great deal? In the three most successful applications—meat tempering, bacon cooking and rubber vulcanization—microwave hardware manufacturers saw these as areas of potentially large sales and became expert in the technology and marketing of those applications, often relying on outside expertise. In donut processing, it was a bakery mix and machinery manufacturer with its own sales force who identified a significant corporate business opportunity and so developed the technology internally so that it could sell microwave proofers and patented donut mix.

Conflicting corporate goals may inhibit this sort of effort. A microwave manufacturer developing a process applicable to ceramics wants to sell as many systems to as many ceramics manufacturers as possible. But limited resources may prevent it from doing so on his own. On the other hand, the ceramics manufacturer will generally only support the development of a proprietary technology which prevents the sale of microwave systems to its competitors. Perhaps a third party is required to spearhead the process development—a component manufacturer whose sales to the ceramics industry would be increased by the adoption of a new microwave process using that component. Sometimes these developments are best done with strategic alliances or assistance from the outside.

Still another factor for the user is the lack of skilled microwave experts. This is a small field. The teaching of microwave power applications at the graduate and undergraduate levels is sparse. So, a company wishing to develop microwave technology may not have the internal resources and will have to go to the outside to a consultant, to a microwave manufacturer or try to hire an expert—but does the project warrant this latter investment?

The fundamental question is "What is the business benefit desired from the microwave system?" If it is to sell many microwave systems that would be the goal of the microwave equipment manufacturer and might be the case of a process equipment manufacturer or perhaps a utility or a processor with many plants and/or processing lines, then the application has an opportunity to lead to many installations. Otherwise, it is likely that the systems will all be custom built and will usually be hybrids of microwave-gas oven or microwave-high velocity hot air oven, etc.

Presence of a "Champion" or "Hero". This is often an individual or small group in a company which forces the development and the adoption of an innovative technology against the desire to maintain the status quo. This can be one of the most important components of innovation.

Presence of "Linker". This is usually an independent expert knowledgeable in both microwaves and user technologies who can bridge the technology gap between microwave hardware and application. The Linker can speak both technology languages and assists in technology transfer. The Linker may also serve an R&D function sometimes in the form of an independent laboratory.

Improving the Likelihood of Success

In order to maximize the chances for success and avoid costly failures of microwave processing applications it is important that processors and equipment manufacturers realistically assess the potential applications. Of utmost importance are in-depth examinations with meaningful answers to the following questions: (7)

- *What is the real problem?* The most basic of all questions and should be followed by examining why other solutions to the problem have been rejected. The approach should be "Why Microwaves?" rather than "Microwaves!" The cost of microwave equipment is usually high and the investment of time and manpower to develop and install a microwave system can be several times that so the problem must be clearly understood and approached.
- *What are the true economics of the situation?* As noted earlier, this is sometimes hard to determine for the old system but can be easily calculated for the microwave equipment. As a rule of thumb, an industrial microwave

system usually has an installed cost of $4,000 to $7,000 per kilowatt (which includes the generator, applicator, conveyor and control system). The required amount of microwave power for the process is easily estimated by standard engineering heat balance equations. To illustrate a quick and dirty case for drying: one kilowatt of microwave will dry 2 to 2 1/2 lbs. of water in an hour. So if a drying operation needs to remove 250 lbs. of water in an hour, it will require about 100 kilowatts at a capital cost of $500,000 or more. It is apparent that capital cost is why this energy source is never used alone to dry huge quantities of product. Another thing to keep in mind is that in order to be practical the intrinsic value of the material should be high preferably, at least, several dollars a pound and the higher the better. That is why microwave drying of pharmaceuticals, where the major ingredients may be valued at thousands of dollars, is far more practical than food drying.

- *Is the effect of the microwaves in the system truly unique?* Microwave tempering and bacon cooking are successful because no other energy source can provide all the unique benefits that microwaves can. But, most often, the proposed systems could be done as well or nearly as well with another cheaper energy source.
- *Would direct and indirect labor changes on a given process be reduced bv the addition of microwave power?* With ever increasing labor costs and need for improved profits microwaves have a unique opportunity among other energy producing processes. They may make it possible to increase the throughput of a present process with little or no increase in floor space and a net decrease in labor on a pound of product produced basis. Increases of 50% to 100% above the present throughput are within reason. Sometimes there are other significant labor savings. A microwave sausage processing system installed over a decade ago saved 16 hours of cleanup labor daily because of better temperature control and ease of sanitation, plus eliminated the fires that occurred daily in the gas fired grill.
- *What is the best combination of energy sources for the new application?* Rarely are microwaves used alone but rather in combination with hot air steam, infrared or other heating means. This will usually reduce the cost of the system since these are usually far less expensive than microwave energy and the combination is often synergistic.

Final Thoughts

In a recent issue of *Scientific American* (8) Mark Weiser stated: "The most profound technologies are those that disappear. They weave themselves into the fabric of everyday life until they are indistinguishable from it." Hot air ovens are a good example—they are used daily in all industries for innumerable heating applications. We no longer say "Wow—a hot air oven!" And yet, that is still the attitude regarding microwave heating. It is still thought of mysterious and magical and, unfortunately, processors expect magical effects which are not likely to occur.

What is necessary for success is sober judgment, a thorough technical understanding and a great deal of laboratory and pilot plant work to determine not only that the process is satisfactory but that all process parameters have been examined. This should be followed by a thorough evaluation of the economic parameters to ensure that there is a positive effect upon the bottom line. Finally, everyone from the potential line operators to the vice presidents of sales and marketing should be interviewed to be sure that no hidden flaws will rear their ugly heads after major investments have been made. Even then, the likelihood of adoption of the microwave process may not be great but if adopted it is likely to be successful.

REFERENCES

1. P.L. Spencer, "Method of Treating Foodstuffs", U.S. Patent 2,495,429 (1950).
2. P.L Spencer, "Prepared Food Article and Method of Preparing", U.S. Patent 2,480,679 (1949).
3. P.L. Spencer, "Means for Treating Foodstuffs" U.S. Patent 2,605,383 (1952).
4. M.R. Jeppson, "The Evolution of Industrial Microwave Processing in the United States", J. Microwave Power Vol 3(1), pp 29-38(1968).
5. B. Krieger, "Why Use Microwave", Cober Food Processing Seminar (1993).
6. G. Freedman, "The Future of Microwave Power in Industrial Applications", J. Microwave Power, Vol 7(4), pp 353-365 (1972).
7. K. Bedrosian, "The Necessary Ingredients for Successful Microwave Applications in the Food Industry", J. Microwave Power, Vol 8(2), pp 173-178 (1973).
8. M. Weiser, "The Computer in the 21 st Century", Scientific American September, (1991).

An Industrial Application of Microwave Heating: Continuous Drying of Inorganic Salts

J.I. Ortigosa

Abstract

A new application of microwave industrial heating is presented. The objective of the project is the design and manufacturing of an inorganic salt drying machine to be used by an industrial chemical manufacturing company in Spain.

Real-world industrial economic requirements, including final quality of product, energy efficiency of the process, and throughput rates are an important condition for the project, and impose very important restrictions upon the design criteria applied to the machine. A mix of direct incidence and multi-mode cavity was selected as the energy application method, and a modular system was developed to allow the adaptation of the machine to variable workload. Safety of personnel operating the machine from microwave leakage was also a consideration for the design, as all the regulations in industrial environments must be applied to the design to permit its operation.

A complete example of industrial application design and manufacturing is presented, covering issues as project specification, microwave and industrial engineering calculations, physical implementation of theoretical designs and final unit qualification and testing results.

Project Requirements

A machine to partially dry a water solution of inorganic salts is needed by an industrial chemical manufacturer in Spain. The water content of the product must be reduced from about 26% of the weight at the input to a maximum of 5% of weight at the output. There are several other requirements to be followed by the design: a continuous process is desired, as the machine will be fed by the output of another machine in the factory; the size of the drier must be minimized due to space restrictions in the production plant; the quality of the final product must pass several controls on crystal size, density and solubility on water; industrial and economic constrains (the investment required is about €250,000, and the desired payback time is 5 years) require a production throughput of 270 kg hour at the input for the process to be profitable enough for the company.

Microwave Energy Applicator

Selection of Heating Process

Conventional and microwave heating techniques were considered for the project. Finally microwave heating was selected because of the following reasons:

- *Simplicity of the process:* The machine will handle the liquid product in a very easy way, therefore simplifying the mechanical design.
- *Quality of the final product:* As the drying process takes place in a very short time and without the use of containers or hot drying surfaces, the quality of the final product is excellent, according to all of the criteria the customer is interested in.
- *Process efficiency and cleanliness:* the designed microwave drying process was tested to be efficient enough to meet the customer's demands, and the secondary exhaust and cooling systems follow the customer's current environmental friendly behavior policy. The machine was also designed for easy and short cleaning and setup cycles, allowing high efficiency in production time.

Applicator Design

The application of the energy required for the evaporation of the water is made via microwave radiation. The selected frequency, 2,450 MHz, is allocated in Spain and in most of the countries in Europe for IMS (Industrial Medical and Scientific) applications, and CIM D'OR has a long experience designing industrial microwave equipment at this frequency.

Due to the fact that the incoming solution has a very good behavior in terms of microwave energy absorption and that the absorbed power can be made large, the designed application consists of a stainless steel rectangular-section long cavity. The material to be dried circulates on a V-shaped teflon moving belt and is sequentially exposed to the microwave energy generated by high power magnetrons through rectangular waveguide apertures. A schematic of the energy application method is presented in figure 1.

Cavity Resonance

Although cavity resonance was not the main objective of its design, more than 125 unloaded modes are resonant in the 50-MHz frequency band around 2,450 MHz, where the magnetron manufacturer allocates the energy generation. We have experimentally found that a high number of modes contributes to an extra degree of power absorption by the material and increases the uniformity of the drying. The main objective of the cavity is to contain and conduct the vapors generated in the drying, which are mainly composed of evaporated water with stain the harsh working environment, while an exhaust system was designed to conduct the gases to the factory's main extraction lines.

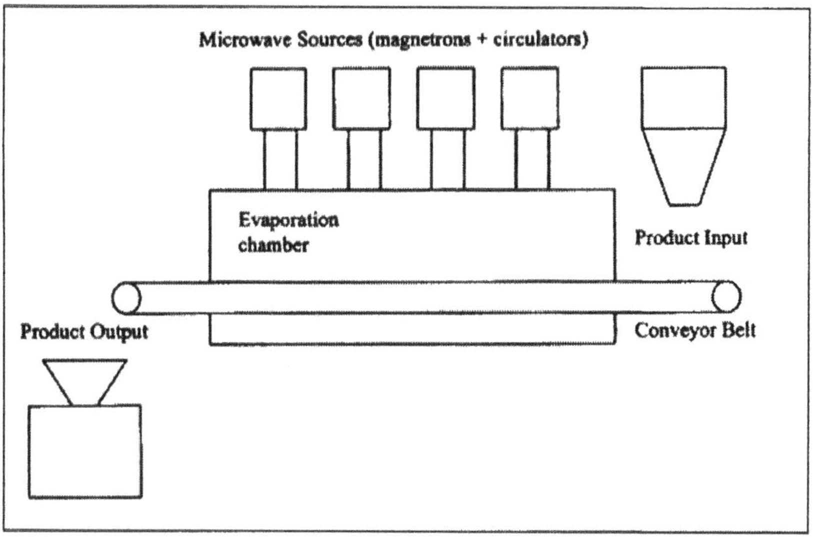

Figure 1. Schematic of the industrial drying machine.

Machine Design and Construction

Determination of Required Microwave Power

Starting from the customer's production throughput requirements, and taking into consideration overall and specific efficiency rates, a determination of the required microwave power must be performed, in order to check the feasibility of the operation and to estimate manufacturing costs.

The basic process to be performed by the machine is the evaporation of the water included in the input mixture. Therefore, the amount of water to be processed will be the starting data for the calculations. The desired production load of wet product at the input is 270 kg/hour with a 26% of weight of water, accounting for a water working load of 70 kg/hour. The temperature of the input product is 70°C. The corresponding output rate, in dry product, is 210 kg/hour including a total of 10 kg/hour of water (4.7% of weight).

The energy spent in the evaporation process can be roughly found as the energy used in the heating of the water from 70°C to the evaporation temperature of 100°C (the machine works at atmospheric pressure) plus the heat required for the evaporation of the water. The total power requirement is 32,500 kcal/hour, equivalent to 37.81 kW. The evaporation process has been considered to have an efficiency of 85% (this efficiency coefficient would cover for any extra power requirement not accounted for, as the heating of the dissolution, etc.), and this increases the total microwave power requirement to 44.48 kW.

Selection of Microwave Sources

The available sources of microwave energy specifically designed and manufactured for industrial heating applications at 2,450 MHz include magnetrons with output powers ranging from less than 500 W to more than 25kW.

The election of the source to be used must take into consideration multiple facts, like the number of units to be used to obtain the required power, the complexity of the cooling and powering systems, and the costs of the isolators required to apply the microwave energy to the process, for instance.

The final decision was to use 1.9 kW continuous wave magnetrons with 2 kW water cooled ferrite isolators as the basic microwave source block. A total of 26 units were used in the machine, accounting for a more realistic total microwave power of 45.5 kW (26 times 1.75 kW).

The selected magnetrons are fed with full-wave single phase rectified power sources that can be individually switched on and off. This disposition allows a power control going from 4% to 100% of maximum power in 4%-increment steps, which is perfectly enough to accommodate the needs of the customer. The use of a large number of microwave sources is also convenient in terms of protection of the production against failures in the equipment, because the stop of one or two of the sources would only imply a small reduction of the production throughput.

Physical and Functional Design

Production and economic considerations have been incorporated to the mechanical design of the machine. The machine has been divided into three sections, each of them with one evaporation chamber. This structure allows modularity of the system and permits future increases of the installed microwave power. Fully accessible evaporation chambers allow for an easy and fast cleaning process. A total of 12 source blocks (each of them composed by a magnetron and a circulator) have been installed in the first section for a faster heating of the incoming product flow. The second section is equipped with 8 source blocks and the third section uses 6 source blocks, although each of them would allow the installation of up to 12. As the efficiency of the sources irradiating the product from the top was experimentally found to be better, the maximum possible number of magnetrons has been mounted in that position.

Safety of Operation

Current Spanish and European regulations on safety of operation of industrial machinery have been applied during the design. The criteria of a maximum microwave radiation leakage level of 2.5 mW/cm^2 at a distance of 5 cm away from the machine was also applied to the design and checked during the final tests of the machine. These minimum radiation levels have been accomplished with the use of microwave filtering at the input and output product flow openings, and also with the use of microwave seals and $\lambda/4$ traps along the edges of the evaporation chamber openings.

Final Product Quality

After the real-size trials performed by CIM D'OR, the quality of the final product was excellent, according to the customer's requirements. Figures 2 and 3 show photographs of the laboratory trials of the drying process. A sample of the inorganic solution to be processed by the drying machine and the same sample after being exposed to the microwave drying process are shown.

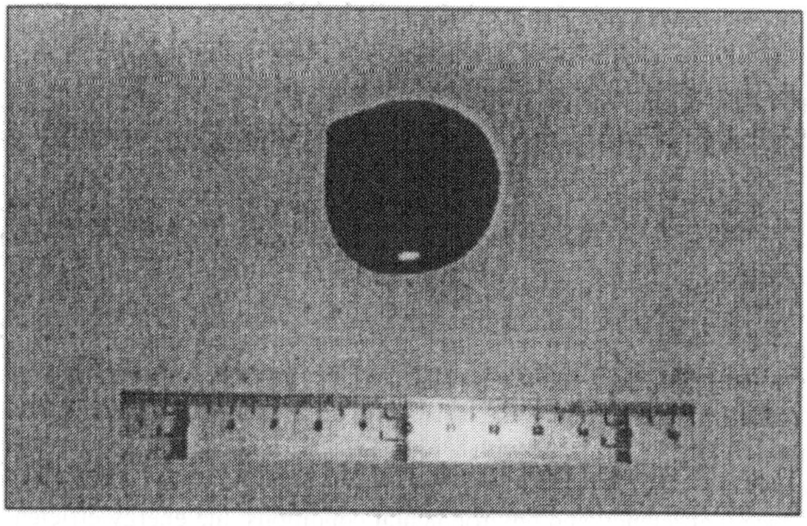

Figure 2. Sample of the inorganic salt solution before drying.

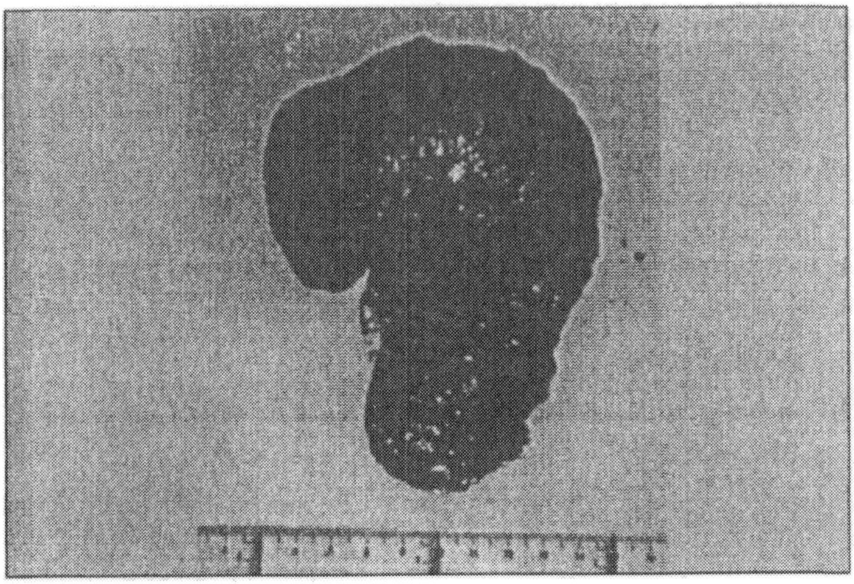

Figure 3. Sample of the result of microwave drying of inorganic salt solution.

Conclusions

An application of microwave heating to the drying of inorganic salts has been presented. An industrial drying machine was designed and manufactured according to the customer's specifications, taking into account key technical, economical and safety aspects. The most important features of the new drying machine are:

- High quality of final product: the most restrictive conditions on product quality and process control were met, allowing the development of a new business for the customer.
- Industry-scale production capability: a production throughput of 270 Kg/hour was accomplished, with a high level of reliability due to the use of a large number of microwave sources.
- Process efficiency and environmental conditions: the energy efficiency and the environmental behavior of the microwave-based drying process was successfully tested against the requirements of the customer.

Scaling Up the Microwave Firing of Ceramics

F.C.R. Wroe

Abstract

EA Technology, through a comprehensive R&D program, is developing new microwave furnace technology focused on the ceramics processing industries. Using a combination of computer modelling, experimentation and feasibility studies, EA Technology has developed processes and procedures for firing large ceramic components. This has achieved the goal of producing uniform microstructures and low thermal stress by careful control of the firing cycle.

Introduction

Energy efficiency surveys [1] carried out in the ceramic industry over the past five years have identified that the specific energy consumption for ceramic components is often more than ten times the theoretical minimum energy required to fire the product. This is due, in part, to heat losses from furnaces, although much work has been carried out to minimize these.

There are two main reasons for the specific energy consumption figures being so high. The first cause is fluctuations in temperature uniformity throughout the kiln. The second is due to the physical nature of ceramic components, whether they are bricks, refractories, industrial ceramics or pottery. When a temperature profile is created within a ceramic component, in either the heating or cooling cycles, it results in thermal stresses being created within the components. The stresses and temperature profile can lead to distortion, an inhomogeneous microstructure and ultimately fracture of the component. To minimize thermal stresses, heating and cooling cycles tend to be long, often with holds at critical temperatures, for example at 574°C in clay-containing bodies due to the alpha to beta quartz inversion.

The aim of the present work program at EA Technology is to investigate the firing of ceramic products such as bricks, pottery, refractories and industrial ceramics, using advanced processing techniques to produce and maintain uniformity of temperature throughout the components and kiln environment.

EA Technology has developed and patented processes and procedures for firing large ceramic components with microwave-only heating [2], and more recently microwave-assisted heating. It has been the development of microwave-assisted firing which has achieved the goal of producing uniform microstructures and low thermal stresses, through the careful control of the firing cycle, using an economically viable process. EA Technology had demonstrated the feasibility of achieving this via microwave-only firing, but as Sections C & D will demonstrate, in terms of energy costs and process control, microwave-assisted firing is the more economically viable process.

Problems Associated with Current Firing Techniques

The manufacture of ceramic products such as bricks, pottery, refractories and industrial ceramics involves heating the ceramic to temperatures in excess of 800°C, a process known as firing. Firing typically involves the use of either direct or indirect gas flames, or electrical resistance elements to radiate heat. Gas firing accounts for 88% of the total firing market. Heat is first transferred to the surface of the ceramic product, primarily by radiative thermal transfer, with minor contributions from thermal conduction and thermal convective transfer. This heat is then passed to the center of the ceramic product by thermal conduction.

Ceramics are poor thermal conductors, particularly if they are not fully dense. Ceramics are usually only 50% dense after the processing stages prior to firing, and only begin to significantly densify at the maximum firing temperature. Therefore when they are being heated to the maximum firing temperatures they are very poor thermal conductors, and can in essence be regarded as thermal insulators.

The poor thermal conduction can result in the center of a ceramic component being at temperatures significantly lower than the surface. The same is true for components which are placed in stacks during firing, as the components at the center can be as much as 100°C lower than those at the surface close to the gas flames.

The physical nature of ceramic components is such that if the temperature difference is too great, cracking and distortion will occur and the product is rendered worthless. To avoid this situation in practice, the rate of heating must be limited, and protracted holds are used at high temperatures to assist with temperature uniformity, therefore firing is usually a slow process. For example, in the brick and refractory industry, it is relatively common to have firing cycles lasting longer than one week. The length of the firing process also results in significant heat losses from the kiln. Consequently there is scope for improved energy efficiency in the firing of ceramics.

Microwave Firing

Microwave processing has been offered as an alternative to conventional kilns. The principal benefit offered by microwave processing is that by depositing the energy directly within the component problems of heat transfer throughout individual components and throughout the kiln can be overcome.

However, despite these claims, recent development work at EA Technology has highlighted a number of potentially serious drawbacks to the process of microwave-only firing which have been the cause of numerous failed attempts to fire realistically sized components resulting in highly non-uniform heating and thermal stress fracture. This has led to significant skepticism within the ceramic industry as to the value of microwave firing. In theory, a material in a sufficiently uniform electromagnetic field can be heated evenly and quickly.

In practice, however, difficulties arise when heating a body to high temperature with only microwave energy. As the temperature of the body rises heat is lost to the surroundings. This then causes an inverted parabolic temperature distribution across the specimen. As the temperature of the body increases, the heat losses become greater, and the magnitude of the temperature difference increases.

Many ceramics, such as alumina, are poor absorbers of microwave energy at low to intermediate temperatures and do not heat quickly in a typical microwave oven. However as they approach a critical temperature, they very rapidly begin to absorb substantial amounts of microwave energy and an effect known as thermal runaway can occur. Furthermore, as the amount of microwave energy absorbed is very dependent on temperature, the surface, from which heat can be lost in a microwave-only situation, will rapidly become cooler than the center of the components, therefore the center will start to absorb the microwave energy in preference to the surface, and the thermal runaway, in effect, becomes self propagating. In extreme cases this can cause the center of the ceramic component to melt while other regions remain at ambient temperature.

The main causes of thermal runaway in ceramic components are:

- Heat losses from the surface of the ceramic component,
- Ceramics are essentially non-uniform in their material properties, which can lead to differential absorption of the microwave energy, and
- Non-uniformity of the distribution of microwaves within the kiln.

Microwave-Assisted Firing

The aim of the current work at EA Technology is to combine radiant surface energy, supplied by conventional gas firing technology, or electric resistance element heating with microwave energy to create conditions of uniform volumetric heating of the product which is essential, as the previous sections have outlined, to produce a uniform temperature throughout the ceramic.

Over the past six months, in conjunction with interested UK and European ceramic companies, we have established that the creation of a uniform temperature will lead to the following benefits identified in the sections below. Statistics do exist on the cost of many of these factors, for example, thermal stresses within the pottery industry can lead to 30% scrap [3] These statistics will assist in making the benefits of this process more quantifiable, and assist with its uptake by the industry.

Thermal Stresses

The creation of a uniform temperature profile throughout the component will reduce the thermal stresses within components and reduce the percentage of scrap due to stress fractures or non-uniformity of microstructure.

Energy Consumption

The total energy consumption will be reduced due to the creation of uniform volumetric firing conditions. Currently the surface has to be heated much higher than the body in order to teach the minimum required temperature. This is even more significant when one considers that most ceramic components are fired in multiples to maximize throughput.

Figure 1 shows the results of computer modelling combined with careful experimentation, the energy costs associated with firing a 95% commercial alumina composition under conventional, microwave-only and microwave-assisted firing conditions. The maximum heating rate obtainable with conventional and microwave-only firing (without the use of susceptors) was 5°C per minute, while firing rates of 35°C per minute were achievable with microwave-assisted firing.

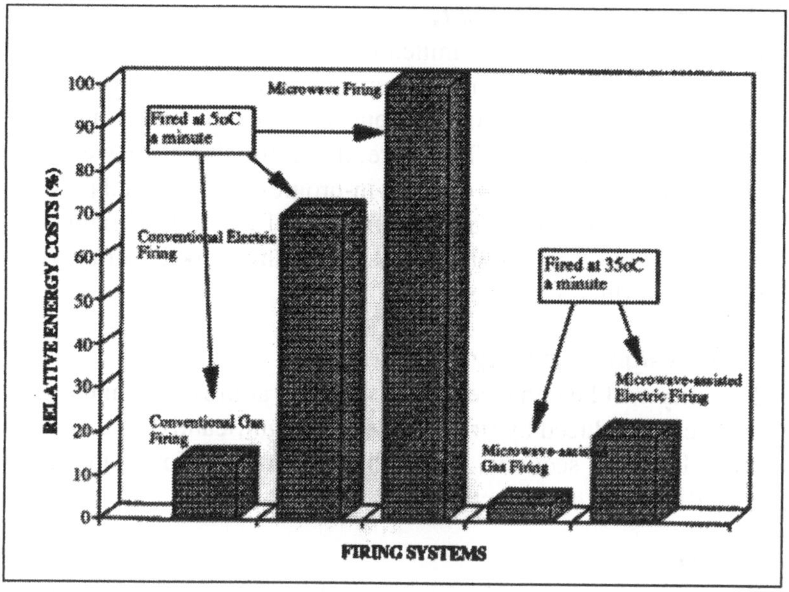

Figure 1. Relative energy costs for the firing of alumina cylinders.

Fast Firing

A uniform temperature profile will permit greater use of fast firing techniques, which are currently limited to simple shapes, which have low aspect ratios, i.e. are wide and thin. One demonstrated example of faster firing has been work carried out on earthenware mugs. The maximum firing rate achievable, consistent with obtaining acceptable quality, in the same conventional kiln was 5°C per minute. In the hybrid kiln we achieved in excess of 20°C per minute, to the same quality standard.

Production Scheduling Flexibility

At present production planning is limited in flexibility because it is not possible to plan to produce the exact current requirements, due to the long firing times and the precise nature of the stacking of components within the kiln to maximize heat transfer and product throughput. Therefore, it is difficult to achieve just-in-time manufacturing and large amounts of work-in-progress are required to satisfy customer requirements. Microwave-assisted firing will allow the flexibility needed to achieve just-in-time manufacturing and reduce the work-in-progress, thereby improving the cash flow of the company.

Component Design Flexibility

Design flexibility will be enhanced, as designs are at present limited because of the thermal stresses induced by firing. These can significantly weaken items with large or irregular cross sections, and high aspect ratios. For example, sanitary ware is limited to 2 cm cross section.

Kiln Furniture

We are beginning to design kiln furniture which balances the requirement to absorb the minimum of microwave energy consistent with the need to prevent thermal shock. This will contribute to energy savings as kiln furniture, at present, absorbs significant amounts of energy. In order to do this, we are continuing our extensive program into identifying microwave-material interactions, because if you do not have a good understanding of these, it is very difficult to design materials for specific processes. In kilns where single-layer firing is adopted, this is eliminated.

Simplification of the Firing Schedule

The firing schedule will be simplified and shortened due to the elimination of holding periods to ensure that the components are all at the same temperature, and the requirement to re-fire due to non-uniform heating will cease to be necessary.

Environmental Benefits

There are also environmentally-friendly aspects to the introduction of fast firing in the clay using industries, apart from the energy savings. Approximately 80% of the companies in the ceramic industry manufacture clay-based products, for example, pottery, house bricks and clay pipes. Clay contains significant quantities of impurities which can be emitted during firing. One example of this is fluorine. In Germany, under the T.A. Luft 1986 Act fluorine emissions are limited to a maximum of 5 mg/m^3 (<0.5 kg/hr). Recent surveys in the UK show that the average fluorine emission rate from kilns firing clay products is 43 mg/m^3

[4], with the maximum recorded emission of 106 mg/m^3. Less than 3% of kilns monitored would conform to these regulations. In the less stringent UK regulations which were first introduced on the 1st January, 1985, and were followed by the Secretary of State for Heavy Clay Industries Guidance Notes in April 1992, which allow 10 mg/m^3 of fluorine, only about 7% of kilns would conform to these regulations.

Fluorides have been singled out because it is accepted that there is technology, scrubbers, that can deal with the problem. These scrubbers contain calcium carbonate, but this presents a further waste disposal problem. In addition, the long term use of scrubbers has still to be evaluated. Secondly, fluorides emitted from firing kilns have been identified as the cause of many local complaints regarding damage to plant life, particularly grapevines in Germany where most of the studies have been performed, and to property, for example, they have caused etching of glass in windows.

Fluorine emissions would be substantially reduced with the use of microwave-assisted firing for two reasons. Firstly the amount of fluorine released is directly related to the time spent above 800°C [5]. The faster throughput will mean that the total time spent will be reduced by up to a factor of 5. Secondly, 90% of the fluorine released is due to water remaining in the clay, and the use of hydrogen-containing fuels. Microwave energy will remove the water prior to the critical temperature of 800°C, and at least 25% of the energy will be microwaves and hence not hydrogen-containing.

To summarize, for the above reasons it is reasonable to expect that the use of microwave-assisted firing in the clay-based ceramic sectors will reduce the fluorine emissions to below the statutory minimum, without the use of scrubbers.

The cost of a scrubber for a kiln is approximately £250,000 with a further £100,000 for continuous monitoring equipment. The estimated cost of retrofitting microwave power supplies to a similar size of kiln is approximately £200,000, and that will stop the problem at source, not attempt to deal with it once it has happened.

Uniformity of Microstructure

In the ceramic industry, the use of radiant heating and the requirement for a high radiant heat flux due to the poor thermal conductivity of the ceramic components or powders, results in higher temperatures at the surface and, in addition to uneven firing, can produce inhomogeneous microstructures. If the ceramic is carbon-containing this results in the surface being effectively sealed due to the faster firing rate, prior to the center being "burnt-out". In the industry this is known as "black-coring" and occurs widely throughout the sanitaryware, brick, pottery, refractory and refractory castable sectors.

Much research has been done into the reasons for the black core and methods of avoiding it's occurrence, with no effective success. The most common partial solution has been to ensure adequate distance is maintained between the radiant heating source and the products. This is not effective in all cases, particularly when alkalis are present, which enhance the firing of the surface at lower temperatures due to the creation of a flux. In addition, it reduces the energy transfer efficiency, especially if there are apertures in the kiln. If the black core appears close to the surface it will detract aesthetically from the product.

It may also cause cracking if the component is in use at high temperatures, and the surface becomes damaged enabling oxygen to enter and cause a build up of carbon monoxide, or dioxide pressure, which can crack the component.

Application to The Calcination of Powders

The establishment of a uniform temperature profile will enable the uniform calcination of powders, either reactive calcination or removal of chemically bonded water or carbonaceous matter. Powder has a worse thermal conductivity than most ceramic components, with the exception of thermal ceramic insulation materials, and can act as an thermal insulator, preventing, or minimizing heat transfer to the center of the container or powder bed. EA Technology has patented the Rotek calciner, which tackles this problem using induction heating of integral mixing blades, but this is limited in temperature to 800°C, and many powders are calcined above this temperature. It also does not provide volumetric heating.

Capital Equipment Cost And The Retrofit Option

The cost of capital equipment is also of prime importance when evaluating the potential benefit offered by a new manufacturing process. In the UK, for a new process to be widely adapted, the return on capital investment should be high, usually a minimum of the return gained on long term high interest accounts, and if the project is considered to carry significant risk, the return must be much greater.

In addition, the pay back period, which can be thought of, in the case of microwave firing, as the time needed to recover the cost of the original investment in microwave kilns from the process savings due to the faster throughput and lower energy consumption, in the UK should be less than 3 years, and preferably less than eighteen months.

Furthermore, it has been demonstrated on a laboratory scale that it is possible to design and install (i.e. retrofit) microwave equipment on most existing kilns, thereby reducing the risk and capital cost involved in designing and installing a completely new kiln system based entirely on microwave firing.

An example of the capital cost associated with installing microwave equipment on an existing furnace as a microwave-assisted retrofit option follows. One recent modelling and experiment exercise estimated the total amount of microwave power required to retrofit a modern gas fast firing 5,860 kW (200 therm) tunnel kiln as 5% in order to achieve uniformity of temperature and increased throughput (The critical temperature range was 500-900°C in the heating cycle). The cost of industrial scale microwave capital equipment is estimated to be between £1785 to £2,500 (£1=US$1.42) per KWh [6], with the higher cost associated with the lower power equipment. This corresponded to an approximate capital cost of £523,000, and gave a pay back time of 13 months, which would have been acceptable to the potential customer.

The equivalent capital cost associated with installing microwave equipment on a microwave-only kiln would be in excess of £10 million. In addition, using the microwave-only option would normally require new kilns to be built, further adding to this cost.

Therefore it can be seen that microwave-assisted firing is the most economically viable microwave firing option. This is also true for microwave drying where microwave energy is usually used along with a conventional form of heat energy, to remove the last few percentages of moisture from a material, at a much reduced capital cost [6].

Summary

In microwave-assisted firing the volumetric heating provided by the microwaves heats the components, while the more conventional radiative heating provided by gas flame or electric resistance heating elements minimizes heat loss from the surface of the components, by providing heat to the surface and its surroundings. This prevents the creation of the temperature profiles associated with both conventional and microwave-only firing.

As thermal stresses are also minimized due to the improvement in uniformity of temperature, components can be heated much more rapidly. Therefore the major energy and process savings will be in the reduction in time associated with the firing cycle, and the increased flexibility as the microwave energy can be directed to where it is required most; or even switched off. An example would be in the firing of clay-based bodies: the critical temperature is 574°C, where cracking on heating and cooling can occur due to the quartz inversion. Microwave energy could be used in this example in either, or both the front and back end of the kilns to provide a uniform temperature profile where it is required the most.

Finally, and possibly most importantly for the future uptake of microwave firing, it has been demonstrated that it is possible to design and install microwave power supplies on existing kilns.

REFERENCES

1. UK Dept. of Energy, Energy Efficiency Office, Energy Efficiency Technology Support Unit (ETSU), Best Practice Program, Guide No. 8, The Firing of Ceramic Tableware, Building 156, Harwell Laboratory, Oxfordshire. UK. November 1990.
2. Brandon, J.R., Samuels, J, and Hodgkin, W.R., Microwave Processing of Materials III, MRS Vol. 269, p 237-244, eds. Beatty, R.L et al, 1992.
3. Twentyman, M.E., Some Factors Affecting the Refire Dunting of Vitreous China: A Review, Brit. Ceram. Trans. & J, Vol. 89 (2), p67, 1990.
4. Amieson, A., Stack Emissions in the Brick Industry, Ceram. Indust. p 61-64, March 1992.
5. Pollutech, Effect on the energy efficiency of tunnel kilns....emission control systems, phase III: final report (OSQ 8500321). Microlog ref. 8703998.
6. Schiffman, RF, Microwave and Dielectric Drying, Chapter 10, The Handbook of Industrial Drying, Institute of Electrical Engineers, U.K. 1992.

Microwave-Assisted Processing—From Laboratory to Production

S. M. Bond

Abstract

Towards the end of the 1980's extensive research efforts were applied to the microwave firing of ceramics. Whilst success was found in this area, technical and economic reasons precluded the use for full-scale ceramic production. The use of susceptors to achieve some form of temperature balance between the surface and the body of the samples appeared to be an uneconomic option. The beneficial combination of radiant (gas or electric elements) and volumetric (microwave) heating in the same furnace, at the same time, was originated and patented. Market forces dictated that, for large scale production, the radiant heating source was gas. Following an extensive development program, microwave-assisted gas firing kilns are now a commercial reality. Laboratory microwave-assisted furnaces are now in use for a diverse range of applications.

Development of Microwave-Assisted Gas Firing

This development was undertaken over a eight year period and involved funding from equipment suppliers, UK electricity industry and government. Forty two ceramic manufacturers from UK, USA, Japan, Germany and other European countries, provided production components for microwave-assisted trials.

The initial furnace developed is a batch operation and can process products up to a maximum of 1650°C. The maximum load is 10kg, with a working zone of 222 x 200 x 300mm. As with all of these furnaces, it can be operated purely as a gas furnace as well as in microwave-assisted mode, hence providing a direct comparison. This furnace is still in operation with over seven years of almost continuous use.

Following the success of the initial furnace, a pilot production shuttle furnace was designed and constructed. With a working zone of 1 x 1 x 1M and loads of up to 1000kg, it was able to demonstrate the advantages of microwave-assisted processing on a meaningful scale. Microwave energy is supplied by two separate generators with a combined maximum output of 85kW. The frequency of the microwaves used is 896 Mhz, being a common industrial standard. The furnace has been running virtually continuously since 1995.

A pilot production microwave-assisted tunnel furnace has been in operation at EA Technology since 1996. Eighteen, high turn down ratio, pulse fired gas burners are fitted to the furnace. The microwave power at 896Mhz is provided via two generators with a total output of 126kW. Fifteen meters long with twenty three cars each capable of carrying a load of 200kg. Working with a twenty four hour shift pattern, the microwave-assisted tunnel furnace has processed products on a pilot production basis, often exceeding 50,000 kg per week. A transmitted temperature monitor system, developed by EA Technology, provided real time measurements of the products core temperatures. This ensured the maximum advantage from each trial period, which was normally twelve days duration.

The tunnel furnace has proven conclusively the benefits and viability of microwave-assisted processing on a continuous basis. It has also demonstrated the scale up from laboratory, through batch, to continuous firing is feasible and predictable.

Product Types

Working with international ceramic manufacturers, the range of products has been extensive. In most cases the green products were removed directly from the production line, thus providing an excellent comparison. The product range included heavy clay pipes, sanitarywave, tableware, advanced ceramics, bricks, refractories, pigments, abrasive components, foundry filters, wall and floor tiles.

Figure 1. Initial microwave-assisted gas firing furnace.

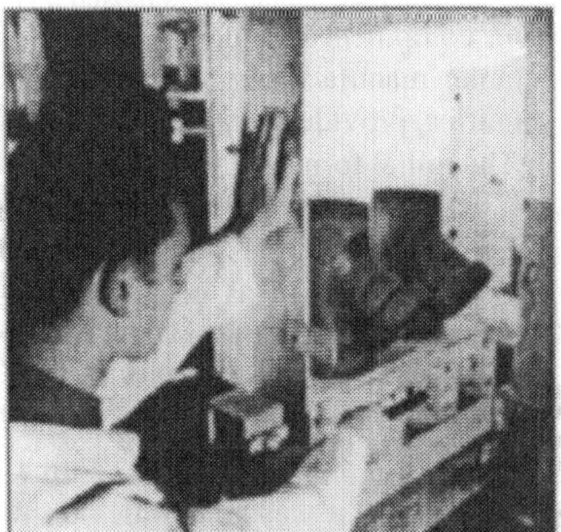

Figure 2. Pilot production microwave-assisted furnace.

Increase in Throughput

One of the prime advantages that microwave-assisted processing brings to ceramic production is the demonstrated throughput increase. Essentially ceramics cannot be processed faster, when using conventional heating, because of the temperature differential from the surface through the body. These are minimized by the combined radiant/volumetric heating approach and subsequently the products can be processed faster. Up to three times faster than conventional has been demonstrated for a wide range of products, without detriment to the product quality. Figure 5 demonstrates the faster throughput, in this instance using the continuous microwave-assisted furnace.

The energy lost from a furnace is a function of time, if products are being processed at significantly faster rates than the heat loss per unit of production is reduced. An independent energy audit, undertaken by Ceram Research, UK, was carried out on a 862 kg batch of clay bricks. The gas only firing required in excess of 43 hours to produce quality products. Microwave-assisted gas firing accelerated the process to 14 hours 15 minutes. The audit demonstrated a 39% reduction in energy use, with the efficiency of the microwave-assisted furnace being 66% higher than the gas mode.

Overall, for all the products types processed, the average energy saving exceeded 35%, however, energy cost savings are dependent on the ratio of gas/electricity costs. In the UK, the cost ratios are such as to still offer cost saving for most ceramic types.

Figure 3. Waveguides for the microwave-assisted tunnel furnace.

Figure 4. Examples of products processed using microwave-assistance.

HF Reductions

Emissions from production furnaces are closely monitored and strict levels are maintained. Hydrogen fluoride is emitted from clay-based products above 850°C. It has been demonstrated for both batch and continuous microwave-assisted furnaces that the reduction in emissions is pro rata with the increased speed of throughput.

Production Microwave-Assisted Furnaces

As with all new technology the uptake is often undertaken under strict confidentiality clauses. To this end, it is impossible to supply the exact location and companies involved. However, production furnaces have and are being supplied in conjunction with traditional furnace manufacturers.

In the field there are batch and continuous microwave-assisted furnaces, with several awaiting build at the present whilst engineering studies and designs are completed.

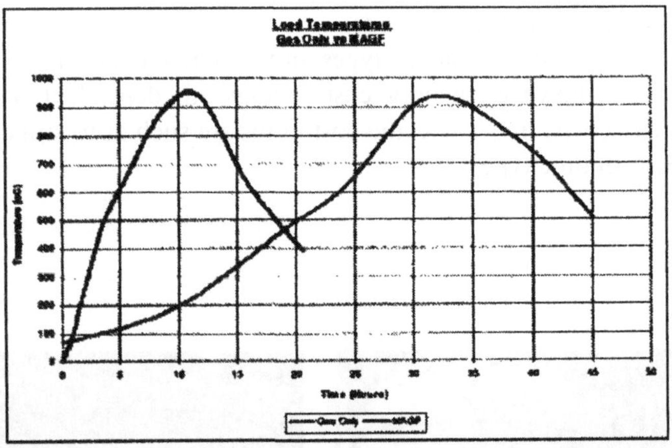

Figure 5. Comparison of microwave-assisted and gas only for heavy clay products energy use.

The engineering required to move from the controlled development furnaces, at EA Technology to production situations has been ongoing. Where unforeseen problems have occurred the availability of pilot production equipment on our site has been essential to their solution.

Microwave-Assisted Processing

Although a concentration on the sintering of ceramics has permitted the early work in this area, the range of uses is now expanding to many other areas, both on a laboratory and production scale.

Figure 6. Rapid drying using microwave/RF-assistance

Figure 7. Expanded slate.

Rapid Drying

Using the basic concept associated with the sintering of ceramics, rapid drying has now been developed using microwave or RF assistance. The speed of throughput is greatly enhanced with a good quality product.

Expanded Slate

Recently EA Technology processed almost 100 tons of slate, expanding it may times for uses in the building and land reclamation areas. This has been so successful that a production unit is planned in the near future.

Microwave-Assisted Laboratory Furnace

EA Technology have been using electric elements/microwave furnaces for almost a decade, to evaluate and develop the technology. Temperatures up to 1700°C are obtainable and can be fitted with dilatometers, fluidized beds, mass balances and controlled atmospheres, all of which can be used in the microwave-assisted mode.

The range of products evaluated successfully, using laboratory microwave-assisted equipment, is constantly growing, including magnets, metal powder components superconductors, electroceramics and solid oxide fuel cell plates.

Summary

The development and production use of a new technology has been achieved. Where difficulties have arisen, they have been systematically evaluated and the solutions pre-tested on our own equipment.

The technology is now expanding into new and existing areas and will become a standard technique. Laboratory furnaces are being used to evaluate benefits and products within a wide range of companies.

Parallam—New Structural Wood Composite

M.T. Churchland

Abstract

The work presented here represents the accomplishments of many talented people from many disciplines. Many of us have devoted our careers to the Parallam® project and we have much more to do before we can rest and reflect on our success lest they cease to be successes. We are in the middle of the difficult task of converting a technical and marketing home run into a profitable business. By July 1987, MacMillan Bloedel had spent over 40 million dollars on Parallam research and development. In July 1987, MacMillan Bloedel committed a further 97 million in capital. These funds were used to build two commercial plants—one in Vancouver, British Columbia and the other in Colbert, Georgia. A third plant has been built in Buckhannon, West Virginia and will start in the third quarter of 1995.

Product

Parallam® PSL. is made from long veneer strands. A photo of strands is shown in Figure 1. Standard plywood technology is used to produce rotary-peeled veneer, typically 102 inches along the grain, and 1/8 to 1/10 inches thick. The veneer is cut into strips about 1/2 inch in width. Most strands are 102 inches long but strands as short as 24 inches are used. We have experimented successfully with many different species, and are largely producing with Douglas fir in Vancouver and Southern pine in Colbert. The two products have strength properties so similar they have the same code allowable strength values.

Phenol formaldehyde resin is used to bind the strands together. The formulation is similar to that used for exterior grade plywood. It is waterproof, chemically stable and emits effectively no free formaldehyde. Because this resin has been used successfully in the plywood industry for over 40 years, we are sure of its long-term reliability.

Parallam® PSL is assembled in the following steps:

- The resin is applied to all surfaces of the strands and penetrates effectively into lathe checks and cracks.

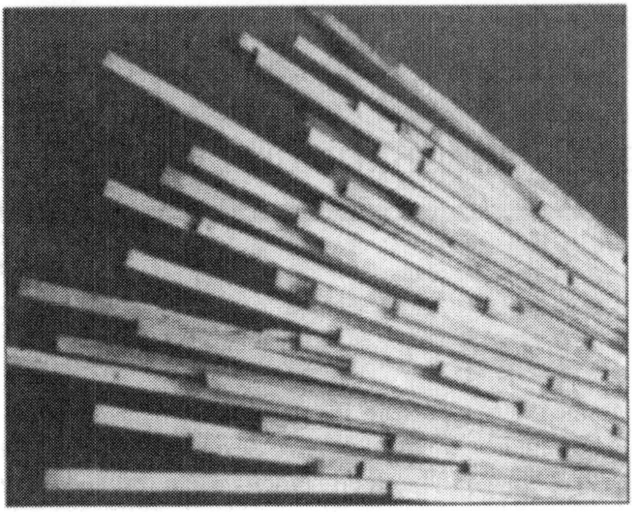

Figure 1. Photograph showing veneer strands used in the fabrication of Parallam®.

- The strands are placed into a trough where each strand is oriented parallel to the product axis with more accuracy than other oriented strand products.
- The ends of the strands are randomly distributed along the lay-up.

The mat of parallel glued strands is continuously formed in this way, compressed to a fixed cross-section and heated with microwave energy. The compression produces only about a ten percent over-density compared to the host species. The process steps will be discussed in more detail later.

Attributes

What we gain over other products and processes relates to usable strength properties and dimensional stability.

First, the product is very strong. The stranding process inherently weeds out knots, severe grain deviations and weak wood. Such weaknesses cause the strand to break up into short lengths. These are automatically discarded. The mat when formed from the remaining strands has evenly distributed very small strand ends. Each of these ends, although theoretically a weakness, has little impact and the inherent strength of each individual strand is fully utilized.

Second, the product is very consistent. Because a single 3.5 inches by 16 inches beam contains over a thousand strands in any given cross-section, the weakness of a single strand or the placement of a strand end does not significantly affect strength. In the testing of full size beams, the beam-to-beam variability in strength and stiffness is less than 10% (coefficient of variation). Solid wood varies from 25% to 50% (coefficient of variation).

Third, the product is very durable. Because of the waterproof resin and the even microwave heating, the product behaves like solid wood in the presence of water. It is strong even when wet and has minimal swelling, effectively recovering both strength and original dimensions on redrying. In a recent 5-year exterior exposure test conducted by the American Plywood Association, Parallam® PSL retained 100 % of its original strength after conditioning back to initial moisture levels.

Fourth, the product properties are controllable. Because the process steps are controllable, the final product can be engineered to ensure the advertised strength properties are met. The combination of strength, consistency, durability and controllability combine to give a range of usable strength properties unparalleled in the industry.

Product Testing

We have continued to evolve Parallam® PSL in subtle ways to improve its strength and durability. In doing so, we have undertaken the most thorough testing program in the history of new wood products. Over 30 different tests of over 50,000 specimens have been done. They include many tests for: bending (MORE, MOE, shear modulus), compression, tension, shear, cyclic aging, etc. They also include duration of load, water absorption, dimensional stability, treatability, fire response, etc..

An important part of our philosophy is to test full size beams or sub-assemblies whenever possible (e.g., trusses). Size effects in structural wood products are not well understood by the industry, so full size testing is very important in predicting serviceability. We have tested significant numbers of 40 feet long, 4-inch by 12-inch beams and 4-inch by 24-inch beams. We have also extensively tested parallel cord roof and floor trusses as well as double pitched roof trusses at similar lengths.

Conversion of Veneer to Billets

The part of process that has consumed most of the research effort for the first 10 years is the conversion of dry veneer into billet. The manufacture of Parallam® PSL billets from dry veneer can be discussed under the following headings:

- Veneer clipping (to 0.5-inch wide strands).
- Defect removal (short strand elimination).
- Glue application (all sides covered).
- Continuous mat formation (parallel strand lay-up).

• Continuous pressing.

• Continuous microwave heating.

I will focus on the last two points—pressing and microwave heating. They form the heart of the process and the two areas really integrate into one machine.

The essential components of the press and microwave windows are shown in Figure 2. The symmetrically opposed steel belts are continuously pulled between the press platens by the 4-drive drums. The continuous compression action on the mat to form the product comes when the material is drawn, by the motion of the steel belts, into the fixed gap formed by the platens. The loose mat of wood strands needed to make the product is about 3 feet thick. The finished product is about 1 foot thick. The principal reason for compressing is to ensure good contact between strands and therefore, good glue bond quality. The wood density is increased only about 10 percent above the host species.

Two features of our process differ from standard continuous presses:

• Sidedams physically constrain the lay-up.

• Microwave energy is continuously applied.

Figure 2 also shows the sidedams installed in the press. The width of most of our production varies between 12 inches and 16 inches, depending on sidedam positioning. The sidedams maintain full supportive forces on the edge of the product as the press belts continuously draw strands into the press. Even though the sidedams are fixed, compressive forces can be in excess of 300 psi.

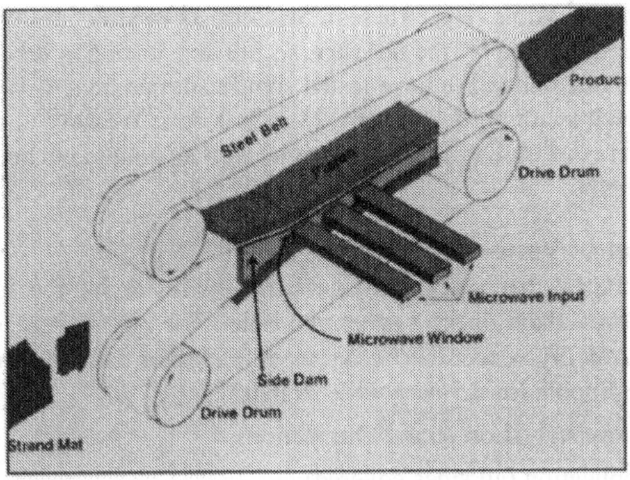

Figure 2. Schematic of the press/microwave set-up.

Unlike particleboard or oriented strand board processes, edge trim waste is virtually non-existent. Further, because the process continuously forms a beam, we can cut the lengths required by the market without stopping and without end trim losses.

The microwave windows are between the microwave waveguides and the sidedams. In our current press, these windows are made of ceramic. The ceramic as part of the sidedam, also holds full compressive forces on the edge of the product. The ceramic is transparent to microwave energy and allows microwave power to penetrate into the product.

Microwaves heat materials like wood (which contain water and various ions) effectively. The choices of wood moisture content and of chemical additives such as the phenolic glue affect the penetration of the microwaves into the product. By controlling these variables, we can control the evenness of heating of the compressed mat.

Our use of microwave energy allows us to evenly heat much larger cross-sections (12" by 20") than would be possible with conventional hot pressing. The high frequency microwaves also allow us to use conventional, well-proven, low cost alkaline phenolic resins. These resins are exterior grade and waterproof. Radio frequency systems used, for example, in wood composite preheating do not work well with phenolics because electrical arcing occurs.

Industry standard hot platen presses can cure product up to about 2 inches thick. Thicker products become impractical because wood is a good thermal insulator and press times increase beyond practical limits. In contrast, our microwave system is currently heating a 12-inch thick product in about two minutes. Further, it heats much more evenly than conventional hot presses resulting in more consistent product properties.

Summary of Product and Process Evolution

The evolution of the product and the process have been intertwined inextricably. The process now in use has very high fiber recovery and is highly automated (low labor cost). Because of this, we have low operating costs. The product has high, consistent strength properties and excellent dimensional stability. Because of this, we gain premium sales value. Further, the product and process are protected by 14 internationally patented inventions.

We have exceeded the original concepts of making a composite wood product equivalent in strength to lumber and now manufacture the highest strength wood-based building material available today. Our strength properties exceed those of machine stress related lumber, glulam beams and most laminated veneer lumber.

Appendix 1

Editors

David E. Clark
Virginia Polytechnic Institute & State University
Department of Materials Science & Engineering
Blacksburg, VA 24060
dclark@vt.edu

Dr. David E. Clark is Professor and Head of the Department of Materials Science and Engineering at Virginia Tech since January 2001. He comes from the University of Florida where he taught ceramic and materials engineering in MSE since 1976. His research interests include glass processing and corrosion, nucleation and crystallization of glass nuclear and hazardous waste remediation and drying and firing of ceramic powders. Most recently, his research has focused on microwave processing of materials, including fundamentals of microwave-material interactions, sintering of ceramics and composites, waste remediation and metal recovery/recycling, self-propagating high-temperature synthesis (SHS) and combustion synthesis of powders and nano-particulates. He is a past-president and trustee of the National Institute of Ceramic Engineers (NICE), a Professional Engineer and currently serves as the chair of the Professional Engineering Committee that writes and grades the ceramic engineering licensing examination. He is a past-chair of the Engineering Ceramics Division, Fellow of the American Ceramic Society, Fellow of the National Institute of Ceramic Engineers, served on the ACerS Board of Directors from 1996 to 2001 and has served as an ABET evaluator for NICE since 1990. Dr. Clark has authored/edited 16 books, over 200 papers and 13 patents.

Diane C. Folz
Virginia Polytechnic Institute and State University
Department of Materials Science and Engineering
Blacksburg, VA 24060
dfolz@vt.edu

Diane Folz is a Senior Research Associate in materials science and engineering at Virginia Tech, where she works primarily in the area of microwave processing of materials. She graduated from the University of Florida in 1987 with a B.S. in materials science and engineering. She stayed at Florida as a project and laboratory manager until January 2001 when she left to join the MSE department at Virginia Tech. Since 1994, Folz has served as the Executive Director of the National Institute of Ceramic Engineers (NICE). She just finished her term as Chair of the Education and Outreach Committee of the American Ceramic Society (ACerS) and is an evaluator for the Accreditation Board for Engineering and Technology (ABET). In January, she began serving a three-year term on the American Association of Engineering Societies (AAES) Board of Governors and she is Chair of Link 180 of the Order of the Engineer. She holds memberships in the Ceramic Education Council, ACerS, Keramos Ceramic Engineering Fraternity, and Epsilon Lambda Chi (Engineering Leadership Circle) and ASM International. She is a Fellow of the NICE and ACerS. Folz has served as editor on 3 books, has more than 25 technical publications and 2 patents. She served as Technical Vice Chair of the Third World Congress on Microwave and Radio Frequency Applications (September 2002, Sydney, Australia) and currently is Chair of the Technical Program Committee for the Fourth World Congress scheduled for November 2004 in Austin, TX.

Carlos E. Folgar
Virginia Polytechnic Institute and State University
Department of Materials Science and Engineering
Blacksburg, VA 24060
cfolgar@vt.edu

Carlos E. Folgar is a Ph.D. graduate student in material science and engineering at Virginia Tech where he has worked as a graduate research assistant in the area of microwave processing of materials since 2002. He graduated from the UCA (Central American University at San Salvador, El Salvador) in 1991 with a B.S. in electrical engineering. Prior to attending Virginia Tech, Folgar worked for 9 years for the food manufacturing industry where he was responsible for designing, building and maintaining packaging equipment. He also was a teaching assistant in electronic engineering at the University Don Bosco of El Salvador for 2 years. He holds memberships in the IEEE, ACerS TMS and ASM-International. His research interests include microwave processing of sol-gel materials, waste remediation and electronic ceramics.

Morsi M. Mahmoud
Virginia Polytechnic Institute and State University
Department of Materials Science and Engineering
Blacksburg, VA 24060
momahmou@vt.edu

Morsi Mahmoud is a Ph.D. graduate student in the Department of Materials Science and Engineering at Virginia Tech. He joined the department in January 2002. He received his B.Sc. in Chemistry from Alexandria University, Alexandria, Egypt in 1991 and his M.Sc degree in materials science in November 1999, Department of Materials Science, Institute of Graduate Studies and Research, also at Alexandria University. He joined the National Research Center, Cairo, Egypt, as a Research Assistant in 1993, and then moved to a newly established research center as an Assistant Researcher for the Advanced Technologies and New Materials Research Institute (IATNM), Mubarak City for Scientific Research and Technology (MCSRT), Alexandria, Egypt. His major research interests include production of glass-ceramic materials from industrial waste, ceramic materials properties and applications and, most recently, microwave processing of materials with special emphasis on crystallization of glass. Mr.Mahmoud has won the 2nd place in the academic poster competition in the 4th World Congress on Microwave and Radio Frequency Applications," from 7-12 November 2004, Austin, Texas, USA. He is a member in at the American Ceramic Society (ACerS), the National Institute of Ceramic Engineers (NICE), Materials Engineering Professional Societies (MEPS) at Virginia Tech, The Minerals, Metals & Materials Society (TMS), and finally in The Materials Information Society (ASM-International).

Contact List for Primary Authors

Prasad A. Apte
Sherritt Inc.
For Saskatchewan
Alberta T8L 2P2, Canada

S. Aravindan
Tokyo Institute of Technology
Dept. of Mechanical & Control Engineering
Tokyo, Japan
aravindhans@hotmail.com

S.B. Bhaduri
Clemson University
School of Materials Science & Engineering
110 Olin Hall, Box 340971
Clemson, SC 29634-0971
Bhaduri@clemson.edu

A. Birman
University of Maryland
Inst for Research in Electronics & Applied Physics
Energy Research Facility (Bldg. 223)
Point Branch Drive
College Park, MD 20742-3511
www.ireap.umd.edu

S.M. Bond
EA Technology
Chester, United Kingdom

John H. Booske (and K. Thompson)
Professor of Electrical and Computer Engineering and Director,
Materials Science Program
University of Wisconsin-Madison
1415 Engineering DriveMadison, WI 53706
booske@engr.wisc.edu

Jon G.P. Binner
IPTME
Loughborough University
Loughborough
LE11 3TU
UK
J.Binner@Lboro.ac.uk

J.B.O. Caughman
Oak Ridge National Laboratory
P.O. Box 2008, M/S 6169
Oak Ridge, TN 37831-6169
caughmanjb@ornl.gov

M.T. Churchland
MacMillan Bloedel Research
925 West Georgia Street
Vancouver V6C 3L2, Canada
604-661-8000

David E. Clark
Virginia Polytechnic Institute and State University
213 Holden Hall
Department of Material Science and Engineering
Blacksburg, VA 24061
dclark@vt.edu

David A. Earl
Alfred University
McMahon Building 354B
Alfred, NY 14802
earlda@alfred.edu

Zak Fathi
Lambda Technologies, Inc.
960 Aviation Parkway, Suite 900
Raleigh, NC 27560
zfathi@microcure.com

Diane C. Folz
Virginia Polytechnic Institute and State University
213 Holden Hall
Department of Material Science and Engineering
Blacksburg, VA 24061
dfolz@vt.edu

Richard S. Garad
Lambda Technologies, Inc.
960 Aviation Parkway, Suite 900
Raleigh, NC 27560

Clark Gellings
Electric Power Research Institute
3412 Hillview Avenue
Palo Alto, CA 94304
EPRI Customer Assistance Center
askepri@epri.com

John F. Gerling
Gerling Applied Engineering, Inc.
P.O. Box 580816
Modesto, CA 95358-0816
jgerling@2450mhz.com

Daniel J. Grellinger
Oak Ridge National Laboratory
Metal and Ceramics Division
Oak Ridge, TN

Walter A. Hendrix
MicroGas Corporation
127 Bellevue Way SE, Suite 200
Bellevue, WA 98004-6229

Ronald M. Hutcheon
Microwaves Properties North
RR#1 Deep River
Deep River, ON K0J 1P0, Canada
Hutcheon@mpn.ca

Uwe Kolberg
Schott Glass
Bernard Krieger
Cober Electronics, Inc.
151 Woodward Avenue
Norwalk, CT 06854
bern@cober.com

Guido Link
Forschungszentrum Karlsruhe GmbH
Institut für Hochleistungsimpuls-und Mikrowellentechnik
Postfach 3640, D-76021
Karlsruhe, Germany
Guido.link@ihm.fzk.de

M.S. Morrow
Microwave Technology, Inc.
200 Compress Street
Chattanooga, TN 37405
mstanmorrow@msn.com

R.E. Newnham
The Pennsylvania State University
251A Material Research Laboratory
University Park, PA 16802
bobnewnham@psu.edu

Juan Ortigosa
CIM D'OR S.A. (Corporació)
Riera Roja 11
Sant Boi de Llobregat
08830, Spain

Robert F. Schiffmann
RFSchiffmann Associates, Inc.
149 West 88th Street
New York, NY 10024
microwaves@juno.com

Willard H. Sutton, Retired
United Technologies Research Center
East Hartford, CT

Mary P. Sweeney
Northwestern University
Department of Material Science and Engineering
Evanston, IL

D.V. Szabfi
Forschungszentrum Karlsruh GmbH
Institut für Materialforschung III
P.O. Box 3640
76021 Karlosruhe, Germany

S. Takayama
Gifu Prefectural Institute for Fusion Science
322-6 Orosi, Toki-Shi
Gifu 509-5292, Japan

Garth V.A. Tayler
Acme Brick Company, Technical Department
P.O. Box 425
Fort Worth, TX 76101-0425

Wayne R. Tinga, Retired
University of Alberta
Department of Electrical Engineering
Edmonton, AB, Canada

Ruth Wroe
EA Technology
Chester, United Kingdom
Ruth.Wroe@capenhurst.com

Appendix 2
Index of Papers

This appendix contains a listing of all papers published by The American Ceramic Society in six volumes on microwave technology. The papers are sorted by topic and the book code shown in brackets [ ].

Book's Codes

Code	Topics
BM	Biomedical
CU	Curing
DP	Dielectric properties-measurement
DR	Drying
EC	Economic and commercialization
EQ	Equipment & Design
FU	Fundamentals of microwaves
GM	Glass-Melting
JO	Joining
MM	Microwave material interaction
MO	Modeling
OF	Other frequencies (not 2.45GHz)
PP	Microwave plasma processing
RE	Review
SA	Safety
SI	Sintering & Annealing
SM	Surface modification
SY	Synthesis
TM	Temperature measurement
WR	Waste remediation and recycling

Code	Publication
1	Microwaves: Theory and Application in Materials Processing (Ceramic Transactions, Vol 21.), *David E. Clark, Frank D. Gac, Willard H. Sutton.*
2	Microwaves: Theory and Application in Materials Processing II (Ceramic Transactions, Vol 36), *David E. Clark, Wayne R. Tinga, Joseph R. Laia, Jr.*
3	Microwaves: Theory and Application in Materials Processing III (Ceramic Transactions, Vol 59), *David E. Clark, Diane C. Folz, Steven J. Oda, Richard Silberglitt.*
4	Microwaves: Theory and Application in Materials Processing IV (Ceramic Transactions, Vol 80), *David E. Clark, Willard H. Sutton, David A. Lewis.*
5	Microwaves: Theory and Application in Materials Processing V (Ceramic Transactions, Vol 111), *David E. Clark, Jon G.P. Binner, David A. Lewis.*
6	Microwave and Radio Frequency Applications, *Diane C. Folz, John H. Booske, David E. Clark, John F. Gerling.*

BM Alina L Hategan, Diana Martin, Alina S Popescu, C Oproiu, R Cramariuc, A Margaritescu, Vasile V Morariu; "Effects of Microwave Irradiation on the Osmotic Properties of Human Erythrocyte Membranes," [4] pp. 193-200.

BM Susan C Hagness, Xu Li, Kristen M Laininger, Michel Okoniewski; "Dielectric Characterization of Human Breast Tissue and Breast Cancer Detection Algorithms for Confocal Microwave Imaging," [5] pp. 85-92.

BM Yoshio Nikawa, Keiji Kawai, Shigeyuki Abisu; "Application of Microwaves to dental Caries Treatment," [5] pp. 435-442.

BM J Mark Elwood; "Epidemiological Studies of Cancers and Exposures to Radio Frequencies," [6] pp. 507-514.

BM Fred Sterzer; "Continuous and Pulsed medical Microwave Devices," [6] pp. 479-486.

CU Ursula Hottong, Jianghua Wei, Ramakrishna Dhulipala, Martin C Hawley; "Microwave Processing of Polyester and Polyester / Glass Composites," [1] pp. 587-595.

CU Zak Fathi, Richard S Garard, J Clemons, Craig Saltiel, R M Hutcheon, Mark T DeMeuse; "Processing and Characterization of a Polymer Matrix Composite Using Variable Frequency Microwave Heating," [3] pp. 441-448.

CU Freddy Y C Boey; "Application of Microwave Curing for the Production of Structural Fiber Reinforced Composite Components Using a High Pressure Autoclave Process," [4] pp. 445-452.

CU S L Mak, R Banks; "Advances in Microwave Curing of Concrete," [6] pp. 377-387.

CU V Adegbite, L Fellows, J Wei, Y Qui, M Hawley; "Uniform Microwave Heating of Polymer Matrix Composites," [3] pp. 149-157.

CU D A Lewis, Susan J Lamaire, Alfred Viehbeck; "Microwave Processing of Polyimide Thin Films for Electronics," [3] pp. 433-439.

CU Albert J Gourdenne; "Polymerization Under Microwaves: Fifteen Years of Experience," [4] pp. 425-436.

CU Elias Siores, Paul Groombridge, Adekunle Oloyede; "Preliminary Investigations into the Use of Microwave Energy for Fast Curing of Adhesively Bonded Joints Formed Using Engineering Thermoplastics," [4] pp. 437-444.

DP Weiguo Xi, W R Tinga; "A High Temperature Microwave Dielectrometer," [1] pp. 215-224.

DP R Moreno, J Requena, J S Moya, J Molla, A Ibarra; "Effect of Powder Characteristics on Dielectric Properties of Alumina Compacts," [1] pp. 225-233.

DP N H Harris, J R Chow, R L Eisenhart, B M Pierce; "Dielectric Properties of Ceramics at Microwave Frequencies," [1] pp. 235-242.

DP R D Hollinger, V V Varadan, V K Varadan, D K Ghodgaonkar; "Free-Space Measurements of High-Temperature, Complex Dielectric Properties at Microwave Frequencies," [1] pp. 243-250.

DP X D Yu, V V Varadan, V K Varadan, D Ghodgaonkar; "In Situ Measurement of the Dielectric Properties of Ceramics with a Single Mode Microwave Heating Device," [1] pp. 251-259.

DP R M Hutcheon, G F Morin, B H Smith, M S de Jong, F P Adams; "The High-Temperature Microwave Response of Industrial Ferrites," [2] pp. 469-481.

DP M Arai, J G P Binner, G E Carr, T E Cross; "High Temperature Dielectric Property Measurements of Engineering Ceramics," [2] pp. 483-492.

DP V F Janas, C J Malarkey,M F Iskander, O Andrade, S Bringhurst,; "Dielectric Property Measurement of Ceramic Bodies," [2] pp. 493-502.

DP Shane Bringhurst, Magdy F Iskander, Octavio M Andrade; "New Metallized Ceramic Coaxial Probe for High-temperature Broadband Dielectric Properties Measurements," [2] pp. 503-510.

DP M Arai, J G P Binner, G E Carr, T E Cross, A L Bowden, N G Evans, M G Hamlyn, R Hutcheon, G Morin, B Smith; "Elevated Temperature Dielectric Property Measurements: Results of a Parallel Measurement Program," [2] pp. 539-546.

DP Gerald J Vogt, Wayne R Tinga, Ross H Plovnick; "Dielectric Property Measurement of Zirconia Fibers at High Temperatures," [3] pp. 227-234.

DP R M Hutcheon, P Hayward, B H Smith, S B Alexander; "High Temperature Dielectric Constant Measurement - Another Analytical Tool for Ceramic Studies?," [3] pp. 235-242.

DP J Batt, W H Sutton, J G P Binner, T E Cross, N R Greenacre, M G Hamlyn, R M Hutcheon, C M Weil; "A Parallel Measurement Program in High Temperature Dielectric Property Measurements: An Update," [3] pp. 243-250.

DP Binshen Meng, J Booske, R Cooper, B Klein; "Measurements of the Complex Permittivity of Low Loss Ceramics at Microwave Frequencies and Over Large Temperature Ranges," [3] pp. 251-258.

DP R S Donnan, R E M Vickers, M Samandi; "Application of FTIR-Reflectance Spectroscopy to Estimate the Temperature-Dependent Dielectric Loss of Engineering Ceramics," [5] pp. 75-83.

DP Stuart O Nelson; "Comparison of Measured and Calculated Powdered Mixture Permittivities," [5] pp. 101-108.

DP T Nakamura, Y Nikawa, F Okade; "Measurement of Microwave Permittivity Using Ferrite Loaded Cavity Resonator," [5] pp. 145-151.

DP Samir Trabelsi, Andrzej W Kraszewski, Stuart O Nelson; "Dielectric Calibration Methods for Industrial Microwave Moisture Sensors," [5] pp. 381-389.

DP I Youngs, T Powers, D Robinson, T Stickland, S G Appleton, A Howard; "The Use of Dielectric Property Measurements in the Selection of a Thermosetting Resin for the Microwave Cure of Filament Wound Composites," [5] pp. 93-100.

DP Sheryl A Barringer, Oya Sipahioglu, Cavit Bircan; "The Dielectric Properties of Fruit, Vegetables and Meat," [6] pp. 57-67.

DP Johanna B Salsman; "Measurement of Dielectric Properties in the Frequency Range of 300 MHz to 3 GHz as a Function of Temperature and Density," [1] pp. 203-214.

DP H W Jackson, M Barmatz; "Method for Calculating and Observing Microwave Absorption by a Sphere in a Single Mode Rectangular Cavity," [1] pp. 261-269.

DP Baiqiang Tian, Wayne R Tinga; "New Self-Oscillating, Self-Heating Microwave Active Dielectrometer," [2] pp. 517-527.

DP R M Hutcheon, F P Adams, M S DeJong; "What is the Best Microwave Absorber?," [3] pp. 215-225.

DP F J Alvin; "The Use of Dielectric Properties in the Design of a Reactor Heated By Microwaves," [5] pp. 119-127.

DP Morris Rimbi, Howard C Reader; "Dielectric Measurement Using an Internally Calibrated Split-Stripline Fixture," [5] pp. 129-136.

DP Marc Regier; "Determination of Volume Fraction and Particle Size in Dispersed Systems by Dielectric Spectroscopy and Principal Component Analysis," [6] pp. 29-37.

DP J L Watkins, C O Hagenlocher, M Barmatz; "Measurement of Cavity Perturbation to Determine the Microwave Absorption of a Sphere in a Single Mode Resonator," [1] pp. 271-279.

DP Ronald Hutcheon, Jozef Mouris, Xiaomei Fang, Deniel A Scola; "Measurements of the High-Temperature Microwave (400 - 3000MHz) Complex Dielectric Constants of Monomers ε-Caprolactam and ε-Caprolactone," [5] pp. 109-118.

DP Serge Lefeuvre, Faycal Azzouz, Magdy Massoud, Gean V Salmoria; "From Dielectric Measurements to Applicators: Drying, Polymer Processing, Sintering," [5] pp. 153-160.

DR Mark Hamann, EIT, CEM; "Variable Frequency Microwave Moisture Leveling of Mouth Swabs," [5] pp. 567-574.

DR D A Earl, D E Clark, R L Schulz; "Microwave Energy Versus Convected Hot Air for Rapidly Drying Ceramic Tile," [3] pp. 449-456.

DR D E Christiansen, W I Unruh; "Use of a TM010 Microwave Cavity at 2.45 GHz for Aerosol and Filament Drying," [1] pp. 597-604.

DR Pavel Kadlec Alexandra Rubecova Andrea Hinkova Jitka Kaasova; "Microwave Treatment of Germinated Yellow Pea," [5] pp. 443-450.

DR I S Balbaa, S J Oda, K E Haque, P D Kondos, R J C MacDonald; "Microwave Reactivation of CIP Spent Carbon," [1] pp. 475-483.

DR Juan I Ortigosa; "An Industrial Application of Microwave Heating: Continuous Drying of Inorganic Salts," [5] pp. 521-526.

EC F C R Wroe; "Scaling Up the Microwave Firing of Ceramics," [2] pp. 449-458.

EC Ruth Wroe; "Microwave - Assisted Firing of Ceramics," [4] pp. 671-678.

EC S M Bond; "Microwave-Assisted Processing - From Laboratory to Production," [5] pp. 489-496.

EC M L Cunha, M W Canto, A Marsaioli Jr; "Drying Pulped Coffee Cherry Beans by Means of Heated Air Assisted by Microwaves," [4] pp. 641-649.

EC Grigori Torgovnikov, Peter Vinden; "Some Aspects of Microwave Applications in the Forest Industry," [4] pp. 659-670.

EC Alexander S Zusmanovskiy, Vadim V Yakovlev; "Application Microwave Energy in Processing of Fish and Marine Products," [4] pp. 679-686.

EC D A Lewis, Stanley J Whitchair, William V Corso, Alfred Vichbeck; "Microwave Processing of Continuous Wide Webs," [4] pp. 651-658.

EC Willard H Sutton; "Microwave Processing: Steps to Successful Commercialization - Panel Discussions," [3] pp. 3-5.

EC R F. Schiffmann; "Commercializing Microwave Systems:Paths to Success or Failure," [3] pp. 7-16.

EC Bernard Krieger; "Commercialization - Steps to Successful Applications and Scale-up: May 1, 1995," [3] pp. 17-21.

EC J F Gerling; "Process Commercialization: Don't Risk Losing the Benefits by Ignoring the Hazards," [3] pp. 23-27.

EC Richard S Garard; "Controlling & Understanding the Variables: Key to Commercializing Microwave Processing of Advanced Materials," [3] pp. 29-32.

EC Muarry J Kennedy; "Commercialization of Microwave Processes," [3] pp. 43-54.

EC Wayne R Tinga, Steven J Oda; "Obstacles and Remedies in Microwave Processing Research and Commercialization - an Open Forum," [3] pp. 575-579.

EC D A Lewis; "World Congress on Microwave Processing," [3] pp. 581-585.

EC Clark Gellings; "Dielectric Heating: EPRI's Perspective on the Market and the Technology," [4] pp. 27-39.

EC Dr Richard Dunham Smith, J P Kesserlring; "Microwave Clothes Dryer," [4] pp. 687-694.

EC B C Langenegger, I O Coker, G V A Tayler; "Technology Transfer of Dielectric Heating Technology in South African Industry," [4] pp. 695-705.

EC Chokichiro Shibata; "Statistics of Microwave and HF Power Application in Japan," [4] pp. 707-712.

EC Walter Van Loock; "A Simple Basis for Decision to Use Microwave and Radio Frequency Energy For Industrial Processing," [5] pp. 497-502.

EC Preston L Rs, Dr. Michael V Vasilik; "Radio Frequency and Microwave Solutions for Industrial Processing," [5] pp. 513-519.

EC Wayne R Tinga; "Microwave and RF - Application, Engineering and Social Issues," [5] pp. 577-585.

EQ Hussamaldin S Sa'adaldin, W Murray Black, Yonglai Tian, Iftikhar Ahmad, Richard Silberglitt; "Microwave Heating of Ceramics in a Single Mode Mitered Bend Cavity," [2] pp. 359-364.

EQ R J Lauf, D W Bible, S R Maddox, C A Everleigh, R J Espinosa, A C Json; "Materials Processing Using a Variable Frequency Microwave Furnace," [2] pp. 571-579.

EQ G J Vogt, A H Regan, A S Rohlv, M T Curtin; "Use of a Variable Frequency Source with a Single-Mode Cavity to Process Ceramic Filaments," [3] pp. 125-132.

EQ J D Katz, D E Rees; "Quasi-Optical Gyrotron Materials Processing at Los Alamos," [3] pp. 141-147.

EQ Frank Demmerle, Werner Wiesbeck, Christian Stenzel; "A New Applicator Design for Microwave Processing of Zeolites," [4] pp. 217-224.

EQ Michael G Hamlyn, Neil A Hart, Nigel G Evans; "A Combined Microwave and Electric Radiant Heating System for Firing Ceramics," [4] pp. 555-562.

EQ Steven M Bradshaw, Deneel Jordean, Howard C Reader, Johan W Gerber, T V Chow Ting; "Developing Microwave Solutions for Industry: A Case Study," [5] pp. 543-550.

EQ Guido Link, Manfred Thumm; "Dilatometer System with Hybrid Heating Technique for Detailed Investigations on Sintering of Ceramics," [6] pp. 301-310.

EQ R A Rudder, R C Hendry, G C Hudson, R J Marlunas, A C Json, L T Thigpen, R S Garard, C A Everleigh; "Diamond Chemical Vapor Deposition (CVD) Via Microwave-Excited Plasma from Water-Ethanol Solutions," [2] pp. 377-384.

EQ L Feher, M Thumm; "Hephaistos-A Novel Automated Microwave Processing System for Carbon Fibre Reinforced Plastics (CFRP)," [6] pp. 291-299.

EQ Magnus Sundberg; "Semi-Automated Design of Microwave Applicators for Sterilization of Packed Foods," [4] pp. 233-240.

EQ B Q Tian, W R Tinga; "A Wide Range Tunable and Matchable High Temperature Applicator," [1] pp. 647-654.

EQ J Asmussen, Jr, B Manring, R Fritz, M Siegel; "Experimental Examination of Material Loaded Cylindrical Applicators and Comparison with Theoretical Models," [1] pp. 655-665.

EQ J F Gerling, Gene Fournier; "Techniques to Improve the Performance of Microwave Process Systems Which Utilize High Q Cavities," [1] pp. 667-674.

EQ Edmund H Moore, Iftikhar Ahmad, D E Clark; "(No Suggestions) Analyzer (MTGA)," [1] pp. 675-681.

EQ Van Nguyen Tran; "An Application Design for Processing Large Quantities of Dielectric Material," [1] pp. 683-691.

EQ A C Metaxas; "Applicators for Industrial Microwave Processing," [2] pp. 549-562.

EQ A C Json, L T Thigpen, R J Espinosa, W A Lewis; "Use of a Variable Frequency Microwave Furnace for Large-Area, Uniform Processing," [2] pp. 563-570.

EQ C McDonough, M Barmatz, H W Jackson; "Applications of the Boltzmann-Ehrenfest Principle to Containerless Microwave Processing in Microgravity," [2] pp. 581-590.

EQ Eugene Eves; "Equipment Supplier's Role in Process Development," [3] pp. 33-39.

EQ L Feher, G Link, M Thumm; "Microwave Raytracing in Large Overmoded Industrial Resonators," [3] pp. 159-166.

EQ M Barmatz, O Iny, T Yiin, I Khan; "Vibration Method for Tracking the Resonant Mode and Impedance of a Microwave Cavity," [3] pp. 167-174.

EQ H Senko, V N Tran; "Control of Microwave Induced Thermal Run-Away Using Temperature Derivative Feedback," [4] pp. 241-250.

EQ V Tessier, L Dubois, M Chive, J P Sozanski, L Prevors; "Non-Invasive Temperature Control by Microwave Radiometry for Industrial Applications," [4] pp. 251-258.

EQ Ed Brown, Tom Tella, Martin Yonnone; "Control Systems and Concepts of Control for Industrial Microwave Equipment," [5] pp. 371-379.

EQ Joseph W Cresko, N Lee Brown, L Myles Phipps; "A Vertical Radio Frequency Design as a Unit Operation for the Continuous Processing of Free-Flowing, Bulk Granular or Particulate Products," [5] pp. 535-541.

EQ J F Gerling; "Waveguide Components and Configurations for Optimal Performance," [5] pp. 559-566.

EQ Andrew Harrison, Graeme Robb, Gavin Whittaker; "Instrumentation for in SITU Diffraction of Materials Under Microwave Radiation," [6] pp. 317-327.

EQ J F Gerling; "A Technical Review of Equipment and Methods for Waveguide Power Measurement in Microwave Heating Applications," [6] pp. 259-268.

EQ R f Schiffmann; "Designing Industrial Microwave Heating Systems for Safe Operation: Batch Ovens," [6] pp. 269-279.

EQ W M Van Loock; "Thermography for Optimizing High Frequency and Microwave Installations," [6] pp. 311-316.

FU R E Newnham, S J Jang, Ming Xu, Frederick Jones; "Fundamental Interaction Mechanisms Between Microwaves and Matter," [1] pp. 51-67.

FU V M Kenkre; "Theory of Microwave Interactions with Ceramic Materials," [1] pp. 69-80.

FU J D Katz, R D Blake, V M Kenkre; "Microwave Enhanced Diffusion?," [1] pp. 95-105.

FU Ralph W Bruce; "Activation Energies for the Dielectric Loss Factor / AC Conductivity of Some Polycrystalline Ceramics," [1] pp. 107-116.

FU E F Sturcken, L E McCurry; "Microwave Enhanced Pyrochemical Reactions of PuO_2, UO_2, and U_3O_8," [1] pp. 117-123.

FU V Fleurov, M Molotskii; "Resonant Excitation of Processes in a Solid by a Microwave Magnetic Field," [5] pp. 57-63.

FU T T Meek , X Zhang, M Rader; "An Analysis of the Individual Grain Structure of an Oxide Heated Using Microwave Radiation and Heated Conventionally," [1] pp. 81-93.

FU I Plazl, D Ravnjak, T Koloini, A Lubej; "Microwave Assisted Anatase - to - Rutile Phase Transition of TiO_2 - Anatase Gels," [5] pp. 207-213.

FU R F Schiffmann; "Principles of Industrial Microwave and RF Heating," [4] pp. 41-60.

FU D E Clark, Willard H Sutton, D A Lewis; "Microwave Processing of Materials," [4] pp. 61-96.

FU J K West, D E Clark; "A Molecular Orbital Model for Absorption of Microwave Energy by Materials," [5] pp. 43-55.

FU Georges De Corte, Philip Yde, Walter Van Loock, Michel Machiels, Deniel Van Dommelen; "Microwaves Energy Applications Design: Development and Evaluation of a Computer-Based Course," [5] pp. 65-72.

FU D Acierno, L di Maio, L Cappetta, M Feo, V Fiumara, D Napoli, V Pierro, I M Pinto, M Ricciardi; "Selected Issues in the Microwave Processing of Polymers," [4] pp. 409-416.

GM Dr Uwe Kolberg; "Microwave Heating of Glass," [5] pp. 527-533.

GM Carol M Jantzen, James R Cadieux; "Characterization of Radioactive Waste Melter Feed Vitrified by Microwave Energy," [1] pp. 441-449.

JO Richard Silberglitt, D Palaith, W Murray Black, H S Sa'adaldin, Joel D Katz, Rodger D Blake; "Investigation of Interlayer Materials for the Microwave Joining of SiC," [1] pp. 487-495.

JO T Y Yiin, V V Varadan, V K Varadan, J C Conway; "Microwave Joining of Si-SiC/Al/Si-SiC," [1] pp. 507-514.

JO Salwan Al-Assafi, Iftikhar Ahmad, Zakarye Fathi, D E Clark; "Microwave Joining of Ceramics," [1] pp. 515-521.

JO Iftikhar Ahmad, Richard Silberglitt, W Murray Black, Hussamaldin S Sa'adaldin, Yong-Lai Tian, Joel D Katz; "Microwave Joining of Dissimilar SiC Ceramics," [2] pp. 409-415.

JO Ronald E Loehman; "Problems and Opportunities in Microwave Joining of Ceramics," [2] pp. 417-430.

JO J Cai, Y L Tian, J K Guo; "Microwave Joining of Bi-based 2223 Superconductor and its Microstructure Features," [2] pp. 439-445.

JO J G P Binner, P A Davis, T E Cross, J A Fernie; "Microwave Joining of engineering Ceramics," [3] pp. 335-346.

JO I Ahmad, R Silberglitt, T A Shan, Y L Tian, R Cozzens; "Microwave-Assisted Pyrolysis of SiC and its Application to Joining," [3] pp. 357-365.

JO J Cai, B S Li, H R Zhuang, J K Guo, Y L Tian; "Microwave Joining of Alumina Both Directly and With 2223 Superconductor Interlayer," [3] pp. 367-372.

JO Mingli Li, Guy O Beale, Yong Lai Tian; "Automatic Control During Microwave Heating of Ceramics," [3] pp. 373-380.

JO R R Di Fiore, D E Clark; "Microwave Joining of Zinc Sulfide," [3] pp. 381-387.

JO A D Cozzi, D E Clark, M K Ferber, V J Tennery; "Apparatus for the Joining of Ceramics Using Microwave Hybrid Heating," [3] pp. 389-396.

JO Iftikhar Ahamad, Richard Silberglitt, Yong-Lai Tian, Joel D Katz; "Microwave Joining of SiC Ceramics and Composites," [4] pp. 455-463.

JO Alex D Cozzi, D E Clark, Mattison K Ferber; "Microwave Joining of Alumina Ceramics," [4] pp. 465-474.

JO Prasad K D V Yarlagadda, C T Soon, A Ahmed, E Siores; "Microwave Welding of Alumina Ceramic Using a ridge Waveguide," [4] pp. 475-482.

JO Y Ma, V K Varadan, VV Varadan; "Enhanced Microwave Absorption in chiral Composite Materials," [1] pp. 523-529.

JO Ruyen Ro, Vasundara V Varadan, Vijay K Varadan; "Parametric Experimental Study of Microwave Absorption in Chiral Composites Materials," [1] pp. 531-538.

JO Arthur C Lind, Frederick C Wear, James E Kurz; "Microwave Heating for Fiber-Placement Manufacturing of Carbon-Fiver Composites," [1] pp. 539-546.

JO R Fritz, J Asmussen, Jr.; "Distribution of Dissipated Power in a Graphite Fiber Reinforced Epoxy Composite Heated in a Microwave Cavity," [1] pp. 547-556.

JO Eric Bescher, J D MacKenzie; "Microwave Heating of Cermets," [1] pp. 557-563.

JO S Aravinden, J Ramkumar, R Krishnamurthy; "Microwave Joining of Al_2O_3 and SiC With Glass Interlayer," [5] pp. 287-294.

JO WR Tinga, J D Xu, F E Vermeulen; "Open Coaxial Microwave Spot Joining Applicator," [3] pp. 347-355.

MM Iftikhar Ahmad, D E Clark; "Effect of Microwave Heating on Solid State Reactions of Ceramics," [1] pp. 605-612.

MM P J Gielisse, H Niculescu, B Roy, P Pernambuco-Wise, J A Crow, G Sykora, R Wahlers; "Applications of Microwaves in High Temperature Superconductor Materials and Devices," [1] pp. 613-621.

MM S A Freeman, J H Booske, R F Cooper, B Meng, J Kieffer, B J Reardon,; "Studies of Microwave Field Effects on Ionic Transport in Ionic Crystalline Solids," [2] pp. 213-220.

MM Iftikhar Ahmad, D E Clark; "Effect of Microwave Heating on the Mass Transport in Ceramics," [2] pp. 287-295.

MM G J Vogt, W P Unruh; "Microwave Hybrid Heating of Alumina Filaments," [2] pp. 297-306.

MM Binshen Meng, J Booske, Reid Cooper, Benjamin Klein; "Microwave Absorption in Ionic Crystalline Ceramics with Various Concentrations of Point Defects," [3] pp. 177-184.

MM S A Freeman, J H Booske, R F Cooper; "Ionic Transport in Sodium Chloride and Silver Chloride During Microwave Irradiation," [3] pp. 185-191.

MM Joachim Maier, Janez Jamnik; "Electrochemical Processes in Ceramics - Overview and Correlation with Microwave Effects," [4] pp. 123-133.

MM J H Booske, R F Cooper, S A Freeman, B Meng, K I Rybakov, V E Semenov; "Thermal and Nonthermal Interactions Between Microwave Fields and Ceramics," [4] pp. 143-151.

MM Garth V A Tayler, Michael Hamlyn, Michael Anderson; "Microwave Firing of Heavy Clay Bodies," [4] pp. 547-554.

MM Garth V A Taylor, Mike Anderson, Mike Hamlyn; "Microwave Firing of Heavy Clay Bodies: the Examination of the Key Variables, Which Influence the Microwave Heating Process," [5] pp. 295-303.

MM D C Dube, M L Lanagan, H Peng, P Sharma, D Agrawal; "Investigations of Dielectric Behavior of Materials in the GHZ Range During Microwave," [6] pp. 157-165.

MM S V Egorov, A G Eremeev, K I Rybakov, V E Semenov, A A Sorokin, S A Gusev; "Microwave Intensity-Dependent Mass Transport Enhancement in Nanostructured Alumina Membranes," [6] pp. 167-174.

MM C C Sorrell, O C Standard, N Ehsani, K R Ratinac, J R Rider, X Q Zheng, A J Ruys; "Precision Microwave / Resistance Heating Study of Zirconia: Absence of Microwave Effect," [6] pp. 175-190.

MM Garth V A Taylor, Mike Anderson, Mike Hamlyn; "Hybrid Microwave Firing of Heavy Clay Products," [6] pp. 389-398.

MM K Thompson, J H Booske, R F Cooper, Y B Gainchandani; "Electromagnetic Induction Heating for Ultra Shallow Junction Formation," [6] pp. 411-421.

MM A Corradi, C Siligardi, P Veronesi, G Marucci, M Annibali, G Ragazzo; "Microwave Irradiation of Asbestos Containing Materials," [6] pp. 445-452.

MM Y Zhang-Steenwinkel, J Beckers, H L Castricum, A Bliek; "Using $La_{1-x}Ce_xMnO_3$ Perovskites as the Active Components for Soot Filter Regeneration by Dielectric Fields," [6] pp. 465-475.

MM S Aravindan, J Ramkumar, S K Malhotra, R Krishnamurthy; "Enhancement of Cutting Performance of Cemented Carbide Cutting Tools by Microwave Treatment," [6] pp. 199-206.

MM Clifford E George, Inchul Jun, Jiayi Fan; "Application of Microwave Heating Techniques to the Detoxification of Contaminated Soils," [1] pp. 459-466.

MM Martin Vala, Jan Szczepanski; "A Microwave Effect: Molecular Level Microwave Study of Water Vapor," [4] pp. 107-114.

MM Jianfeng Lou, T Alan Hatton, Paul E Laibinis; "Mixing Rules for the Dielectric Properties of Solvent Mixtures at Microwave Frequencies," [4] pp. 115-122.

MM R N Gedye; "The Question of Non-Thermal Effects in the Rate Enhancement of Organic Reactions by Microwaves," [4] pp. 165-172.

MM Christine M Wehlburg, Martin T Vala, Jan Szczepanski, Mark Moore; "Infrared Spectroscopic Analysis of Matrix Isolated Tetraethylorthosilicate (TEOS): Understanding the Effect of Microwave Irradiation on the Sol-Gel Process," [4] pp. 185-192.

MM R P Currier, D J Devlin; "Ceramic Matrix Composites by Microwave Assisted CVI," [2] pp. 387-396.

MM M T Churchland; "PARALLAM - New Structural Wood Composite," [3] pp. 63-68.

MM Kristie S Leiser, D E Clark; "Microwave Behavior of Silicon Carbide / High Alumina Cement Composites," [5] pp. 267-275.

MM D Jaglin, B Vaidhyanathan, J Binner, C Prentice, B Shatwell; "Densification Mechanism in SiCf / SiC Composites by Microwave Enhanced Chemical Vapour Infiltration," [6] pp. 231-240.

MM Tetsushi Yamamoto, Yujii Wada, Shozo Yanagida; "Preparation of Metal Oxides by Polyol Method Under Microwave Irradiation," [6] pp. 103-110.

MM M C Chow, A N Ma; "Processing of Fresh Palm Fruits Using Microwaves," [6] pp. .

MM Peter C S Carter, Nguyen Tran, Peter Hutchings; "Microwave Technology for Heating Large Diameter Logs for Sapstain Treatment and Stress Relief," [6] pp. 341-348.

MM Stuart O Nelson; "Microwave and Radio-Frequency Power Applications in Agriculture," [6] pp. 331-339.

MM G Torgovnikov, P Viden; "Innovative Microwave Technology for the Timber Industry," [6] pp. 349-356.

MM A D Cozzi, Z Fathi, D E Clark; "Crystallization of Sol-Gel Derived Barium Aluminosilicate in a 2.45GHz Microwave Field," [2] pp. 317-324.

MM A Boonyapiwat, D E Clark; "Nucleation and Crystal Growth of Li2O - 2SiO2 in a Microwave Field," [3] pp. 505-514.

MM A Boonyapiwat, Z Fathi, D C Folz, D E Clark; "Repair of Glass by Sol-Gel Coating Using Either Conventional or Microwave Heating," [2] pp. 341-349.

MM Thorsten Gerdes, Monika Willert-Porada, Philip Bahke, Rudolf Burghardt, Ingo Kleinwachter; "Microwave-Assisted Recovery of Platinum," [6] pp. 453-463.

MM E B Ripley, T L White, P A Eggleston; "Direct Microwave Coupling to Metals at Elevated Temperatures," [6] pp. 241-246.

MM J T Strack, I S Balbaa, B T Barber; "Microwave Regeneration of Activated Carbon for Gold Processing Applications," [3] pp. 99-105.

MM S Bradshaw, T V Chow Ting Chan, N Rowson, R Greenwood, S Kingman, K Jackson; "Microwave Assisted Comminution of Ores at Economically Viable Energy Inputs," [6] pp. 247-256.

MM L A Dissado, R M Hill; "What Form of Motion Gives Rise to Dielectric Response at Microwave Frequencies?," [4] pp. 99-106.

MM J R Thomas, Jr, J M Curtis Jr, Curtis Stern, W A Davis; "Controlled Microwave Heating of thermal Runaway Materials," [5] pp. 345-354.

MM Jorg Rauß, Dieter Bathen, Henner Schmidt-Traub, Michael Hoffmeister, Andreas Binder; "Multicomponent desorption of Fixed-Bed-Adsorbers by Microwaves," [5] pp. 427-434.

MM R Roy, D K Agrawal, J I Cheng; "New First Principles of Microwave-Material Interaction. Discovering the Role of the H Field, and Anisothermal Reactions," [5] pp. 471-485.

MM Edmund H Moore, D E Clark, Ronald Hutcheon; "Removal of Polymethyl Methacrylate Binder from Alumina Using Conventional and/or Microwave Heating," [2] pp. 325-332.

MM William V Corso, Jovan Mijovic; "Comparison of Microwave and Thermal Reaction Kinetics Via In-Situ FTIR Spectroscopy," [4] pp. 173-183.

MM J W Cresko; "Innovative Application of Dielectric Heating Technologies to Concurrently Reduce Energy Usage and Emissions in Polymer Manufacturing Processes," [6] pp. 367-376.

MO K S Nikita, G S Stamatakos; "Analysis of the Interaction Between a Layered Spherical Human Head Model and a Linear Antenna," [5] pp. 35-42.

MO E C Fear, M Okoniewski; "Image Formation Algorithms with Automatic Skin Detection for Confocal Microwave Imaging," [6] pp. 487-494.

MO Huawei Zhao; "FDTD Method for Modeling the Electromagnetic Field Effect on the Human Body in MRI," [6] pp. 495-505.

MO M F Iskander, O Andrade, R Smith, S Lamoreaux, Jes Asmussen, Jr, H Kimrey; "Computational Techniques in Modeling and Quantifying Microwave Interactions with Materials," [1] pp. 141-158.

MO Xiang Dong Yu, Vasundara V Varadan, Vijay K Varaden; "Modeling Microwave Heating of Ceramics," [1] pp. 167-176.

MO Gory A Kriegsmann; "Microwave Heating of Ceramics," [1] pp. 177-183.

MO J H Booske, R F Cooper, I Dobson, L McCaughan; "Models of Nonthermal Effects on Ionic Mobility During Microwave Processing of Crystalline Solids," [1] pp. 185-192.

MO G A Kriegsmann, P Varatharajah; "Formation of Hot Spots in Microwave Heated Ceramic Rods," [2] pp. 221-228.

MO J D Newman, L Walsh, R Evans, T Tholen, O M Andrade, M F Iskander, K J Bunch, H Kimrey; "Experimental Validation of Numerical Simulations of the Microwave Sintering Process," [2] pp. 229-237.

MO Brian J Reardon, J Kieffer, J H Booske, Reid F Cooper; "Analysis of the Microwave Sintering Effect in NaCl Using Molecular Dynamic Simulations," [2] pp. 239-246.

MO Rajiv K Singh, J Viatella, Zak Fathi, D E Clark; "Thermal Analysis of Microwave Processing of Ceramics," [2] pp. 247-255.

MO G A Kriegsmann; "Mathematical Models of Microwave Heating: Bistability and Thermal Runaway," [3] pp. 269-277.

MO Daniel J Skamser, Jeffrey J Thomas, Hamlin M Jennings, D Lynn Json; "Microwave Heating of a Changing Material," [3] pp. 289-296.

MO J R Thomas, Jr; "Importance of Dielectric Properties in Modeling of Microwave Sintering of Ceramics," [3] pp. 297-304.

MO A Birman, B Levush, Y Carmel, D Gershon, D Dadon, L P Martin, M Rosen; "Modeling of Multi-Frequency Microwave Sintering of ZnO Ceramic," [3] pp. 305-312.

MO Michael T K Koh, Rajiv K Singh, D E Clark; "Temperature Distribution Considerations During Microwave Heating of Ceramics," [3] pp. 313-322.

MO Liqiu Zhou, Gang Liu, Jian Zhou, Jiping Cheng; "Finite Element Computer Model of Microwave Heated Ceramics," [3] pp. 323-332.

MO Gory A Kriegsmann, Jonathen H C Luke; "Microwave Joining of Two Long Hollow Ceramic Tubes: A Combined Asymptotic and Numerical Analysis," [5] pp. 27-34.

MO Gory A Kriegsmann; "Microwave Heating of Silicon Wafers in Cylindrical TM Cavities: A Mathematical Model," [6] pp. 47-56.

MO J R Thomas, Jr; "Heat Transfer Model of Microwave Enhanced Catalysis," [4] pp. 397-406.

MO B S Tilley, G A Kridgsmann; "Microwave-Enhanced CVI Processing: A Moving Interface Model," [5] pp. 3-10.

MO N A Zharova, K I Rybakov, V E Semenov, S V Egorov; "Computer Simulation of Millimeter-Wave Sintering of Ceramic and Composite Materials," [5] pp. 11-18.

MO Ingolf Meier, Johann B de Swardt; "Pattern Synthesis for Microwave Heating Applications," [5] pp. 19-26.

MO Claude Lorenson, Christine Gallerneault; "Numerical Methods for the Modeling of Microwave Fields," [1] pp. 193-200.

MO Wayne R Tinga; "Microwave Material Interactions and Process Design Modeling," [2] pp. 29-43.

MO Magdy F Iskander; "Modeling the Microwave Process — Challenges and New Directions," [2] pp. 167-187.

MO H W Jackson, M Barmatz, P Wagner; "Transient Temperature Behavior for a Sphere Heated by Microwaves," [2] pp. 189-199.

MO E B Manring, J Asmussen, Jr.; "Numerical Model for the Modes of a Lossy, Coaxially Loaded Cylindrical Cavity," [2] pp. 201-212.

MO Y L Tian, W M Black, H S Sa'adaldin, I Ahmad, R Silberglitt; "Dynamic Model for Electromagnetic Field and Heating Patterns in Loaded Cylindrical Cavities," [3] pp. 261-268.

MO H W Jackson, M Barmatz, P Wagner; "Steady State Temperature Profile in a Cylinder Heated by Microwaves," [3] pp. 279-287.

MO Nguyen Tran, Andrew Piotrowski; "Advances in the Modeling of Microwave FR and Hot Air Drying of Materials," [4] pp. 201-213.

MO J Haala, F Demmerle, W Wiesbeck; "Analysis and Optimization of Microwave Heating Applicators Using Finite Difference Time Domain," [4] pp. 225-231.

MO A C (Ricky) Metaxas; "Recent Developments in Radio Frequency Technology," [4] pp. 269-276.

MO Vadim V Yakovlev; "Comparative Analysis of Contemporary Electromagnetic Software for Microwave Power Industry," [5] pp. 551-558.

MO B Liu, T R Marchant, I W Turner, V Vegh; "A Comparison of Semi-Analytical and Numerical Solutions for the Microwave Heating of a Lossy Material in a Three-Dimensional Waveguide," [6] pp. 17-27.

MO P Kopyt, M Celuch-Marcysiak, W K Gwarek; "Microwave Processing of Temperature-Dependent and Rotating Objects: Development and Experimental Verification of FDTD Algorithms," [6] pp. 7-16.

MO Vyacheslav V Konarov, Vadim V Yakovlev; "Computational Analysis of a Tapered Slotted Waveguide as a Radiating Element," [6] pp. 39-45.

MO Nguyen Tran; "Successful Microwave Processing Applications in Industry," [6] pp. 359-366.

OF Ralph W Bruce, Arne W Fliflet, Richard P Fischer, D Lewis III, Barry A Bender, Gan-Moog Chow, Roy J Rayne, Lynn K Kurihara, Paul E Schoen; "Millimeter - Wave Processing of Alumina Compacts," [4] pp. 287-294.

OF G Link, W Bauer, A Weddogen, H J Ritzhaupt-Kleissl, M Thumm; "MM-Wave Processing of Ceramics," [4] pp. 303-311.

OF S Miyake, Y Setsuhara, S Kinoshita, M Kamai, T Ohmae, N Abe, S Sano; "Study of Ceramic Heating and Sintering by High-Power Millimeter-Wave Radiation Within JWRI at Osaka University," [4] pp. 313-320.

OF Yurii V Bykov, Anatolii G Eremeev, Vladislav V Holoptsev, C Odemer, Hans-J Ritzhaupt-Kleissl, Anatolii I Rachkovskii; "Sintering of Piezoceramics Using Millimeter-Wave Radiation," [4] pp. 321-328.

OF Richard S Garard, Zak Fathi, Billy Wei; "Materials Processing Via Variable Frequency Microwave Irradiation," [3] pp. 117-124.

OF Peter L Jones; "RF Heating, an Old Technology with a Future," [4] pp. 261-268.

OF Jean-Paul Bernard; "The Development of 50 ? RF Technology in Industry," [4] pp. 277-283.

OF Thomas W Hardek, Wayne D Cooke, Joel D Katz, William L Perry, Daniel E Rees; "Intense High-Frequency Gyrotron-Based Microwave Beams for Material Processing," [4] pp. 295-302.

PP Hunghai Su, Guy E Salinas, Steve Kim, Vinayak P Dravid, D Lynn Json; "Mechanical Properties of Alumina Sintered in a Microwave Induced Oxygen Plasma," [3] pp. 205-212.

PP D V Szabo, D Vollath, W Arnold; "Microwave Plasma Synthesis of Nanoparticles: Application of Microwaves to Produce New Materials," [5] pp. 217-224.

PP A Dittmar, H Kosslick, A Bruckner, D L Hoang; "Microwave Plasma-Assisted Processing of Nanoscaled Supported Metal Oxide Catalysts," [6] pp. 147-153.

PP Ji Tian-Ren; "Recent Research and Development on Microwave Power Applications at the University of Electronic Science and Technology of China," [6] pp. 399-408.

PP Ward C Roman, Willard H Sutton, Denise A Tucker, Barbara Walden, Fred A Otter, Michael T McClure; "Microwave Plasma Processing of Diamond Coatings for Aerospace Applications: Deposition, Characterization, and Performance Evaluation," [4] pp. 375-386.

PP R W Boswell, A Ellingboe, A W Degeling, M Lieberman, J Derouard; "The Transition From Capacitive to Inductive to Wave Sustained RF Discharges," [4] pp. 367-374.

RE Dieter Vollath; "Some Activities of Microwave Processing of Ceramics in Germany," [2] pp. 147-156.

RE Yaw A Owusu, Peter J Gielisse; "Microwave Application in Precision Investment Casting Operations," [2] pp. 307-315.

RE S J Oda; "An Overview of Microwave Applications Research in Canada," [1] pp. 3-15.

RE D L Johson; "Microwave Processing of Ceramics at Northwestern University," [1] pp. 17-28.

RE D E Clark, D C Folz; "Microwave Processing Activities at the University of Florida," [1] pp. 29-34.

RE M F Iskander, O Andrade, A Virkar, R Smith, S Lamoreaux, C Cheng, C Tanner, R Knowlton, K Mehta, H Kimrey,; "Microwave Processing of Ceramics at the University of Utah — Description of Activities and Summary of Progress," [1] pp. 35-48.

RE Willard H Sutton; "Key Issues in Microwave Process Technology," [2] pp. 3-18.

RE Isidoro Enrico Campisi; "Results of the Workshop on Microwave-Absorbing Materials for Accelerators (MAMA): A Personal View," [2] pp. 19-28.

RE Wayne R Tinga, Willard H Sutton; "Key Issues in Microwave Processing, Research and Commercialization - An Open Forum," [2] pp. 45-49.

SA D Greeop; "Determination of Plutonium in Urine Using Stopped Flow Microwave Sample Preparation," [4] pp. 341-348.

SA Rajender S Varma; "Microwave-Assisted Reactions Under Solvent-Free 'Dry' Conditions," [4] pp. 357-365.

SA J M Osepchuk; "Radiofrequency Safety Issues in Industrial Heating Systems," [1] pp. 125-137.

SA R F Schiffmann; "Fires in Microwave and RF Heating Systems: Causes and Prevention," [5] pp. 355-362.

SA Van Nguyen TRAN; "Two Successful Industrial Microwave Recycling Applications," [5] pp. 503-512.

SI Yong-Lai Tian; "Practices of Ultra-Rapid Sintering of Ceramics Using Single Mode Applicators," [1] pp. 283-300.

SI D S Patil, B C Mutsuddy, James Gavulic, M Dahimene; "Microwave Sintering of Alumina Ceramics in a Single Mode Applicator," [1] pp. 301-309.

SI M A Janney, C L Clahoun, H D Kimrey; "Microwave Sintering of Zirconia-8 MOL % Yttria," [1] pp. 311-318.

SI Arindam De, Ifitkhar Ahmad, E Dow Whitney, D E Clark; "Microwave (Hybrid) Heating of Alumina at 2.45 GHz: I. Micro Structural Uniformity and Homogeneity," [1] pp. 319-328.

SI Arindam De, Ifitkhar Ahmad, E Dow Whitney, D E Clark; "Microwave (Hybrid)) Heating of Alumina at 2.45GHz: II. Effect of Processing Variables, Heating Rates and Particle Size," [1] pp. 329-339.

SI Y Fang, D M Roy, R Roy; "Rapid Sintering of Hydroxyapatite Ceramics by Microwave Processing," [1] pp. 349-356.

SI A D Cozzi, D K Jones, Z Fathi, D E Clark; "Micro structural Evolution of YBa2Cu3O7-x Using Microwave Energy," [1] pp. 357-364.

SI Mary P Sweeney, D Lynn Json; "Microwave Plasma Sintering of Alumina," [1] pp. 365-372.

SI C E Holcombe, N L Dykes; ""Ultra" High-Temperature Microwave Sintering," [1] pp. 375-386.

SI T N Niegs, M K Ferber, J O Kiggans, K L More, C M Hubbard, D W Coffey; "Microstructure Development During Microwave Annealing of Dense Silicon Nitride," [1] pp. 411-420.

SI X D Yu, V V Varadan, V K Varadan,; "Application of Microwave Processing to Simultaneous Sintering and Joining of Ceramics," [1] pp. 497-505.

SI D S Patil, B C Mutsuddy, James Gavulic; "Microwave Sintering of Al_2O_3:ZrO_2 Ceramics," [1] pp. 565-575.

SI Yong-Lai Tian, Bao-Shun Li, Jian-Lin Shi, Yao-Ping Xu, Jing-Kun Guo, Dong-Sun Yen; "Microwave Sintering of Y2O3(3%) - ZrO2(TZI)," [1] pp. 577-584.

SI M A Willert-Porada, S Vodelgel; "Metallorganic and Microwave-Processing of Ceramic Powders and Compacts," [1] pp. 631-638.

SI Mark A Janney, Mitchell L Jackson, Hal D Kimrey; "Microwave Sintering of ZrO2 - 12 MOL% CeO2," [2] pp. 101-108.

SI Y Fang, D K Agrawal, D M Roy, R Roy; "Microwave Sintering of Calcium Strontium Zirconium Phosphate Ceramics," [2] pp. 109-114.

SI H K Varma, S Ananthakumar, P Perumal, P Prabhakar Rao, K G K Warrier, A D Damodaran; "Microwave Processing of Rare Earth Oxide Doped ZnO Varistors," [2] pp. 115-122.

SI A J Pigram, R Freer, M G Hamlyn, A L Bowden; "Microwave Sintering of Ni-Zn Ferrites," [2] pp. 123-131.

SI D Lynn Json, Daniel J Skamser, Mark S Spotz; "Temperature Gradients in Microwave Processing: Boon and Bane," [2] pp. 133-146.

SI T N Tiegs, K L Ploetz, J O Kiggans, R L Yeckley; "Crystallization of Grain Boundary Phases in Silicon Nitride with Low Additive Contents by Microwave Annealing," [2] pp. 259-267.

SI J O Kiggans, T N Tiegs, H D Kimrey, C E Holcombe; "Processing of Complex Sintered Reaction-Bonded Silicon Nitride Parts by Microwave Heating," [2] pp. 269-276.

SI R D Blake, J D Katz; "Microwave Sintering of Large Alumina Bodies," [2] pp. 459-465.

SI Prasad S Apte, William D MacDonald; "Microwave Sintering Kilogram Batches of Silicon Nitride," [3] pp. 55-62.

SI F C R Wroe, A T Rowley; "Microwave Enhanced Sintering of Ceramics," [3] pp. 69-76.

SI Yu Bykov, A Eremeev, V Flyagin, V Kaurov, A Kufatin, A Luchinin, O Malygin, I Plotnikov, V Zapevalov; "The Gyrotron System for Ceramics Sintering," [3] pp. 133-140.

SI M Willert-Porada; "The Role of Space Charge for Microwave Sintering of Oxides," [3] pp. 193-203.

SI L P Martin, M Rosen, D Dadon, D Gershon, B Levush, Y Carmel; "Characterization of Microwave Processed ZnO," [3] pp. 399-406.

SI R W Bruce, A W Fliflet, D Lewis III, R Rayne, G M Chow, P E Schoen, B A Bender, L K Kurihara, A K Kinkead; "Microwave Sintering of Nanocrystalline Ceramics," [3] pp. 407-414.

SI Yoon Chang Kim, Sang Cheol Koh, Do Kyung Kim, Chong Hee Kim; "Effect of Microwave Heating on the Sintering of Silicon Nitride-Doped with Al_2O_3 and Y_2O_3," [3] pp. 415-422.

SI Jiping Cheng, Yi Fang, Dinesh K Agrawal, Rustum Roy; "Densification of Large Size Alumina Objects by Continuous Microwave Sintering," [3] pp. 457-463.

SI D J Grellinger, J H Booske, S A Freeman, R F Cooper; "Processing Parameters Influencing the Sintering of Alumina in a Conventional and 14 GHz Microwave Furnace," [3] pp. 465-471.

SI Ki-Yong Lee, Eldon D Case, Jes Asmussen Jr, Marvin Siegel; "Sintering of Alumina Ceramics in a Single Mode Cavity Under Automated Control," [3] pp. 473-480.

SI J B O Caughman, D J Hoffman, F W Baity, M A Akerman, S C Forrester, M D Kass; "Sintering of Ceramics Using Low Frequency RF Power," [3] pp. 481-488.

SI M D Kass, M A Akerman, F W Baity Jr, J B O Caughman, S C Forrester, R A Lowden, E D Brewer, W J Bruchey, E J Rapacki, L H Tressler; "Influence of Microwave-Annealing on the Mechanical Properties of Alumina Tiles," [3] pp. 489-496.

SI K I Rybakov, V E Semenov; "Analysis of the Influence of Temperature Nonuniformity on Ceramics Sintering," [3] pp. 497-504.

SI Rustum Roy, D Agrawal, J P Cheng, M Mathis; "Microwave Processing: Triumph of Applications-Driven Science in WC-Composites and Ferroic Titanates," [4] pp. 3-26.

SI J P Calame, Y Carmel, D Gershon, K Rybakov; "Electric Field Intensification in Spherical Neck Ceramic Microwstructures During Microwave Sintering," [4] pp. 135-142.

SI M Willert-Porada; "A Micro structural Approach to the Origin of "Microwave Effects" in Sintering of Ceramics and Composites," [4] pp. 153-163.

SI Masayuki Hirota, Manuel E Brito, Kiyoshi Hirao, Koji Watari, Motohiro Toriyama, Takaaki Nagaoka; "Microwave Sintering of Silicon Nitride with Rare Earth Sesquioxide Additions," [4] pp. 515-522.

SI S Rhee, G Link, M Thumm; "Dilatometric Measurements of Nanoscaled Ceramics in a 30 GHZ Millimeter Wave Field," [5] pp. 137-144.

SI M Sato, T Mutoh, S Ito, K Ide, T Inoue, K Asaki, O Motojima, M Fujiwara, S Takeyama, M Mizuno, S Obata, T Shimade, K Satake; "Insulation Blankets for Microwave Sintering of Traditional Ceramics," [5] pp. 277-285.

SI S Takeyama, M Mizuno, S Obata, T Shimade, K Satake, M Sato, T Mutoh, S Ito, K Ide, T Inoue, K Asaki, O Motojima, M. Fujiwara; "Sintering of Traditional Ceramics by Microwaves (84GHz and 2.45GHz)," [5] pp. 305-312.

SI M Mizuno, T TaKeyama, S Obata, T Hirai, T Shimade, K Satake, M Sato, T Mutoh, S Ito, T Inoue, K Asaki, O Motojima; "Analysis of Microwave Sintered Porcelain," [5] pp. 313-319.

SI Cheng-Yu Hsieh, Chun-Nan Lin, Hung-Jia Chen, Shyan-Lung Chung, Jiping Cheng, Dinesh K Agrawal; "Microwave Sintering of a Combustion Synthesized AIN Powder," [6] pp. 221-230.

SI Aimin Chang, Jianwen Zhuang, Wen Yang, Yanrong Li; "Microwave Sintering of Large-Sized NTC Thermistor Ceramics," [6] pp. 191-198.

SI M Sato, T Mutoh, T Shimotuma, K Ida, O Motojima, M Fujiiwara, S Takayama, M Mizuno, S Obata, K Ito, T. Hitai, T Shimada; "Recent Development of Microwave Kilns for Industries in Japan," [6] pp. 281-289.

SI B Vaidhyanathan, K Annapoorani, J Binner, R Raghavendra; "Microwave Sintering of Integrated Passive Devices," [6] pp. 423-430.

SI Y H Kao, J D MacKenzie; "Magnetite as a Sintering Aid for Microwave Consolidation of Soda-Lime Glass," [1] pp. 341-348.

SI Y Fang, D K Agrawal, D M Roy, R Roy, J Cheng; "Microwave Sintering of Hydroxyapatite-Based Composites," [2] pp. 397-407.

SI M D Mathis, H S Dewan, D K Agrawal, R Roy; "Microwave Processing of Ni-Al_2O_3 Composites," [2] pp. 431-438.

SI T Gerdes, M Willert-Porada, K Rodiger, H Kolaska; "Guidelines for Large Scale MW-Processing of Hardmetals," [3] pp. 423-431.

SI R Borchert, M Willert-Porada; "Pressureless Microwave Sintering of Metal-Ceramic Functionally Gradient Materials," [4] pp. 491-498.

SI J Buckley, R Bryant, M Long, A Buchman, J R Gleason; "Rapid Microwave Processing and Production of Carbon-Carbon Composites," [4] pp. 499-506.

SI M Willert-Porade, H S Park; "Heating and Sintering of Steel Powders with Microwaves at 2.45 GHz Frequency - Relation Between Heating Behavior and Electrical Conductivity," [5] pp. 459-470.

SM Z Fathi, I Ahmad, J H Simmons, D E Clark, A R Lodding; "Surface Modification of Sodium Aluminosilicate, Glasses Using Microwave Energy," [1] pp. 623-629.

SM A Lunk, A Neuffer, I Scheible, H Schmidt, I I Vinogradov; "Progress in the in SITU Control of Low Temperature Plasma Processing," [5] pp. 231-240.

SM M S Morrow, D A Schechter, H A Huey, Q S Wang; "Annealing and Self-Healing of Microwaved Ceramics," [5] pp. 329-334.

SM Yu V Bykov, S V Egorov, A G Aremeev, K I Rybakov, V E Semenov, A A Sorkin, S A Gusev; "Study of Microwave-Enhanced Mass Transport in Nanostructured Alumina Membranes," [5] pp. 335-342.

SM Z Fathi, D C Folz, D E Clark, R Hutcheon; "Surface Modification of Sodium Aluminosilicate Glasses Using Microwave Energy II," [2] pp. 333-340.

SY S N Kumar, A Pant, R R Sood, J Ng-Yelim, R T Holt; "Production of Ultra-Fine Silicon Carbide by Fast Firing in Microwave and Resistance Furnaces," [1] pp. 395-402.

SY J O Kiggans, C A Holcome, C R Hubbard, R R Steele, H D Kimrey, T N Niegs; "Characterization of Silicon Nitride Synthesized by Microwave Heating," [1] pp. 403-410.

SY A K Singh, P Mehta, A I Kingon; "Synthesis of Nonoxide Ceramic Powders by Nonthermal Microwave Plasma," [1] pp. 421-429.

SY M Willert-Porada; "Reaction-rate Controlled Microwave-Processing of Ceramic Materials," [2] pp. 277-286.

SY P Komarenko, D E Clark; "Microwave Susceptor Materials in the $(Mg_xNi_{1-x})O$ System for High Temperature Use in Air," [2] pp. 351-358.

SY M D Mathis, D K Agrawal, R Roy, R H Plovnick; "Microwave Synthesis of Salons," [3] pp. 533-540.

SY M D Roth, R R Di Fiore, D C Folz, D E Clark, C Wehlburg, J Szczepanski, M T Vala; "Vibrational Spectroscopy of Silica Sol-Gel Components During Microwave Irradiation," [3] pp. 549-556.

SY M D Mathis, D K Agrawal, R Roy, R H Plovnick, R M Hutcheon; "Microwave Synthesis of Aluminum Titanate in Air and Nitrogen," [3] pp. 557-564.

SY J G P Binner, N A Hassine, T E Cross,; "Microwave Synthesis of Non-Oxide Ceramic Powders," [3] pp. 565-572.

SY H Kosslick, H L Zubowa, U Lohse, H Landmesser, R Fricke, J Caro; "Microwave Processing of Zeolites," [4] pp. 523-537.

SY M Cirakoglu, W A Irisbrey, J R Jokisaari, S Bhaduri, S B Bhaduri; "Microwave Synthesis of Calcium Tungstate," [5] pp. 173-180.

SY S B Bhaduri, J G Huang, A H Zhou, S Bhaduri, W R Tinga; "Auto Ignition Synthesis and Microwave Sintering of Nanocrystalline Ceramics," [5] pp. 181-188.

SY Paolo Veronesi, Cristina Leonelli, Anna BoNamertini Corradi; "Restoring of Catalytic Alumina Porous Spheres by Microwave Treatments," [5] pp. 249-256.

SY Cristina Leonelli, Federica Bondioli, Cristina Siligardi, Paolo Veronesi, Gian Carlo Pellacani; "Synthesis of Oxide Pigment Powders by Microwave Treatments," [5] pp. 321-327.

SY T Schubert, M Willert-Porade; "Synthesis of n-TiO2 for Photovoltaic Applications by Colloidal Microwave Processing," [5] pp. 419-425.

SY Christian Gerk, Monika Willert-Porade; "Laser Assisted Microwave Plasma (LAMP) - A New Tool in Ceramic Processing," [5] pp. 451-458.

SY Elodie Michel, Thierry Caillot, Denis Chaumont, Didier Stuerga; "Flash Synthesis of Titanium Dioxide Colloids," [6] pp. 99-102.

SY S Miyake, T Matsumoto, A Niinou, H Saito, K Numata, S Baba; "Application of Mirror-Confinement-Type ECR Plasma Sputtering and Microwave Radiation to the Low Temperature Synthesis of Crystallized High-K Material Thin Films," [6] pp. 139-145.

SY Wang Jian-hua, Wang Sheng-gao; "Synthesis of Carbon Nanotubes by Microwave Plasma Chemical Vapor Deposition at Low Temperature," [6] pp. 79-87.

SY R Gedye, K Westaway, F Smith; "The Application of Microwaves to Rapid Chemical Synthesis," [3] pp. 525-531.

SY Steven L Suib, Elizabeth Vileno, Qiuhura Zhang, Carolina Marun, L Daniel Conde; "Microwave Induced Chemical Reactions in Synthesis and Catalysis," [4] pp. 331-339.

SY Federica Bondioli, Anna Maria Ferrai, Cristina Leonelli, Cristina Siligardi, Tiziano Manfredini; "Conventional and Microwave Preparation of the Al_2O_3 / Cr_2O_3 (SS) Pink Pigment," [4] pp. 483-490.

SY Hyung Chun Kim, Hee Young Kim, Seong Ihl Woo; "Ultrapyroloysis of Chlorodifluoromethane in a Fluidized Bed Reactor Heated by Microwave Energy," [4] pp. 593-600.

SY S Kataoka, D T Tompkins, M A Anderson, M A Zion, W A Zeltner, J H Booske; "Microwave-Assisted Heterogeneous Photocatalytic Oxidation," [5] pp. 225-229.

SY Jeffrey K S Wan, M Catherine Depew; "the Potential of Microwave Induced Catalytic Process: A Recent Revisit," [5] pp. 241-247.

SY M Hajek, Marilena T Radoiu; "Microwave Induced Catalytic Reactions in Liquid Phase," [5] pp. 257-264.

SY C Mazzocchia, A Kaddouri, G Modica, R Nannicini; "Fatty Acid Methyl Esters Synthesis from Triglycerides Over Heterogeneous Catalysts in Presence of Microwaves," [6] pp. .

SY G Bond, J A Gardner, J G McGuire; "Microwave Catalysis for the Production of Fine chemicals and Pharmaceuticals," [6] pp. 131-138.

SY F Bauer, M Willert-Porada; "Microwave Assisted Synthesis of Hybrid Organic-Inorganic Membranes for Fuel Cell Applications," [6] pp. 119-129.

SY T Matsummura-Inoue, Y Yamamoto, N Yoshikawa; "Efficient Microwave Synthesis of Luminescent Metal Complexes," [6] pp. 111-118.

SY Hiromitsu Kozuka, J D Mackenzie; "Microwave Synthesis of Metal Carbides," [1] pp. 387-394.

SY M Willert-Porada, B Fisher, T Gerdes; "Application of Microwave Heating to Combustion Synthesis and Sintering of Al_2O_3-TiC Ceramics," [2] pp. 365-375.

SY T Yiin, M Barmatz; "Microwave Induced Combustion Synthesis of Ceramic and Ceramic-Metal Composites," [3] pp. 541-547.

SY Yan Yin, Jon Binner, Tom Cross; "Microwave Assisted Chemical Vapour Infiltration for Ceramic Matrix Composites," [4] pp. 349-356.

SY Ch Gerk, C W Schmidt, A Niesenhaus, M Willert-Porada; "Microwave Interaction with Emulsions and its Application to the Synthesis of Nanostructured Powders and Composites," [4] pp. 387-395.

SY Ki-Yong Lee, Eldon Dcase, Jes Asmussen Jr.; "Microwave Binder Burn-Out for Batch Processing of Al_2O_3, Al_2O_3 / SiC Platelet, and Al_2O_3 / ZrO_2 Particle Powder Compacts," [4] pp. 539-546.

SY R H G A Kiminami, M R Morelli, D C Folz, D E Clark; "Microwave Combustion Synthesis of Al_2O_3 / SiC Powders," [5] pp. 189-196.

SY Duangduen Atong, D E Clark; "Microwave - Induced Combustion Synthesis of Al2O3 - TiC Composite Powder," [5] pp. 197-206.

SY R Tap, M Willert-Porada; "Synthesis of Composite Powders and Coating of Fibres by Combined CVD-PECVD in a Microwave Heated Fluidised Bed Reactor," [6] pp. 89-97.

SY Yuji Wada, Shozo Yanagida; "Synthesis of Nano-Sized Inorganic Compounds Induced by Microwave Irradiation," [6] pp. 71-78.

SY Duangduen Atong, D E Clark; "The Effect of Reaction Parameters on Microwave - Induced Combustion Synthesis of Al_2O_3 - TiC Composite Powders," [6] pp. 207-220.

SY J P Zhong, Z Fathi, G P LaTorre, D C Folz, D E Clark; "Microwave Densification of Porous Silica Gel," [2] pp. 157-164.

SY Andre Louly, Laurence Perreux, Alain Petit; "Solvent-Free Microwave Assisted Organic Synthesis," [5] pp. 163-172.

SY J Jacobs, L H L Chia, F Y C Boey; "Microwave Radiation in Polymerization: Its Effect on the Molecular Weight of PMMA, PMA, and PS," [4] pp. 417-424.

TM Daniel J Grellinger, Mark A Janney; "Temperature Measurement in a 2.45GHz Microwave Furnace," [2] pp. 529-538.

TM G Darby, D E Clark, R Di Fiore, R Schulz, Attapon Boonyapiwat, D Roth; "Temperature Measurement During Microwave Processing," [3] pp. 515-522.

TM R S Donnan, M Samandi,; "A Critical Assessment of thermometric Accuracy During Microwave Sintering of Engineering Ceramics," [5] pp. 363-370.

TM Keith Thomlson, J H Booske, Reid F Cooper, Yogesh B Fianchandeni, Shiaoling Ge; "Temperature Measurement in Microwave-Heated Silicon Wafers," [5] pp. 391-398.

TM Jason Mershon; "Accurate High Temperature Measurements in Microwave Environments," [1] pp. 641-464.

TM H E Berek, E M Jensen; "Electric Field Strength/Power Density and Temperature Measurements in an electromagnetic Field," [2] pp. 511-516.

WR Terry Strack, Maura Ryan, Rodney Dobson, Mark Taitz; "Microwave Treatment of Biomedical Waste Provides a Cost-Effective Alternative to Incineration," [4] pp. 611-617.

WR J Amouroux, K Coulibaly, M F Renou-Gonnord, H Lancelin, I Dudragne, P Adam; "Plasma Treatment of Halocarbons - Diagnostic and Waste Destruction Applications," [4] pp. 565-575.

WR Dieter Bathen, Henner Schmidt-Traub; "Regeneration of Exhausted Adsorbents by Microwaves," [4] pp. 577-584.

WR S M Bradshaw, E J van Wyk, J B de Swardt; "The Benefits of Microwave Regeneration of CIP Granular Activated Carbon," [4] pp. 585-592.

WR Edward R Peterson; "Crude Oil Emulsion Separation rate Enhancement with High Frequency Energy," [4] pp. 601-610.

WR R Varma, S P Nandi; "Oxidative Degradation of Trichloroethylene Adsorbed on Active Carbons: Use of Microwave Energy," [1] pp. 467-473.

WR R L Schulz, D C Folz, D E Clark, G G Wicks; "Microwave Destruction/Vitrification of Electronic Components," [2] pp. 81-88.

WR R L Schulz, D E Clark, R M Hutcheon, G G Wicks; "Microwave Processing of Simulated Nuclear Waste Glass II," [2] pp. 89-97.

WR E F Sturcken; "The Use of "Self Heating" Ceramics as Crucibles for Microwave Melting Metals and Nuclear Waste Glass," [1] pp. 433-440.

WR R L Schultz, Z Fathi, D E Clark, G G Wicks; "Microwave Processing of Simulated Nuclear Waste Glass," [1] pp. 451-458.

WR T R Krause, J E Helt; "Applications of Microwave Radiation in Environmental Remediation Technologies," [2] pp. 53-59.

WR Jospeh Dovits, Rolf-Erhard Schmitt, Volker Friehmelt; "Microwave Processing of Geopolymer-Cement-Based Waste Forms," [2] pp. 61-72.

WR S J Oda; "Microwave Applications to Process Nuclear and Hazardous Waste," [2] pp. 73-79.

WR G G Wicks, D E Clark, R L Schulz, D C Folz; "Microwave Technology for Waste Management Applications Including Disposition of Electronic Circuitry," [3] pp. 79-89.

WR J A Pearce, D E Daniel, Jana Boerigter, Andres Zuluaga; "Radio Frequency Enhanced Soil Vapor Extraction for in Situ Remediation," [3] pp. 91-98.

WR R L Schulz, D C Folz, D E Clark, C J Schmidt, G G Wicks; "Microwave Treatment of Emissions from the Destruction of Electronic Circuitry," [3] pp. 107-114.

WR D E Clark, D C Folz, R L Schulz, A Boonyapiwat, R R Di Fore, G Darby, K Leiser, R M Hutcheon; "Applications of Microwave Processing in Ceramics and Waste Remediation," [4] pp. 507-514.

WR W M Van Loock; "Overview of Microwave and High-frequency Energy for Hazardous Waste Processing," [4] pp. 619-626.

WR G G Wicks, D E Clark, R L Schulz; "Microwave Technology for Waste Management Applications: Treatment of discarded Electronic Circuitry," [4] pp. 627-637.

WR R L Schulz, G G Wicks, D E Clark; "Microwave Remediation of Emissions Resulting From Treatment of Electronic Components," [5] pp. 401-409.

WR George G Wicks, Bebecca L Schulz, D E Clark; "Microwave Remediation of Hazardous and Radioactive Wastes," [5] pp. 411-417.

WR R L Schulz, D E Clark, D C Folz; "Microwave Waste Remediation: An Update on Applications and Current Research," [6] pp. 433-443.

Printed and bound by CPI Group (UK) Ltd, Croydon, CR0 4YY

15/09/2026

14962410-0003